AF531248

BIOCHEMISTRY OF NUCLEIC ACIDS

BIOCHEMISTRY OF NUCLEIC ACIDS

By

Dr. G.P. Garg

DISCOVERY PUBLISHING HOUSE PVT. LTD.

NEW DELHI-110 002

Published by:
Tilak Wasan
DISCOVERY PUBLISHING HOUSE PVT. LTD.
4383/4B, Ansari Road, Darya Ganj
New Delhi-110 002 (India)
Phone : +91-11-23279245; 23253475; 43596065
E-mail: discoverybooksindia@gmail.com
discoverypublishinghouse@gmail.com
namitwasan9@gmail.com
web : www.discoverypublishinggroup.com

***Reprinted:* 2019**

***First Edition:* 2010**

ISBN: 978-81-8356-624-7

Biochemistry of Nucleic Acids

Printed at:
Infinity Imaging Systems
Delhi

Preface

Biochemistry today has made spectacular progress in unraveling the mysteries of animate nature. This progress has allowed us to gain deeper insight into the principles of vital activity and has to a very significant extent stimulated the development of applied disciplines, especially medicine. The present title *"Biochemistry of Nucleic Acids"* is intended for those who wish to understand living organisms, especially man. Biochemistry is essential for this purpose, but it would be almost impossible for a student to survey on his own the massive body of existing knowledge, constantly augmented by a remarkable torrent of brilliat discoveries. The purpose of the book, then is to organize our knowledge into something that can be comprehended in a relatively short time and still convey a reasonable complete picture of the chemical structure and function of man. Readability without sacrifice of coverage has been a prime goal. An important device in gaining that goal is to keep attention constantly focused on function, with repeated use of rationalization to show that the chemical facts are not isolated, but part of a whole.

Biochemistry has two major goals as a fundamental science, it treats the vital functions from the standpoint of physical chemistry, and as an applied discipline, it points out practical applications for the wealth of scientific knowledge it has acquired. The dual purpose has been, as far as possible, take into account in the writing of this book. We have also tried to summarize our pedagogical experience in teaching biochemistry to students specializing in medicine and pharmacy. The material of this book has been organized according to the principle of functionally in order to trace the close relationship between the functions of the living organism and the structures of its constituents molecules as well as the chemical and physico-chemical process in which they are involved.

The aim of this book is to present a core of biochemical knowledge that is desirable for undergraduate and postgraduate students and also those involved in the field of medical, microbiology, biotechnology and pharmaceutical. Every attempt has been made to keep abrest of the advances in the subject and at the same time to include the fundamentals.

To make the work more comprehensive and informative, the author has consulted many authoritative books, research journals, abstracts, monographs etc. He is grateful to all those great scholars whose work are cited or substantially reproduced.

There can be no claim to originality except in the manner of treatment and much of the information has been obtained from the books and scientific journals available in the different libraries.

The author expresses his thanks to his friends and colleagues whose continue inspirations have initiated him to bring out this book.

The author expresses his gratitude to Mr. Wasan and Staff of M/s Discovery Publishing House Pvt. Ltd., for their whole hearted co-operation in the publication of this book.

In the mean time, the author will remain sincerely responsible for any shortcomings of the book and be grateful to the readers for their suggestions and constructive criticism for the continuous betterment of the book. He takes this opportunity to appeal to the readers to send their suggestions straightway to his publisher.

Author

CONTENTS

CHAPTER

1 The Nucleic Acids

The phosphorus- and nitrogen-containing materials that came to be known as nucleic acids were first isolated from cells around 18'70 by Friedrich Miescher but were long regarded as something of a curiosity. Nevertheless, the structures of the monomer units, the *nucleotides,* were established by 1909 and the correct *polynucleotide* structure of the chains of DNA and RNA was proposed by Levene and Tipson in 1935.

The nucleotides are made up of three parts:

1. One of the *pyrimidine* or *purine "bases"* uracil, cytosine, adenine, or guanine. All four of these bases are present in RNA, while DNA contains thymine instead of uracil. Atoms in the bases are numbered 1 – 6 or 1 – 9.
2. A *sugar,* either D-ribose or D-2-deoxyribose. Carbon atoms in sugars are numbered 1′ – 5′.
3. *Phosphoric acid*

Although the biological synthesis is indirect, we can imagine that nucleotides are formed from these parts by elimination of two molecules of water as indicated in Fig. 1.1. In nucleic acids the nucleotides are combined through phosphodiester linkages between the 5′-hydroxyl of the sugar in one nucleotide and the 3′-hydroxyl of another. Again, we can imagine that these linkages were formed by the elimination of water.

The structures of a pair of short polynucleotide strands in DNA are shown in Fig. 1.1. That of a segment of double-helical DNA is shown in Fig. 1.3 and, that of a transfer RNA is shown in Fig. 1.30 & 1.31. Despite the fact that Levene had deduced correct structure for polynucleotides, he was thrown off the trail of a deeper understanding by the roughly equal amounts of the four bases found in either DNA or RNA. He assumed that nucleic acids must be regular repeating polymers for which there was no obvious biological function. It was not until 1944 that there was concrete evidence that DNA carried genetic information. However, it was James Watson and Francis Crick's recognition of the double-helical structure of DNA.

In 1953 and the mechanism of replication that this structure implied that captured the imagination of biologists and chemists alike and paved the way for the present-day explosion of knowledge of DNA, RNA, and of the encoded proteins. Watson and Crick proposed that DNA is a double helix of two antiparallel polynucleotide chains. The structure was deduced from model building together with knowledge of the X-ray diffraction data of Maurice F. Wilkins and Rosalind Franklin on artificially formed DNA fibers. An additional key piece of information was the discovery by Erwin Chargaff that *in all double-stranded DNA the content of adenine equals that of thymine and the content of guanine equals that of cytosine.* The most significant feature of the proposed structure was the pairing of bases in the center through hydrogen bonding.

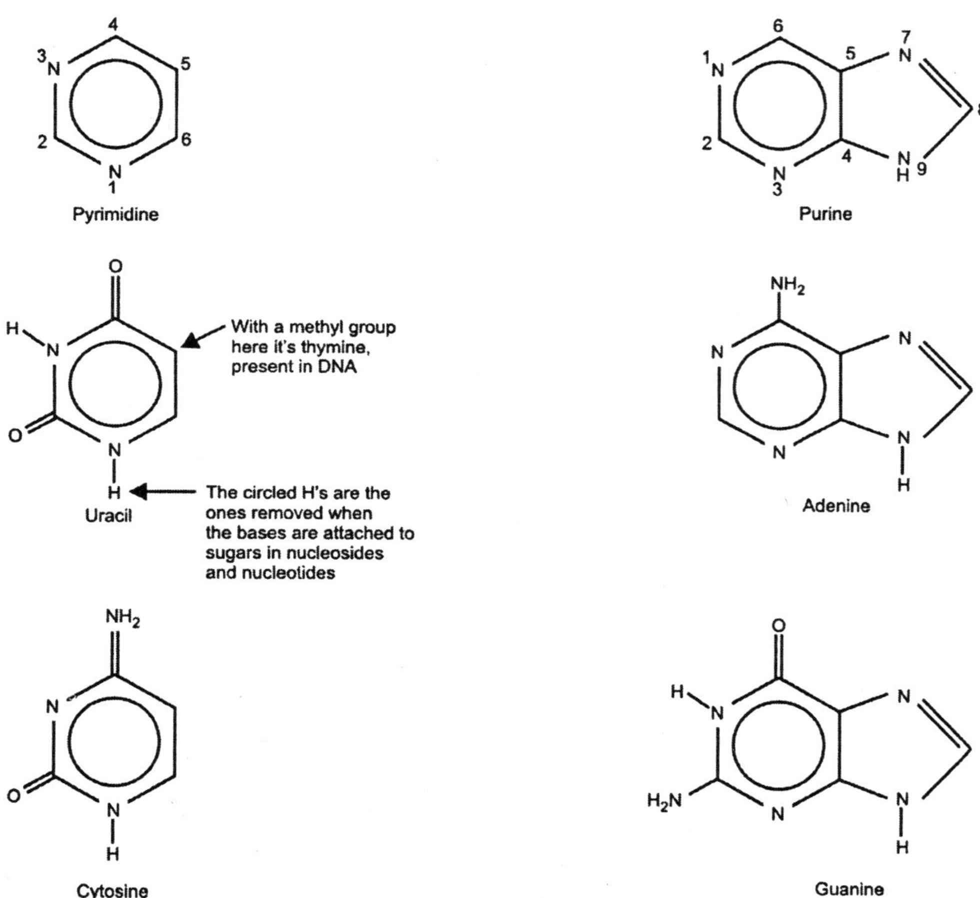

Fig. 1.1 Structures of the major pyrimidine and purine bases of DNA and RNA.

The pairs and triplets of hydrogen bonds could form in the manner shown only if adenine (A) was paired with thymine (T) and cytosine (C) with guanine (G) at every point in the entire DNA structure. Thus, *the nudeotide sequence in one chain is complementary to but not identical to that in the other chain.* It was apparent almost immediately that the sequence of bases in a DNA chain must convey the encoded genetic information. The complementarity of the two strands suggested a simple mechanism for replication of genes during cell divisions. The two strands could separate and a complementary strand could be synthesized along each strand to give two molecules of the DNA, one for each of the two cells.

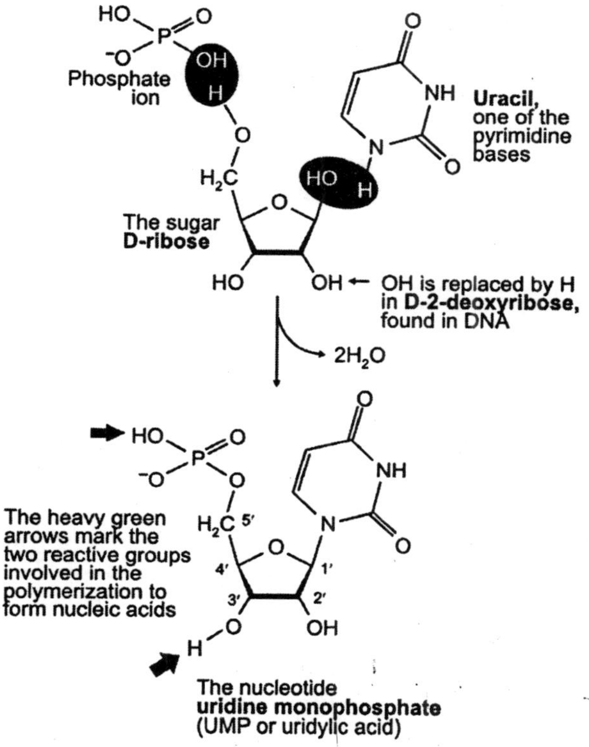

Phosphate ion

Uracil, one of the pyrimidine bases

The sugar **D-ribose**

OH is replaced by H in **D-2-deoxyribose,** found in DNA

$2H_2O$

The heavy green arrows mark the two reactive groups involved in the polymerization to form nucleic acids

The nucleotide **uridine monophosphate** (UMP or uridylic acid)

...(1.1)

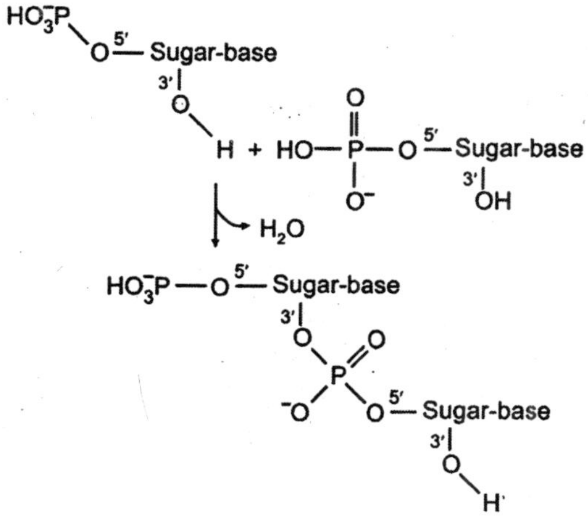

HO_3^-P—O—5′ Sugar-base —3′ O—H + HO—P(=O)(O⁻)—O—5′ Sugar-base —3′ OH

H_2O

HO_3^-P—O—5′ Sugar-base —3′ O—P(=O)(O⁻)—O—5′ Sugar-base —3′ O—H

...(1.2)

The essential correctness of the concept has been proved. Two extremely important developments came in the 1970s: (1) Methods were found for cutting and rejoining DNA fragments and for *cloning* them in bacteria and (2) ways were devised for rapid determination of *nucleotide sequences.* The application of these techniques is now providing startling advances in biology and medicine.

In 1970 we knew virtually nothing about the sequences of nucleotides in genes but today, we know the sequences for many thousands of genes of all types. By the 1980s sequences had been established for hemoglobin, γ-globulins, collagen, and for many enzymes. An example is the gene sequence for mitochondrial aspartate aminotransferase. Its coding regions consist of 1299 pairs of mononucleotides, the bases being paired in double helical form. Complete DNA sequences are known for numerous viruses including the 9740-base pair (bp) DNA provirus form of the RNA virus that causes AIDS and for bacterial viruses such as T7 bacteriophage (39, 936 bp). Also determined in the 1980s were sequences of human mitochondrial DNA (16, 598 bp) and of chloroplast DNA from the tobacco plant (155, 844 bp). In 1995 the first complete sequences of bacterial genomes were obtained. These were followed by sequences of many other bacterial genomes, including the 4.2×10^6 bp *E. coli* genome.

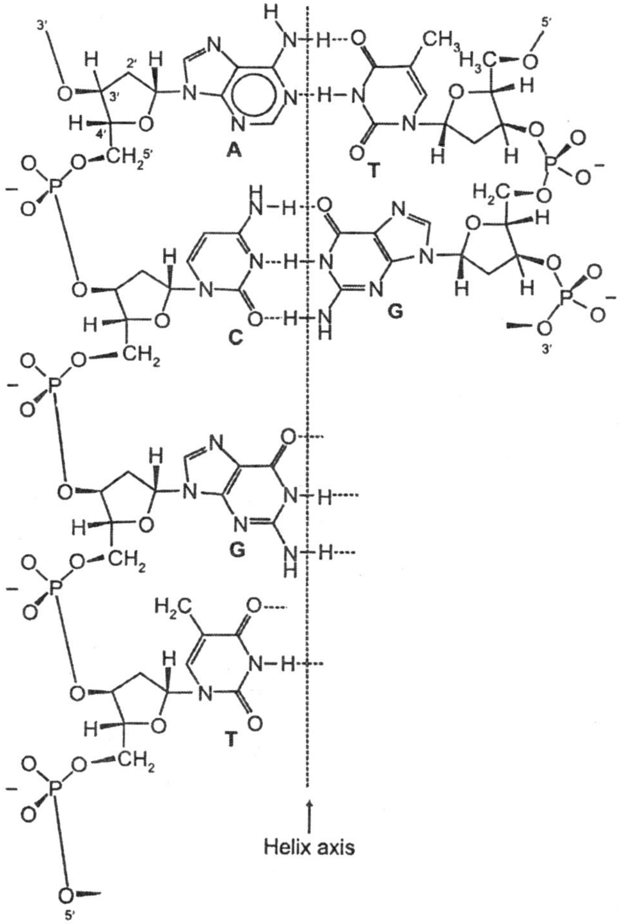

Helix axis

Fig. 1.2. A distorted (flattened) view of the Watson-Crick structure of DNA showing the hydrogen-bonded base pairs.

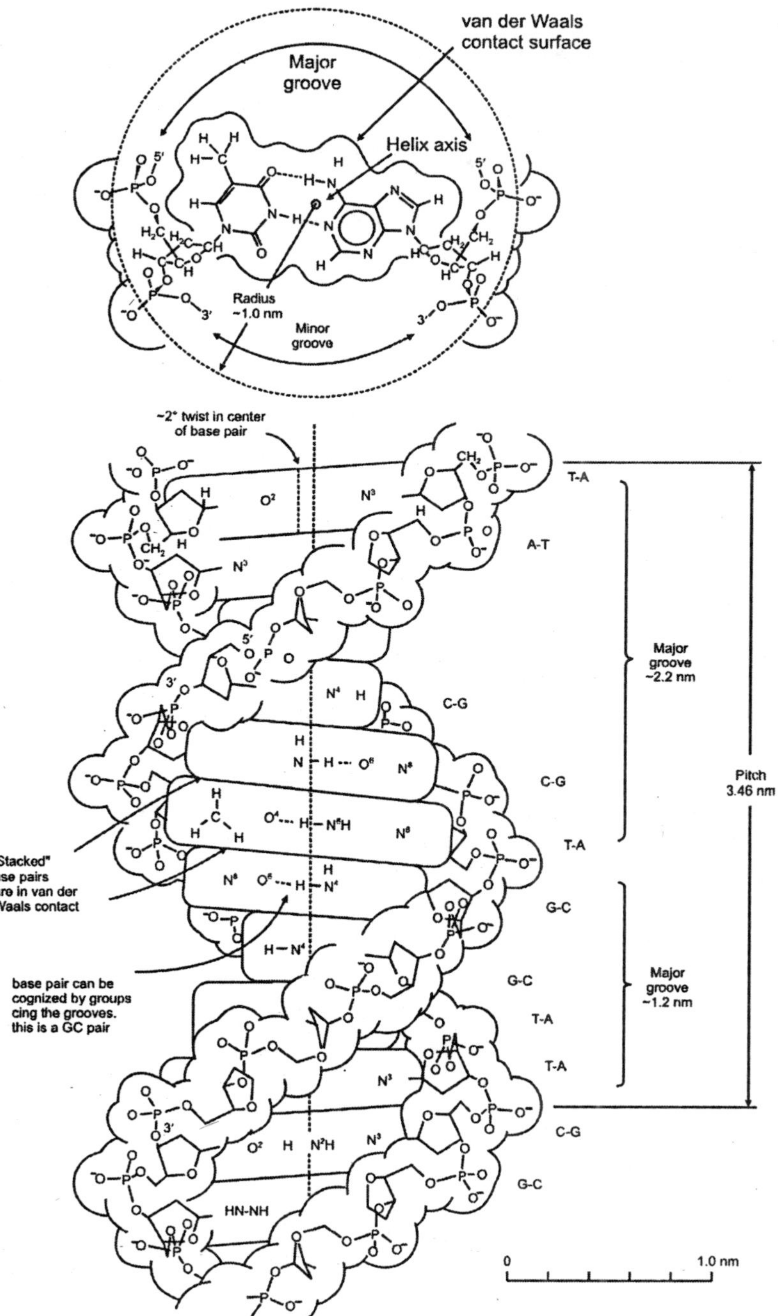

Fig. 1.3. The double-helical structure of DNA. The structure shown is that of the B form. The major and minor grooves, are marked.

Sequencing of the 16 chromosomes of the 12.07 Mbp genome of yeast containing ~6300 genes was completed in 1996 and that of the 97 Mbp genome of the nematode *Caenorhabditis elegans* in 1998. By 2000 the sequence of the 180 Mbp genome of the fruit fly *Drosophila melanogaster* was largely completed, and most sequences of the 3×10^9 bp human genome were known. Cloning of DNA has not only provided an essential step in sequence determination but also has given birth to a new industry devoted to producing proteins from genes cloned in bacteria, yeast, or other cells.

Human insulin and the antiviral protein interferon were two of the first proteins produced in this way. Methods now in use permit us to introduce at will alterations at any point in a DNA sequence. We are able to locate and study the genes responsible for many genetic defects.

The first successes in using virus-like particles to carry new pieces of DNA into the cells to help correct these defects have been reported. Our ability to breed new varieties of plants and microorganisms has been enormously enhanced. We can foresee the production of artificially designed enzymes to conduct many industrial chemical processes. These are among the many reasons for the excitement today in the fields of nucleic acid chemistry and molecular genetics.

STRUCTURE AND CHEMISTRY OF NUCLEOTIDES

Names and Abbreviations

The purine and pyrimidine ring compounds found in nucleic acids are known as "bases," even though some of them have almost no basic character. *Nucleosides* are the N-glycosyl derivatives of the bases with ribose or 2-deoxyribose. The *nucleotides* are phosphate esters of nucleosides. Similar names are applied to related compounds such as adenosine triphosphate (ATP) that are not present in DNA or RNA. The names of the principal nucleotides from which the nucleic acids are formed are given in Table 1.1. The novice may find these confusing! Even worse than the names in the table is *hypoxanthine* (Hyp), which is derived from adenine by replacement of its $-NH_2$ group with –OH and tautomerization:

$$-\overset{\displaystyle OH}{\overset{|}{C}}=N- \longrightarrow -\overset{\displaystyle O}{\overset{||}{C}}-\overset{\displaystyle H}{\overset{|}{N}}-$$

The nucleoside formed from hypoxanthine and ribose is known as *inosine* (Ino or I) and the corresponding nucleotide as *inosinic acid*. Further substitution at C-2 of –H by –OH and tautomerization yields *xanthine* (Xan). Its nucleoside is xanthosine (Xao, X). A similar hydroxylation at C-7 converts xanthine to *uric acid,* an important human urinary excretion product derived from nucleic acid bases.

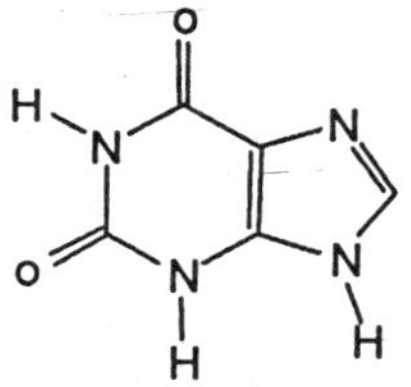

Xanthine

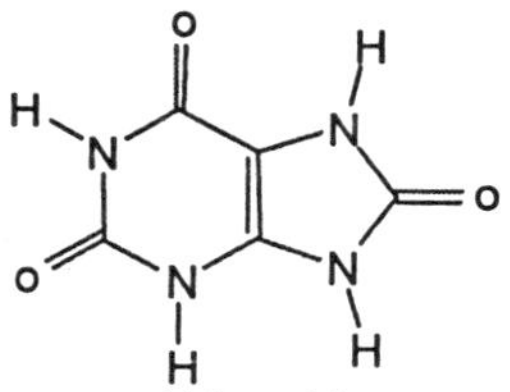

Uric acid

Table 1.1. Names of Pyrimidine and Purine Bases, Nucleosides, and 5′-Nucleotides

A. Nucleotide units of RNA (abbreviations in parentheses)[a]

Base :	**Uracil (Ura)**	**Cytosine (Cyt)**	Adenine (ade)	**Guanine (Gau)**
Nucleoside:	**Uridine (Urd or U)**	**Cytidine (Cyd or C)**	Adensoine (Ado or A)	**Guanosine (Guo or G)**
5'-Nucleotide:	**Uridine 5′-phosphate or 5′-uridylic acid (Urd-5'-*P* or UMP)**	**Cytidine 5′-hosphate or 5′-cytidylic acid (Cyd-5′-*P* or CMP)**	Adenosine 5′-phosphate or 5′-adenylic acid Ado-5′-*P* or AMP)	**Guanosine 5′-phosphate or 5′-guanylic acid (Guo-5′-*P* or GMP)**

B. Nucleotide units of DNA

These contain 2-deoxyribose and the nucleosides and nucleotides are called deoxyadenosine (dAdo or dA), deoxyadenosine 5′-phosphate (dAMP), etc.

DNA contains thymine (Thy) rather than uracil. The deoxyribose derivatives are thymidine (dThd or dT) and thymidine 5′-phosphate. The ribose derivatives of thymine are the nucleoside ribsoylthymidine (Thd) and ribsoylthymidine 5′-phoshpate (Thd-5′-*P*).

C. Abbreviations used in describing polynucleotide sequences

U,T,C,A,G	Uracil, thymine, cytosine, adenine, guanine
Y or Pyr	Pyrimidine (T or C)
R or Pur	Purine (A or G)
M	Amino base (A or C)
K	Keto base (G or T)
S	Strongly pairing (G or C)
W	Weakly pairing (A or T)
H	Not G (any other base)
B	Not A
V	Not T or U
D	Not C
N	Any base

[a] Isomers of the 5′-nucleotides, in which the phosphate is attached to the oxygen on C-3′, are the 3′-nucleotides. Care must be taken to avoid ambiguity. The simple abbreviations UMP, CMP, AMP, and GMP always refer to the 5′-nucleotides.

Nucleic acid structures are abbreviated in several ways. For example, the sugar rings may be portrayed by vertical lines. The abbreviations A, C, U, T, and G for the individual bases or Pu (purine) and Py (pyrimidine) are placed at the upper ends of the lines, and slanted lines with P in the centers represent the 3′ – 5′ phosphodiester linkages in a polynucleotide.

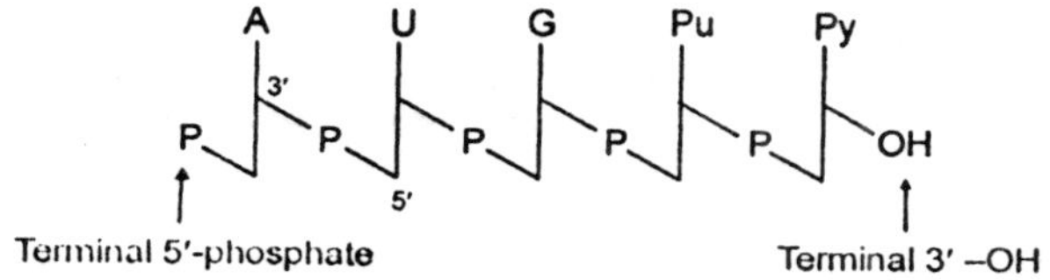

A U G Pu Py

3′

P P P P P OH

5′

Terminal 5′-phosphate Terminal 3′ –OH

The same structure can be further abbreviated

	5′ end	pApUpGpRpY	3′ end
or	5′ end	AUGRY	3′ end

Here purine is abbreviated R and pyrimidine Y. By convention the 5′ end of a polynucleotide is ordinarily placed to the left in these formulas.

Lengths of nucleic acid chains are usually given as a number of bases or kilobases (kb). For double-stranded DNA (dsDNA) the length is given as base pairs (bp), kilobase pairs (kbp), or megabase pairs (Mbp). However, in most places, including this book, the abbreviations kb and Mb will be used for a length of DNA whether single or double stranded. For double-stranded DNA, one strand, usually the *coding strand,* from which the amino acid sequence can be read using the code in Table 1.4 or 1.5 has the 5′ end at the left while the complementary strand has the 3′ end at the left, *e.g.*,

```
      Ala   Tyr   Gly   Pro   Phe   Cys   Gln   Termination
5′  G C C T A T G G A C C T T T C T G C C A G T G A  3′
3′  C G G A T A C C T G G A A A G A C G G T C A C T  5′
```

ACID – BASE CHEMISTRY AND TAUTOMERISM

The ionized phosphate groups of the polymer "backbone" give nucleic acid molecules a high negative charge. For this reason DNA in cells is usually associated with basic proteins such as the *histones* or *protamines* (in spermatozoa), with polycations of amines such as *spermidine* ($H_3^+NCH_2CH_2CH_2CH_2\,NH_2^+CH_2CH_2CH_2NH_3^+$), or with alkaline earth cations such as Mg^{2+}. If the pH of a solution containing double-helical DNA is either lowered to ~3 or raised to ~12, the two strands unravel and can be separated. Over the entire range of pH the alternating sugarphosphate backbone of the polymeric chains remains negatively charged. However, depending on the pH, the bases can be protonated or deprotonated with a resultant breaking of the hydrogen bonds that hold the pairs of bases together. Pyrimidines and purines, which contain the $-NH_2$ group, are weakly basic.

The cationic protonated conjugate acid forms of cytidine, adenosine, and guanosine have pK_a values of 4.2, 3.5, and 2.7, respectively. Similar values are observed for the 5′-nucleotides. In these compounds it is not the $-NH_2$ group that binds the proton but an adjacent nitrogen atom in the ring.

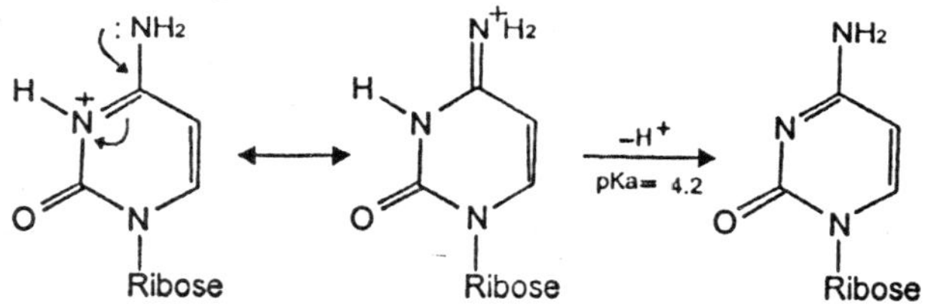

...(1.3)

We can understand this if we recognize that the bases have substantial aromatic character. In aniline (aminobenzene) electrons are withdrawn from the amino group into the aromatic ring with a strong decrease in basicity of this $-NH_2$ group (the pK_a is 4.6). Similarly, electrons are withdrawn from the NH_2 groups of cytosine, adenosine, and guanine into the pyrimidine and purine rings as is indicated by the small curved arrows on the left-hand structure of Fig. 1.3.

The effect is even stronger than in aniline, largely because of the presence of the nitrogen atoms in the rings. In cytosine it is primarily N-3 that serves as the electron acceptor. As a consequence this nitrogen becomes more basic than the $-NH_2$ group and is the major site of protonation. However, as is indicated in Eq. 1.3 the positive charge on the cation is shared by resonance with the exocyclic amino group. Adenosine is similar to cytosine in its acid-base chemistry; N-1, adjacent to the $-NH_2$ group, is the principal site of protonation.

A tautomer of the cation protonated at N-3 is formed in smaller amounts. Guanosine is electronically more complex, being protonated mainly at N-7 and to a lesser extent at N-3. This can be understood in terms of electronic interaction with the adjacent oxygen as indicated in the resonance structure to the right in the following diagram:

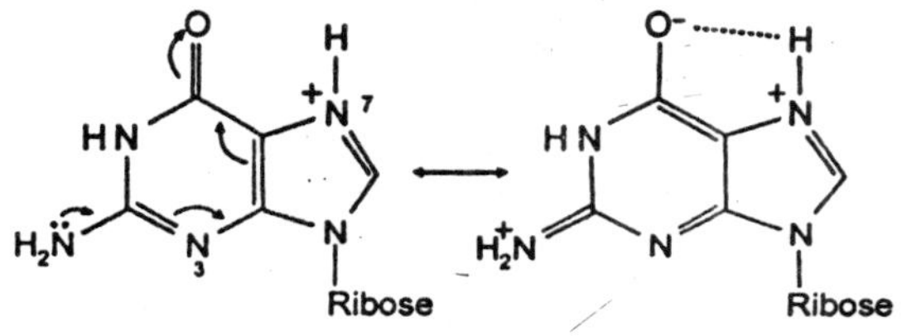

HN H $^+$N 7 $H_2\ddot{N}$ N 3 N Ribose ⟷ O$^-$ H $^+$N HN $H_2\overset{+}{N}$ N N Ribose

Under basic conditions the proton on N-3 of uridine or thymine or on N-1 of guanosine can dissociate with a pK_a of ~9.2. These "bases" are actually weak acids!

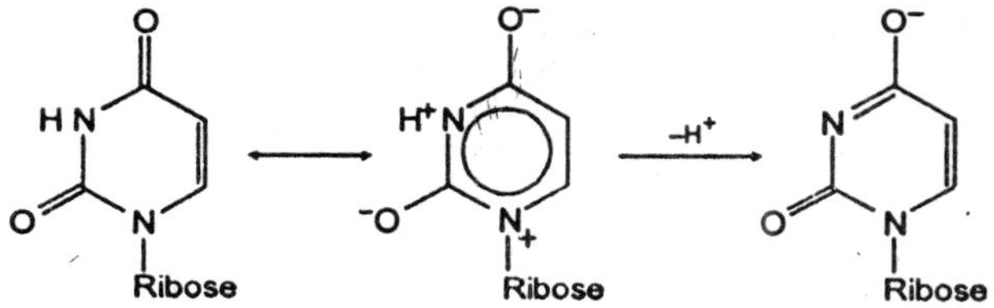

O HN O N Ribose ⟷ O$^-$ H$^+$N $^-$O N$^+$ Ribose $\xrightarrow{-H^+}$ O$^-$ N O N Ribose ...(1.4)

Look at the "Kekule" resonance structure shown in the center. The two negative charges flanking the dissociable proton are sufficient to hold it firmly to the ring at low pH. The proton is half dissociated only when the pH is raised to 9.2. At still higher values of pH (pK_a ~12.4) a proton dissociates from a ribose hydroxyl group.

The tautomeric forms shown in Fig. 1.1 predominate. However, it is possible that minor tautomers such as the following are sometimes preferentially bound into active sites of enzymes where the dielectric constant may be low and where the geometrical arrangement of amino acid functional groups may favor protonation on oxygen rather than nitrogen.

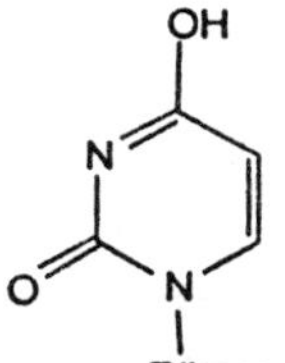

OH N O N Ribose

A minor tautomer of uridine

Absorption of Ultraviolet Light

Nucleic acids strongly absorb ultraviolet light of wavelengths below about 300 nm, with an absorption maximum at ~260 nm and a stronger one below 200 nm. This property is biologically important because of the resultant induction of mutations, a natural result of exposure to sunlight. In the laboratory the same property is useful in identification and quantitative analysis of nucleotides. The ultraviolet spectra of four nucleosides are shown in Fig. 1.5. Notice the changes that accompany protonation or deprotonation of the ring. The absorption bands, which are related to those of benzene, provide another indication of the partial aromatic character of the bases.

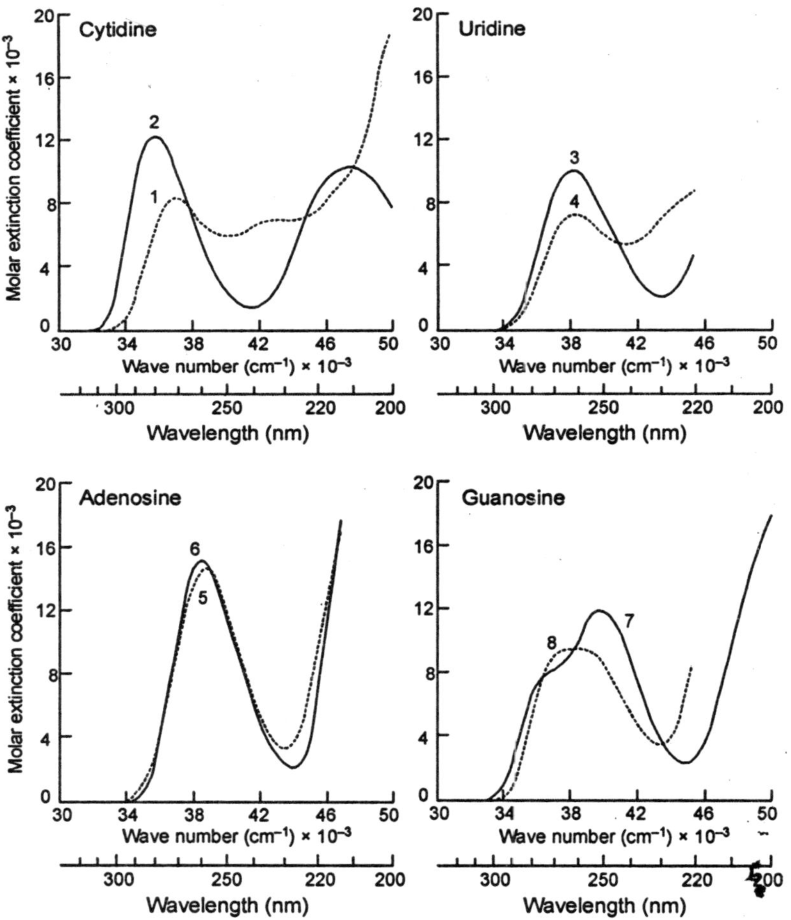

Fig. 1.4. Near ultraviolet absorption spectra of cytidine, uridine, adenosine, and guanosine. 1. Monoprotonated form of cytidine (for which pK_a = 4.2). 2. Neutral form (pH ~ 7) of cytidine. 3. Neutral form of uridine (for which pK_a = 9.2). 4. Monoanionic form of uridine. 5. Monoprotonated form of adenosine (pK_a = 3.5). 6. Neutral form of adenosine. 7. Neutral form of guanosine (pK_a = 9.2). 8. Monoanion of guanosine.

Chemical Reactions of the Bases

The purines and pyrimidines are relatively stable compounds with considerable aromatic character. Nevertheless, they react with many different reagents and, under some relatively mild conditions, can be completely degraded to smaller molecules. The chemistry of these reactions is complex and is made more so by the fact that a reaction at one site on the ring may enhance the reactivity at other sites. The reactions of nucleic acids are largely the same as those of the individual nucleosides or nucleotides, the rates of reaction are often influenced by the position in the polynucleotide chain and by whether the nucleic acid is single or double stranded. The reactions of nucleosides and nucleotides are best understood in terms of the electronic properties of the various positions in the bases. Most of the chemical reactions are nucleophilic addition or displacement reactions.

Positions 2, 4, and 6 of pyrimidine bases are deficient in electrons and are therefore able to react with nucleophilic reagents. The 6 position is especially reactive toward additions, while the 2 position is the least reactive. The corresponding electron-deficient positions in the purine bases are 2, 6, and 8. These positions, which are marked by asterisks on the following structures, have electrophilic character in all of the commonly occurring pyrimidines and purines.

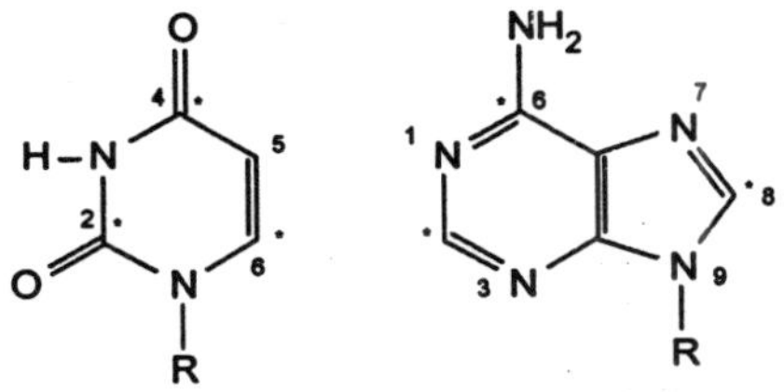

All of the oxygen and nitrogen atoms in the pyrimidines, as well as the 5 position of the ring, have nucleophilic character and can therefore react with electrophilic centers of various reagents.

Base Pairs, Triplets, and Quartets

The purine and pyrimidine bases are the "side chains" of the nucleic acids. The polar groups that are present in the bases can form hydrogen bonds to other nucleic acid chains, *e.g.*, in

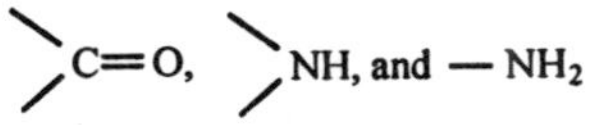

$$>C{=}O, \quad >NH, \text{ and } -NH_2$$

the base pairs of the DNA double helix, and also to proteins. Fig. 1.6 shows the shapes and the hydrogen bonding groups available in the bases. The number of both electron donor groups and proton donors available for hydrogen bonding is large and more than one mode of base pairing is possible.

The base pairs proposed by Watson and Crick are shown in Fig. 1.2 and again in Fig. 1.7. While X-ray diffraction studies indicate that it is these pairs that usually exist in DNA, other possibilities must be considered. For example, Hoogsteen proposed an alternative A-T pairing using the 6-NH_2 and N-7 of adenine. Here the distance spanned by the base pair, between the C-1′ sugar carbons, is 0.88 nm, less than the 1.08 nm of the Watson-Crick pairs. Duplexes of certain substituted poly (A) and poly (U) chains contain only Hoogsteen base pairs and numerous X-ray structure determinations have established that Hoogsteen pairs do occur in true nucleic acids Fig. 1.7 shows the structures of two Hoogsteen pairs. In each case the purine component has formed a Watson-Crick pair with a third base to give *a base triplet*. The first of these triplets may be designated T*A•T, where * represents the Hoogsteen hydrogen bonding and • represents the Watson-Crick bonding. The

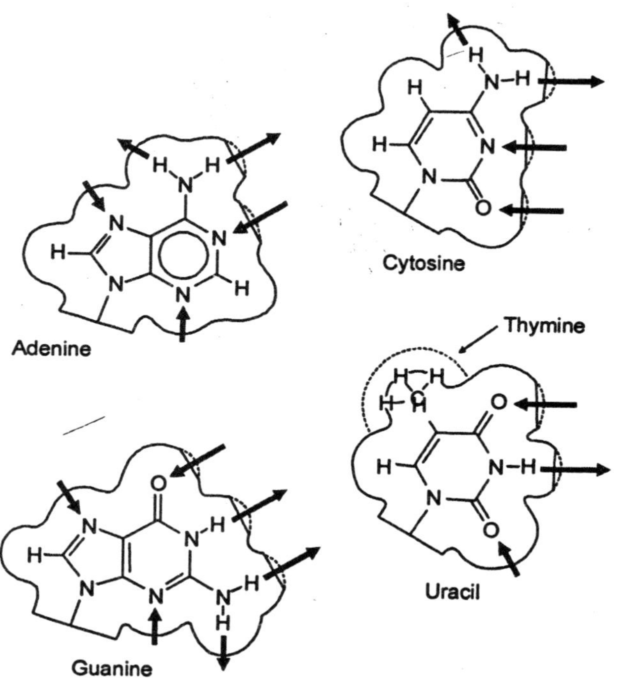

Fig. 1.5. Outlines of the purine and pyrimidine bases of nucleic acids showing van der Waals contact surfaces and some of the possible directions in which hydrogen bonds may be formed. Large arrows indicate the hydrogen bonds present in the Watson-Crick base pairs. Smaller arrows indicate other hydrogen bonding possibilities. The directions of the green arrows are from a suitable hydrogen atom in the base toward an electron pair that serves as a hydrogen acceptor. This direction is *opposite,* to that in the first edition of this book to reflect current usage.

second triplet in Fig. 1.7. C^{+}*G•C, can form only with the N-1 protonated form (low pH form) of cytosine. *Reversed* Watson-Crick or Hoogsteen AU or AT pairs are formed if the 2-carbonyl rather than the 4-carbonyl of the U or T makes a hydrogen bond with the amino group of adenine.

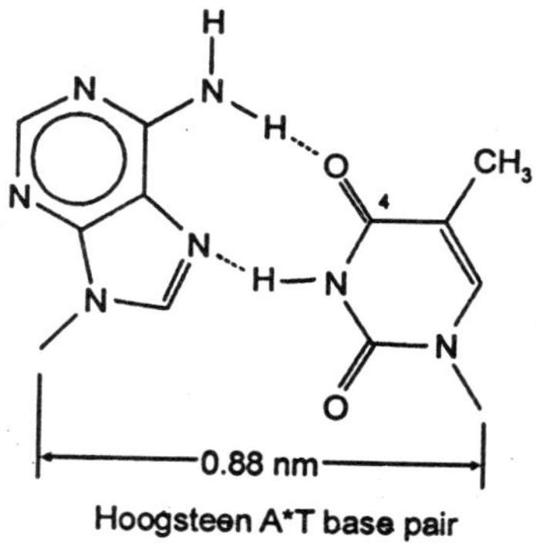

0.88 nm

Hoogsteen A*T base pair

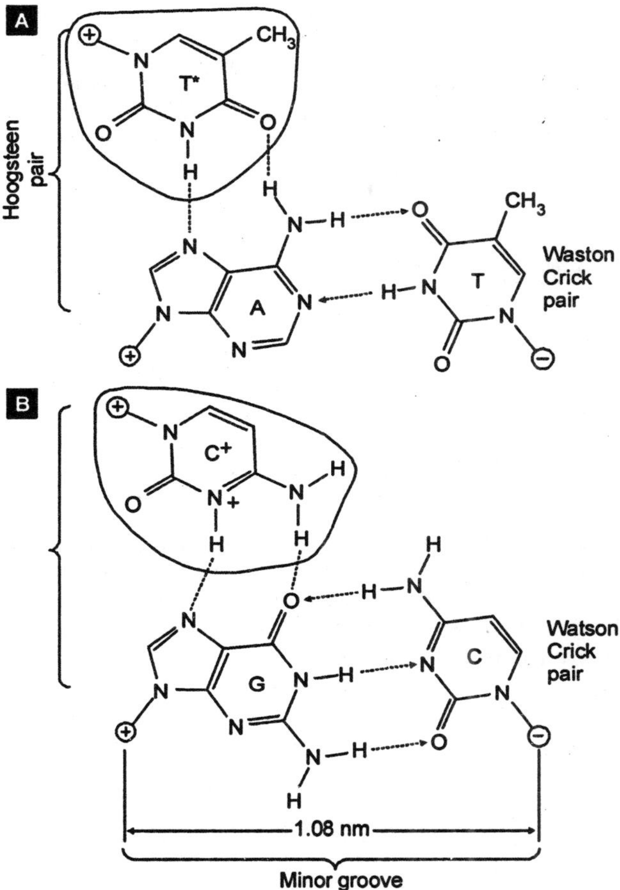

Fig. 1.6. Two base triplets that form in triple-stranded DNA and involve both Watson-Crick and Hoogsteen base pairing. (A) The triplet T*A•T, where the T (marked T*) of the third strand is hydrogen bonded as a Hoogsteen pair (.........) to an adenine of a Watson-Crick AT pair (whose hydrogen bonds are indicated (– – –). (B) *The* triplet C⁺*G•C, where *C⁺* is cytosine in its N-1 protonated (low pH) form. The Watson-Crick strands are antiparallel, as indicated by the ⊕ and ⊖ signs. The third strand may have either orientation, but when it contains largely pyrimidines it is parallel to a purine-rich strand.

Because of the resulting arrangement of the ribose rings the base pairs cannot fit into the ordinary Watson-Crick double-stranded DNA structure. A reversed Watson-Crick pair can also be formed between G and C but with only two hydrogen bonds, while a reversed Hoogsteen pair can form only if a minor tautomer of cytosine is used. Hoogsteen pairs were first observed in nature in transfer RNA molecules. These molecules contain mostly Watson-Crick base pairs but there are also two reversed Hoogsteen pairs. One of them, between U8 and A14, is invariant in all tRNAs studied.

Hoogsteen pairing also occurs in four-stranded DNA, which has important biological functions. A *G quartet* from a DNA tetraplex held together by Hoogsteen base pairs is shown in Fig. 1.8.

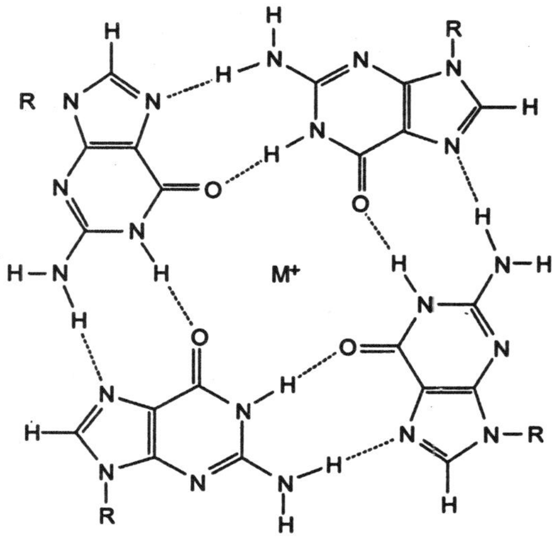

Fig. 1.7. A guanine quartet held together by Hoogsteen base pairing. This structure is found in the telomeres at the ends of linear chromosomes. Four segments of DNA, each of which may be part of a single folded strand, give structures in which four or more of these G quartets are stacked one above the other. Monovalent ions, usually K^+ or Na^+, are bound in the center, although not always in the plane of the bases.

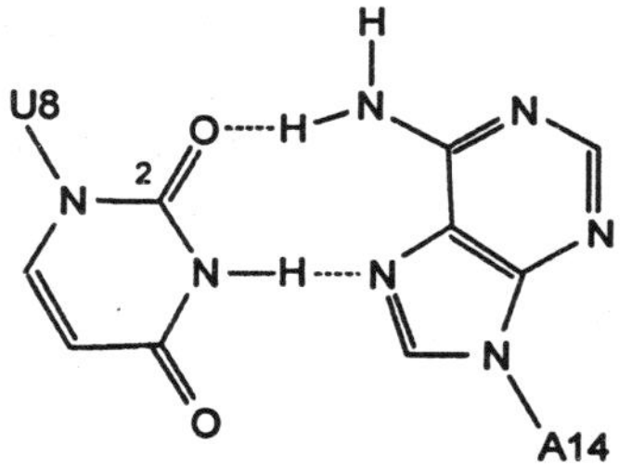

A reversed Hoogsteen base pair present in tRNA molecules

Another pairing that occurs in tRNAs allows guanine to pair with uracil, *e.g.,* G4 with U69. This was originally proposed to account for codon-anticodon interactions between tRNA molecules and messenger RNA. It is commonly called *wobble pairing* because the uracil must wobble away from its orientation in the normal Watson-Crick pair.

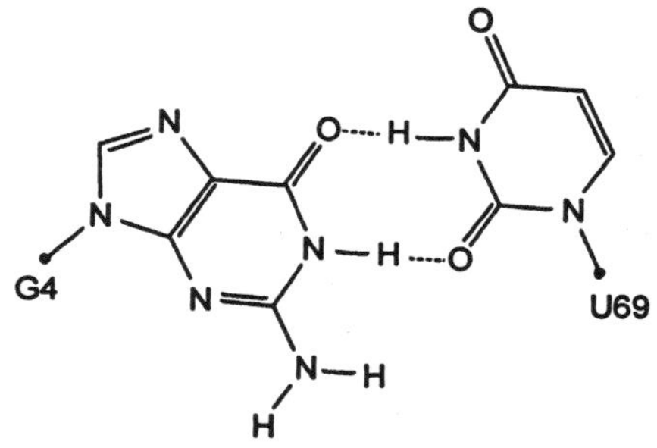

A "wooble" base pair

Strengths of Base Pairs

How strong are the bonds between pairs of bases in DNA? The question is hard to answer because of the strong interaction of the molecules with polar solvents through hydrogen bonding and hydrophobic effects. Some insight has come from studies of the association of bases in non-polar solvents. Thus, 1-cyclohexyluracil forms a dimer involving either hydrogen bonding or stacking, but the association is weak with the Gibbs energy of formation $\Delta G_f^\circ = 5$ kJ m ol^{-1}.

When the same compound was mixed with 9-ethyladenine, a base-paired complex formed between the two compounds with a formation constant over ten-fold greater than that for the dimmer. When the circled hydrogen atom in Eq. 1.5 was replaced by $-CH_3$, which blocked the pairing K_f fell below 1 kJ/mol ($\Delta G^\circ > 0$). The difference in ΔG_f in the two cases was only 7 kJ mol^{-1}. Many other estimates have been made. The small energies summed over the many base pairs present in the DNA molecule help provide stability to the structure.

Stacking of Bases

The purines and pyrimidines of nucleic acids, as well as many other compounds with flat ring structures and containing both polar and nonpolar regions, are sparingly soluble in either water or organic solvents. Molecules of these substances prefer neither type of solvent but adhere tightly to each other in solid crystals. Both experimental measurements and theoretical computations suggest that hydrogen bonding is the predominant force in the pairing of bases in a vacuum or in nonpolar solvents. However, in water stacking becomes important.

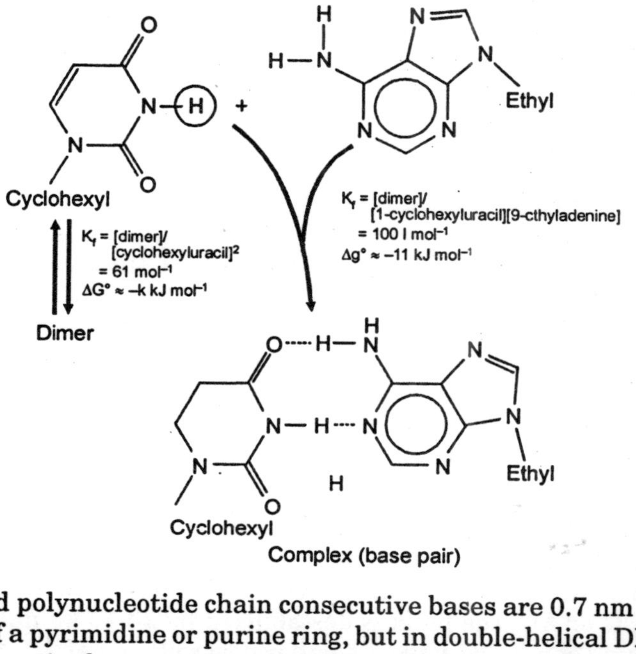

Cyclohexyl

K_f = [dimer]/[cyclohexyluracil]^2 = 61 mol^{-1}
$\Delta G^\circ \approx -k$ kJ mol^{-1}

Dimer

Ethyl

K_f = [dimer]/[1-cyclohexyluracil][9-cthyladenine] = 100 l mol^{-1}
$\Delta g^\circ \approx -11$ kJ mol^{-1}

Cyclohexyl
Ethyl
Complex (base pair)

...(1.5)

In a fully extended polynucleotide chain consecutive bases are 0.7 nm apart, twice the van der Waals thickness of a pyrimidine or purine ring, but in double-helical DNA of the B type, the distance between consecutive base pairs is only 0.34 nm. They are touching. One effect of stacking is a decrease in the expected intensity of light absorption. The molar extinction coefficient of a solution of double-helical DNA or RNA is always less by up to 20 – 30% than that predicted from the spectra of the individual nucleosides. Because both hydrogen bonding and stacking are involved, the thermodynaimcs of base pairing in nucleic acids is complicated.

The hydrophobic parts of exposed bases tend to induce an ordering of the surrounding water molecules and therefore a decrease in their entropy. However, hydrogen bonding of the polar groups of the bases to the solvent causes a decrease in water structure. This is greater than the increase in structure around the hydrophobic regions and the stacking of bases leads to a net decrease in entropy.

The entropy change ΔS for addition of a base pair to the end of a double-stranded RNA helix in a hairpin loop such as that displayed in Fig. 1.9, ranges from –0.05 to –0.15 kJ/degree per base pair. The enthalpy change ΔH tends to be small and positive for association of alkyl groups in water and nearly zero for association of aromatic hydrocarbons. However, ΔH is distinctly negative for association of heterocyclic bases. This has also been attributed to a decrease in the ordering of solvent around the bases as a result of exclusion of water.

Attraction or repulsion of partial charges on the polar groups comprising the purine and pyrimidine bases may also be an important factor. For addition of a base pair to an RNA helix, the change in enthalpy, ΔH, varies from about –24 to –60 kJ/ mol. Since $\Delta G = \Delta H - T\Delta S$, the net result is a negative value of ΔG, a "hydrophobic effect" that favors association of bases. Substantial efforts have been made to estimate quantitatively the Gibbs energies of formation of helical regions of RNA molecules in hairpin stem-loops such as that of Fig. 1.9. Table 1.2 the observed increments in ΔG_f° of such a helix upon addition of one base pair at the end of an existing helix. Addition of an AU pair supplies only –4 to –5 kJ to ΔG_f°. The exact amount depends upon whether an A or a U is at the 5′ end in the existing helix. If an AU pair is added to the helix terminating in CG or GC, about –9 kJ/mol is added.

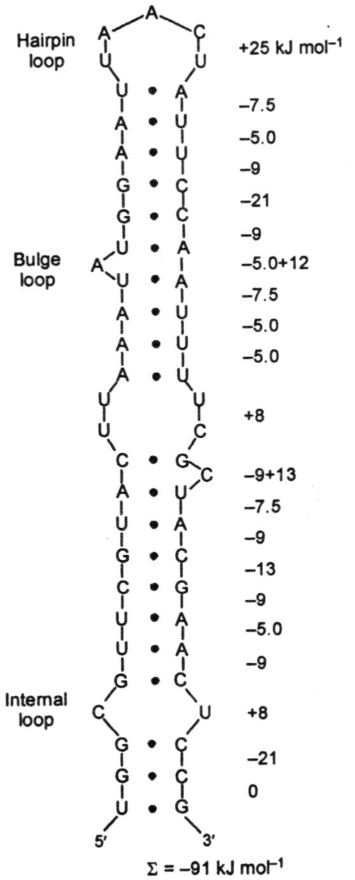

Fig. 1.8. The contribution of base-paired regions and loops to the Gibbs energy of a possible secondary structure for a 55 base fragment from R17 virus.

Larger increases in $-\Delta G_f^\circ$ result from addition of GC pairs, which contain three hydrogen bonds between the bases versus the two in AU pairs. UG pairs provide a very small amount of stabilization to an RNA double helix, while the presence of unpaired bases has a destabilizing effect. The most stable hairpin loops contain four or five bases. Depending upon whether the loop is "closed" by CG or AU, the helix is destabilized by 20-30 kJ mol. "Bulge loops," which protrude from one side of a helix, have a smaller destabilizing effect. An example of the way in which Table 1.2 can be used to estimate the energies of formation of a loop in a straight-chain RNA is illustrated in Fig. 1.9. Similar analysis of base pairing in DNA can also be done.

Table 1.2. Gibbs Energies of Formation ΔG_f° at 25°C for Addition of One Base Pair to an Existing RNA Helix.

Base pair at end of existing helix	*Base pair added*	ΔG_f° *(kcal mol^{-1} ± 10%)*	ΔG_f° *(kJ mol^{-1} ± 10%)*
A • U	A • U	−1.2	−5.0
U[a] • A	A • U	−1.8	−7.5
C or G[b] • • G C	A • U	−2.2	−9
C • G	G • C	−3.2	−13
G • C	C or G • • G C	−5.0	−21
G • U	U • G	−0.3	−1
Hairpin loops of 4 or 5 bases			
Closed by GC		+5	+21
Closed by AU		+7	+29
Bulge loops			
1 base		+3	+12
4–7 bases		+5	+21

[a] All base pairs in the table are oriented as follows : 5′—A→3′
3′←U—5′

[b] ΔG_f° is the same for U•A or added to an A•U end.

[c] ΔG_f° is the same for C•G or G•C added to an A•U end.

Tautomerism and Base Pairing

Tautomerism has an interesting relationship to the formation of the pairs and triplets of hydrogen bonds in DNA or RNA. Each base exists predominately as one preferred tautomer, but at any moment a very small fraction is present as less stable tautomers. Watson and Crick suggested that this fact may be responsible for the occurrence of some mutations.

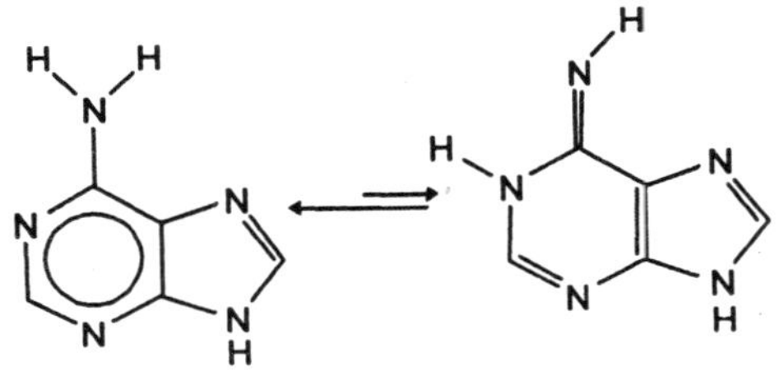

A B

Thus, tautomer B of Eq. 1.6 would not be able to pair with thymine, its proper pairing partner, but could pair with cytosine. Similarly, tautomer B of uracil in Eq. 1.4 could pair with guanine instead of its proper partner adenine. If a similar event occurred to an AT pair during gene replication an incorrect copy of the gene, differing in a single "code letter," would be formed.

However, because the tautomerism is affected so strongly by the environment the extent to which it may cause mispairing while replication enzymes act is uncertain. Both bases in a pair could be tautomerized by synchronous transfer of protons in two parallel hydrogen bonds. However, theoretical calculations predict a high energy barrier to this process. Proton transfer, which can also be induced by light, has been studied on a femtosecond scale. After photochemical transfer of one proton a second is transferred within a few picoseconds.

Conformations of Nucleotides

The furanose ring of ribose or deoxyribose is flexible and can be interconverted smoothly among an infinite number of envelope (E) and skew or twist (T) conformations. However, there are limits set by steric and anomeric effects. Conformations are often described by staring which atom in an envelope conformation lies mostly out of the plane of the other four atoms. If this atom lies aboce the ring, *i.e.*, toward the base, the ring conformation is known as *endo;* when below the ring, it is known as *exo*. The C(2′)-*endo*(^{2}E) and C(3′)-*endo* (^{2}E) conformations are most commonly approximated in nucleotides and nucleic acids.

The C(3′)-*exo* conformation is designated E_3, the twist conformation C(2′)-*endo*-C(3′)-*exo* as $^2T_3$, etc. A conformation can be specified more precisely by the five torsion angles ν_0 to ν_4 (which have also been designated τ_0 to τ_4). All of the envelope and twist conformers can be interconverted readily. The interconversions can be imagined to occur in a systematic way by *pseudorotation*, a rotation of the pucker around the sugar ring. In a commonly used convention a pseudorotation phase angle P ranges from 0° to 360° as the pucker moves *twice* around the ribose ring to restore the original conformation. *P* is taken as 0° for the symmetric $^3T_2$ twist, –18° for E_2, +18° for ^{3}E, +54° for ^{4}E, +198° for E_3, etc. The five torsion angles ν_j, for $j = 0 - 4$ as defined in the foregoing diagram, are related to *P* as follows:

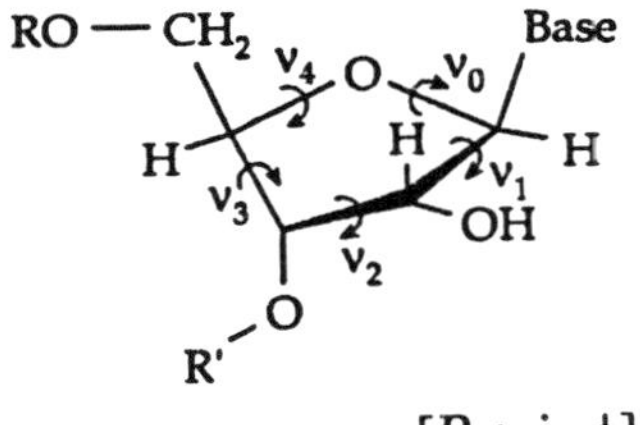

RO — CH2 Base O H H H OH O R′ ν_0 ν_1 ν_2 ν_3 ν_4

$$\upsilon_j = \upsilon_{max} \cdot \cos [P + j \cdot \phi]$$

where ϕ = 720°/5 = 144°. The maximum torsion angle, ν_{max}, is about 40°. Conformational alterations of ribose and deoxyribose rings can occur within polynucleotides and are of biochemical importance.

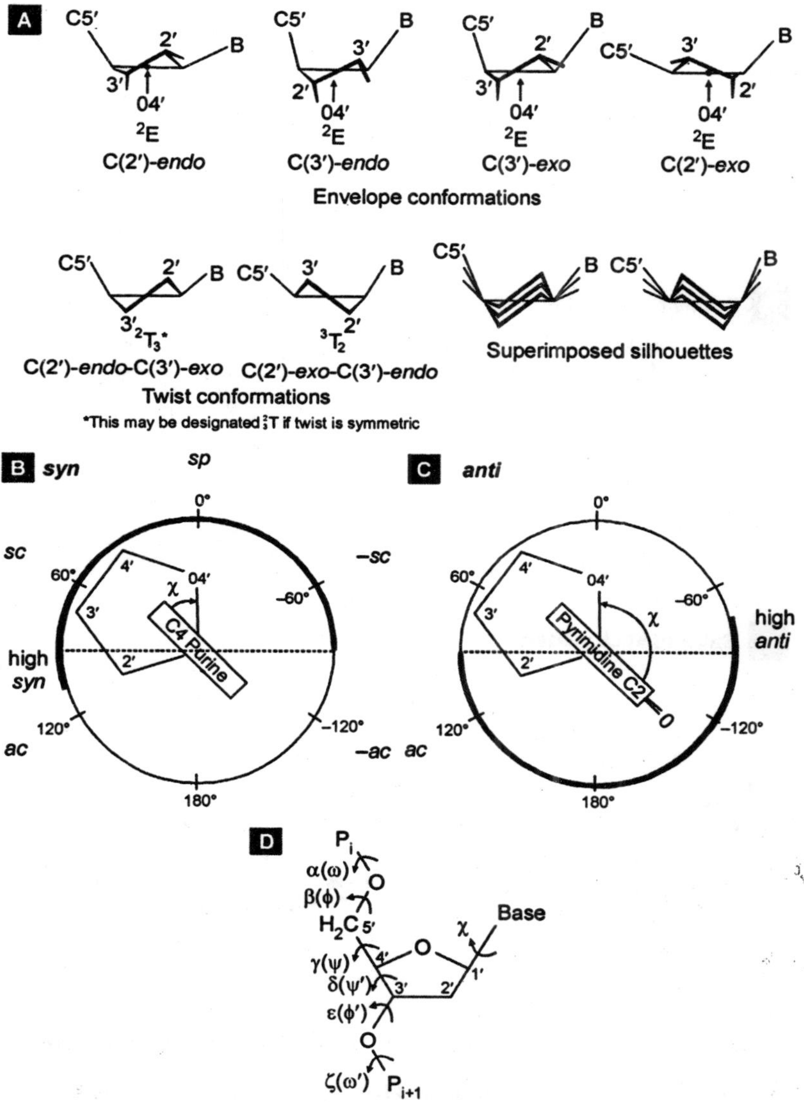

Fig. 1.9. Conformational properties of nucleosides. (A) Representations of several conformations of a ribose or deoxyribose ring in a nucieoside. (B) View down the N–C axis joining a purine base to the sugar in a nucieoside in a *syn* conformation. The atom marked C4 is in the six-membered ring, which protrudes further over the sugar ring. The angle χ is measured from 0° to ±180°. *Syn* conformations are those for which χ falls in the region of the heavy semicircle. (C) An *anti* conformation of a pyrimidine nucieoside. Other *anti* conformations fall in the region of the heavy semicircle. (D) Labeling of the conformational angles of the main chain in a polynucleotide.

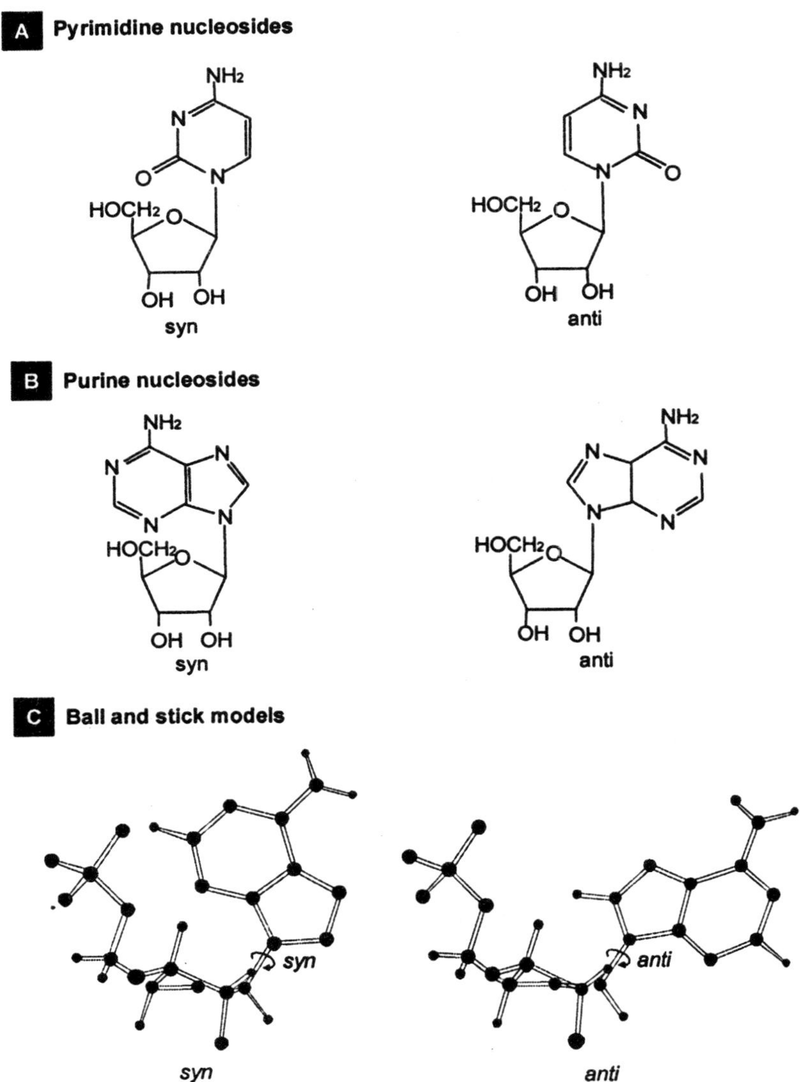

Fig. 1.10. Vies of a nucleoside in *syn* and *anti* conformations.

An interesting consequence of changes in ring conformation is that the distance between the C(5′) and N atoms attached to the sugar ring of a nucleoside may vary by as much as 0.05 nm. The orientation of a base with respect to the sugar is specified by the angle χ. In one convention, its zero value is taken as that in which the N(1) – C(2) bond of a pyrimidine or the N(9) – C(4) bond of a purine is *cis* to the C(1′) – O(4′) bond, [sometimes called the C(1′) – O(1′) bond]. However, other definitions have been used.

Typical values of χ for nucleotides and nucleosides vary between –75° and –165°. In these *anti* conformations the CO and NH groups in the 2 and 3 positions of the pyrimidine ring (or in positions 1, 2 and 6 of the purine ring) are *away* from the sugar ring, while in the *syn* conformations they lie over the ring. *Anti* conformations are more often present than *syn* in nucleic acids. An additional five torsion angles are needed to specify the backbone conformation of a polynucleotide.

According to the convention adopted by the IUB the six angles are designated α, (β, γ, δ, ε, and ξ, as is indicated. In an older but much used convention, starting at any phosphorus atom the angles ω, ϕ, and ψ specify the next three torsion angles as one moves toward the 3′ end of the chain, while ω', ϕ', and ψ', specify the angles lying toward the 5′ end (these are shown in parentheses in Fig. 1.10D. Notice that $\delta(\psi') = \nu_3 + 120°$.

DOUBLE HELICES

The B Form of DNA

This form of DNA, whose structure is depicted in Fig. 1.3 and 1.12, stable at high humidity and is thought to approximate mat of most DNA in cells. If we look directly at the axis of the double helix and perpendicular to one of the base pairs, and ignore the fact that the base pair is asymmetric, we see that the nucleotide unit in one chain is related to the nucleotide unit lying across from it in the opposite plane by a two-fold axis of rotation (*dyad axis*). This symmetry element, which arises from the antiparallel arrangement of the chains, makes the DNA molecules from the outside look nearly identical whether viewed from one end or the other – and whether viewed as a model by the human eye or through contact with an enzyme which might act on the molecule.

Actually, *the two chains are not identical,* and the genetic information can be read off from the functional groups exposed in two *grooves* in the surface of the helix. The broader groove in the B form, which is referred to as the *major groove,* is about 0.85 nm deep and 1.1 – 1.2 nm wide when allowance is made for the van der Waals radii of the atoms. In some other forms of DNA the major groove is narrow, but it can always be identified by the larger of the arcs that can be drawn between the two N–C bonds of the nucleosidic linkages in a nucleotide pair. The *minor groove* or narrow groove, which is defined by the smaller are between the two N–C fonds, is ~ 0.75 nm deep and 0.6 nm wide in B-DNA. The diameter of the double helix of B-DNA, measured between phosphorus atoms, is just 2.0 nm.

The rise per turn, the *pitch,* is 3.4 nm. There are about ten base pairs per turn (9.7 and 10.6 in two different crystal forms). Thus, the rise per base pair is 0.34 nm, just the van der Waals thickness of an aromatic ring. It is clear that the bases are stacked in the center of the helix. A 1000-bp (1-kb) gene would be a segment of DNA rod about 340 nm long, about 1/40 the length of the molecule in the electron micrograph. As is appropriate for the cell's master blueprint, DNA in the double helix is stable.

Factors contributing to this stability are (1) the pairs and triplets of hydrogen bonds between the bases; (2) the van der Waals attraction between the flat bases which stack together; (3) the fact that on the outside of the molecule are many oxygen atoms, some negatively charged, which are able to form strong hydrogen bonds with water, with small ions, or with proteins that surround the DNA; and (4) the ability to form superhelices. Nevertheless, the long DNA chains present in our chromosomes are frequently broken and an elaborate system of repair enzymes is needed to preserve the reliability of this master code for the cell. Today it is generally accepted that most DNA exists in nature as a double helix resembling the B form.

However, doubts were expressed as recently as 1980. The reason for the uncertainty lay in the fact that X-ray data from the stretched "paracrystalline" DNA fibers used in earlier studies are much less precise than those from true crystals. The X-ray data for B-DNA fibers could also be interpreted in terms of an alternative "side-by-side" structure which would permit easier separation of strands during replication. However, numerous high-resolution X-ray structure determinations on single crystals of synthetic DNA fragments have confirmed the double-helical structure and the presence of Watson-Crick base pairs. Similarly, fragments related to the double helices of RNA have been crystallized and the structures determined to atomic resolution.

The right-handed helical structure of DNA in solution has also been confirmed by independent methods based on electron microscopy, scanning probe microscopy, fluorescence resonance energy transfer, and computation. Dickerson and associates discovered, that B-DNA has a "spine" of water in the minor groove of regions rich in A-T base pairs. Two water molecules per base pair are hydrogen bonded to form a long chain. Half of the water molecules in the chain also form hydrogen bonds to oxygen and nitrogen atoms that are exposed in the minor groove while the others hydrogen-bond to a second chain of water molecules, forming a ribbon of hydration.

Additional water binds to polar groups in the major groove. The hydration pattern is largely *local, i.e.,* each base has characteristic hydration sites. Some of the hydration sites may be occupied partially by the nomovalent ions, Na^+ or K^+, depending upon the medium. Bound divalent metal ions are also sometimes seen in X-ray structures. However, most cations are thought to remain mobile.

Other Double-Helical Forms of DNA

The B form of fibrous DNA is stable under conditions of high (~93%) humidity but at 75% humidity it is converted into *A-DNA* in which the base pairs are inclined to the helix axis by about 13° and in which the ribose rings are primarily C3′-*endo* rather than C2′-*endo*. There are 10.9 base pairs per turn. In the A form the major groove is very deep (~1.35 nm) and narrower (~ 0.27 nm) than in the B form and extensively hydrated. The minor groove is wider (~1.1 nm) and shallower (~0.28 nm) than in the B form.

The crystal structure for a form intermediate between A and B has also been reported. Paracrystalline forms of DNA known as B′, C, and D have also been observed and others have been proposed. The most interesting additional form, *Z-DNA,* was discovered by X-ray studies of the alternating oligodeoxy-ribonucleotides d(CpGp-CpGpCpG) and d(CpGpCpG). The helix of Z-DNA is *left-handed.* The repeating unit consists of two Watson-Crick base pairs, the backbone following a zigzag pattern. There are 12 base pairs per turn. The cytosine groups have the usual *anti* conformation but the guanosine groups are *syn.* Some of the ribose rings have the C2′-*endo* conformation characteristic of B-DNA but some are C3′-*endo.* The alternating CpG sequence is not essential for Z-DNA formation.

However, alternating *syn* and *anti* conformations are important and purines assume the *syn* conformation more readily than do pyrimidines. The major groove of Z-DNA is shallow allowing it to accomodate bulky substituents at C8 of purines or C5 of pyrimidines. Such substituents favor Z-DNA: The B form is the most hydrated and most stable form of DNA under conditions of high humidity but even in solution it can be converted to A-DNA and

Z-DNA by a high concentration of NaCl. This is presumably because the salt dehydrates the DNA.

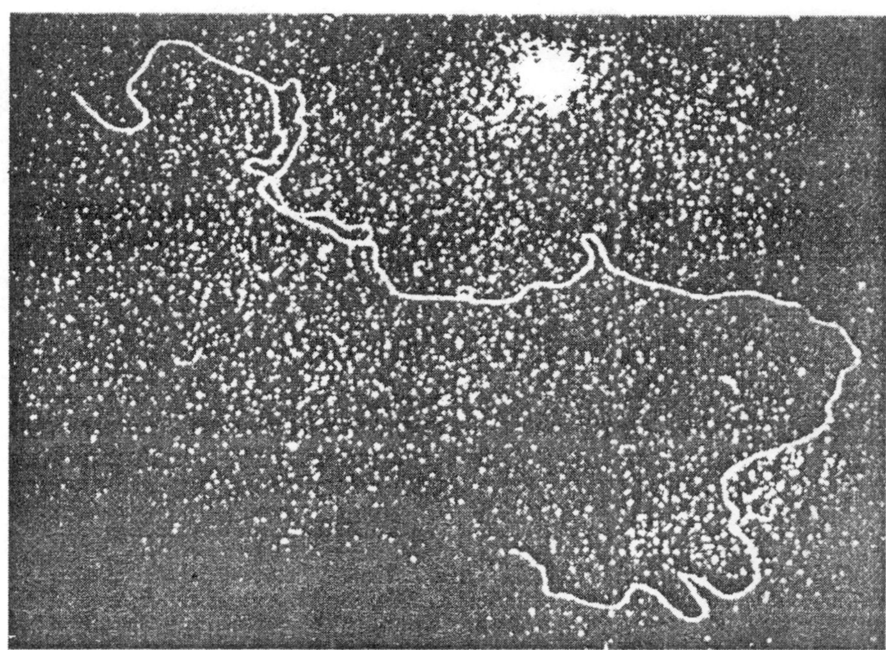

Fig. 1.11. Electron micrograph of a DNA molecule (from a bacterial virus bacteriophage T7) undergoing replication. The viral DNA is a long (~14 μm) duplex rod containing about 40,000 base pairs. In this view of a replicating molecule an internal "eye" in which DNA has been duplicated is present. The DNA synthesis was initiated at a special site (origin) about 17% of the total length from one end of the duplex. The DNA was stained with uranyl acetate and viewed by dark field electron microscopy.

Saenger and coworkers pointed out that the oxygen atoms of successive phosphate groups in the polynucleotide backbone of B-DNA are at least 0.66 nm apart too far apart to be bridged by a water molecule. However, the phosphates are individually hydrated. On the other hand, in A-DNA and Z-DNA the oxygen atoms on successive phosphates are as close as 0.53 and 0.44 nm, respectively. This allows one H_2O molecule to bridge between two phosphates, stabilizing these forms in environments of low humidity. This may be one factor that affects the B to A and Z transitions. In the narrow Z-DNA helix repulsion between negative charges on the phospho groups of opposite strands is strong.

By shielding these charges high salt concentrations also help to stabilize Z-DNA. Since the Z form of DNA is favored in regions rich in G-C pairs, it is reasonable to expect that it may occur in nature. Antibodies have been prepared which bind specifically to regions of DNA in the Z form and have been used to identify many such regions. Genetic studies in *E. coli* have also provided strong evidence that left-handed DNA sequences are formed *in vivo* in that bacterium. Z-DNA-forming sequences have also been found in the *Halobacterium* genome. Segments of Z-DNA may occur in control regions called *enhancers and* Z-DNA may also form behind RNA polymerase molecules that are moving along a gene while synthesizing mRNA.

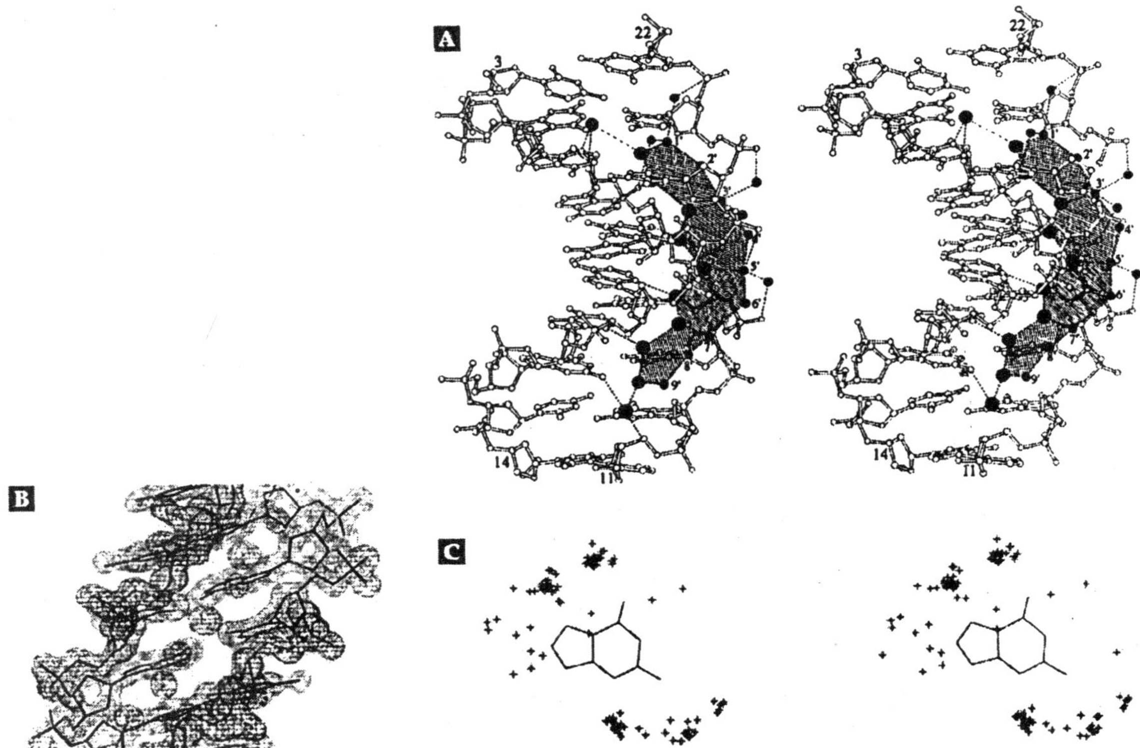

Fig. 1.12. (A) Stereoscopic drawing showing two layers of water molecules that form a "spine" or "ribbon" of hydration in the minor groove of B-DNA. The inner layer is shown as larger filled circles; water molecules of the outer layer are depicted with smaller dots and are numbered. Hydrogen bonds are shown as dashed lines. (B) Electron density map. (C) Stereoscopic representation of the superimposed electron densities of 101 water molecules observed to hydrate 14 guanine rings found in 14 B-DNA molecules for which high-resolution X-ray structures were available. Positions of 101 water molecules within 0.34 nm from any atom of the 42 guanines are plotted.

The associated negative supercoiling of the DNA could cause it to assume the Z form. This region of Z-DNA can, in turn, be a site for interaction with specific proteins. In the Watson-Crick structure the two strands are *antiparallel,* an essential for replication. However, stable segments of double-stranded DNA with *parallel* strands can also be formed and may occur in specialized regions of the genome.

The Conformational Flexibility of Double Helices

Local variations in the sequence of nucleotides affect the conformation of a DNA molecule and it is clear that the helix is not uniformly coiled throughout the entire length. While most helix segments probably have a right-handed twist others may be left-handed. Most DNA is probably in the B form but there are segments in the A form. These may arise from formation of hybrid duplexes with RNA, which assume the A conformation and are also favored by certain base sequences. Rules for predicting the DNA conformation from the nucleotide sequence have been proposed.

Table 1.3. Helix Parameters for Three Types of DNA

Parameter	*Form of DNA or RNA (A-form)*		
	B	*A*	*Z*
Helical twist, degrees mean	28–42 36 ± 4	16–44 33 ± 6	GC – 51 ± 2 CG –8.5 ± 1
Base pairs per turn	10.0 (9.7–10.6)	11-12	12
Helix rise per base pair, nm	0.34 ± .04	0.29 ± .04	CG 0.35 ± 0.2 CG 0.41 ± .02
Base inclination, degrees	–2.0 ± 5	13 ± 2	8.8 ± .7
Propeller twist, degrees	12 ± 5	6 ± 5	4.4 ± 3
Base roll degrees	–1.0 ± 5	6 ± 5	3.4 ± 2
Predominant conformation of deoxyribose	C2′-*endo*	C3′-*endo*	C C2′-*endo* G C3′-*endo*
Depth of grooves (nm)			
Major	0.85	1.35	very shallow
Minor	~0.75	~0.28	very deep
Width of grooves (nm)			
Major	1.1 – 1.2	0.27	broad
Minor	0.6	~1.1	narrow

In the simplest case we consider each pair of adjacent nucleotides in the double helix. There are 16 possible pairs in one chain. These can be designated as in the following examples: (AA, TT), (CG, CG), and (AG, CT). The first two letters within the parentheses represent the sequence (from 5′ to 3′) in one chain while the second pair of letters represent the sequence (again from 5′ to 3′) in the complementary chain. The rules state that (AA, TT) or (TT, AA) repeated in a sequence will stabilize the B form of DNA. Repetition of (CC, GG) or (GG, CC) will favor conversion to the A form.

Repetitions of (CG, CG) favor the Z form, especially if an alternating sequence of purines and pyrimidines is present throughout the (G + C)-rich region. Even within the regular B structure the torsion angles χ and δ variable. Their changes are highly correlated. As χ ranges from –140° to –90°, ζ varies between 80° and 160°. Other pairs of torsion angles are also correlated. Besides allowing for changes between B, A, and Z conformations, this flexibility of the DNA helix together with cooperativity with adjacent base pairs may allow transmission of conformational effects for some distances along a DNA helix.

Supercoiling, also affects the helical conformation. DNA can be stretched into yet another form or forms. Application of a force of 65 – 70 piconewtons (pN) stretches B-DNA by 70%. This "overstretched DNA" may also be important biologically.

Rotational and Translational Movements of Bases

In considering the conformational flexibility of a polynucleotide it is useful to define the parameters associated with movement of base pairs or individual bases. *A long axis* is established for each base pair in a structure for which X-ray data are available. Making use of these axes, we can move along the helix and measure the *angle of twist* from one base pair to the next. While the mean value of the twist for B-DNA is 36°, values of 28 – 42° have been observed in structures of oligonucleotides determined by X-ray analysis. The *base tilt* which is nearly zero for B-DNA, is less variable. Within each base pair there is also a *propeller twist* around the long axis. It averages about 12° for B-DNA.

In addition, the whole base pair may *roll, i.e.,* be rotated around its long axis by several degrees. These motions relieve steric interferences in the center of the helix, for example, those that would arise between purines in adjacent base pairs if the same helix parameters were imposed on each base pair. A more detailed analysis reveals the large range of motions, mostly small. The mathematics needed to deal with computer-based modeling of polynucleotide structures has been developed. Its application showed that there is a strong correlation between twist and roll. Typical average values of base tilt, propeller twist, and roll are given in Table 1.3 B-, A-, and Z-DNA.

Bends and Bulges

If we overlook the ridges and grooves on their surfaces the DNA structures, straight rods. However, real DNA rods are crooked and may contain distinct bends. The existence of *bent DNA* in nature was first discovered from study of fragments enzymatically cut from kinetoplast DNA of tropical parasites. The bent fragments moved anomolously slowly during electrophoresis in polyacrylamide gels of small pore size. Evidently, the bent shape impedes movement through the pores. Since the initial discovery, regions of bent DNA have been found in origins of replication and other specialized sites.

Bending of DNA because of local sequence variations may also be important to the positioning of nucleosomes. Several causes for DNA bending have been identified. For example, a sharp 26° bend is expected at the junction between B- and A-DNA segments. The presence of a thymine photodimer may cause a 30° bend. Some naturally occurring bent DNAs contain repeated (A+T)-rich sequences. This suggested that the sequence (AA, TT), containing adjacent thymine methyl groups, can be thought of as a wedge, consisting of both roll and tilt components. Such a wedge repeated at intervals averaging 10.5 bp along the helix might cause the helix to bend.

However, computer modeling experiments suggest that thymine methyl groups do not distort the DNA helix. The bending may be a result of the greater tendency for AT pairs to roll and to enhance the van der Waals and electrostatic attractions between the 2′-O of cytosine and the 2′-NH_2 of guanine in the CG base pairs of bent DNA. Bending can occur in both (A+T)- and (G+C)-rich regions and is often induced by binding to proteins or protein complexes that act in replication, transcription, and recombination. Various errors are made during replication or recombination of DNA. Incorporation of an incorrect nucleotide will cause a mismatched base pair in which proper hydrogen bonds cannot be formed.

Most of the mispaired bases that result from mistakes in replication are removed by repair processes. Those that remain can often assume alternative pairings that distort the helix only

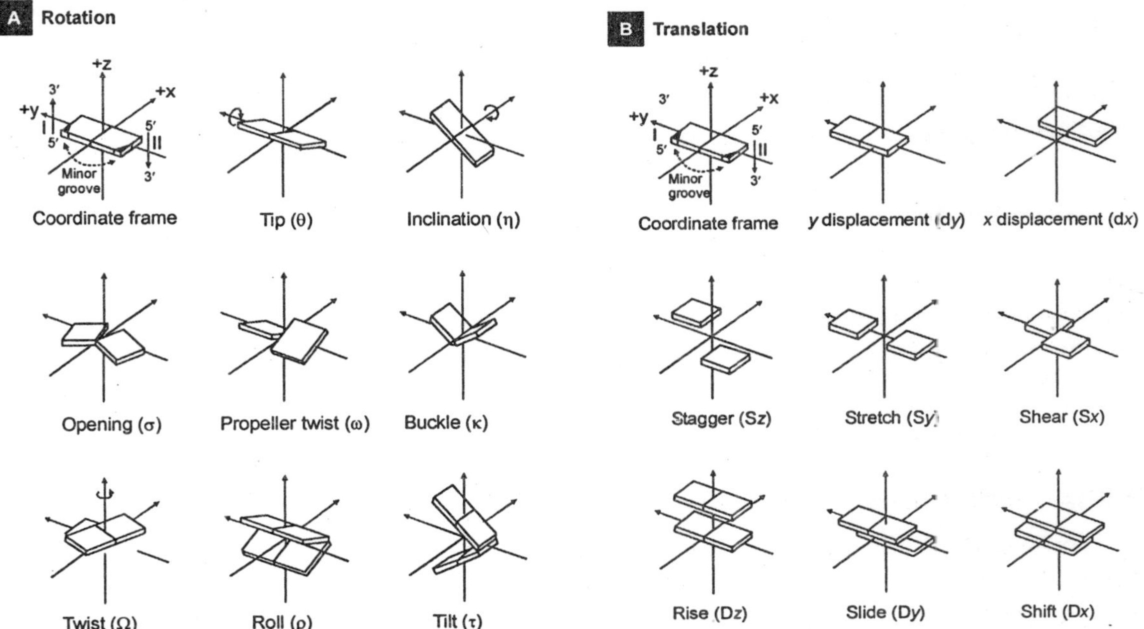

Fig. 1.13. (A) Drawing illustrating various rotational movements of bases in polynucleotides. *Upper two rows:* rotations of two bases of a pair. *Lower row.* rotations involving two successive base pairs. (B) Translational movements. *Upper two rows:* involving two bases of a pair. *Lower row:* two successive base pairs.

slightly. For example, a G-T wobble pair fits readily into an oligodeoxyribonucleotide double helix. Even a bulky G-A base pair causes little perturbation of the helix. Incorporation of an extra nucleotide into one strand of the DNA will create a bulge in the helix. NMR spectroscopic studies on bulged oligonucleotides have shown that the extra base can be stacked into the helix causing a sharp bend. However, some oligonucleotides with mismatched bases crystallize as straight helices with the extra nucleotide looped out. Mismatched base pairs tend to destabilize helices.

Interactions with Ions

Because each linking phospho group carries a negative charge (two charges per base pair, the behavior of polynucleotides is strongly affected by cations of all kinds. The predominant small counterions within cells are K^+ and Mg^{2+}. They are attracted to the negative charges on the polynucleotide backbone and, although they remain mobile, they tend to occupy a restricted volume. Some may bind in well-defined locations. Because of the presence of these positive ions the interactions of nucleic acids with cationic groups of proteins are strongly affected by the salt concentration. Organic cations compete with the simple counterions K^+ and Mg^{2+}.

Among these, the polyamines are predominant. Crystal structures have revealed that spermine binds across the deep grooves of tRNA and the major groove of B-DNA. Spermine also binds into the deep groove of A-DNA interacting by hydrogen bonding with bases in GTG sequence in both strands. It binds tightly to CG-rich sequences in the minor groove of Z-DNA, where it tightens the structure and shortens the helix. At higher concentrations of DNA, as occur in cell nuclei, spermidine induces the conversion of the DNA into liquid crystalline phases. Heavier metal ions and metal complexes can find sites on nitrogen atoms of the nucleic acid bases.

Examples are the platinum complex *cisplatin* and the DNA-cleaving antibiotic *neocarzinostatin*. Can metals interact with the π electrons of stacked DNA bases? A surprising result has been reported for intercalating complexes of ruthenium (Ru) and rhodium (Rh). Apparent transfer of electrons between Ru (II) and Rh (III) over distances in excess of 4.0 nm, presumably through the stacked bases, has been observed, as has electron transfer from other ions. Stacked bases are apparently semiconductors.

THE TOPOLOGY AND DYNAMICS OF NUCLEIC ACIDS

Rings, Catenanes, and Knots

While a DNA molecule may exist as a straight rod, the two ends are often covalently joined. Thus, the chromosomes of *E. coli* and of other bacteria are single closed circles. Circular DNA molecules are also found in mitochondria, chloroplasts, and many viruses. Further complexity arises from the fact that the circles of DNA are sometimes interlocked in chainlike fashion (*catenated*). An unusual example of this phenomenon is the presence of thousands of small catenated DNA circles in the single mitochondrion of a trypanosome. Sometimes circular DNA is *knotted*. Knots and catenanes often appear as intermediate forms during replication and recombination, especially involving circular DNA. Methods have been devised for synthesis of even very complex DNA knots. Let's look briefly at the topology of knots.

The three simple knots shown here have a chirality beyond that of the nucleotide units. This can be expressed by indicating the sign of each node in the knot:

Trefoil knot (–)

Figure-8 knot

Trefoil knot (+)

The node is negative if the crossing is like that in a left-handed supercoil and positive if like that in a right-handed supercoil:

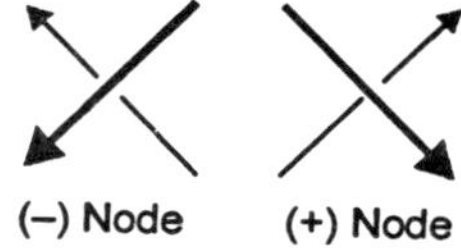

(–) Node (+) Node

Supercoiled DNA

Double-stranded DNA in solutions of low salt content usually assume the B-DNA conformation with 10.4 – 10.5 base pairs per turn. If the two ends are joined the resulting covalently closed circular DNA will be "relaxed." However, there are topoisomerases that act on this form of DNA by cutting both strands, holding the ends, and twisting the two chains. The energy source for the process is provided by cleavage of ATP. *DNA gyrase* untwists relaxed circular dsDNA one turn at a time and reseals the cut ends. A *reverse gyrase* from certain bacteria twists the relaxed DNA more tightly. In both cases the change causes the DNA to form *superhelical turns.* These may be either *solenoidal* or *plectonemically interwound* (as a twisted thread). The geometric and topological properties of closed supercoiled DNA molecules may be described by three quantities: The *linking number* (*Lk* also called the winding number, α), the *twist* (*Tw*), and the *writhe* (*Wr*). If a segment of double helical DNA were laid on a flat surface and the ends were joined to form a relaxed circle both *Lk* and *Tw* would equal the number of helical turns in the DNA. The writhe *Wr* would be zero.

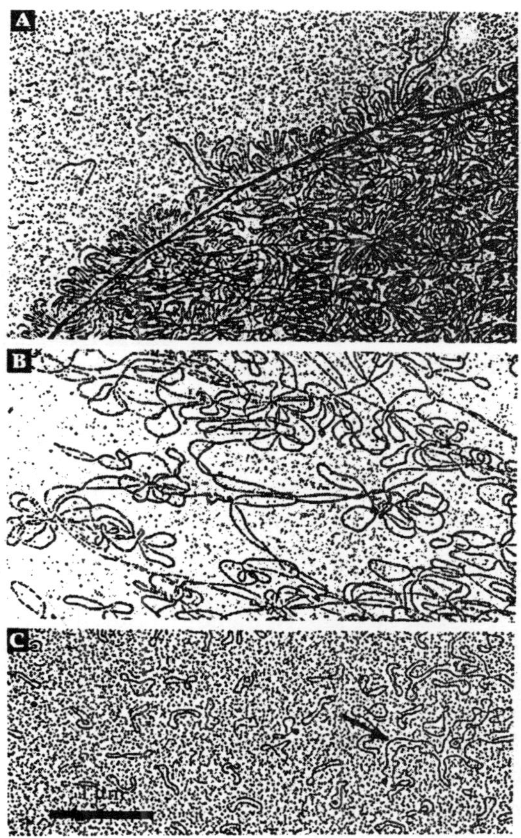

Fig. 1.14. (A) Electron micrograph of the network of catenated DNA circles in the mitochondrion of the trypanosome *Crithidia fasciculata.* (B) and (C) The same network after treatment with a *topoisomerase* from bacteriophage T4 that catalyzes a decatenation to form individual covalently closed circles. Five times as much enzyme was added in (C) as in (B). Two sizes of circles are present. Most are "minicircles", each containing about 2300 bp but a smaller number of larger ~35-kb "maxicircles" are also present. One of these is marked by the arrow.

The linking number is a topological property. It has an integral value which is unchanged if the DNA molecule is distorted. It can be changed only for DNA with open ends. When the ends are joined the linking (winding) number is constant unless one or both chains are cleaved. However, twist and writhe are geometric properties, which can change according to Eq. 1.8, while the *Lk* remains constant. The twist is related to the number of helical turns while the writhe is related to the number of superhelical turns.

$$Lk = Tw + Wr \qquad ...(1.8)$$

However, *Tw* and *Wr* do not usually have integral values. It is hard to define the relaxed state for a circular DNA. For example, changing the ionic composition will alter *Tw* and *Wr* according to equation 1.8. Both *Lk* and *Tw* are taken as positive for a right-handed *toroidal* (solenoidal) supercoil or a left-handed interwound twist.

For relaxed B-DNA, *Lk* is equal to the total number of base pairs in the circle divided by 10.5, *Wr* = 0, and *Tw* = *Lk*. Both *Lk* and *Tw* are positive in the right-handed B- and A-DNA forms, but *Tw* is negative for Z-DNA. In closed circular DNA the value of *Wr* is usually negative, the secondary structure being a fully formed Watson-Crick helix but with right-handed interwound superhelical turns or left-handed toroidal superhelical turns. The helix is said to be *underwound* ($Lk < Tw$). Some of the topological properties or doublestranded DNA can be demonstrated by twisting together two pieces of flexible rubber tubing whose ends can be joined with short rods to form a closed circle.

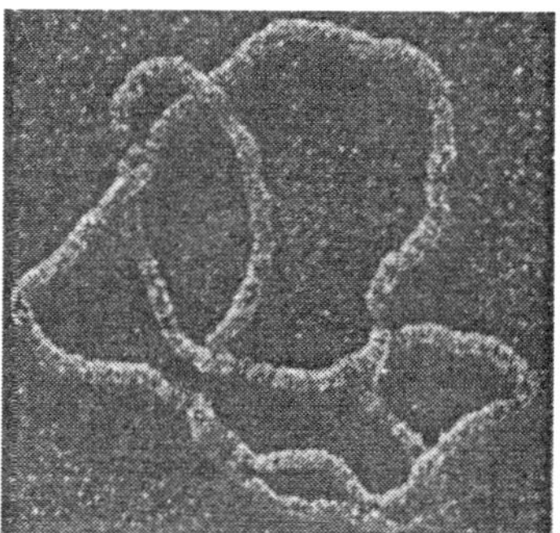

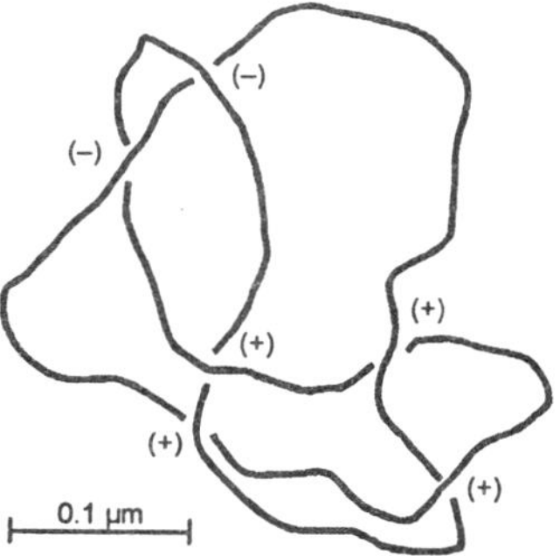

Fig. 1.15. Electron micrograph of a six-noded knot made by the Tn3 resolvase which is involved in movement of the Tn3 transposon from one location to another within the genome. Putative six-noded knot DNA was isolated by electroelution from an agarose gel. The knots, which are nicked in one strand, were denatured to allow the nicked strand to slide away and leave a ssDNA knot. This was coated with *E. coli* recA protein to greatly thicken the strand and to permit the sign of each node (designated in the tracing) to be seen.

Twist the tubing in a right-handed fashion as tightly as possible without causing supercoiling ($Lk = Tw$; $Wr = 0$). If the ends are now joined the circle will be relaxed. Now twist one turn tighter before joining the ends. A right-handed toroidal supercoil will be formed ($\Delta Wr = 1$; $\Delta Lk = 1 + Tw$; *Tw* is the same as before). If twisting is continued until several supercoils appear before the ends are joined, two interconvertible forms result. One form has right-handed toroidal supercoils and the other left-handed interwound supercoils. If relaxed circular DNA is unwound by one turn by cutting and resealing one chain, a single right-handed interwound supercoil will be formed ($\Delta Wr = -1$, $Lk = Tw-1$).

On the other hand, if the two chains in a closed relaxed helix are pried apart, as happens during intercalation *Lk* must remain constant, *Tw* will decrease, and *Wr* will increase with appearance of left-handed supercoiling. During replication of DNA pairing of some bases associated with the replication apparatus is prevented.

Upon release of the constraint the newly replicated DNA forms base pairs and becomes supercoiled. Since *Lk* is constant in circularly closed DNA a change of one turn of B-DNA ($Tw = 1$) into a turn of Z-DNA ($Tw = -1$) will cause the writhing number *Wr* to change by –2. Conversely, if the writhing number is forced to change by –2, a turn of Z-DNA may develop somewhere in a suitable (G + C)-rich region of the DNA.

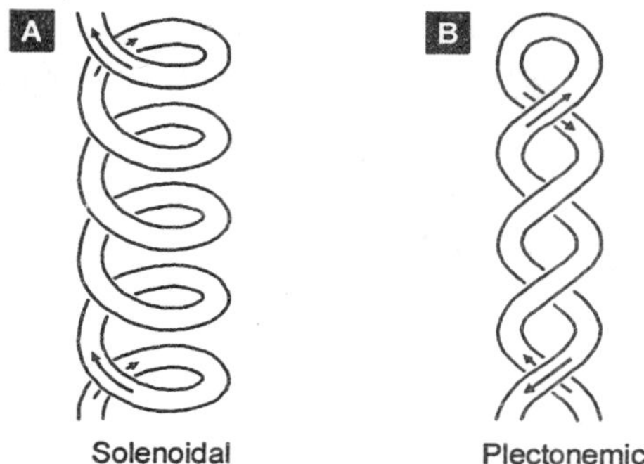

Fig. 1.16. Two forms of supercoiling of a DNA duplex: (A) Solenoidal. (B) Plectonemically interwound. These are both negatively supercoiled, as can be deduced from the arrows at the nodes, but the solenoid is left-handed and the plectonemic form right-handed.

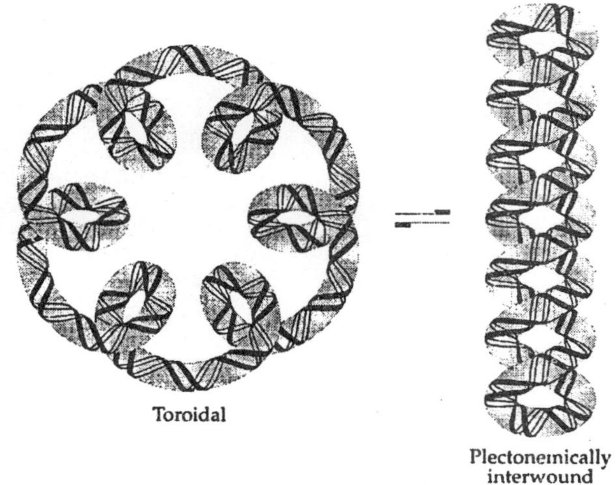

Fig. 1.17. Topological equivalence of toroidal (solenoidal) and plectonemically interwound forms of a circular DNA. These two forms have a constant value of the linking number *Lk* (or α), the twist *Tw*, and writhing number *Wr*.

Some of the information about supercoiled DNA can be summarized as follows:

Relaxed circular DNA	*Lk = Tw, Wr = 0*
Underwound circular DNA	*Lk* < *Tw*; *Wr* is negative; right-handed interwound supercoils or left-handed toroidal supercoils may be present. Alternatively *Lk* = *Tw* and left-handed Z-DNA regions may appear.
DNA with intercalated molecules	DNA is partially untwisted; *Tw* is lowered; *Wr* is increased with decrease in number of negative supercoils. As *Tw* approaches *Lk* the DNA becomes relaxed.

The *superhelix density* of a DNA molecule is often expressed as $\sigma = Wr/Tw \approx$ number of superhelical turns per 10 bp. In most naturally occurring circular DNA molecules σ is negative, a typical value being –0.05 (~5 negative superhelical turns per 1000 bp). The presence of superhelices in circular DNA molecules can be recognized readily by its effect upon the sedimentation constant of the DNA. Naturally occurring supercoiled DNA from polyoma virus sediments rapidly but after nicking of one of the strands of the double helix by brief exposure to a DNA-hydrolyzing enzyme the resulting relaxed form of the molecule sediments more slowly.

Supercoiling lowers the viscosity of solutions of DNA and increases the electrophoretic mobility and may also be recognized by electron microscopy. Naturally occurring or artificially prepared supercoiled DNA molecules can often be separated by electrophoresis into about ten forms, each differing from the other by one supercoiled turn and by ΔLk: = ± 1.

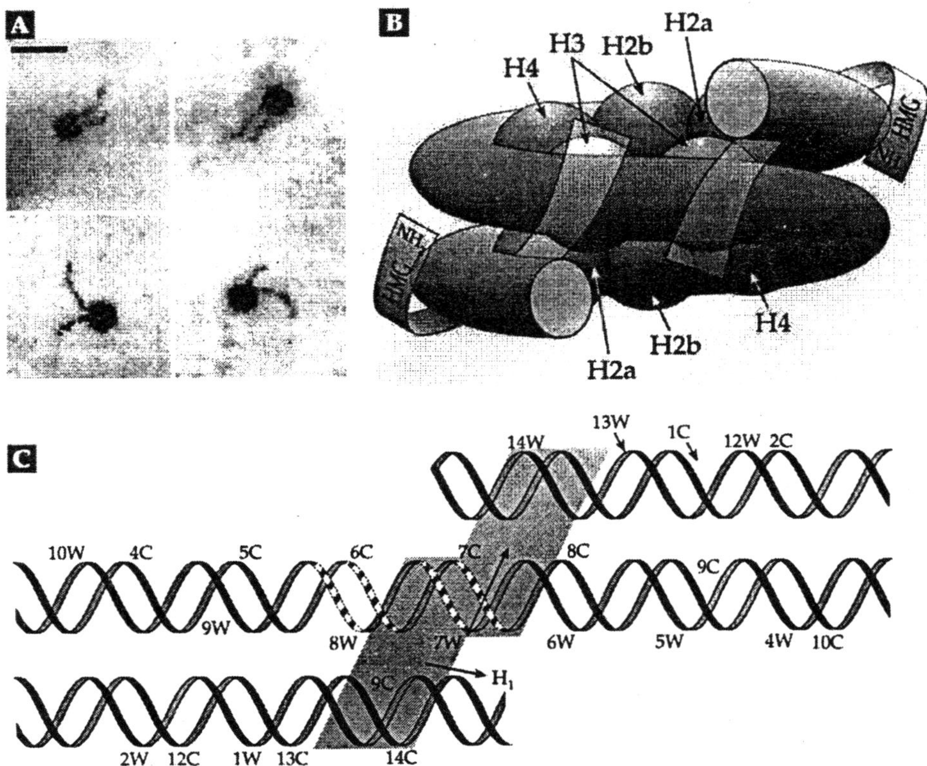

Fig. 1.18. Nucleosomes. (A) Electron micrographs of individual nucleosomes reconstituted from 256-bp DNA fragments and separated proteins. (B) Model of a nucleosome core. The 1.75-turn (145-bp) DNA superhelix winds around the histone octomer which consists of two subunits apiece of histones H2A, H2B, H3, and H4. In addition, two elongated molecules of proteins HMG-14 or HMG-17 are indicated. (C) Schematic radial projection of the double-helical DNA showing areas protected from cleavage by hydroxyl radicals by the bound proteins. The shaded areas are those protected by HMGs. The zigzag lines near the dyad axis indicate the most prominent regions of protection.

The relative amounts of these *topological isomers* form an approximately Gaussian distribution. The isomers apparently arise as a result of thermal fluctuations in the degree of supercoiling at the time that the circles were enzymatically closed. Why is DNA in cells supercoiled? One effect of supercoiling is to contract the very long, slender double helices into more compact forms. In eukaryotic cells much of the DNA exists in *nucleosomes*. Each bead-like nucleosome consists of a core of eight subunits of proteins called *histones* around which an ~ 140 bp length of DNA is coiled into two negative, left-handed toroidal superhelical turn.

There is some spacer DNA between nucleosomes. Otherwise a nucleosome providing two superhelical turns per 200 bp would produce a superhelix density of –0.1, twice that observed. A detailed analysis of the geometric properties of DNA wrapped around nucleosomes or other protein particles has been developed. Nucleosome formation is thought to protect DNA and to keep it in a more compact state than when it is fully active and involved in transcription. Nucleosomes, are also important in the regulation of transcription. Even though nucleosomes as such are absent, bacterial DNA also has a superhelix density of –0.05, apparently a result of interaction with other proteins such as the histonelike HU. Interaction with smaller molecules can also affect supercoiling. The presence of naturally supercoiled DNA and a variety of topoisomerases suggests that the control of DNA supercoiling is biologically important.

In fact, topoisomerases play essential roles in both replication of DNA and in transcription of genes. Supercoiling requires energy and topoisomerases that induce supercoiling must provide energy, *e.g.*, by cleavage of ATP. Conversely, supercoiled DNA can be a source of energy for biological processes. It has been estimated that the ΔH, ΔS, and ΔG per mole for formation of a single superhelical turn are 35.9 kJ, 68 J/°K, and 14.6 kJ, respectively.

A reduction in superhelix density is accompanied by a decrease in Gibbs energy and can therefore be coupled to other processes that have positive values of ΔG. An example is conversion of (G + C)-rich regions into the Z form of DNA, which is favored by negative supercoiling. Supercoiling affects binding of various proteins to DNA as well as intercalation (discussed in the next section) and formation of cruciform structure. The folding of DNA into compact forms, such as that in chromosomes, is also influenced by supercoiling. In the absence of nucleosomes supercoiled DNA may assume the plectonemic form.

Segments of the resulting rods may aggregate side-by-side in a liquid crystalline state. The relatively high cation concentration within cells favors this transformation. Polyamines such as spermidine are especially effective in promoting aggregation of DNA, in formation of Z-DNA, and possibly in facilitating cooperative processes that require that two DNA molecules interact with each other.

Intercalation

Flat, aromatic, hydrophobic rings are often able to insert themselves between the base pairs of a DNA duplex. Such *intercalation* is observed for many antibiotics, drugs, dyes, and environmental pollutants. Among them are proflavine, ethidium bromide, actinomycin, hycanthone, and daunomycin. Hycanthone, employed in the treatment of schistosomiasis, is one of the most widely used drugs in the world. Since intercalating agents can be mutagenic, such drugs are not without their hazards. Intercalation is often used to estimate the amount of negative Supercoiling of DNA molecules.

Varying amounts of the intercalating agent are added, and the sedimentation constant or other hydrodynamic property of the DNA is observed. As increasing intercalation occurs, the secondary turns of DNA are unwound (the value of *Tw* decreases). Each intercalated ring causes an unwinding of the helix of ~26°. Since for a closed covalent duplex the value of *Lk* in Eq. 1.8 constant, the decrease in *Tw* caused by increased intercalation leads to an increase in the value of *Wr*, which is usually negative for natural DNA. When sufficient intercalation has occurred to raise *Wr* to zero, a minimum sedimentation rate is observed. Addition of further intercalating agent causes a positive supercoiling. When the "replicative form" of DNA of the virus ϕX174, a small circular molecule containing ~ 5000 bp, was treated with proflavine the binding of 0.06 mol of proflavine per mole of nucleotides reduced *Wr* to zero.

From this it was estimated that σ = 0.055, corresponding to –27 superhelical turns at 25°, pH 6.8, ionic strength ~0.2. Changes in temperature, pH, and ionic environment strongly influence supercoiling. In general σ becomes less negative by ~3.3×10^{-4} per degree of temperature increase. For example, the observed value of σ for ϕX174 DNA was –0.059 at 15°C and –0.040 (–20 superhelical turns) at 75°C at an ionic strength of ~0.2. The exact ways in which intercalating substances can fit between the base pairs of nucleic acids are being revealed by X-ray diffraction studies of complexes with nucleosides, dinucleotides, and other oligonucleotides.

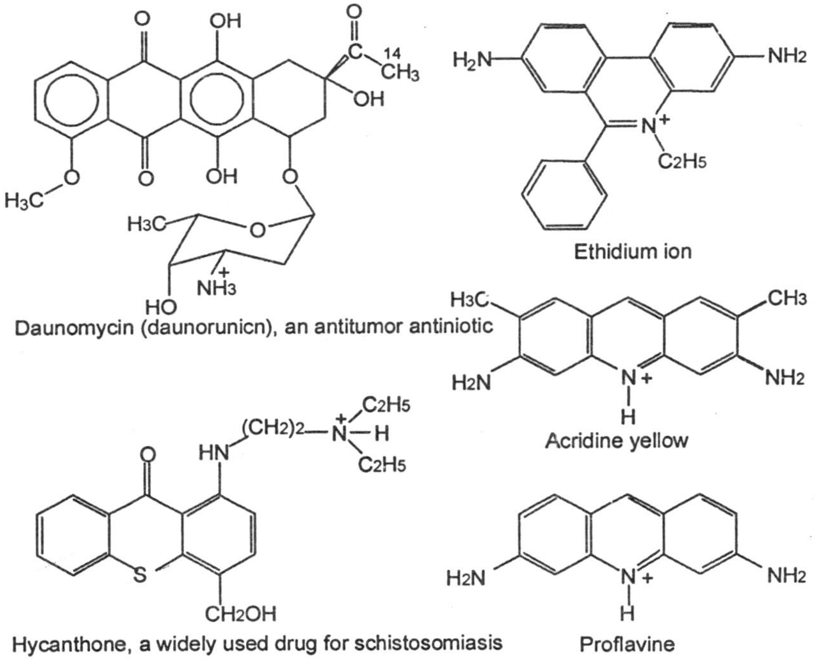

Daunomycin (daunorunicn), an antitumor antiniotic

Ethidium ion

Acridine yellow

Hycanthone, a widely used drug for schistosomiasis

Proflavine

Fig. 1.19. Structures of some substances that tend to "intercalate" into DNA structures.

The structure of a complex in which daunomycin is intercalated between two GC base pairs in DNA is shown as in Fig. 1.23. Daunomycin and other related anthracycline antibiotics also have an amino sugar ring that binds into the minor groove of the DNA, providing both electrostatic stabilization and hydrogen bonding. Substituents on this aliphatic ring also hydrogen bond to DNA bases. Does intercalation of flat molecules into nucleic acid chains have a biochemical function?

Aromatic rings of amino acid side chains in proteins designed to interact with nucleic acids may sometimes intercalate into nucleic acid helices serving a kind of "bookmark" function. Changes in superhelix density caused by such intercalation may be important in the orderly handling of DNA by enzymes within cells.

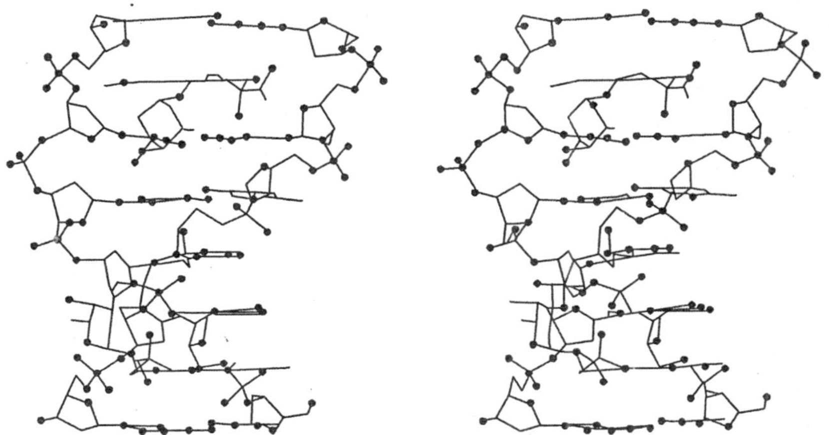

Fig. 1.20. Stereoscopic drawing showing a molecule of daunomycin intercalated between two base pairs in a molecule of double-helical DNA, d(CGTACG). Nitrogen and oxygen atoms are shown as dots. Both daunomycin and adriamycin (doxorubicin; 14-hydroxy-daunomycin) are important but seriously toxic anticancer drugs.

Polynucleotides with Three or Four Strands

Some nucleotide sequences in DNA favor the formation of a regular triple-helical (*triplex*) structure. This is possible when there are long stretches of adjacent pyrimidines having any sequence of C and T in one strand of double-helical DNA. The other strand of the DNA will contain the correct purines for formation of Watson-Crick base pairs. With such a structure it is always possible to add a third strand of a polypyrimidine using Hoogsteen base pairing. Triads formed in this way contain either two T's and one A or two C's and one G. In the latter case one of the C's must be protonated to allow formation of the pair of hydrogen bonds. These triplex structures can be formed only for stretches of DNA containing all pyrimidines in one strand (a *homopyrimidine* strand) and all purines in the other (a *homopurine* strand). A poly (AAU) triplex can also be formed, using the hydrogen bond pattern of the first triplet in Fig. 1. Triple-stranded synthetic polynucleotides of these types were prepared by Felsenfeld and others as early as 1957. The third strand in a triplex can be either parallel or antiparallel to the homopurine strand. The third strand is either homopurine or a mixture of purine and thymine.

The triplets are G•C*C, A*A•T, or T*A•T and are formed by Watson-Crick and *reversed Hoogsteen* pairing. Recently, there has been a renewed interest in DNA triplets because of their occurrence in natural DNA, their possible importance in genetic recombination, and the potential for design of powerful inhibitors of replication and transcription that function via triplex formation.

H-DNA

Strands of DNA contain many homopyrimidine "tracts" consisting of repeated sequences of pyrimidines, *e.g.*, d(T-C)$_n$, which may be abbreviated more simply as (TC)$_n$. In this example the complementary strand would contain the two purines in the repeated sequence (GA)$_n$. At low pH, where protonation of the cytosine rings occurs, or in negatively supercoiled DNA, the two strands of the repeating sequence may separate, with the (TC)$_n$ strand folding back to form a triple helix in which the base triplets have the hydrogen bonding pattern of Fig. 1.7. The resulting structure, which is shown in Fig. 1.28 known as *H-DNA*. A variety of relatect "nodule" and looped forms of DNA can also be formed.

R-DNA

A different type of triplex DNA may be formed during genetic recombination. A Watson-Crick duplex is brought together by one or more proteins with a single strand that is, for at least a considerable distance, an exact copy of one of the strands of the duplex. It is within such a triplex that cutting of a strand of the duplex takes place to initiate recombination. Can a triplex structure containing two identical chains be formed? Possible base triplets include the following C•G*G triplet which occurs in a crystalline oligonucleotide structure. This triplet contains a variation on.

Hoogsteen pairing and is related to the first triplet in Fig. 1.7. The other triplets needed for the proposed *R-DNA* recombination intermediates are G•C*C, T•A*A, and A•T*T. While there is keen interest in R-DNA the formation of a stable intermediate triplex is still uncertain.

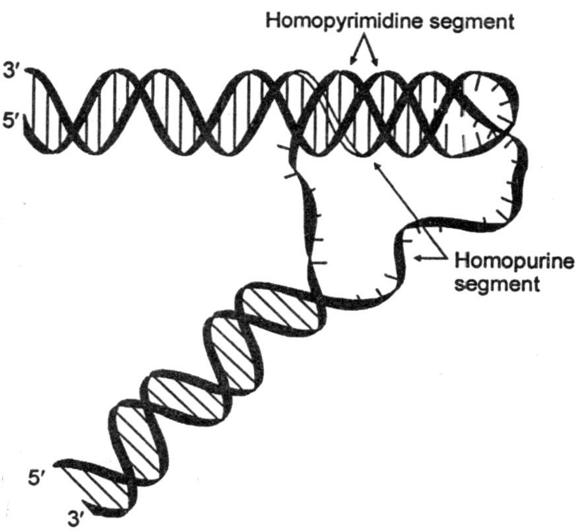

Fig. 1.21. Proposed structure of H-DNA which can be formed when a homopyrimidine segment and its complementary homopurine segment separate as a result of protonation of the cytosine rings or of negative supercoiling Stress.

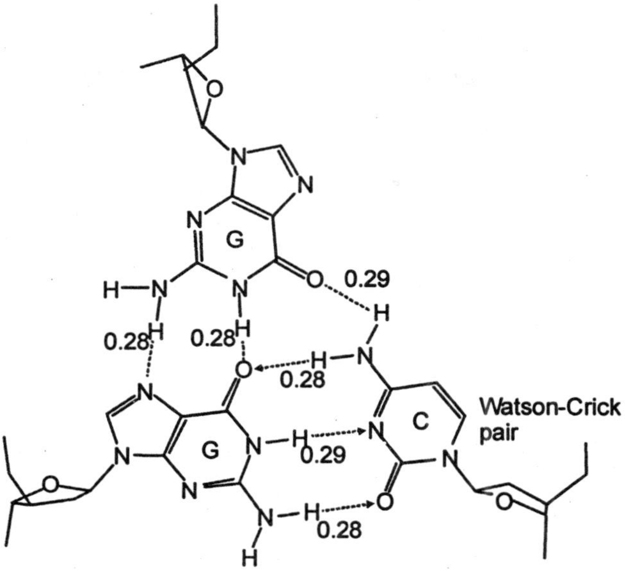

G
0.29
H—N
0.28
0.28
0.28
C
Watson-Crick pair
G
0.29
0.28

Designed third Strands

If triplex DNA segments can form naturally it should be possible to design oligonucleotides that will bind into the major groove of a DNA duplex to form a triplex at a specific location or locations in the genome. Dervan and associates are studying this approach systematically. Such oligonucleotides may be modified chemically to provide stronger binding to targeted locations and may prove to be useful therapeutic agents. Among these are synthetic phosphoramidates in which the phosphodiester linkage between nucleosides is replaced by 3′–NH-PO_3^-–O-5′. The resulting oligonucleotide is resistant to digestion by phosphodiesterases present in cells. Its geometry favors binding to A-DNA or RNA.

Another oligonucleotide mimic, dubbed PNA (peptide nucleic acid), contains monomer units of the following type and is able to form a triple helix: Yet another approach is to synthesize a "hairpin polyamide" that contains pyrrole, hydroxypyrrole, and imidazole groups in a sequence that favors tight binding to a specific dsDNA sequence. Some possible geometries for triplex DNA are illustrated in Fig. 1.25. These are based on computer-assisted modeling.

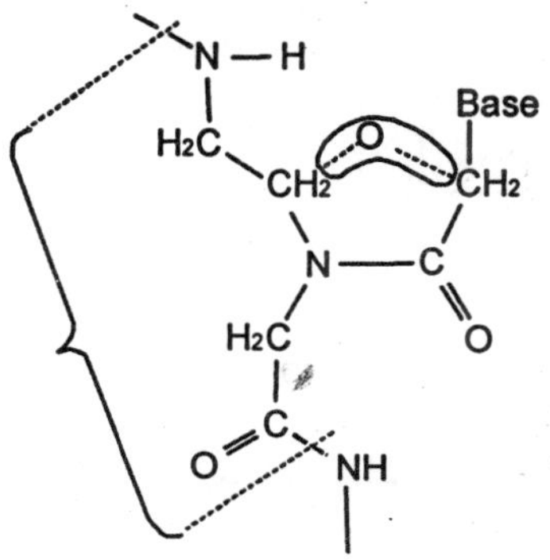

N—H
Base
H_2C
CH_2
CH_2
N—C
O
H_2C
C
O
NH

Tetraplex (Quadruplex) Structures

The ends of linear chromosomes, the *telomeres*, have unusual nucleotide sequences repeated hundreds or thousands of times. There is usually a *guanine-rich strand* running 5′ to 3′ toward the end of the chromosome and consisting of sequences such as TTAGGG in vertebrates, TTTTGGGG in the ciliate *Oxytricha,* and TG^{1-3} in *Saccharomyces cerevisiae.* The complementary strand is cytosine rich. The guanine-rich strand is longer than its complement, "overhanging" by about 2 repeat units. The thing that has attracted most attention to telomeric DNA is the unusual structure of the G-rich strand that was signaled by the first NMR studies.

Subsequent investigation revealed the presence of G quartets which are apparently stacked in folding patterns such as the following:

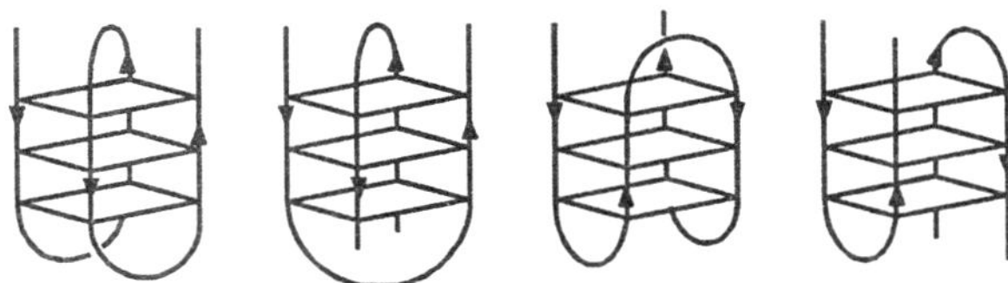

These structures are stabilized by the presence of univalent cations, K^+ being more effective than Na^+, Li^+, Rb^+, or Cs^+. A protein isolated from *Tetrahymena* binds specifically to the G4-DNA regions. The presence of 5-methylcytosine (m^5c) also stabilizes quadruplex structures. The oligoribonucleotide UG_4U forms G_4 quartets so stable that it takes days at 40°C in D_2O for the hydrogen-bonded NH protons forming the quartet to be exchanged for ^2H. Tetraplex structures have also been observed for G-rich repeating sequences associated with the human *fragile X syndrome.*

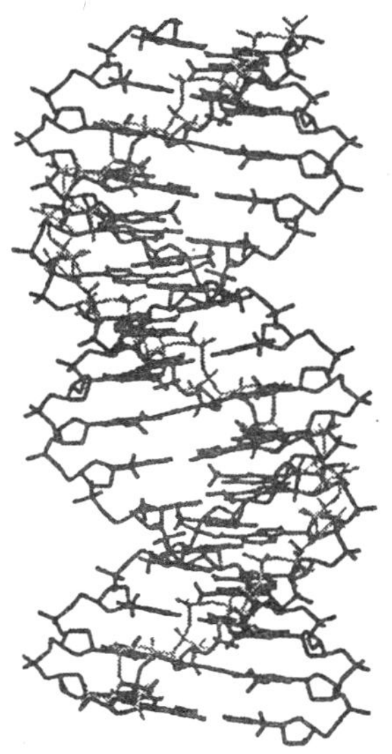

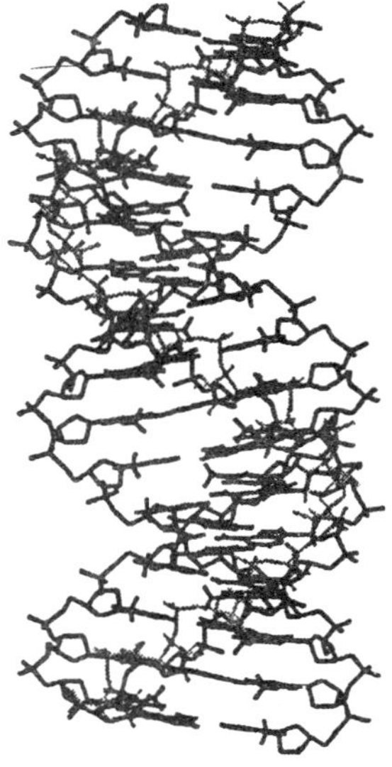

Fig. 1.22. Two superposed stereoscopic diagrams illustrating the lowest energy triple helical conformations of one pyrimidine, one purine, and one Hoogsteen DNA strand (in black). Superposed is an RNA purine strand in gray.

This is the most common cause of inherited mental retardation and appears to arise as a result of the presence of an excessive number of repeats of the trinucleotide sequence $(CGG)_n$. For normal persons $n = 60$ or less; for healthy carriers n may be as high as 200 but for sick individuals it may be much higher. The structure in solution, as determined by NMR spectroscopy is shown in Fig. 1.26. Another variant of four-stranded DNA, which arises from cytosine-rich DNA, contains C•CH⁺ pairs such as the following at low pH. In sequences such as $d(TC_5)$ and $d(C_3T)$ these C•CH⁺ pairs are intercalated.

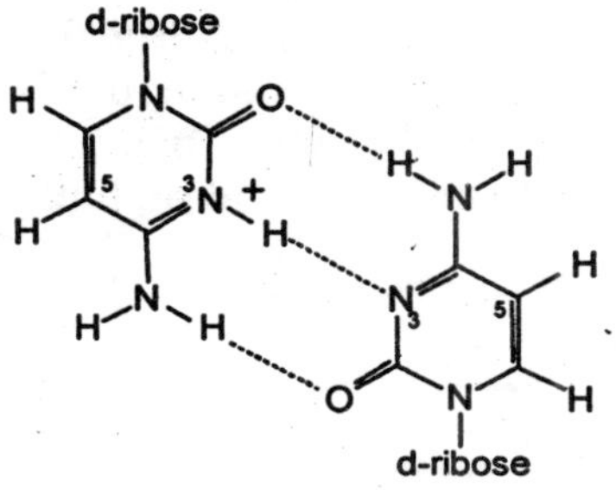

Hemiprotonated d(CH⁺• C)

This intercalated DNA (*I-DNA*) may provide an alternative conformation for some telomeric sequences. An I-DNA motif has also been identified in oligonucleotides from the DNA of human centromeres. The seemingly unusual forms of DNA described in this section may represent only a fraction of the naturally occurring DNA structures of biological significance.

Junctions

Special structural features may be found at junctions between different types of DNA, *e.g.*, between A-DNA and B-DNA. However, the most interesting junctions are *branched*. For example, Fig. 1.28 shows a four-way junction in which all or the bases form Watson-Crick pairs. This junction is better known as a *Holliday junction* because it was proposed by Holliday in 1964 as an intermediate in genetic recombination. As shown of the top of Fig. 1.28 the junction is formed from *two homologous DNA duplexes*. These are identical except for the boxed and shaded base pairs. The ends of the first duplex are marked I and II and those of the second III and IV.

The Holliday junction appears to arise by cleavage of one strand of each duplex with rejoining of the strands as indicated by the green

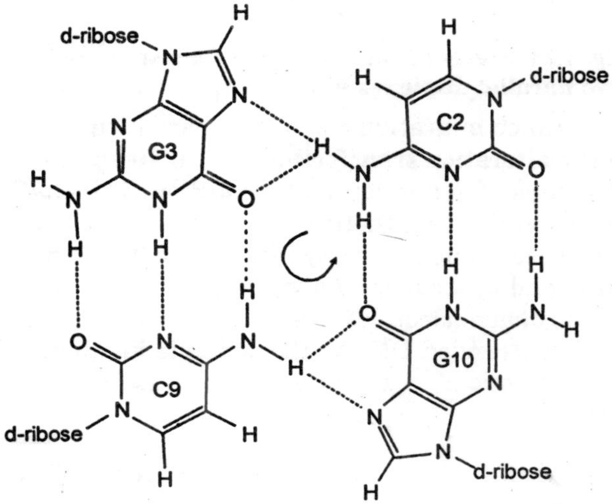

Fig. 1.23. Structure of a G•C*G•C* tetrad present in a quadruplex structure formed by the oligonucleotide d(GCG-CTTTGCGC) in Na⁺-containing solution.

arrows. Rotation gives the untwisted Holliday junction structure. The postulated three-dimensional structure of the junction is indicated in Fig. 1.28B. An important characteristic of Holliday junctions formed from homologous duplexes is that they can move by a process called *branch migration.*

Because of the twofold symmetry of the branched structure the hydrogen bonds of one base pair can be broken while those of a new base pair are formed, the branch moving. Notice that, in this example, the non-homologous base pairs TA and GC have become *mispaired* as TG and AC after branch migration. More significantly, the junction may be cut by a *resolvase* at the points marked by the green double-headed arrow. If the strand break is then resealed, and the DNA strands are replicated, the boxed base pair in duplex I-II will have been transferred into a strand replicated from III-IV and genetic recombination will have been accomplished.

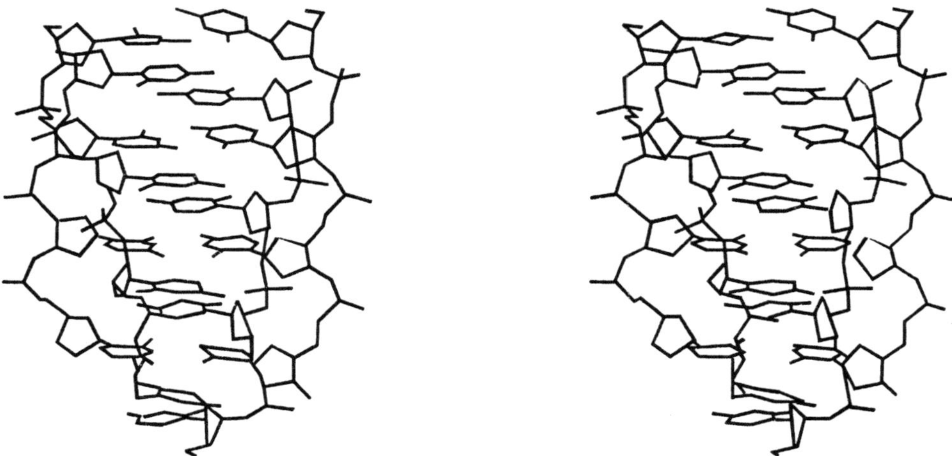

Fig. 1.24. Stereoscopic view of a four-stranded intercalated DNA or I-DNA formed from d(C_4). Two parallel duplexes with C•CH^+ pairs are intercalated into each other.

Branch migration can occur over much longer distances than are indicated in this figure, so the alteration transferred may be far from the site of the initial cleavage and whole genes or groups of genes can be transferred. The Holliday junctions formed during recombination are mobile, but synthetic *immobile Holliday junctions* can be synthesized by using nonhomologous base sequences or by locking the junctions. This has permitted careful physical study of these and other more elaborate synthetic junctions. With suitable choices of base sequences for the oligonucleotides from which they are made, such junctions will assemble spontaneously. Double-stranded DNA with immobile junctions is a very suitable construction material on a "nanochemical" scale. It has been assembled into knots, rings, cubes, and more complex polyhedra.

RIBONUCLEIC ACIDS (RNA)

The best known forms of RNA are: (1) the long chains of messenger RNA (mRNA), which carry genetic messages copied from DNA to the ribosomes where proteins are made; (2) the much shorter transfer RNAs (tRNAs) which participate in reading the genetic code, correctly placing each ammo acid in its sequence in the proteins; and (3) ribosomal RNAs (rRNA), which provide both structural material and a catalytic center for peptide bond formation. In addition there are numerous small RNAs that function in the splicing and editing of mRNA, processing of tRNA precursors, methylation of ribosomal RNA, transfer of proteins across membranes, and replication of DNA.

The genomes of many viruses consist of RNA. There are doubtless additional as yet undiscovered types of RNA. Unlike DNA, which exists largely as double helices, the single chains of RNA can fold into complex forms containing many bulges and loops of the sort depicted in Fig. 1.9. These loops are closed by double-stranded *stems* which have the A conformation. The B conformation is impossible because of the presence of the 2′-hydroxyl groups on the ribose rings in RNA. Even one ribonucleotide in a 10-nucleotide oligomer prevents formation of the B structure. The 2′-OH groups not only keep the RNA in the A form but also engage in hydrogen-bond formation.

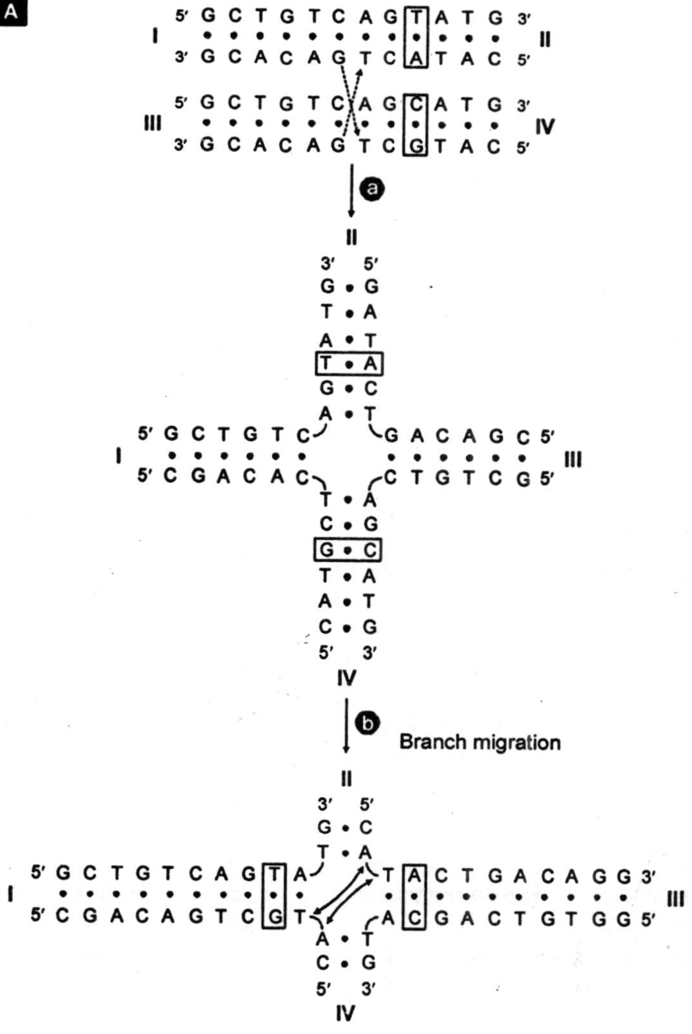

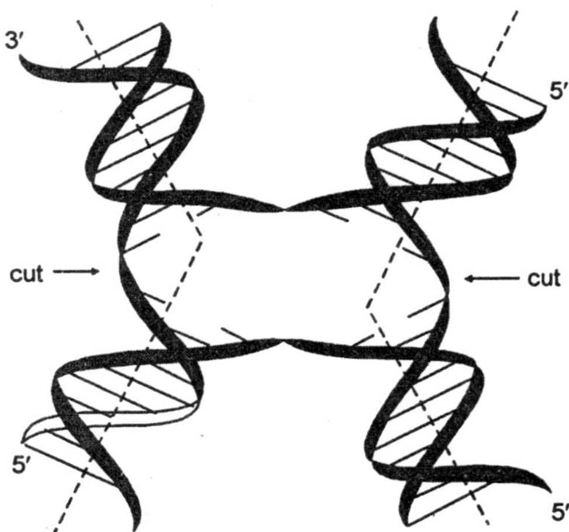

Fig. 1.25. (A) Abbreviated reaction sequence for formation of a four-way Holliday junction between two homologous DNA duplexes. In step *a* strands are cut and rejoined with movement of the strands to a roughly antiparallel orientation. The resulting structure is thought to resemble that shown below the four-stranded representation. In step *b* branch migration takes place, separating the nonhomologous base pairs TA and CG and causing mismatched pairs which will be subject to repair. (B) Proposed three-dimensional structure.

Hydrogen bonds may form between the 2′-OH and the oxygen atom in the next ribose ring in the 3 → 5′ direction. The – OH groups also hydrogen bond to water molecules which form a network within the minor groove. These bound water molecules, in turn, can bond to associated protein and to other atoms of the complex loops found in RNA molecules. The 2′–OH groups also act as ligands for divalent metal ions in some tRNAs and in some RNA catalytic sites. Transient *hybrid DNA-RNA double helices* also exist within cells and they too usually have the overall shape of A-DNA. However, the minor groove is intermediate in width between that expected for the A and B forms.

RNA Loops and Turns

Like polypeptides, polynucleotide chains have preferred ways of bending or turning. The loops at the ends of the hairpin turns of RNA molecules sometimes consist of a trinucleotide such as UUU, but are usually larger. In tRNA there are typically seven bases that do not participate in regular Watson-Crick pairing. The tetranucleotide *UUCG* is frequently present, and the sequence 5′-GGAC*UUCG*GUCC forms an unusually stable hairpin. Other tetranucleotides, such as UGAA, CCCG (also found in DNA loops), GCAA, and GAAA, occur often. The latter are members of a larger group of loop structures with the consensus sequence GNRA, where *N* is any nucleotide and R is a purine.

These sequences are very common in highly folded structures of ribosomal RNAs. Until recently high-resolution X-ray structures were available for only a few tRNAs and ribozymes. To help remedy this deficiency the structures of a great variety of oligonucleotide stem/loop (hairpin) structures are being determined, most by NMR spectroscopy. The sharp turns in the loops involve mostly rotation about the two torsion angles around the phosphorus atom of the third, from the 5′ end, of the seven nucleotides.

Base bulges on stems not only introduce kinks and bends in RNA stems but also provide well-defined hydrogen-bonded binding sites for proteins. Numerous branched three-way and more complex junctions provide other important motifs in folded RNA. Among the new RNA structures are those of RNA-antisense RNA pairs in *"kissing"* hairpin complexes.

Another interesting complex folding pattern in RNA is the *pseudoknot,* a structural feature that has been identified in many RNA sequences. A pseudoknot can be formed if nucleotide sequences favorable to formation of two short RNA stems are overlapped. After stem 2 in this drawing is formed (step *a)* additional base pairing can lead to formation of stem 1 (step *b).* The base pairs of the two stems can stack coaxially to form the pseudoknot (step *c*). For the formation of base-paired stems the RNA must contain antiparallel sequences that allow Watson-Crick or wobble (GU) base pairing in the stems. This requires certain relationships in the sequences of the DNA in the genes that encode these molecules. Because of the base-pairing requirements, some of the bases in the stems protrude as bulges and fail to form pairs. Often, it is possible to find more than one reasonable structure, each having some bases unpaired.

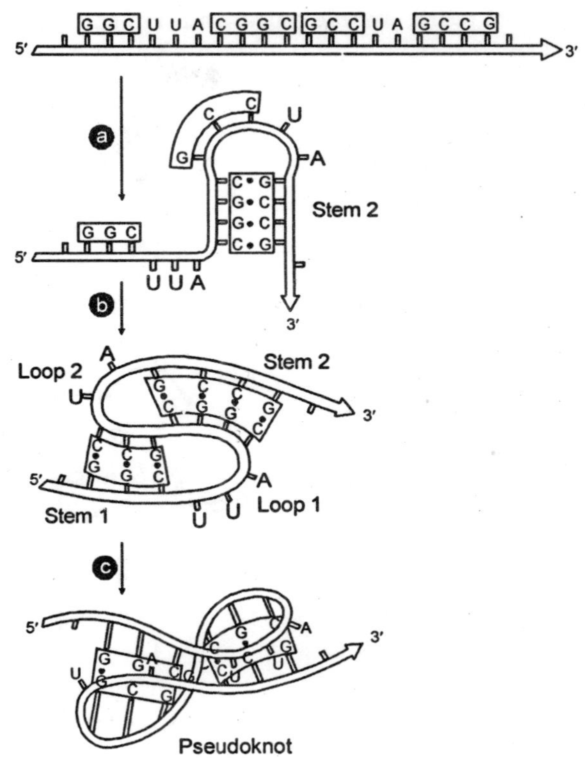

Fig. 1.26. Formation of a pseudoknot in an RNA chain.

Transfer RNAs

In all tRNAs the bases can be paired to form "clover-leaf" structures with three hairpin loops and sometimes a fourth. This structure can be folded into the L-shape shown in Fig. 1.31. The structure of a phenylalanine-carrying tRNA of yeast, the first tRNA whose structure was determined to atomic resolution by X-ray diffraction. An aspartic acid-specific tRNA from yeast, and an *E. coli* chain-initiating tRNA, which places N-formyl-methionine into the N-terminal position of proteins, have similar structures. These molecules are irregular bodies as complex in conformation as globular proteins.

Numerous NMR studies show that the basic structure is conserved in all tRNAs. However, animal mitochondrial tRNAs often lack some of the usual stem-loop "arms" as well as the invariant nucleotides in the dihydrouridine and TψC loops. At the bottom of the structure, the *anticodon,* a triplet of bases having the correct structures to permit pairing with the three bases of the codon specifying a particular amino acid in this case phenylalanine. While tRNAs consist largely of loops and stems containing Watson-Crick base pairs, they also contain Hoogsteen pairs, wobble pairs, and triplets such as the following.

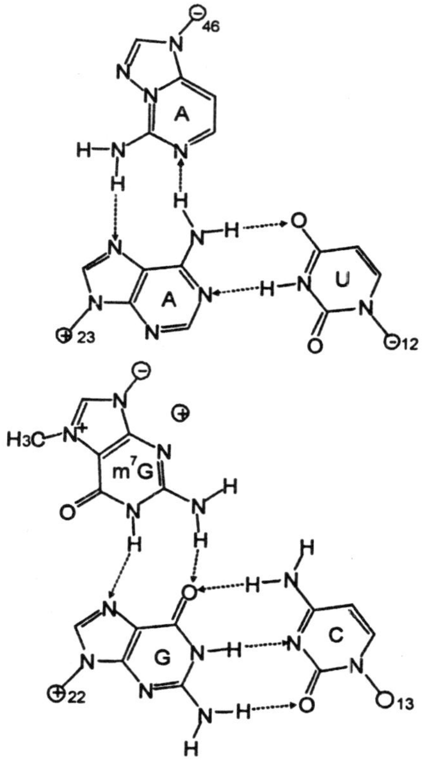

The first of these contains a Watson-Crick A•U pair with a second A bound to form a reversed Hoogsteen A*A pair. The second contains a G•C Watson-Crick pair with a Hoogsteen N-7-guanine*G (m^7G*G) pair. Among the complex base associations present in tRNAs are some that also involve hydrogen bonding to the 2′-hydroxyl groups of ribose rings and to at least one of the phosphate groups. There are over 100 internal hydrogen bonds, a large proportion of which are relatively invariant among the known tRNAs.

Properties and Structures of Some Other RNA Molecules

A few specialized RNA molecules are listed in Table 1.4. There are many others. In addition, there are thousands of different mRNAs within a cell. The most abundant RNA molecules in cells are those of the ribosomes.

Ribosomes consist of two elaborate RNA-protein subunits, a large subunit with sedimentation constant ~30S in bacteria and ~ 40S in eukaryotes. The small subunit contains 16S or 18S RNA and the large subunit 23S or 28S as well as smaller 5S and 5.8S RNAs. A proposed three-dimensional structure of a bacterial 16S ribosomal RNA and the corresponding ribosomal subunit with its 21 proteins are shown in Fig. 1.32. It might seem impossible that the folding pattern of the RNA was deduced correctly before an X-ray structure was available.

However, a *phylogenetic approach,* the comparison of nucleotide sequences among several species, suggested that the stem structures of rRNAs are highly conserved. This fact, together with a variety of other chemical evidence, allowed the prediction of the structure shown as well as many characteristics of the rRNA of the large subunit. Culminating decades of effort, complete structures of bacterial ribosomes were established by 2000 and the peptidyl transferase center was identified as a ribozyme.

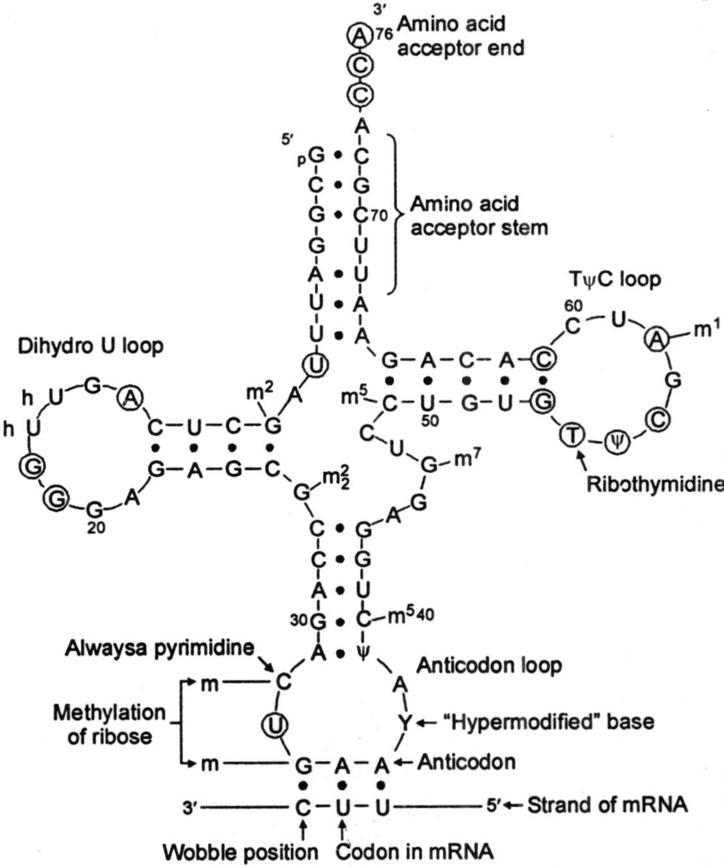

Fig. 1.27. Schematic cloverleaf structure of a phenylalanine-speclfic transfer RNA (tRNA^{Phe}) of yeast. The dots represent pairs or triplets of hydrogen bonds. Nucleosides common to almost all tRNA molecules are circled. Other features common to most tRNA molecules are also marked. The manner in which the anticodon may be matched to a codon of mRNA is indicated at the bottom.

Modified Nucleotide

The picture of DNA or RNA as chains of only four kinds of nucleotides is not quite accurate. DNA contains a significant number of methylated bases among which are 5-*methylcytosine* (5mC or m^{5}C) and 6-methyladenine (6mA). The former is regularly present in the nuclear

DNA of higher animals and plants. In mammals from two to seven percent of the cytosine is methylated. It is likely that these methylated bases mark special points in the genetic blueprint.

Table 1.4. Some Specialized RNA Molecules.

Kind of RNA	*Number of nucleotides*
Transfer RNAs	60 – 85
Ribosomal RNAs	
5S	~120
5.8S (rat)	158
16S (*E. coli*)	1542
18S (rat	1874
23S (E. coli)	2904
28S (rat)	4718
Telomerase guide RNA	159 (*Tetrahymena*)
M1 RNA of Ribonuclease P	350 – 410 377 (*E. coli*)
Tetrahymena Intron ribozyme	413
Viroid hammerhead ribozyme	~55
Signal recognition particle 7S RNA	295 (human)
Small nuclear RNAs	65 – 1200
RNA-editing guide RNA	~60
Thermotolerance factor (G8 RNA)	~300
Viroid RNA	240 – 380
Virus MS2 genome	3569

Methylation, which is accomplished after the synthesis of the polynucleotide, may block the expression of certain genes. This appears to happen when one of the two X-chromosomes becomes inactivated in cells of females. Methylation may also be used *to imprint* certain genes, *i.e.* to mark them as coming from a specific one of the two parents. Another function of methylation is to protect DNA from attack by enzymes (restriction endonucleases) formed in response to invading viruses.

Some viruses, notably the bacteriophage of the T-even series that attack *E. coli*, have developed their own protective devices. They contain 5-*hydroxymethylcytosine* (HOMeC) in place of cytosine. The extra hydroxyl groups provided in this fashion often carry one or two glucose units in glycosidic linkage. A bacteriophage attacking *Bacillus subtilis* substitutes hydroxymethyluracil for uracil and 5-dihydroxypentyluracil for thymine, and phage W14 of *Pseudomonas acidovomns* substitutes the 5-methyl of thymine with $-CH_2-NH-(CH_2)_4-NH_3^+$. The modifications carried out on RNA molecules are more varied and more extensive than those of DNA. Sixty or more modification reactions are known for tRNA, with the number and the extent of modification depending upon the species.

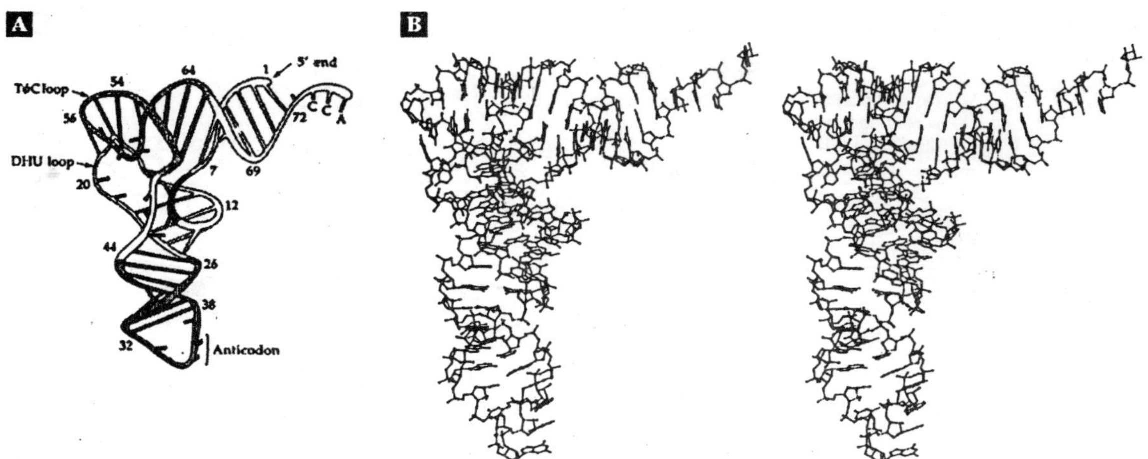

Fig. 1.28. The three-dimensional structure of a phenylalanine-specific transfer RNA of yeast. (A) Perspective diagram of folding of polynucleotide chain. The ribose phosphate backbone is drawn as a continuous cylinder with bars to indicate hydrogen-bonded base pairs. The positions of single bases are indicated by rods which are intentionally shortened: The TψC arm is heavily stippled, and the anticodon armis marked by vertical lines. Tertiary structure interactions are illustrated by black rods. (B) Stereoscopic view of the structure of yeast tRNA as revealed by X-ray crystallography. The acceptor stem with the protruding ACCA sequence at the 3′ end is to the right. The anticodon GAA is at the bottom right side of the drawing. The guanine ring is clearly visible at the very bottom. The middle adenine of the anticodon is seen exactly edge-on as is the "hypermodified" base Y, which lies just above the anticodon. Its side chain is visible at the back of the drawing. Preceding the anticodon on the left (5′ side) are two unpaired bases (C and U). The 2′-hydroxyl groups of the cytosine and of the guanosine in the anticodon are methylated. Moving far up the anticodon stem one can see two groups of base triplets which utilizg both Watson-Crick and Hoogsteen types of base pairing.

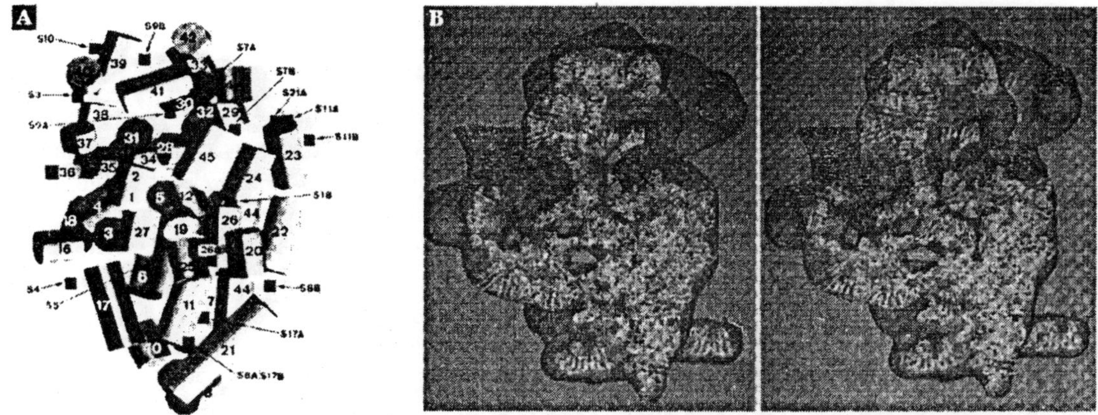

Fig. 1.29. (A) A three-dimensional computer graphics model proposed by Brimacombe *et al.* for the single chain of *E. coli* 16S ribosomal RNA. The helices are depicted as cylinders, which are all connected. The small dark squares denote the positions of artificially formed RNA-protein crosslinks, marked with the appropriate protein number. For proteins exhibiting more than one crosslink site (*e.g.*, 517), the sites are denoted A or B, in each case A being the site nearer to the 5′ terminus of the 16S RNA. (B) Stereoscopic view of tentative atomic model of 16S RNA in the SOS ribosomal subunit. The viewing direction is different from that in (A). From Mueller and Brimacombe.

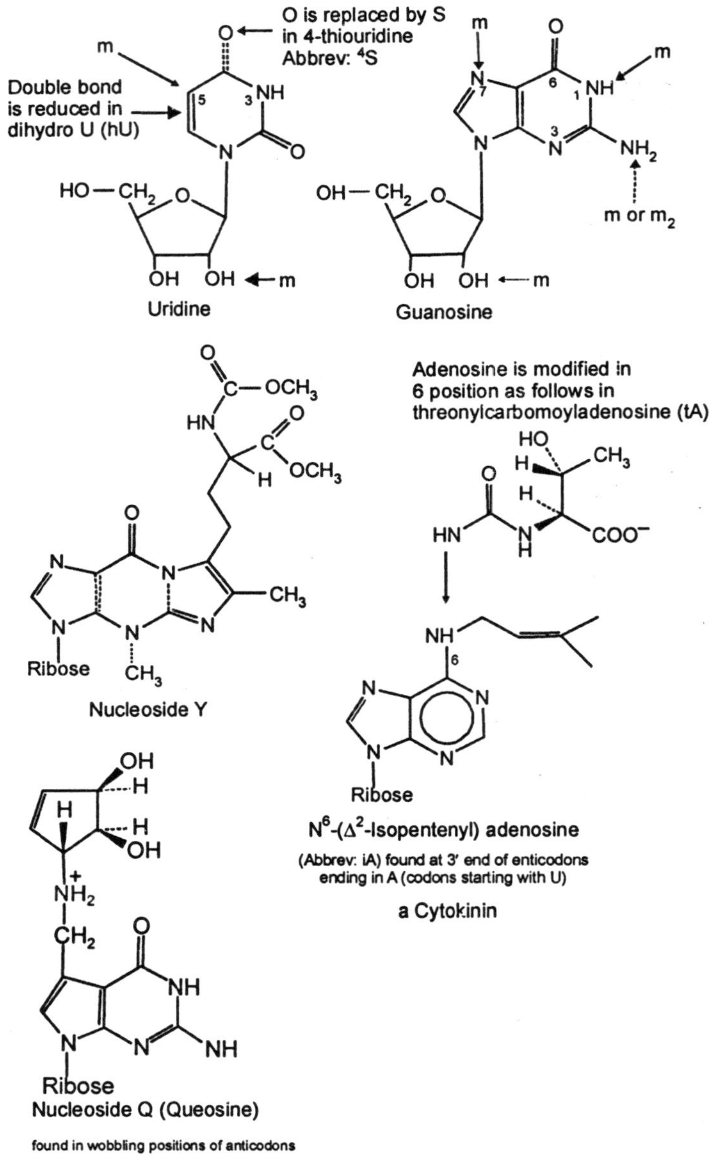

Fig. 1.30. Structures of some nucleosides containing modified bases and found in tRNA molecules. Positions where methylation may occur are designated m. Some other abbreviations are e (ethyl), ac (acetyl), and f (formyl).

Structure of some of the modified bases are indicated in Fig. 1.33, Uridine can be methylated either on the base or on the 2′-hydroxyl of the sugar. Methylation at the 5 position of uridine yields *ribothymidine*. Cytidine can be modified in the same positions. Reduction of the 5, 6 double bond of uridine gives *dihydrouridine* (hU). Replacement of the oxygen at position

4 by sulfur gives 4-*thiouridine* (^{4}sU). Positions in the guanosine structure that can be methylated are also indicated in Fig. 1.33. The symbol *m* is commonly used to designate methylation in nucleic acid bases; m_2, indicates dimethylation, *e.g.*, 6, 6-dimethyladenine is abbreviatd $m_2{}^6$ A. A remarkable transformation is that of uridine into pseudouridine (ψ). Pseudouridine is formed by enzymatic rearrangement of uridine in the original transcript. It can form a base pair with adenine in the same manner as does uracil. Pseudouridine is found not only in tRNA but also in several places in both large and small ribosomal RNA subunits.

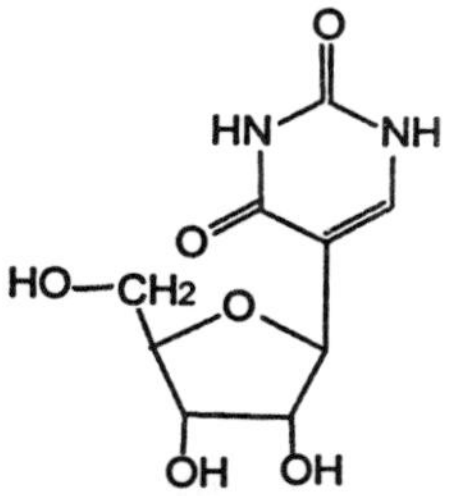

O
HN NH
O
HO—CH2 O
OH OH

Pseidouridine (ψ)

For example, it is present at position 516 in the *E.coli* 16S RNA, at a specific position in the 23S RNA, and at many more locations in eukaryotic rRNA. The bases called "Y" and "Q" are highly modified guanines are also shown as Fig. 1.33. Q is found at the 5′ end of some anticodons in the "wobble position". Two *hypermodified* adensines. The N^6-isopentenyladenosine is found at the 3′ end of the anticodons that pair with codons starting with U. This compound is also a plant hormone, a *cytokinin*. Another highly modified purine, threonylcarbamoyladenine, occurs adjacent to the end of anticodons pairing with codons starting with A. The function of these hypermodified bases is uncertain, but they appear to promote proper binding to ribosomes.

The modifications are often not absolutely essential for function. Another source of modified bases in both DNA and RNA is spontaneous or "accidental" alteration. Nucleic acids encounter many highly reactive and mutagenic materials including hydroxyl radicals, formed from Q_2, and are able to convert guanine rigns into 7,8-dihydro-8-oxoguanine. Other reactive and carcinogenic compounds can form adducts with nucleic acid bases.

RNA Aptamers

Ellington and Szostak synthesized a random "Pool" of ~10^{15} different oligodexyribonucleotides, each ~100 nucleotides in length. They "amplified" these using the polymerase chain reaction and prepared a mixture of the corresponding RNAs by in *vitro* transcription. From the ~10^{13} different sequences still present they selected individual oligonucleotides by affinity chromatography on columns that contained well-defined immobilized ligands such as organic dyes. They called the selected RNAs *aptamers*. Their approach is being used to find RNA sequences that bind to such ligands as ATP, FMN, the bronchodilator theophylline, ammoglycoside antibiotics, arginine, etc. Many of the selected aptamers bind their ligands very tightly and studying them may shed light on interactions of RNA with proteins and on the catalytic activities of RNA.

THE GENETIC CODE

The general nature of the genetic code was suggested by the structure of DNA. Both DNA and proteins are linear polymers. Thus, it was logical to suppose that the sequence of the bases in DNA codes for the sequence of amino acids. There are only four bases in DNA but 20 different amino acids in proteins at the time of their synthesis. It is obvious that each amino acid must be specified by some combination of more than one base. While 16 pairs of bases are possible, this is still too few to specify 20 different amino acids, therefore, it appeared that at least a triplet group of three nucleotides would be required to code for one amino acid.

Sixty-four (4^3) such triplet *codons* exist. Simplicity argues that the genetic blueprint specifying amino acid sequences in proteins should consist of consecutive, nonoverlapping triplets. This assumption turned out to be correct, as is illustrated by the DNA sequence for a gene shown in Fig. 1.5. In addition to the codons that determine the sequence of amino acids in the protein, there are *stop codons* that tell the ribosomal machinery when to terminate the polypeptide chain. One methionine codon serves as an *initiation codon* that marks the beginning of a polypeptide sequence.

One of the valine codons sometimes functions in the same way. How does a cell read the cbde? A key step is the positioning of each amino acid on the ribosome in proper sequence. This is accomplished by the pairing of codons of messenger RNA with the anticodons of the appropriate transfer RNA molecules as is indicated at the bottom. Each tRNA carries the appropriate "activated" amino acid at its 3′ end ready to be inserted into the growing peptide.

The "Reading Frames"

It is immediately obvious that there are three ways of reading the genetic code in mRNA depending upon which nucleotide is used to start each codon. For example, in the following mRNA sequence either codons GCA, CAG, or AGC could be selected as first.

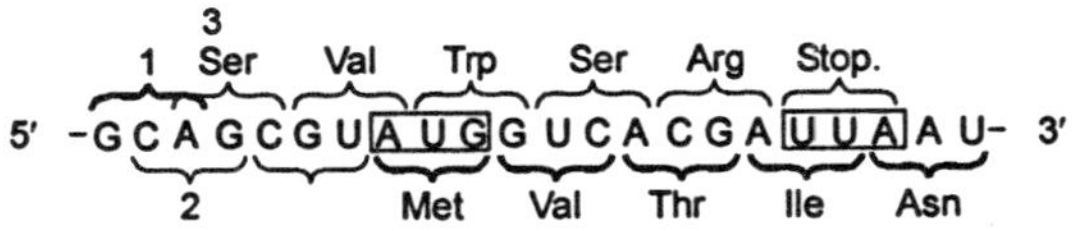

3
1 Ser Val Trp Ser Arg Stop.
5′ -G C A G C G U A U G G U C A C G A U U A A U- 3′
2 Met Val Thr Ile Asn

These codons define the three reading frames or phases in which the code may be read. Here, the term *frame* does not designate a single codon, although frame does designate a single exposure in a motion picture film.

Reading frame designates which of the three possible sets of codons we are using. In the foregoing sequence the codons in reading frames 2 and 3 are labeled. Reading frame 2 contains the initiation codon AUG (shaded) which could mark the beginning of an encoded protein sequence. Reading frame 3 contains a termination (stop) codon which, when the mRNA transcript is read by ribosomes, will terminate polypeptide synthesis. It may be in the position shown but not have any real function.

However, it could represent the end of the coding sequence that is marked if genes for the two proteins overlap a little at the ends, a situation that actually occurs in nature. A reading frame in a specified part of a DNA sequence is said to be *"open"* if there is an initiation codon preceded by suitable regulatory signals (an *operator* region). This means that it *could* encode a protein. The reading frame of a sequence is open until the next termination codon. Recently another usage has appeared. Many writers refer to *an open reading frame* as a segment of DNA in which any one of the three reading frames is open.

Table 1.5. The Genetic code

Amino acid	*Codons*	*Total number of codons*
Alanine	GCX	4
Arginine	CGX, AGA, AGG	6
Asparagines	AAU, AAC	2
Aspartic acid	GAU, gac	2
Cysteine	UGU, UGC	2
Glutamic Acid	GAA, GAG	2
Glutamine	CAA, CAG	2
Glycine	GGX	4
Histidine	CAU, CAG	2
Isoleucine	AAU, AUC, AUA	3
Leucine	UUA, UUG, CUX	6
Lysine	AAA, AUG	2
Methionine (also initiation codon)	AUG	1
Phenylalanine	UUU, UUC	2
Praline	CCX	4
Serine	UCX, AGU, AGC	6
Threonine	ACX	4
Tryptophan	UGG	1
Tyrosine	UAU, UAC	2
Valine (GUG is sometimes an initiation codon)	GUX	4
Termination	UAA (*ochre*)	
	UAG (*amber*)	
	UGA	3
Total		64

Another complexity in the reading of genetic messages arises because splicing of RNA may sometimes cause a shift in the reading frame. For example, a mRNA being transcribed from the sequence in reading frame 2 in the foregoing example may skip over a nucleotide part of the time to form an RNA in which the first part is encoded by reading frame 2, and the second by reading frame 3 (a + 1 frame-shift). Alternatively, a nucleotide could be read twice with a – 1 frameshift with the sequence of reading frame 1 for the latter part of the mRNA. Frameshifts can also occur during protein synthesis as the mRNA is being read. In the present example we have examined the sequence in mRNA.

Table 1.6. The Sixty-Four Codons of the Genetic Code.

5′–OH Terminal base	*Middle base*				*3′–OH Terminal base*
	U(T)	*C*	*A*	*G*	
U(T)	Phe	Ser	Tyr	Cys	(U(T)
	Phe	Ser	Tyr	Cys	C
	Leu	Ser	Terms[c]	Term[d]	A
	Leu	Ser	Term	Trp	G
C	Leu	Pro	His	Arg	U(T)
	Leu	Pro	His	Arg	C
	Leu[a]	Pro	Glan	Arg	A
	Leu	Pro	Gln	Arg	G
A	Ile	Thr	Asn	Ser	U(T)
	Ile	Thr	Asn	Ser	C
	Met[b]	Thr	Lys	Arg	G
G	Val	Ala	Asp	Gly	U(T)
	Val	Ala	Asp	Gly	C
	Val	Ala	Glu	Gly	A
	Val[b]	Ala	Glu	Gly	G

[a] The codon CUA (CTA) encodes threonine and the codon AUA (ATA) methionine in mammalian mitochondria.

[b] Initiation codons. The methionine codon AUG is the most common starting point for translation of a genetic message but GUG can also serve. In such cases it codes for methionine rather than valine.

[c] The "termination codon" UAA (TAA) encodes glutamine in *Tetruhymena*.

[d] The termination codon UGA (TGA) encodes tryptophan in mitochondria and selenocysteine in some contexts in nuclear genes.

In the DNA there are two strands. One is the *coding strand* (also called the nontranscribing or nontranscribed strand), which has a sequence that corresponds.to that in the mRNA and the one that is given in Fig. 1.4. The second antiparallel and complementary strand can be called the *template strand* or the noncoding, transcribing, or transcribed strand. The mRNA that is formed is sometimes referred to as a *sense strand.* The complementary mRNA, which corresponds in sequence to the noncoding strand of DNA, is usually called *antisense DNA.*

Variations

Is the genetic code "universal" or does it vary from one organism to another? Studies with bacteria, viruses, and higher organisms including humans have convinced us that the code is basically the same for all oganisms. However, there are some variations. For example, in mitochondria of both humans and yeast the codon TGA is not a termination codon but represents tryptophan. In mammalian mitochondria CTA represents threonine rather than leucine, and ATA encodes methionine instead of isoleucine.

These differences in the code are related to the fact that mitochondria contain their own piece of DNA. It encodes not only several proteins but also tRNA molecules whose anticodon structures are altered to accommodate the changed meanings of the codons of the mitochondrial DNA and mRNA. Variations in the code for cytoplasmic proteins have been found. In *Tetrahymena* and other ciliates the codon TAA represents glutamine rather than being a termination codon. A few proteins, including some in the human body, contain *selenocysteine,* the selenium-containing analog of cysteine. Selenocysteine is encoded by termination codon TGA.

However, even though TGA is occasionally used in this way, it serves as a termination codon for most proteins within the same cells. Thus, the *context* in which the codon TGA occurs determines how it is read by the ribosomal machinery.

Palindromes and Other Hidden Messages

The sentence "Madam, I'm Adam" reads the same either forward or backward. Such sentences, known as *palindromes,* are infrequent in the English language. However, most DNA contains many palindromes, sequences of base pairs that read the same in forward and reverse directions. Consider, for example, the gene that specifies the sequence of nucleotides for the tRNA molecules of Fig. 1.30. It is a double-stranded DNA segment in which one strand has a sequence identical to that of Fig. 1.30, except for the substitution of T for U and for ψ (pseudouridine) and for the lack of methylation and other base modifications.

The second strand is the exact complement. Fig. 1.34 shows that part of this gene (residues 49 – 76) that corresponds to the 3′ end of the tRNA molecule. This DNA segment could exist in a second conformation having a loop on each side of the molecule. The stems of the two loops in this *cruciform* conformation are identical and symmetrically disposed around the center of the molecule. If we overlook the seven nucleotides in the center of the loop, the message in the stems reads the same in both directions, as is indicated by the green arrows. Palindromes are often imperfect. Here the two stems in the cruciform structure are related by an exact twofold rotational symmetry but the loops at the ends of the stems are not.

Unpaired bases may bulge at various points in double-stranded stems of longer palindromes. These imperfect palindromes in the DNA are responsible for much of the tertiary structure of the various kinds of RNA. The tertiary structure, in turn, often determines the interaction of the RNA with enzymes and other proteins. Special properties may be observed for palindromes containing homopurine tracts in one strand, and therefore homopyrimidine tracts in the other.

If two identical palindromes of this type occur close together it is possible that the pyrimidine-containing strand of one can join with a hairpin loop of the other to form a triplex base structure. A related triplex structure may be formed when inverted repeat sequences occur within a homopyrimidine tract in one chain of DNA. These have been called *mirror-repeats* or *H-palindromes* to distinguish them from true palindromes. Each base triplex structure contains one set of Watson-Crick hydrogen bonds and one set of Hoogsteen hydrogen bonds.

The triplexes of H-DNA are all either TAT or CGC^+, where one C is protonated. Segments of H-DNA might block transcription, and proteins that bind to the triplex H-DNA may be involved in transcriptional regulation. Until rather recently there had been little to indicate that DNA actually assumes cruciform conformations in cells.

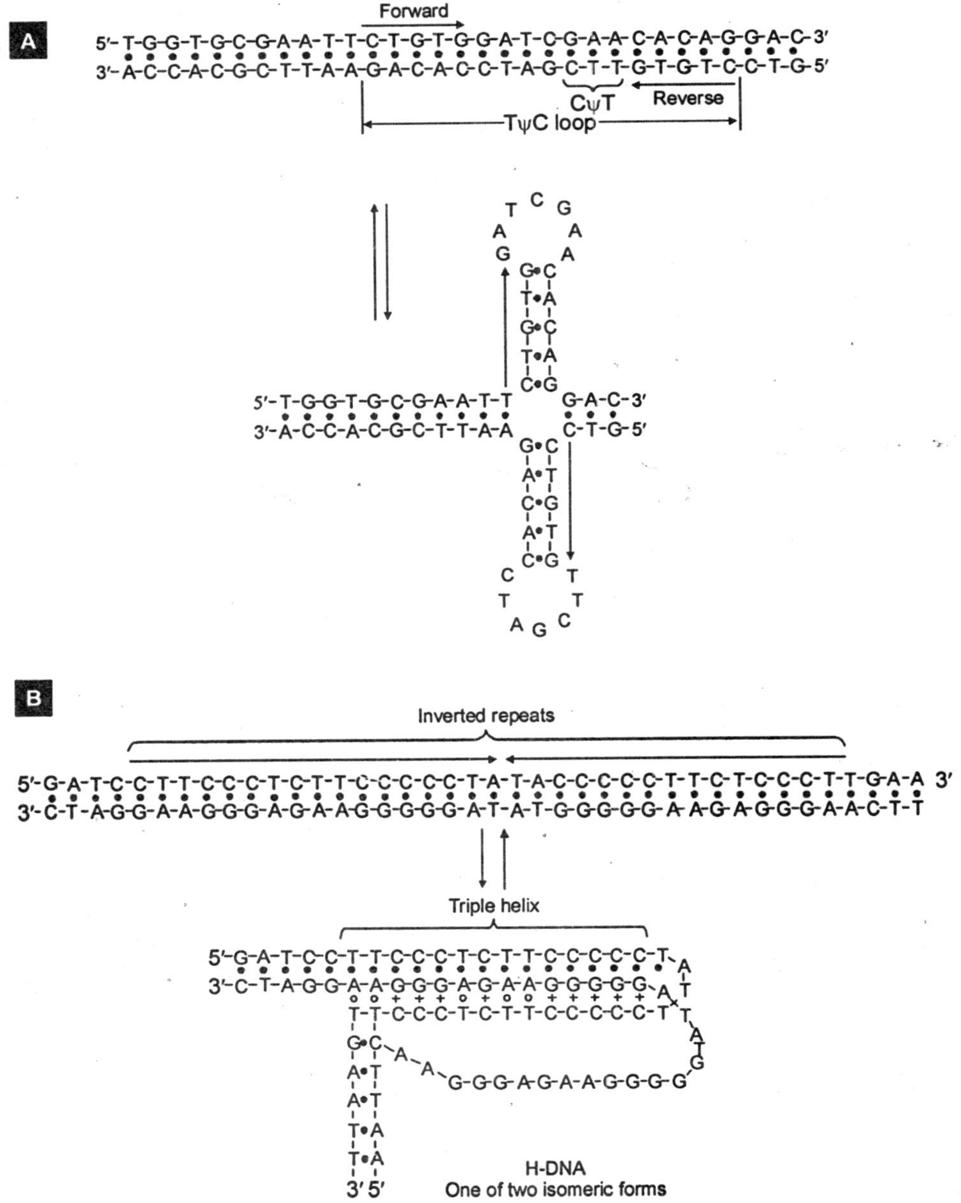

Fig. 1.31. (A) Two conformations of a segment of the yeast phenylalanine tRNA gene. The segment shown codes for the 3′ end of the tRNA molecule. The TψC loop. (B) Formation of H-DNA, proposed for a sequence in plasmid pGG32. The major element of the structure is the triplex, which is formed from the Watson-Crick duplex (•) associated with the homopyrimidine loop through Hoogsteen base pairing (°, +). One of the two possible "isomeric" forms is shown.

However, strong experimental evidence suggests that some cruciform structures do form naturally. Their formation from palindromic DNA [like the formation of Z-DNA from (G + C)-rich sequences] is a way of relieving torsional strain induced by supercoiling. Whether or not cruciform structures occur frequently within cells, there is no doubt that palindromic sequences are of great importance in the interaction of nucleic acids with symmetric dimeric and tetrameric protein molecules such as the gene represser protein. DNA contains numerous other protein binding sites which are not palindromes but whose sequences represent additional encoded information.

The RNA Transcripts likewise contain sequences that direct the catalytic machinery involved in splicing, that bind to ribosomal proteins, that control rates of transcription, and that cause termination of transcription.

The Base Composition of DNA and RNA

The nucleotide composition of DNA is surprisingly variable. The sum of the percentage cytosine plus the percentage guanine (C + G) for bacteria varies from 22 to 74%. That for *E. coli* is 51.7%. Among eukaryotic organisms, the range is somewhat narrower (28 to 58%; for humans, 39.7%). The fact that bacterial DNA molecules are more varied than those of higher organisms is not surprising. The prokaryotes have evolved for just as many more million of years as have we. Perhaps because of their simpler structure and rapid rate of division, nature has done more experimentation with genetic changes in bacteria than in people. Comparisons of the C + G content of organisms have been used as a basis for establishing genetic relatedness.

However, since thymine is especially susceptible to photochemical alteration by ultraviolet light, bacteria with a high (C + G) content may have evolved in environments subject to strong sunlight or high temperatures, whereas those with a low (C + G) content have developed in more protected locations.

INTERACTION OF NUCLEIC ACIDS WITH PROTEINS

Most functions of DNA and RNA are dependent on proteins. Cells contain a vast array of polypeptides that bind to nucleic acids in many specific ways. We have seen that the histones hold supercoiled DNA in the form of nucleosomes eukaryotic nuclei and similar proteins bind to and protect the double-stranded DNA of bacteria. Sperm cells "package" a large amount of DNA into a small space with the help of small arginine-rich proteins called *protanunes*. Cells always have some single-stranded DNA segments as well as single-stranded DNA binding (*SSB*) proteins.

Among the latter are proteins from *E. coli* and from viruses. Specialized proteins bind to DNA sequences in telomeres and centromeres. A large number of proteins interact with RNA in ribosomes, spliceosomes, and other complexes. A host of enzymes, which are described elsewhere in the book, act on DNA and RNA. They include hydrolytic nucleases, methyltransferases, polymerases, topoisomerases, and enzymes involved in repair of damaged DNA and in modifications of either DNA or RNA. While most of these enzymes are apparently proteins, a surprising number are *ribozymes,* which consist of RNA or are RNA-protein complexes in which the RNA has catalytic activity.

The Helix-Turn-Helix Motif

Much current interest in DNA-protein interactions is focused on regulatory processes. In prokaryotes the initiation of transcription of a large fraction of the genes is blocked by the binding of proteins known as *repressers,* While their structures are varied, one large group of repressers have DNA-binding domains with a similar helix-turn-helix architectural motif. They bind with high affinity to specific control regions which contain palindromic DNA sequences such as the following one, which defines a binding site for the *E. coli try* (tryptophan) *represser:*

5′-T-C-T-A-C-T-A-G-T-T-A-A-C-T-A-G-T-A-C-3′
3′-C-A-T-G-A-T-C-A-A-T-T-G-A-T-C-A-T-G-T-5′

These repressor proteins form dimers joined through a rigid central domain with a pair of arms, each containing two helices that form the helix-turn-helix motif. One helix of each pair fits into the major groove of the DNA helix and forms a "reading head" that carries a specific arrangement of amino acid side chains able to locate the symmetric nucleotide sequences flanking the twofold axis of the palindrome. In the case of the *trp* repressor, these are the ACTAGT hexanucleotides (marked by the arrows in the preceding structure).

The interaction is depicted in Fig. 1.35 the repressor has a high affinity for the DNA only if one molecule of L-tryptophan is bound to each subunit at a specific site near the DNA helix. Binding of the tryptophan causes a-conformational change, which is pictured in Fig. 1.35. If tryptophan is absent the repressor binds to the palindromic DNA only weakly and transcription of genes needed for tryptophan biosynthesis occurs freely. However, if tryptophan accumulates within the cell it binds to the repressor molecules causing them to bind firmly to DNA and prevent transcription. There are at least three of these palindromic sequences in *E. coli,* each one regulating set of genes (*operons*) involved in tryptophan synthesis.

In contrast to the effect of tryptophan, the tryptophan analog indole-3-propionate, which lacks the amino group of tryptophan, *derepresses* the same operons. It also binds to the *trp* repressor, but with the indole ring flipped over by 180° so that its carboxylate group contacts the phosphate groups of the DNA repelling them through both electrostatic and steric effects. The picture in Fig. 1.35 is an, oversimplification.

In fact, at some binding sites (operator sites) more than one dimeric repressor binds in a tandem fashion. The helix-turn-helix motif is also found in many other proteins. One of these is the bacterial *lac* (lactose) *represser* which controls the *lac* operon and for which the terms operator, promoter, repressor, and operon were first introduced. Some bacterial viruses, such as *phage lambda* (λ) of *E. coli,* encode repressers that allow the virus to reside in the bacteria without immediately destroying them. The λ repressor and other closely related proteins also utilize the helix-turn-helix motif as do some proteins that *activate* transcription.

Other DNA-Binding Motifs

Nobody knows how many different DNA-binding structures may be discovered. However, most of those that are designed to recognize specific DNA sequences have some part that fits into the major groove of B-DNA. Usually the DNA structure must be in a specific form: B-, A- or Z-, but it may sometimes be bent or distorted. The DNA recognition motifs in the proteins may consist of helices, P strands, or loops.

Leucine Zipper Proteins

Several transcription factors have the leucine zipper structure. Related structures include those of the transcription activators c-Jun and c-Fos and of the yeast transcription factor GCN4.

The latter is a dimeric protein in which the C-terminal halves of the two monomers form the helices of the leucine zipper. The helices, which are continuous for over 60 residues, fan out to interact with the DNA double helix.

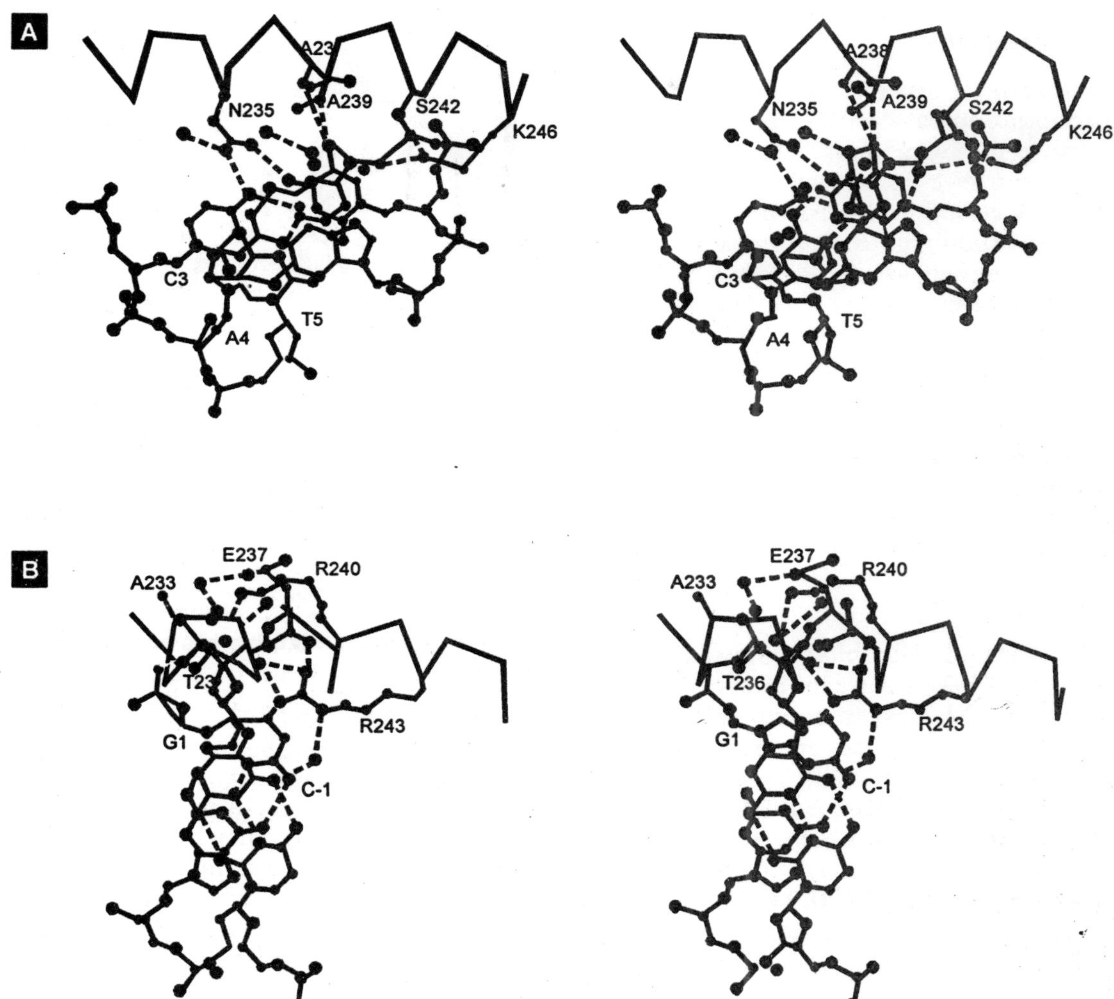

Fig. 1.32. Stereoscopic diagrams showing some of the interactions between an N-terminal helical domain of the yeast transcription factor GCN4-bZIP, a leucine zipper protein, and a specific palindromic DNA binding site:

–5 –1 1 5
5′–A T G A C G T C A T

(The bases are numbered outward from the central C and G.) The small solid spheres are water molecules. Notice the water mediated interactions of the basic arginine and lysine side chains with the nucleic acid bases and also the interaction of R240 and R243 (in B) with a backbone phosphate. The overall structure of the protein is similar to that of another lecucine zipper.

A 19-residue basic domain of each protein helix crosses the major groove of the DNA with side chain groups interacting with the DNA.

Zinc Fingers

Another large group of transcription factors contain a bound zinc ion in a "finger" motif. The Zn^{2+} is held by two cysteine $-S^-$ groups present on a loop of P structure and by two imidazole groups present on an α helix. Amino acid sequences found in many proteins that regulate transcription tend to form zinc fingers and to interact with DNA. These were recognized first in the transcription factor TFIIIA from *Xenopus laevis* in 1985. This protein contains a zinc finger motif repeated nine times. X-ray structures are known for proteins with three and five zinc fingers.

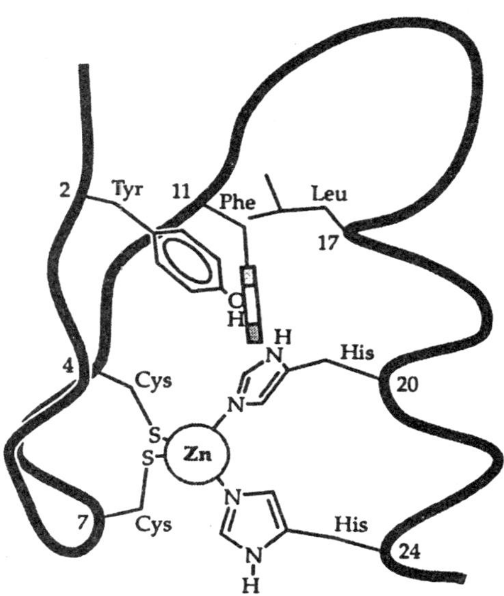

Fig. 1.33. Three-dimensional structure of a zinc finger. This si formed by the binding of Zn^{2+} to the following sequence in a protein:

$$\phi X C X_{2-4} C X X X F X_5 L X X H X_{3,4} H$$

β → turn → β → turn → β helix

Here ϕ is a hydrophobic amino acid and X may be any amino acid.

Beta Ribbons

An antiparallel double-stranded β ribbon can fit into the major groove of DNA and form hydrogen bonds and other interactions with the DNA. This is the basis for specific recognition of the operator sequence for the E. *coli met* (methionine) *repressor* as is illustrated in Fig. 1.39. The same binding motif is employed by some other repressors and also by the abundant bacterial DNA-binding protein *HU*. It has also been utilized in designing the previously mentioned hairpin polyamide DNA-binding compounds.

The Winged Helix Family

A group of large protein transcription factors contain an N-terminal DNA-binding domain with the striking winged helix motif shown in Fig. 1.40. It occurs in proteins from a wide range of organisms from yeast to human.

RNA-Binding Proteins

Among the many proteins that bind to RNA molecules are the aminoacyl-tRNA synthetases, a variety of other well known enzymes, and various proteins with dual functions of catalysis and regulation of translation. A widely found RNA-binding *ribonucleoprotein domain* (RNP domain), occurs in hundreds of proteins including many of the RNA processing proteins. This 70–90 residue αβ module binds an RNA strand against a β sheet surface. Another ~70-residue protein motif binds ds RNA.

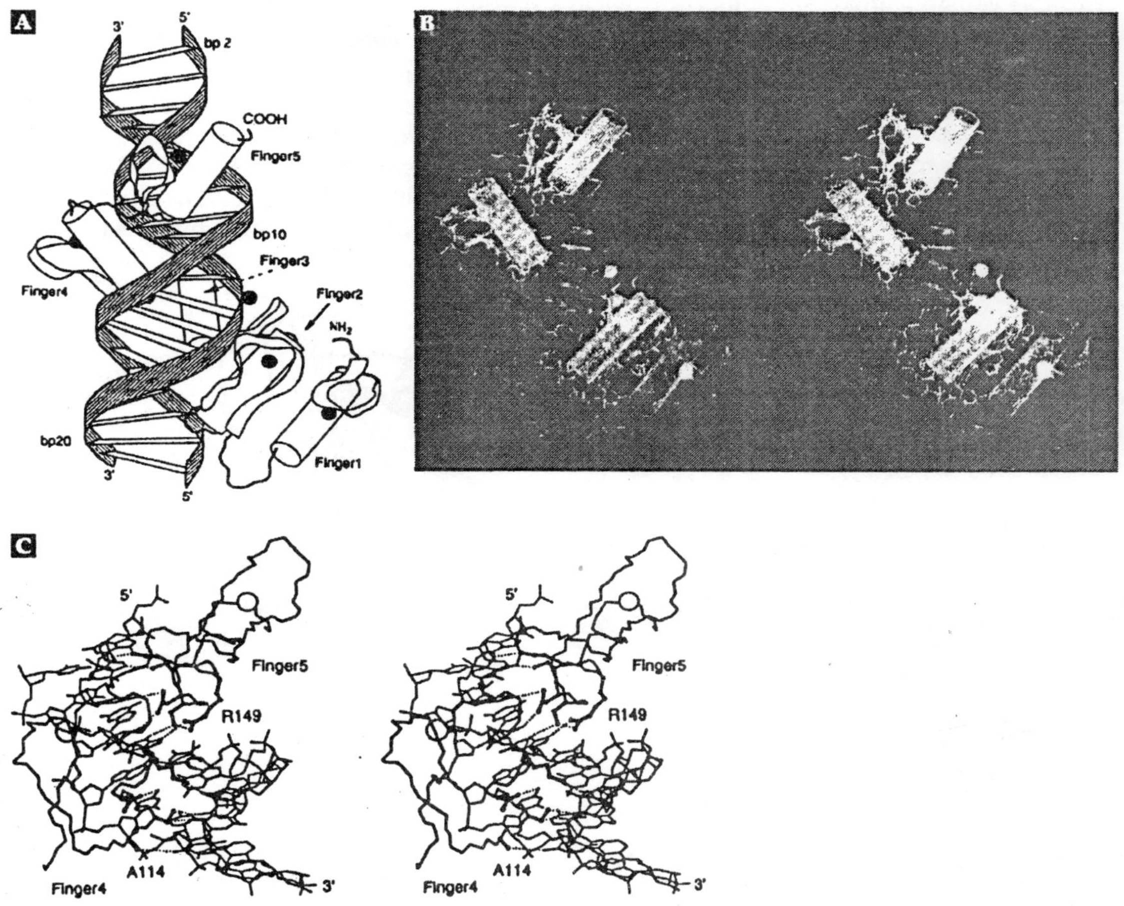

Fig. 1.34. (A) Sketch of a complex of a five-zinc finger protein bound to a high-affinity site in DNA. This protein is a human oncogene called GL1. Its gene is amplified in many tumors, but it is similar to many other transcription factors. The cylinders and ribbons highlight the a helices and β sheets. Cobalt ions, which replaced the Zn^{2+} in the crystals, are shown as spheres. (B) Stereoscopic view of the complex in a similar origentation. (C) Steresocopic view emphasizing the contacts of side chains from fingers 5 and 6 with the DNA. The view is similar to that in (A) and (B) but the structure has been tilted back.

VIRUSES AND PLASMIDS

Attacking every living thing from the smallest mycoplasma to human beings the nucleoprotein particles known as viruses have no metabolism of their own. However, they "come alive" when the nucleic acid that they contain enters a living cell. Viruses are significant to us not only because of the serious disease problems that result from their activities but also as tools in the study of molecular biology. A mature virus particle or *virion* consists of one or more nucleic acid molecules in a protein coat or *capsid* usually of helical or icosahedral form. The capsid is made up of morphological subunits called *capsomers*. They can sometimes be seen clearly with the electron microscope.

The capsomers in turn are usually composed of a number of smaller protein subunits. Some of the larger viruses are surrounded by membranous envelopes. Others, such as the T-even *bacteriophage* which attack *E. coli* have extraordinarily complex structures. Most viruses contain a genome of either double-stranded DNA or single-stranded RNA, but some small viruses have single-stranded DNA and others have double-stranded RNA. The number of nucleotides in a virus genome may vary from a few thousand to several hundred thousand

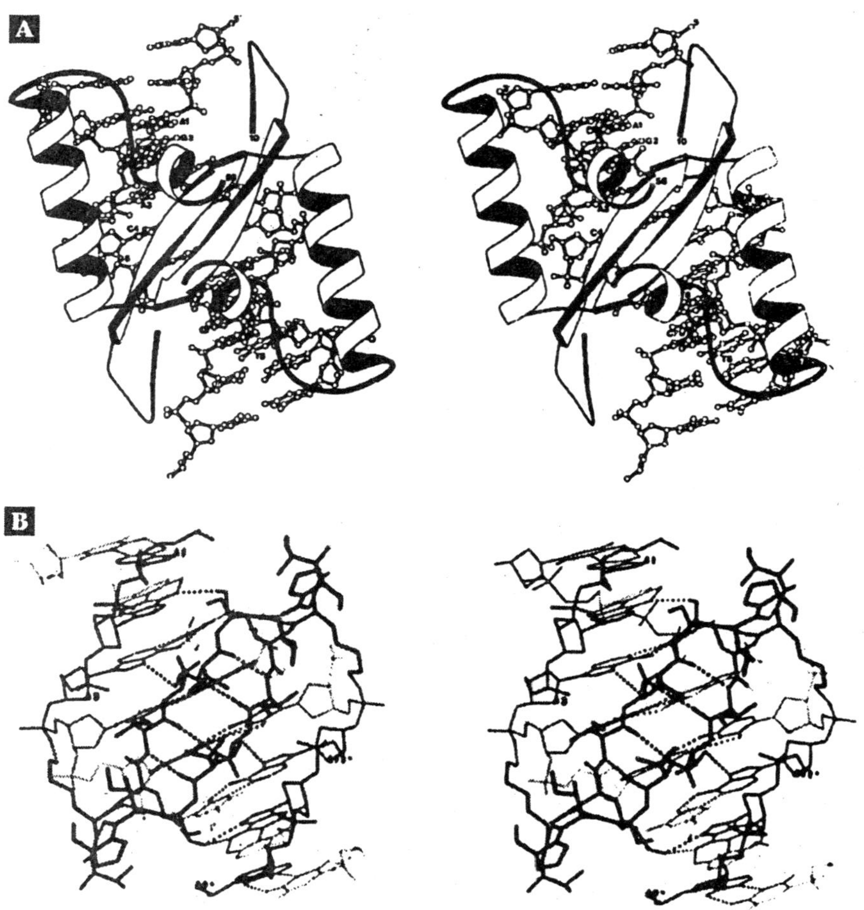

Fig. 1.35. (A) Stereoscopic ribbon representation of the *E. coli* methionine represser-operator complex. Two subunits form a dimer with a double-stranded antiparallel β ribbon that fits into the major groove of the DNA in the B form. One strand is shaded more darkly than the other. (B) View of the β ribbon and its interactions with the DNA. Notice the direct hydrogen bonds from amino acid side chains of lysine and threonine residues to bases in the specific palindromic sequence AGACGTCT.

and the number of genes from 3 to over 200. Sometimes the nucleic acid molecules within the virion are circular, but in other cases they are linear.

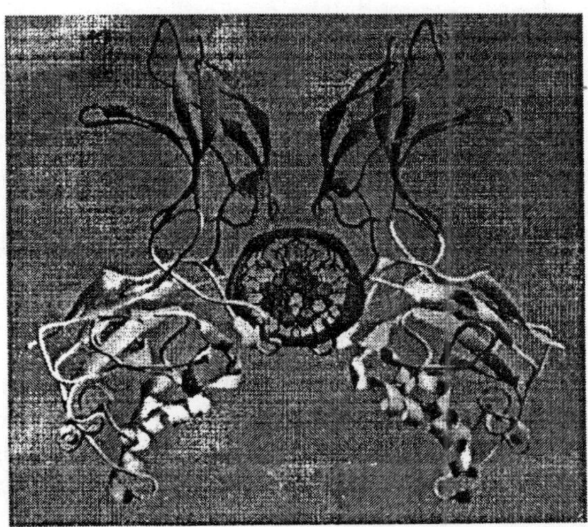

Fig. 1.36. Structure of a protein known as transcription factor NF-KB bound to its DNA target. Each subunit of the dimeric protein contains two (J barrel domains. The loops at the ends of the barrels interact with the DNA in the center.

Table. 1.7 lists a few of the known types of viruses as well as some individual viruses. The size of the genome, in kilobases (kb), or kilobase pairs (kbp) is indicated. The number of genes in a virus is often somewhat more than one per kbp of DNA. A vast amount of information is available about these infectious particles. Only a few reference sources are cited here.

Viruses with Single-Stranded DNA

The very small *helical bacteriophages* of the family, such as *fd, fl,* and M13 resemble thin bacterial pili but each virus particle contains a molecule of single-stranded circular ~ 6400 nucleotide DNA of $M_r \sim 2 \times 10^6$ which encodes only ten proteins. Bacteriophage ϕX174, an *icosahedral* DNA-containing virus 25 nm in diameter, is only three times as thick as the thinnest cell membrane. Its DNA contains just 5386 nucleotides. The similar bacteriophage G4 contains 5577. It is remarkable that such tiny viruses are able to seize control of the metabolic machinery of the cell and turn it all in the direction of synthesis of more virus particles.

There are only 11 known proteins encoded by the genes of ϕX or G4. Three genes encode the three kinds of protein subunits of the virus coat. Sixty copies of each are needed, as are 12 copies of a "pilot" protein. Eight of the genes are spaced closely together, occupying most of the DNA. The other three genes are embedded within some of the first eight but in different reading frames. In one short region of the G4 chromosome the same nucleotides are part of three different genes, using all three possible reading frames. In addition to their own genes, these small viruses make use of many components of the cell that they infect. A large group of animal viruses, the *parvoviruses,* are similar in size and architecture to bacteriophage ϕX174. Canine parvovirus, first identified in 1978, is now endemic.

Childhood *fifth disease* is also caused by a parvovirus. When these single-stranded DNA viruses infect cells a double-stranded *replicative form* of DNA arises by synthesis of the complementary *negative strand* alongside the original positive DNA strand. Many copies of the replicative form are then synthesized. The negative strands of the replicative forms serve as templates for synthesis of numerous new positive strands that are incorporated into the progeny viruses. The whole process may take only 20 minutes. Some parvoviruses are unable to reproduce unless the cell is also infected by a larger adenovirus. While most plant viruses contain dsRNA, the *geminiviruses* which cause a number of plant diseases, contain single-stranded DNA. The virus particle consists of a fused pair of incomplete icosahedra, evidently containing a single ~2500-bp DNA strand. Replication may require coinfection with two virus particles of differing sequence.

Viruses with Double-Stranded DNA

One of the smallest viruses containing double-stranded DNA is the 3180-nucleotide human *hegatitis B virus*. It infects millions of people throughout the world causing chronic hepatitis and often liver cancer. The circular DNA is surrounded by an envelope consisting of proteins, carbohydrate, and a lipid bilayer. Many icosahedral viruses also contain dsDNA. Among them are the *papovaviruses,* some of which cause warts and others malignant rumors. Much studied by biochemists is the 5386-nucleotide *simian virus 40* (*SV40*), a monkey virus capable of inducing tumors in other species. Closely related is the *polyomavirus* of the mouse and the *BK virus* of humans, also suspected of causing cancer. The *papillomaviruses* cause warts and perhaps cancer and the larger (70 nm diameter) *adenoviruses* cause respiratory infections. Some 32 types infect humans.

Table 1.7. Characteristics of Some Individual Viruses and Groups of Viruses

Type of genome[a] and group or individual virus name	*Shape[b]*	*Diameter[c] (nm)*	*Masses in daltons × 10^{-6} Total*	*DNA or RNA*	*Thousands of bases or base pairs (kilobases)*
DNA, single-stranded					
Bacteriophages *fd, f1,* M13	H	~6 × ~880 (length)	17.6	2.1	6.4
Basteriophages φ174, G4	I	25	6.2	1.8	5.4 – 5.6
Parvoviruses	I	18 – 25		1.8	5.5
Geminiviruses	I (fused pairs)				
DNA, double-stranded					
Hepatitis B virus	I (enveloped)			2.1	3.18
Papoviruses	I	45–55			
SV40 (monkey)	I		17.6	3.5	5.22
Polyoma (mouse)	I	45	23.6	3.3	4.96
BK virus (human)	I			3.4	4.96
Papilloma (human wart)	I	56		5.3	8.0
Bacteriophage φ29	T			12	19.3
Adenoviruses	I	70		20 – 30	30 – 45
Bacteriophage Mu				25	38

Type of genome[a] and group or individual virus name	*Shape[b]*	*Diameter[c] (nm)*	*Masses in daltons × 10^{-6} Total*	*DNA or RNA*	*Thousands of bases or base pairs (kilobases)*
Bacteriophage T7	T(short)				39.9
Bacteriophage P22	T			28.5	43.2
Bacteriophage λ	T			3.2	48.6
T-even bacteriophage	T	100 × 800 (head)	215	130	166
Baculoviruses of insects	I	70 – 130			
Herpesviruses	I	100			
	core	78	~1000	80-120	80-140
	envelope	150 – 200			
Pox viruses, *e.g.*,					
Small pox, vaccinia (cowpox)	C	160 × 250	~4000	150 – 240	240 – 300
Cauliflower mosaic virus	I				8.0
RNA, single-stranded					
Unsheathed					
Potato spindle tuber viroid				0.116	0.30
Hepatitis delta virus					1.7
Sheathed, *plus-strand*					
Tobacco necrosis satellite	I	18	1.7	0.4	1.20
Small bacteriophages					
R17, MS2, Qβ	I	23 – 26	3.6 – 4.0	1.2 – 1.5	3.5 – 4.5
Picornaviruses	I		8.4	2.6	7.9
Polioviruses	I	27	6.4	2	6.1
Rhinoviruses	I	27 – 30	7 – 8	2.2 – 2.8	6.7 – 8.5
Turnip yellow mosaic virus	I	28	5.0 – 6.0	2.0	6.1
Tobacco mosaic virus	H	18 × 300	40	2.2	6.7
Togaviruses	I	20 – 40		4	11
Negative-strand virus					
Influenza virus	I	80 – 100	200	2.0	6.1
Bullet-shaped virus					
Rhadbovirus	C	20 × 130			
Retroviruses	I	80 – 100		7 – 10	20 – 30
RNA, double-stranded					
Reoviruses	I	55 – 60		11 – 12	16 – 18
Mobilivirus	I (enveloped)	38			
Rotavirus	Wheel-shaped	70			
Leishmanina RNA virus					5.28

[a] Complete nucleotide sequences are known for most of these viruses.
[b] Shapes are indicated as I, icosahedral; H, helical; T, a tailed phage; C, complex.
[c] The second dimension given for some helical and complex viruses is the length (nm).

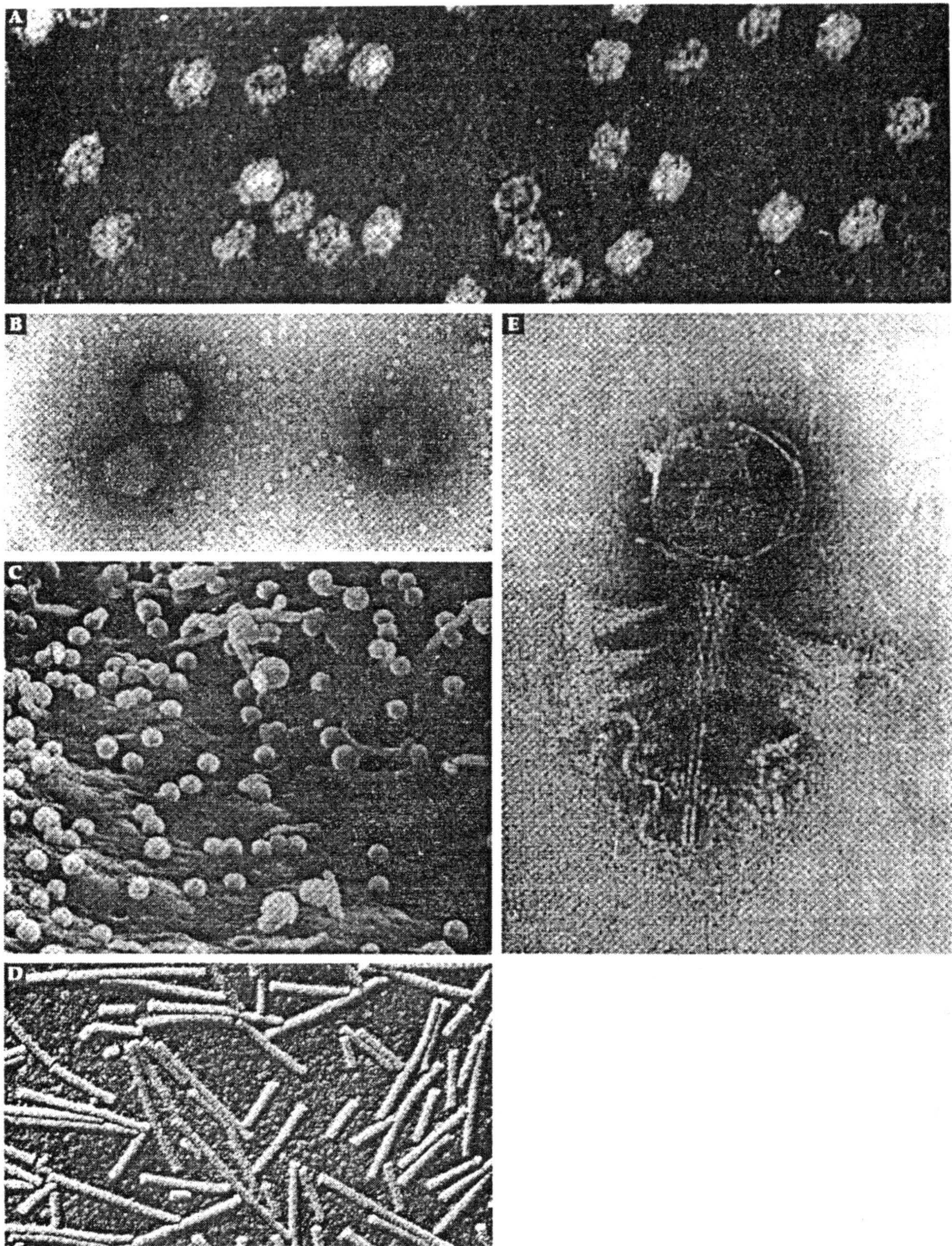

Fig. 1.37. Electron micrographs of some viruses: (A) Bacteriophage Lϕ7, which resembles ϕX174. (B) Adenovirus, (C) Cytomegalovirus particles being released from a fibroblast. Tobacco mosaic virus. (E) A tailed bacteriophage.

Many of the details of eukaryotic transcription were first studied using the adenovirus. The very large *herpesviruses* are enveloped by a lipid-containing membrane. Among them are herpes simplex viruses, which infect human mucous membranes and varicella-zoster virus, which causes chicken pox and shingles. Other herpesviruses include cytomegaloviruses another common human pathogen, and the Epstein-Barr virus, which causes mononucleosis and is suspected of causing cancer. Another herpesvirus has been associated with multiple sclerosis. The very large icosahedral *baculoviruses* cause polyhedroses in insects.

One that infects the fly *Tipula* measures 130 nm in diameter. *Poxviruses* are also large and complex. The tailed bacteriophages range in size from the small ~29-kDa phage P22 to the very complex T-even phage. Some plant viruses also contain dsDNA. One of these, the cauliflower mosaic virus is transmitted by aphids. It has proved useful as a gene-transfer vehicle for genetic engineering.

Viruses Containing RNA

Several plant diseases including *the potato spindle tuber disease* are caused by *viroids,* molecules of single-stranded RNA only 240 – 380 nucleotides in length with a folded structure. Such an RNA could code for a protein containing only about 100 amino acids. However, the known 359-nucleotide sequence of the potato spindle tuber virus contains no AUG initiation codon. It seems impossible that the virus carries any gene for a protein. Conserved features of viroid sequences suggest a close relationship to the intervening DNA sequences known as type I introns. Whatever its genetic message, a plant viroid causes the plant cell to replicate many copies of the viroid molecule, which may then be transmitted to other plants by aphids, or on the surface of tools, by humans.

A larger 1678-nucleotide viroid-like RNA (*hepatitis delta virus*) has been identified in human patients with severe chronic hepatitis and who were also infected with hepatitis B virus. The ssRNA in this virus does encode at least one protein. One of the smallest of the encapsulated RNA-containing viruses is the *satellite tobacco necrosis* virus. It replicates only when the plant is also infected with the larger tobacco necrosis virus. The satellite virus, whose three-dimensional structure is known from X-ray diffraction studies, contains a 1200-nucleotide strand of RNA which encodes a 195-residue protein. Sixty copies of the latter are assembled around the RNA in an icosahedral array to form the virion. The structure of the similar satellite tobacco mosaic virus has also been described in detail. RNAs of the small bacteriophages *f*2, R17, MS2 and Qβ contain 3500 – 4500 nuceotides and are enclosed in icosahedral shells made up of 90 identical subunits.

Intially only three genes were evident but a fourth small gene in a different reading frame has since been found in at least one of them. The RNA molecules in these and other *positive-strand* viruses serve as mRNA within the host cells. Another group of small RNA viruses are the *picornaviruses* (picoRNA, meaning very little RNA). Many of these icosahedral viruses of 15-30 nm diameter attack humans. Among them are the *enteroviruses* including the *polioviruses* the *hepatitis. A virus* the *coxsackieviruses,* and some of the *echoviruses.*

A second class ot picornaviruses include *rhinoviruses* which cause the common cold. More than 100 types are known. The *foot-and-mouth disease* virus, which attacks the cloven-footed animals, and the *Mengo virus,* which can cause a fatal encephalitis in mice, are also picornaviruses. The three-dimensional structures of polio virus, human rhinoviruses, Mengo virus, and many other eicosahedral viruses are known. The *togaviruses*, which are a little larger than the picornaviruses, have an icosahedral core surrounded by a lipid membrane. Yellow fever and rubella (German measles) are both caused by togaviruses.

Other togaviruses, such as *Sindbis virus* and *Semliki Forest virus* have become important in biological research. A large number of icosahedral RNA viruses of diameter 28-30 nm attack plants, causing diseases such as tomato bushy stunt, southern bean mosaic, or turnip yellow mosaic.

Best known of the helical RNA viruses is the *tobacco mosaic virus.* Its genome contains 6395 nucleotides as linear ssRNA. Many strains are known. Related viruses cause cucumber green mottle and other plant diseases. Large viruses of 80 –100 nm diameter bearing 8 – 10 spikes at the vertices of the icosahedra cause influenza, mumps, measles, and related diseases. The internal structure must be complex. Only 1% of the virus is RNA, and that consists of several relatively small pieces. These are *negative strand viruses* whose RNA is of the opposite polarity to the mRNA. The latter must be formed by transcription from the negative strand.

The viruses carry their own RNA polymerase for this purpose. Of even more complex structure are the bullet-shaped *rhabdoviruses* which cause rabies and vesicular stomatitis. The diameter of these viruses is 65 – 90 nm and the length 120 – 500 nm. The internal structure includes a helical arrangement of nucleoprotein. They also are negative strand viruses. Among the *retroviruses* are types B and C *oncoviruses* which induce malignant tumors in mammals and birds and the *human immunodeficiency virus* (HIV), the apparent causative agent of AIDS. Their RNA functions in a surprising way. Each virion contains a *reverse transcriptase,* an enzyme that transcribes copies of circular dsDNA copies from the one or two mRNA-like molecules that make up the virus genome.

Following action of the reverse transcriptase, one of the transcribed DNA circles becomes covalently spliced into the host's own cellular DNA. There it remains permanently as *a provirus.* RNA molecules transcribed from the provirus serve as mRNA for virus-encoded proteins and also as the genomes for new virus particles. Double-stranded RNA is unusual in nature but constitutes the genome of the *reoviruses.* The RNA of these viruses fragments into segments upon infection. One member of the group is thought to be the cause of acute diarrhea of infants.

Viruses without Nucleic Acid?

The cause of the slow, fatal neurological disease of sheep known as *scrapie* has been a mystery for many years. Similar human diseases include *kuru* and *Creutzfeldt-Jakob* disease. Scrapie can be transmitted by injection and this has permitted isolation of the apparent infective agent, a 27- to 30-kDa hydrophobic protein particle which is devoid of DNA or RNA. Prusiner suggested the name *prion* (protein-aceous infectious particle) for the scrapie agent. However, mRNA for the prion is present in normal as well as infected brains, and protein produced in mouse cells from cloned prion genes did not cause scrapie infections. Therefore, there was doubt about the causative agent for the disease. The prion concept is now generally accepted. There are still some who are looking for a nucleic acid component.

Life Cycles

Viruses have many modes of life. They enter cells in various ways. Some enter through coated pits from which they are taken into lysosomes via endocytosis. Others are literally injected into the cells. Within cells some viruses are assembled in the nucleus, some in the cytoplasm, and some in membranes. The typical life cycle of a virus leads to rapid formation of large numbers of progeny. Within 20 minutes after entrance into a bacterial cell, a bacteriophage can induce the formation of 100 – 200 new bacteriophage particles. One of the bacteriophage genes encodes a protein that is also synthesized by the host and which induces lysis of the cell membrane and destruction of the the cell.

Many animal viruses destroy cells in a similar fashion. *Temperate bacteriophage,* the best known being phage λ, have a very different life cycle. Their DNA usually becomes integrated at a specific point into the genome of the bacterium. Only rarely is an infected cell lysed. The retroviruses that attack mammals and birds have a similar characteristic. Their DNA is also integrated into the host genome. Some viruses that usually produce lysis of cells, *e.g.,* SV40, adenoviruses, herpes viruses, and hepatitis B virus, can occasionally be integrated into the DNA of the host.

If such integration occurs in the middle of a gene, that gene will be mutated. This is one way in which such viruses may induce cancers. One of the most important results of integration of viral DNA into the host genome is that the integrated genes are replicated as part of the genome and are transmitted from one generation to the next. Among these are the cancer-causing *viral oncopenes* (*v-onc*).

While viruses are important causes of cancer in some animals, relatively few human cancers are thought to result directly from the action of viruses. However, the Epstein-Barr virus, which causes mononucleosis, can sometimes be integrated into epithelial cells of nasal regions and can evidently cause cancer. The same virus appears to be responsible for Burkitt's lymphoma, a common cancer in certain areas in Africa.

Plasmids and Transposable Genetic Elements

In addition to their chromosomal DNA, bacteria often carry extra small pieces of DNA as permanent parts of their genome. These *plasmids* (sometimes called *episomes*), which are about the size of the DNA of viruses, replicate independently of the host chromosomes. Each bacterial cell usually contains more than a single copy of the plasmid. For example, the "colicinogenic" plasmid ColE1, that infects *E. coli* is a circular piece of DNA of molecular mass 4.2×10^6 Da. Over 20 copies are normally found per cell but in the presence of a suitable concentration of the drug chloramphenicol the number may rise to 1000 – 2000. Plasmids carry a variety of genes which are often useful to bacteria.

Some proteins encoded by plasmid genes confer drug resistance to a bacterium. Some are antibiotics. For example, a protein encoded by a gene in plasmid ColE1 is toxic to other strains of *E. coli*. Some plasmids carry genes for enzymes needed for the oxidation of hydrocarbons. Some plasmids contain genes for the *restriction endonucleases* which have become essential to present-day molecular biology and genetic engineering. As with some viruses, the DNA of many plasmids can become integrated into the genome of the host.

An example is provided by the large 62-kDa plasmids known as *sex factors*. They contain genes encoding the protein subunits of the sex pili and can become integrated into the bacterial chromosome. Bacteria containing integrated sex factors are "male" and are able to transfer genes not only of the sex factor but also of virtually the entire bacterial genome into other susceptible bacterial cells. This provides bacteria with the means for sexual reproduction. The transfer of DNA between the bacteria may occur via the sex pili. In this respect the sex factors are similar to viruses such as M13 that also appear to gain entrance to bacteria via sex pili. Integrated viruses are also related to *transposable. genetic elements* (transposons). These are segments of DNA that allow genes to move from place to place within the chromosomes.

METHODS OF STUDY

Many of the methods are directly applicable to nucleic acids. A few additional methods will be considered in this section.

Isolation and Separation of Nucleic Acids

RNA is often extracted from lysed cells or tissues, separated ribosomes, mitochondria, plastids, or nuclei by warming with aqueous phenol and a detergent such as sodium dodecyl sulfate (SDS). Proteins are denatured by this treatment and are dissolved by the phenol, while RNA remains in the lighter aqueous layer. Depending on the conditions DNA may either remain in the aqueous layer or be removed. Various precipitation and extraction procedures may be used to separate the RNA in the aqueous layer from polysaccharides, from DNA (if present), and from their components.

DNA may be extracted from cells or nuclei as a nucleic acid-protein complex using 1 M NaCl. The protein can then be denatured with an organic solvent, by detergents, or by phenol. It is desirable to digest proteins away with a nonspecific protein-hydrolyzing enzyme such as proteinase K. After removal of proteins DNA is often precipitated with cold ethanol. During isolation of RNA, bentonite (a type of clay) or other inhibitors of ribonuclease are often added. For the same reason, chelating agents that complex metal ions needed for the action of deoxyribonucleases are used to protect DNA. Care is necessary to avoid shearing of the very long, narrow strands of DNA. Even rapid pipeting of solutions will cause such breakage. Extracted RNA molecules may be separated from each other centrifugation in a sucrose gradient.

Fragments of DNA are purified in the same way or by equilibrium centrifugation in *CsCl gradients*. Concentration gradients in the dense salt solution are stable, and the sharpness of banding of particles is ensured by use of a high centrifugal field. Single-stranded DNA may be separated from double-stranded DNA, and DNAs of differing G + C content can be separated. The latter separation is based on differences in buoyant densities ρ in CsCl which are approximately shown in Eq. 1.9.

$$\rho = 1.660 + 0.098 \text{ (mole fraction C + G)}$$

One of the most important methods for separating either RNA or DNA mixtures is zone electrophoresis through polyacrylamide or agarose gels. The separated bands may be visualized by scanning in ultraviolet light or by fluorescence of intercalated dyes such as ethidium bromide. This method is being displaced to some extent by HPLC using DEAE type ion exchange columns for small lengths of DNA including plasmids. The procedure called pulsed *field electrophoresis* makes it possible to isolate very large pieces of DNA, up to several million base pairs in length.

The separation is carried out in agarose gels, through which the long DNA rods must move in a snakelike fashion. The current is delivered in a pulse and then, after a period of a second to several minutes, a second pulse in a different direction, usually at 90° to the first. The procedure is repeated many times. The size of the DNA seems to affect the time required to reorient the molecules and to start moving in the second direction. Intact DNA from small chromosomes can be separated. To prevent breakage of the DNA by shearing, intact cells are suspended in liquid agarose and allowed to gel into a block about 2 × 5 × 10 mm in size. The block is treated with enzymes and detergents to lyse the cells and to remove all protein and RNA. The block, containing the residual DNA molecules, is then embedded in the electrophoresis gel. Other methods of DNA separation include chromatography on hydroxylapatite and gel filtration.

Hydrolysis and Analysis

Both DNA and RNA are easily broken down by r acid-catalyzed hydrolysis. Thus, heating at 100°C for one hour in 12 M $HClO_4$ is sufficient to hydrolyze nucleic acids to their constituent bases. However, for analysis of RNA it is better to heat in 1 N HCl for 1 h at 100°C. The products are adenine, guanine, cytidine-5′-phosphate, and uridine-5′-phosphate. As is suggested by this distribution of products, the glycosylamine linkages to purines are more labile than those to pyrimidines.

The linkages are also less stable in DNA than in RNA. A procedure based on these differences and useful in sequencing by the Maxam-Gilbert method, is to leave DNA overnight in the cold at pH 2 to cleave off all of the purine bases. The resulting polymer is known as an *apurinic* acid. In alkaline solutions RNA is hydrolyzed to a mixture of 2′- and 3′-nucleotides.

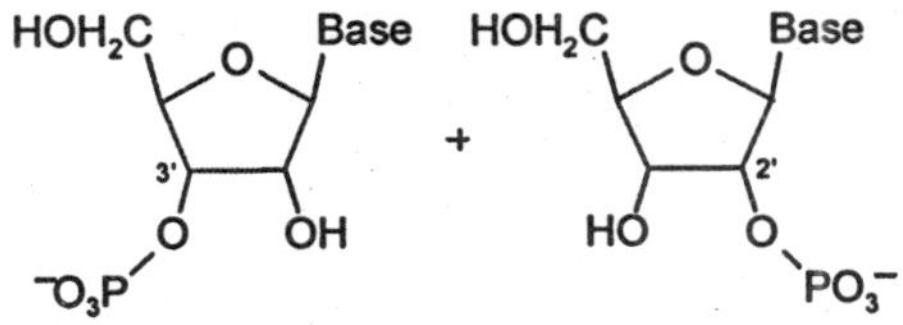

The mechanism involves participation of the free 2′-OH of the ribose groups and formation of cyclic 2′, 3′-phosphates and is similar to that of pancreatic ribonuclease. Because deoxyribose lacks the free 2′-OH, the phosphodiester linkages in DNA are quite stable in base. Hydrolytic cleavages of nucleic acids by the enzymes known as *nucleases* are of great practical value. Pancreatic ribonuclease, an *endonuclease*, cuts a chain adjacent to a pyrimidine in nearly random fashion, leaving phospho groups attached to the 3′ position in the nucleotide products. *Exonucleases* cleave from the ends of chains. For example, the phosphodiesterase of snake venom cleaves from the 3' end, which must have a free 3′-OH group, to give 5′-nucleotides.

On the other hand, the phosphodiesterase from spleen has the opposite polarity, cleaving chains from the 5′ end to give 3′-nucleotides. Similar variations in specificity are found among enzymes that cleave DNA. For example, *pancreatic DNAase I,* which cleaves preferentially between adjacent purines and pyrimidines, yields 5′ mononucleotides whereas DNAase II gives 3′-mononucleotides. Various hydrolytic cleavage reactions of polynucleotides are summarized in Fig. 1.43. The most striking specificity in DNA hydrolysis is displayed by the *restriction endonucleases.* These fussy catalysts cleave only at points within or close to a defined sequence of several nucleotides in double-stranded DNA. For example, the enzyme EcoR I cuts only at the following palindromic sequence:

$$\begin{array}{c} \quad\downarrow \\ 5'-G\ \ A\ \ A\ \ T\ \ T\ \ C-3' \\ \bullet\ \ \bullet\ \ \bullet\ \ \bullet\ \ \bullet\ \ \bullet \\ 3'-G\ \ T\ \ T\ \ A\ \ A\ \ G-5' \\ \qquad\qquad\qquad\uparrow \end{array}$$

The cuts in the two strands are made at the points indicated by the arrows.

This one endonuclease will cut almost any DNA into long pieces averaging about 5000 base pairs each. These pieces can in turn be cleaved by other restriction endonucleases to form smaller fragments. Since there are about 2400 of these enzymes known, with 188 different specificities, it is possible to cut any piece of DNA down to a size of 100 – 500 base pairs, ideal for sequencing. Each fragment has known sequences at the two ends. Some restriction enzymes cleave outside their specific recognition sequence. Some recognize 16-nucleotide palindromes and cut at rare sites. It is sometimes desirable to cut a large DNA molecule at only a few points.

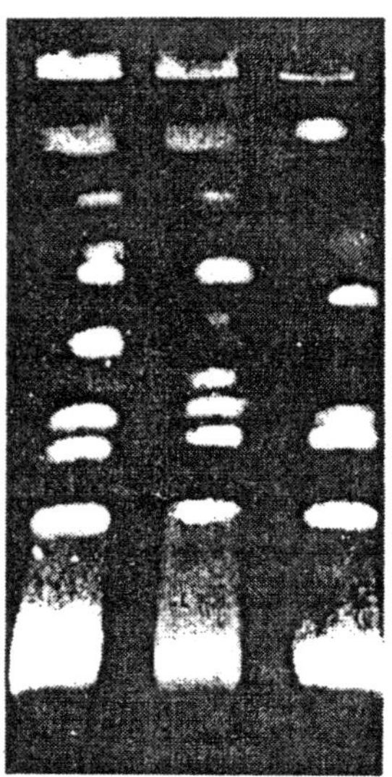

Fig. 1.38. Intact DNA from the chromosomes of three strains of the malaria parasite Plasmodium falciparnm, ranging from 750 Kb to 5 Mb, separated by pulsed-field gel electro-phoresis.

One approach is to protect most sites of a restriction enzyme's action by methylating them while protecting the desired cleavage site, for example by a represser protein or by a PNA molecule of specific sequence designed to "clamp" the site chosen for protection. Ribozymes have been engineered to be as specific or more specific than endonucleases. Other new approaches are being developed. The base composition of either RNA or DNA can be determined after hydrolysis catalyzed by 98% formic acid at 175°C for 30 min or by 12 M perchloric acid at 100°C for 1 h.

The bases can then be separated by ion exchange cnromatography on a sulfonated polystyrene resin. RNA can be hydrolyzed to a mixture of nucleoside 2′- and 3′-phosphates by 0.3 M NaOH at 37°C for h 16 and DNA can be hydrolyzed to nucleotides enzymatically. The negatively charged nucleotides can then be separated by ion exchange chromatography on a quaternary base-type resin.

Periodate cleavage and reduction of the resulting dialdehydes by $[^3H]NaBH_4$ to trialcohols allows introduction of a radioactive label. Alternatively, the dialdehydes can be reductively alkylated by an amine plus $NaCNGH_3$. The total content of RNA + DNA in tissues may be estimated from the phosphorus content or by color reactions of the sugars.

These reactions depend upon dehydration to furfural or deoxyfurfural by concentrated sulfuric acid or HCl. Furfural formed from RNA reacts with orcinol (3, 5-dihydroxytoluece) and ferric chloride to produce a green color useful in colorimetric estimation of RNA. A similar reaction of DNA with diphenylamine yields a blue color.

Quantitative determination of over 90 free nucleotide compounds found within cells can be accomplished by thin layer chromatographic procedures on as few as 10^6 bacterial cells (~2 μg) labeled by growth in a $^{33}P_i$-containig medium.

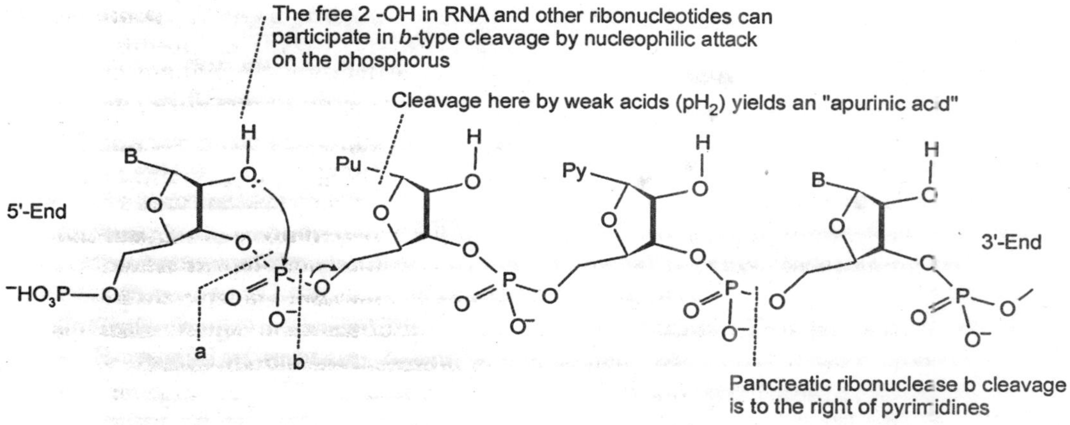

A. Clevage at point *a* is catalyzed
   1. Throughout the molecule by endonucleases
      Pancreatic deoxyribonuclease I
   2. Only at the 3′ end by exonucleases
      Venom diesterase, nonspecific, attacks DNA and RNA. A free 3′-OH is essential

B. Cleavage at point *b* is catalyzed
   1. Randomly throughout the molecule by endonucleases and by bases (nonenzymatically)
      Pancreatic ribonuclease cleaves only to the right of a pyrimidine-containing nucleotide
      Ribonuclease T1 of *Aspergillus oryzae* cleaves to the right of a guanine-containing residu (3′-guanlyate) in ssRNA
      Ribonuclease T2 of *Aspergillus oryzae* cleaves to the right of an adenine-containing residue (3′-adenylate) in ssRNA
      Pancreatic deoxyribonuclease (DNase) II
      Micrococcal DNase
   2. Only at the 5′ end by exonucleases
      Bovine spleen phosphodiesterase hydrolyzes both polyribo-and polydeoxyribonucleotides

Fig. 1.39. Some hydrolytic cleavage reactions of polynucleotides. Reactions of both RNA and DNA are included.

Characteristic Reactions of the Bases and Backbone

Reactions of Nucleophiles

A number of nucleophilic reagents add reversibly at the 6 position of pyrimidines. Thus, bisulfite adds to uridine Hydroxylamine ($HONH_2$) adds in a similar fashion to give a compound with –HNOH in the 6 positions. Sodium borohydride ($NaBH_4$), which can be viewed as a donor of a hydride ion (H^-), reduces uridine to the 5, 6-dihydro derivative. This presumably occurs by attack of the hydride ion at position 6 in a manner analogous to the reaction of bisulfite in Eq. 1.10.

Cytidine reacts in the same way, but the bisulfite addition compound is unstable. These C5-C6 adducts of cytidine all have a greatly enhanced reactivity at C4, presumably because of the lessened aromatic character of the ring. Cytidine is slowly deaminated by base, presumably as a result of attack by hydroxyl ion on the electrophilic center at C4 and subsequent elimination of NH_3.

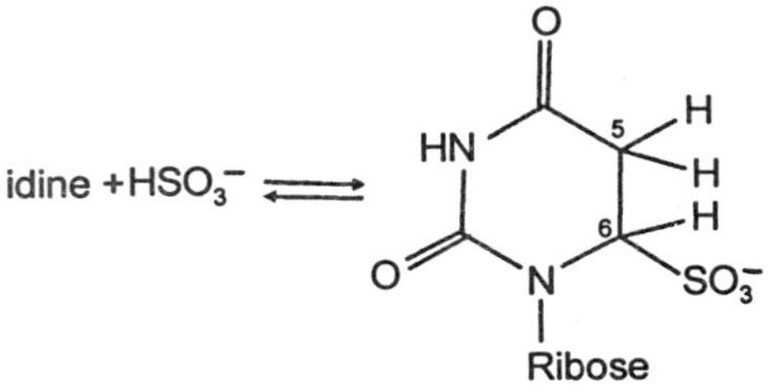

...(1.10)

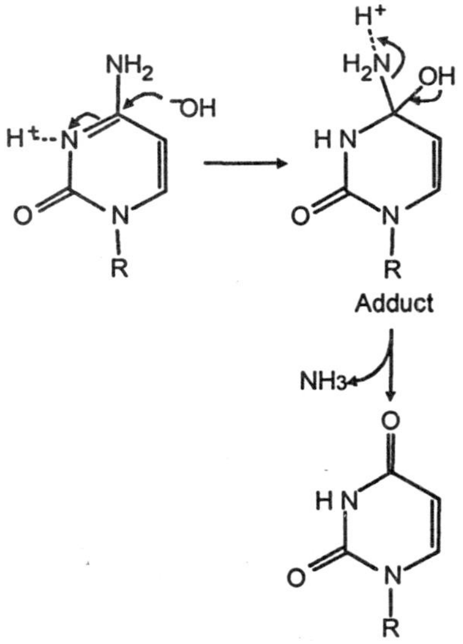

...(1.11)

The reaction is catalyzed by buffer salts and by bisulfite and hydroxylamine. Catalysis probably occurs, at least in part, as a result of addition of these nucleophiles at the 6 position to form compounds with increased nucleophilic reactivity at C4. Hydroxylamine and methoxyamine (NH_2OCH_3) participate in reactions parallel to that of the hydroxyl ion in Eq. 1.11. Products contain $-NHOH$ or $-NH-OCH_3$ in place of $-NH_2$ but tautomerize to the more stable forms shown in Eq. 1.12. Similar substitution reactions occur with other amines.

The C6 adduct with hydrazine can undergo ring cleavage. The initial product then undergoes β elimination, leaving ribosylurea or deoxyribosylurea. The same reaction can be carried out on intact strands of DNA and is widely used in determination of nucleotide sequences. The conversion of 5-hydroxymethylcytosine to the 5-methylenesulfonate (5-CH_2SO_3-) by reaction with bisulfite should also be mentioned. This is a nucleophilic displacement on the electron-deficient methylene group of this base.

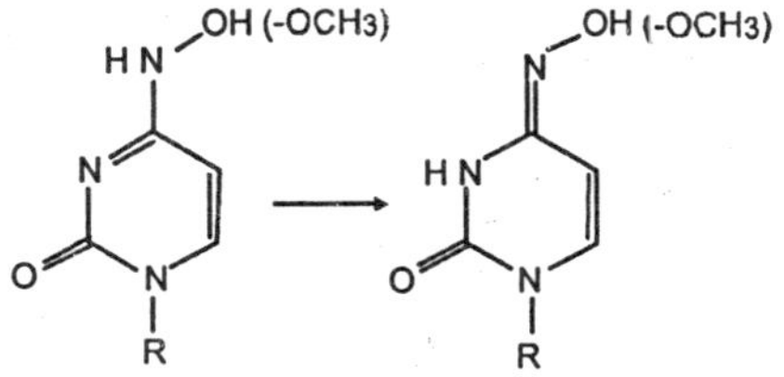

...(1.12)

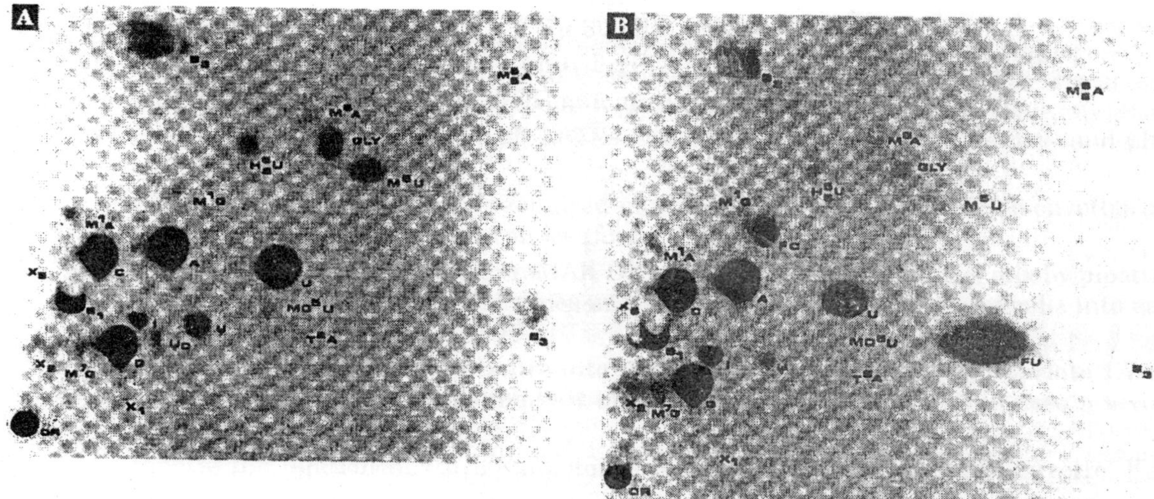

Fig. 1.40. Fluorographs of 2′3′-[^{3}H] nucleoside trialcohols from *Bacillus subtilis* grown in the absence (A) and presence (B) of 5-fluorouracil. About 2.3 nmol of nucleosides from each sample was chromatographed and exposed to X-ray film for 90 h. at –80 °C. Or (origin) B_1, B_2, and B_3 contain unidentified material present in a reaction mixture lacking RNA. Abbreviations used. FU, 5-fluorouridine; FC, 5-fluorocytidine; U, uridine; C, cytidine; G, guanosine; A, adenosine; 1, inosine; m′G, 1-methylguanosine; m^{7}G, 7-methylguanosine; m′A, 1-methyladenosine; m^{6}A, 6-methyladenosine; m$^6_6$A, 6,6-dimethyladenosine; t^{6}A, N-[9-(β-D-ribofuranosyl) purin-6-yl carbamoyl] threonine; H$^5_6$U, 5,6-dihydrouridine; ψ, pseudouridine; $ψ_D$, decomposition product of ψ, m^{5}U, 5-methyluridine (ribosylthymine); mo^{5}U, 5-methoxyuridine; N′, a nucleoside trialcohol obtained by reduction of a nucleoside dialdehyde with [^{3}H]NaBH$_4$; FU-5 and FU-20 samples correspond to tRNAs from cells grown at that final concentration of 5-fluorouracil in μg/ml.

β elimination

Ribosylurea ...(1.13)

Reactions with Electrophilic Reagents

Reactions of nucleic acids with the simplest electrophile, the proton, have already been discussed. Somewhat similar are the reactions by which metal ions bind at many sites on both the bases and the phosphate groups of the backbone. An important reaction is the deamination of amines by dilute *nitrous acid*. This reagent, by a complex mechanism, converts the amino groups of cytidine, adenosine, and guanosine to hydroxyl groups; hydroxy compounds tautomerize to the corresponding amides. Cytidine reacts more rapidly than does adenosine which in turn reacts more rapidly than guanosine.

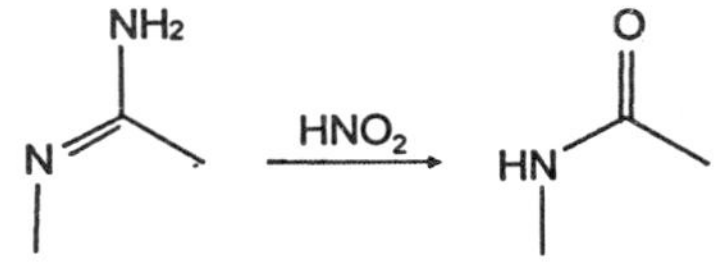

$$\text{N=C(NH}_2\text{)} \xrightarrow{HNO_2} \text{HN–C(=O)} \qquad ...(1.14)$$

The reaction converts cytosine into uracil and adenine into hypoxanthine. The changes are mutagenic because during replication the modified bases of the DNA pair differently than do the original bases. Guanine is converted to xanthine but this is not likely to be highly mutagenic. Nitrous acid can also convert uridine to 5-nitrouridine. The amino groups of the bases react reversibly with aldehydes but to a lesser extent than do the more strongly basic amino groups of the amino acids.

Form aldehyde forms adducts containing either one or two molecules of the aldehyde.

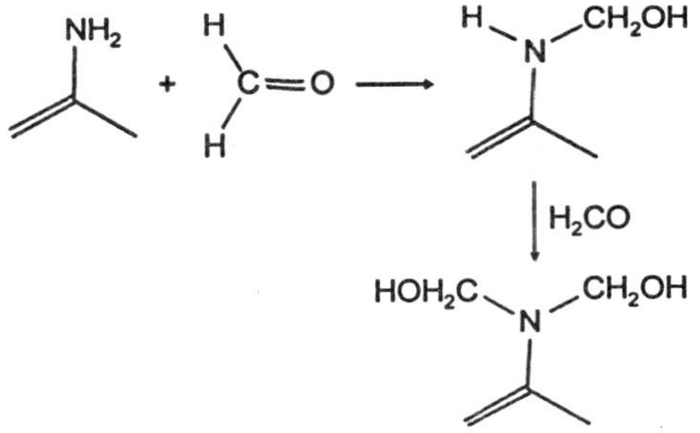

$$\text{C(NH}_2\text{)} + H_2C{=}O \longrightarrow \text{C–N(H)CH}_2\text{OH} \xrightarrow{H_2CO} \text{C–N(CH}_2\text{OH)}_2 \qquad ...(1.15)$$

These are reversible reactions. A more nearly irreversible crosslinking can occur by elimination of water between one of these products and a nucleophilic group in another base. A dicarbonyl reagent that is widely used because of its specificity toward guanine is *kethoxa*.

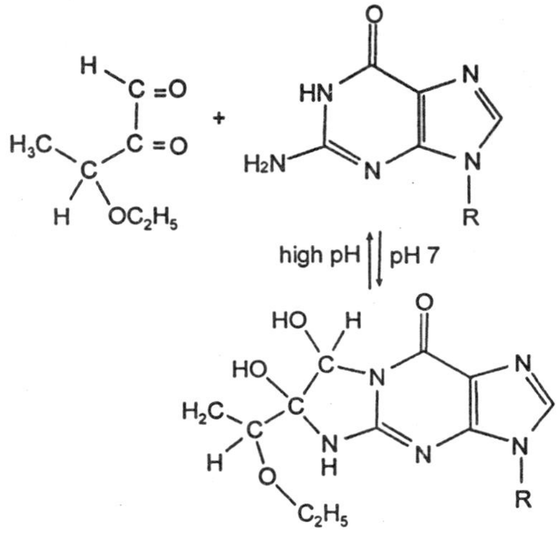

$$\text{H–C(=O)–C(=O)–CH(CH}_3\text{)(OC}_2\text{H}_5\text{)} + \text{guanine (N-R)} \underset{pH\ 7}{\overset{high\ pH}{\rightleftharpoons}} \text{adduct} \qquad ...(1.16)$$

Formation of the cyclic product is a consequence of the presence of the adjacent amino and NH groups in the guanine ring. Pyrimidines undergo *halogenation* at position 5, while guanine reacts at position 8. Adenine is quite unreactive. Elemental halogens or a variety of other halogenating reagents may be used. Of special

$$Br_2 + \text{cytidine} \rightarrow \text{5-Br-cytidine} + H^+ + Br^- \quad ...(1.17)$$

value is iodination with ^{131}I or ^{125}I, by which a high level of radioactivity may be introduced into nucleic acids. *Alkylation* reactions are not only of use in structural studies but also provide the basis for the action of a large class of mutagenic compounds.

Treatment of a nucleoside, nucleotide, or nucleic acid with an alkyl iodide or a dialkylsulfate converts residues of guanosine to an N^7-alkyl-guanosine. Reaction occurs at other nitrogen atoms as well as the oxygen atom of the base and of the ribose ring to a lesser extent.

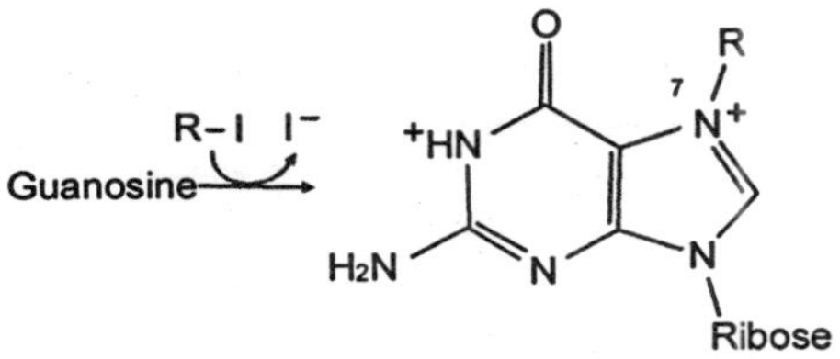

R–I I⁻
Guanosine → (N^7-R guanosinium: $^+$HN, H₂N, N, N$^+$–R at position 7, N–Ribose) ...(1.18)

Adenine is alkylated preferentially at N-1 and cytosine and thymine at the corresponding position (N-3) almost exclusively. Uridine and thymidine react very slowly. Adenine is also alkylated at N-3, N-7 and at the exocyclic N-6. Other alkylating reagents include the powerful mutagens dialkylnitrosamines and alkylnitrosoureas.

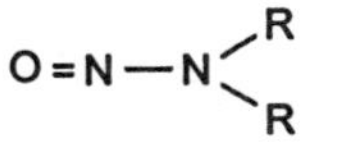

$$O{=}N{-}NR_2$$

Dialkylnitrosamines

Epoxides alkylate by a nucleophilic displacement reaction that opens the ring.

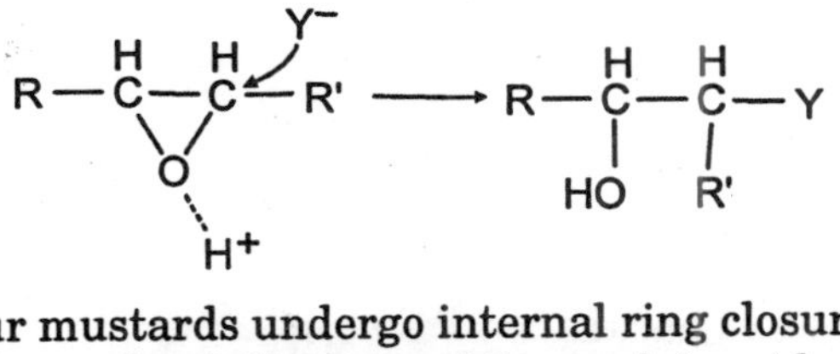

$$R{-}CH{-}CH{-}R' \ (\text{epoxide, O}\cdots H^+;\ Y^- \text{ attack}) \longrightarrow R{-}CH(OH){-}CH(R'){-}Y \quad ...(1.19)$$

The nitrogen and sulfur mustards undergo internal ring closure to an iminium ion which can then open by attack of a nucleophilic atom of the nucleic acid.

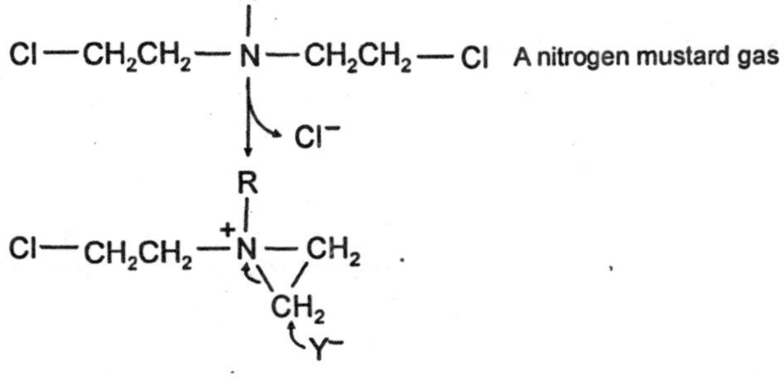

$$Cl{-}CH_2CH_2{-}N(R){-}CH_2CH_2{-}Cl \quad \text{A nitrogen mustard gas}$$

$$\downarrow -Cl^-$$

$$Cl{-}CH_2CH_2{-}\overset{+}{N}(R)\text{(cyclic with }CH_2{-}CH_2) \leftarrow Y^- \quad ...(1.20)$$

Other alkylating agents react through nucleophilic addition to a carbon-carbon double bond. Thus, acrylonitrile reacts with the nitrogen or oxygen atoms of nucleic acid in the same manner as does the SH group in Eq. 1.25. The water-soluble carbodiimides react as in the first step of Eq. 1.10, form adducts of the following type:

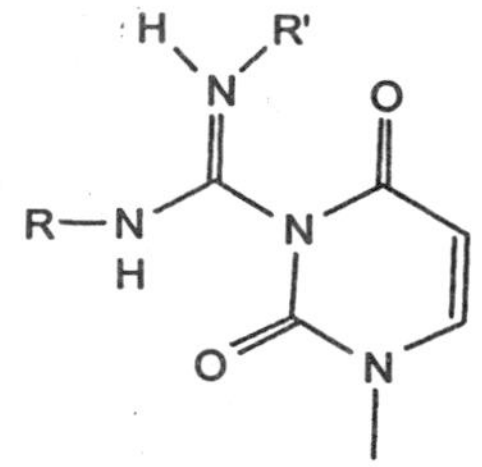

There are many other alkylating agents which often display widely varying reactivity and specificity toward particular nucleic acid bases and particular nucleotide sequences. A striking effect of alkylation of guanine in nucleic acids is the labilization of the N-glycosyl linkage to the ribose or deoxyribose.

This effect can be understood in terms of the induction by resonance of a partial positive charge on the nitrogen of the glycosyl linkage. As is indicated by the small arrows on the right-hand structure, the positive charge assists in an elimination reaction that produces an oxycarbocation. The latter can then react with a hydroxyl ion from water.

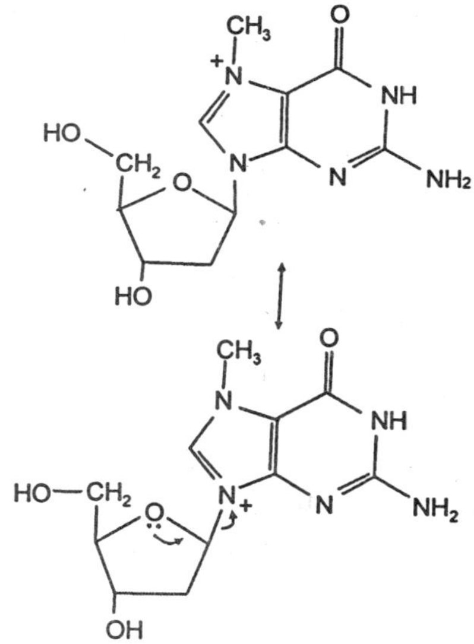

Reactions Causing Cleavage of the Sugar-phosphate Backbone

Treatment of DNA with 16 – 18 M hydrazine leads to the destruction of the pyrimidine rings. The reaction can be made somewhat specific for cytosine by carrying it out in the presence of a high concentration of chloride. The remaining polymer, an apyrimidinic acid, contains residues of ribosylurea. These undergo an aminocatalyzed displacement and a β elimination sequence that cleaves the polynucleotide chain.

Hydration of the aldehyde and several tautomerization steps are involved in step *e* of this equation. This reaction is very useful in sequence determination. Notice that "tracts" of purine nucleotides remain intact after this treatment. A similar base-catalyzed reaction sequence can be used to displace N^7-methylguanine and to cleave the polynucleotide. Ethylnitrosourea, in its reaction with purines, is useful as a structural probe of RNA.

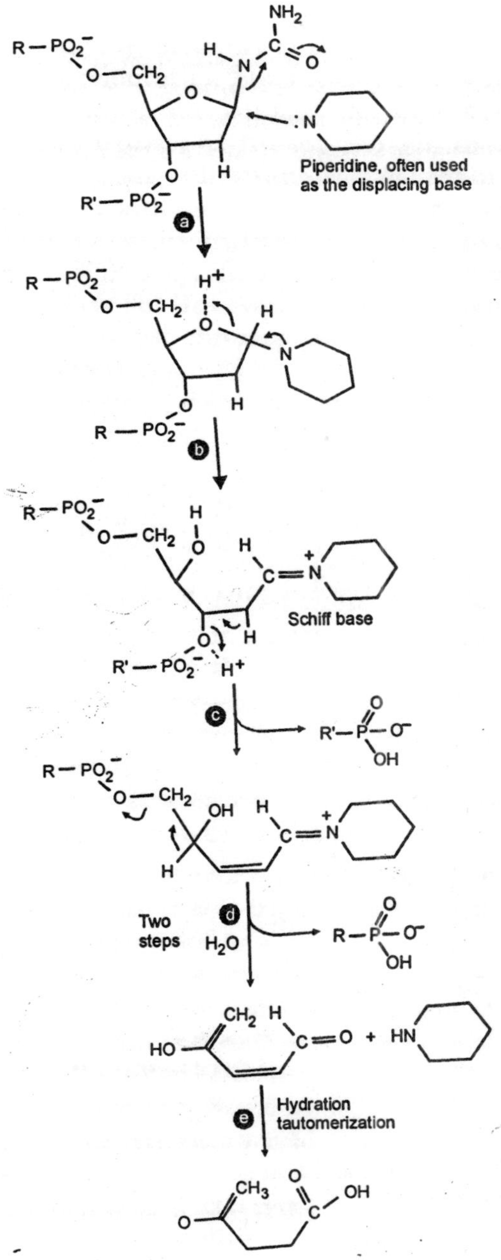

Piperidine, often used as the displacing base

Schiff base

Two steps H_2O

Hydration tautomerization

...(1.21)

Melting, Hybridization, and Polynucleotide Probes

Like proteins, nucleic acids can undergo denaturation. The strands of the double helix of DNA are separated and the double-stranded regions of RNA molecules "melt." Denaturation can be accomplished by addition of acids, bases, and alcohols or by removal of stabilizing counter ions such as Mg^{2+}. The product is a random coil and denaturation can be described as a helix → coil transition.

Denaturation of nucleic acids by heat, like that of proteins, is cooperative and can be described by a characteristic *melting temperature*. A plot the optical absorbance at 260 nm (the wavelength of maximum light absorption by nucleic acids) versus temperature is known as a *melting curve*. The absorbance is lower, by up to 40%, for native than for denatured nucleic acids. This *hypochromic effect* is a result of the interaction between the closely stacked bases in the helices of the native molecules.

The melting temperature T_m is taken as the midpoint of the increase in absorbance. As the percentage of G + C increases, the nucleic acid becomes more stable toward denaturation because of the three hydrogen bonds in each GC pair. T_m increases almost linearly with increases in the G + C content. In the "standard" citrate buffer (0.15 M NaCl + 0.015 M sodium citrate, pH 7.0) Eq. 1.22 holds. The exact numerical relationship depends strongly upon the ionic composition and pH of the medium.

$$\%(G + C) = 2.44\,(T_m - 69.3);\ T_M \text{ in } °C \qquad ...(1.22)$$

The curves, in Fig. 1.45 appear simple, but using newer apparatus and plotting the first derivative of the melting curve yields a complex pattern that depends on the sequence of bases. Complete denaturation of DNA leads to separation of the two complementary strands.

If a solution of denatured DNA is cooled quickly, the denatured strands remain separated. However, if the temperature is held for some time just below T_m (a process known as *annealing*), the native double-stranded structure can be reformed. An important tool for studying DNA has been the measurement of the *kinetics of reassociation* of separated strands of relatively short DNA fragments. Because itdepends upon the concentration of two separated strands, reassociationbeys second-order kinetics and Eq. 1.23 which is readily derived by integrating, [A] = [B] = C from time 0 to *t:*

$$C/C_0 = 1/(1 + kC_0t) \qquad ...(1.23)$$

The initial concentration of denatured DNA, C_0, is related in this way to the concentration C of DNA remaining dissociated at time *t*. A plot of the fraction of molecules remaining single-stranded versus the logarithm of C_0t, is a convenient way of displaying data. As indicated in Fig. 1.46B, the value of C_0t increases in direct proportion to the

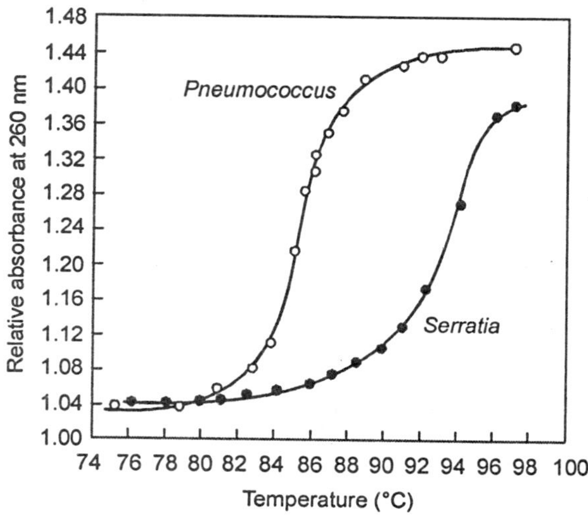

Fig. 1.41. A melting curve for DNA molecules from two different sources.

length of the DNA chain in the genome, but it is very much decreased if the sequence of bases is highly repetitive [poly(T) and poly (A)]. The slope of the plot at the midpoint gives an indication of the heterogeneity of the DNA fragments in a solution. Denatured DNA fragments can sometimes reassociate with DNA from a different source to form *hybrid duplexes*.

Such double helices, in which one strand comes from one strain of an organism and the other strand from a genetic variant of the same organism or from a different species, are known as *heteroduplex*. Some mutations consist of *deletions* or *additions* of one or a substantial number of bases to a DNA chain. Heteroduplexes prepared from DNA of such mutants hybridized with that from a nonmutant strain have normal hydrogen-bonded Watson-Crick base pairs for the most part. However, they may have single-stranded loops in regions where long deletions or additions prevent complementary base pairmg. Hybridization measurements have been used in many studies of *homolog* of nucleic acids from different species.

A nucleic acid is cut (*e.g.*, by sonic oscillation) into pieces of moderate length (~1000 nucleotides) and is denatured. The denatured DNA fragments are mixed with denatured DNA of another species. Nucleotide sequences that are closely similar between species tend to hybridize, whereas sequences that are drastically different between two species. One way to do such an experiment is to immobilize the

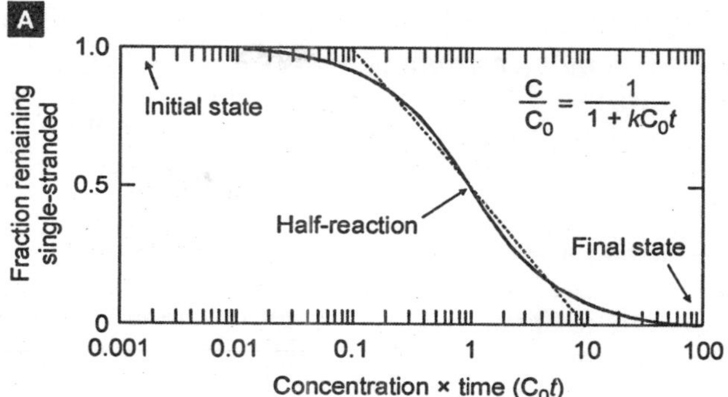

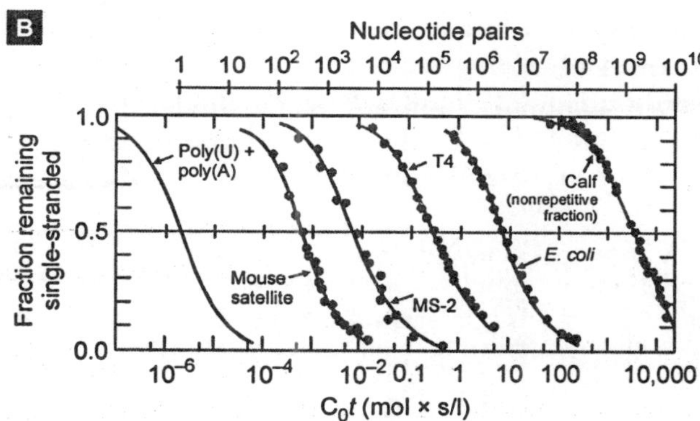

Fig. 1.42. Reassociation curves for DNA, (A) Time course of an ideal, second-order reaction to illustrate the features of the log C_0t plot. The equation represents the fraction of DNA which remains single-stranded at any time after the initiation of the reaction. For this example, *k* is taken to be 1.0, and the fraction remaining single-stranded is plotted against the product of total concentration and time on a logarithmic scale. (B) Reassociation of double-stranded nucleic acids from various sources. The genome size is indicated by the arrows near the upper nomographic scale. Over a factor of 10^9, this value is proportional to the C_0t required for half-reaction. The DNA was sheared, and the other nucleic acids are reported to have approximately the same fragment size (about 400 nucleotides, single-stranded). Correction has been made to give the rate that would be observed at 0.18 M sodium-ion concentration. The temperature in each case was optimal, *i.e.*, ~30°C below the melting temperature T_m. The extent of reassociation was established by measuring optical rotation (calf thymus DNA), ribonuclease resistance (MS-2), or hypochromicity.

long-chain denatured DNA from the one organism by embedding it in an agar gel or by absorbing it onto a nitrocellulose filter. The DNA fragments from the second organism are passed through a column containing "beads" of the DNA-containing agar or through the filter with adsorbed DNA. Pairing of fragments with complementary sequences occurs and such paired fragments are retained while strands that do not pair pass on through the column (or filter). Both DNA hybrids and *DNA-RNA hybrid duplexes,* are very important to present day genetic research.

Molecules of mRNA that represent canscripts of a particular gene will hybridize only with one of the two separated strands of DNA for that gene. A major use of hybridization is to locate a gene or other DNA or RNA sequence by means of a *synthetic probg.* This is a small piece of DNA or RNA which is labeled in some way, *e.g.,* with a radioisotope such as ^{3}H, ^{32}P, or ^{125}I. Alternatively, the probe may carry a highly fluorescent dye or a "tag" that can be recognized by a specific antibody.

An example of the latter is the use of the vitamin biotin and the specific binding protein *avidin.* Related procedures employ labeling with *digoxigenin* and often employ chemiluminescent detection. Several methods for preparation of probes are in use. Some are enzymatic but the direct chemical synthesis of oligonucleotide probes is probably used the most. One of the first methods devised torrnaking a highly radioactive DNA probe is called *nick translation.*

A piece of dsDNA, *e.g.,* a "restriction" fragment cut from a larger piece of DNA by restriction endonucleases, is selected. A small amount of pancreatic DNase I is added. It creates "nicks" in which one strand has been cut and some nucleotides have been removed leaving a gap. Now the DNA is incubated with DNA polymerase I (pol I) and a mixture of the four mononucleotide triphosphates, the precursors of biological synthesis of DNA. Usually a ^{32}P or ^{35}S label with high radioactivity is present in one of the nucleotide triphosphates as indicated in the following structure.

The polymerase fills the gap, adding nucleotides to the exposed 3′ end of the nicked chain. *E. coli* pol I has a second enzymatic activity which allows it to digest a polynucleotide chain from the 5′ end. Thus, as synthesis proceeds at the 3′ side the nick is "translated". Probes may also consist of DNA copied from mRNA. This is known as *cDNA* and is also widely used to determine indirectly the sequences of mRNA molecules.

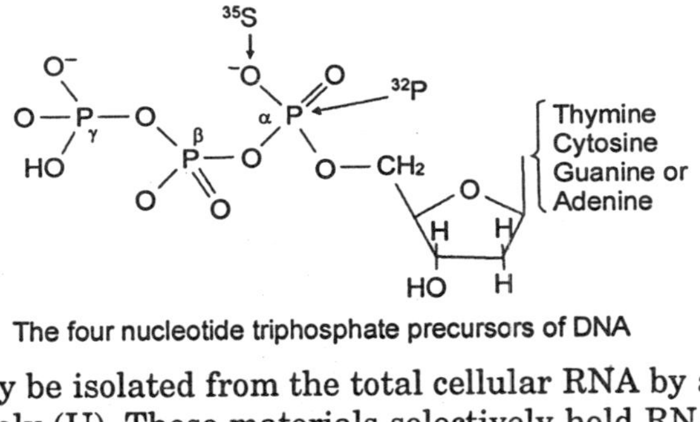

Thymine
Cytosine
Guanine or
Adenine

The four nucleotide triphosphate precursors of DNA

Messenger RNA may be isolated from the total cellular RNA by affinity chromatography on bound poly (dT) or poly (U). These materials selectively hold RNA with the poly (A) tails characteristic of most eukaryotic mRNA. Another source of mRNA is polyribosomes (polysomes), which are "reading" mRNA and actively making proteins. An antibody to the

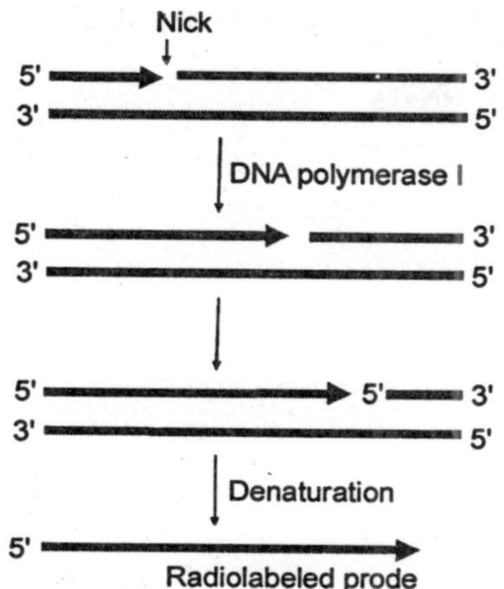

...(1.24)

protein for which mRNA is desired will often bind to the protein chains being synthesized and precipitate the polysomes. The mRNA can be recovered and used as a template for cDNA. Synthesis of cDNA, usually in radiolabeled form is accomplished with reverse *transcriptase*, the enzyme from retroviruses that synthesize a DNA-RNA hybrid from ssRNA.

A short oligo (dT) primer is usually hybridized to the 3′ poly (A) tail to initiate synthesis. Reverse transcriptase also has ribonuclease (RNase H) activity and will digest away the RNA. If desired, synthesis of the second strand can be carried out by a DNA polymerase to give a complete DNA duplex. Many gene sequences have been deduced from cDNA copies. Often the most practical approach to obtaining a DNA probe is synthesis of a mixture of short oligonucleotides, often in radioactive form as described in the next section.

The "redundancy" in the genetic code, *i.e.*, the existence of two or more codons for most amino acids, presents a problem in designing an oligonucleotide probe based on amino acid sequence information. Examination of Table. 1.5 suggests part of the solution. Whereas only Met and Trp have single unambiguous codons nine amino acids have only two codons each. We should try to find an amino sequence that contains Met and Trp and as many of the nine others as possible. We should avoid sequences that contain Ser, Leu, or Arg because each has six codons. We can then make a mixture of oligonucleotides, using the various coding possibilities.

Mixtures of as many as 1024 (2^{10}) oligonucleotides have been used. One of these may bind tightly and specifically to the desired DNA segment. Instead of such a complex mixture it may be more useful to incorporate a modified base at the most ambiguous positions. For example, inosine, which occurs in the wobble position in anticodons, can pair with A, C, or T. Substitution of 2-aminoadenosine can cause a probe to bind more tightly because a third hydrogen bond will be present in each AT pair. Another important procedure is labeling ends of ucleotides.

Most often the 5′ end is labeled with a radioisotope or by covalent attachment of a fluorescent dye. For example, a *polynucleotide kinase* can be used to transfer a radioactive γ-phospho group from ATP to the 5′ end of a polynucleotide that has a free 5′-OH group.

Synthesis of Oligonucleotides and Polynucleotides

Efficient solid-phase methods of synthesis analogous to those for polypeptides have been devised. The pioneering work was done by H.G. Khorana, who made the first synthetic gene and later synthesized a gene for the visual pigment rhodopsin. Several synthetic approaches have been developed. Currently the most popular method involves the use of phosphate esters.

Most nucleophilic groups of the monomers are derivatized with removable blocking groups. For example, N-4 of cytosine and N-6 of adenosine may carry benzoyl groups. The 5′-OH of each nucleotide is blocked by a di-*p*-anisylphenylphenylmethyl (also called dimethoxytrityl, DMTO) group. The 3′-OH is converted to one of a number of activated derivatives such as the following *N, N*-diisopropylamino phosphines. Solid-phase synthesis is usually done on a silica support with a covalently attached succinamide.

The first nucleotide at the 3′ end of the chain to be synthesized is attached by an ester linkage to the bound succinamide. The 5′-protecting group is removed in step *b* and the 5′-OH reacts with the activated phosphine of the second nucleotide. Steps *b* and *c* are then repeated as often as necessary to complete the chain. The finished polynucleotide can be removed from the solid support, the cyanoethyl groups removed from the phosphorus atoms by β elimination and all of the other blocking groups removed by treatment with concentrated NH_3.

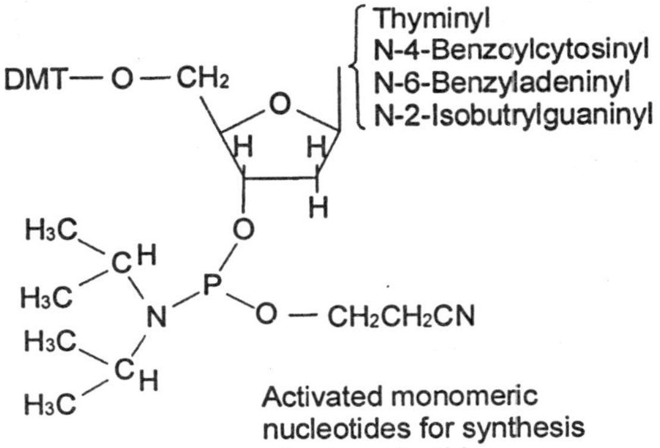

DMT—O—CH_2

Thyminyl
N-4-Benzoylcytosinyl
N-6-Benzyladeninyl
N-2-Isobutrylguaninyl

H_3C CH H_3C N P O—CH_2CH_2CN H_3C CH H_3C

Activated monomeric nucleotides for synthesis of polynucleotides

The whole procedure has been automated. If a large piece of DNA is needed several oligonucleotides can be joined end to end enzymatically. For example, a functional 17-bp gene for the 53-residue human epidermal growth factor was synthesized by joining ten oligonucleotides of lengths 11 – 59 bp. DNA is often synthesized enzymatically. Cloned sequences of synthetic DNA can also be transcribed to produce *polyribonucleotides* of any desired sequence. New nonenzymatic methods for RNA synthesis have also been devised.

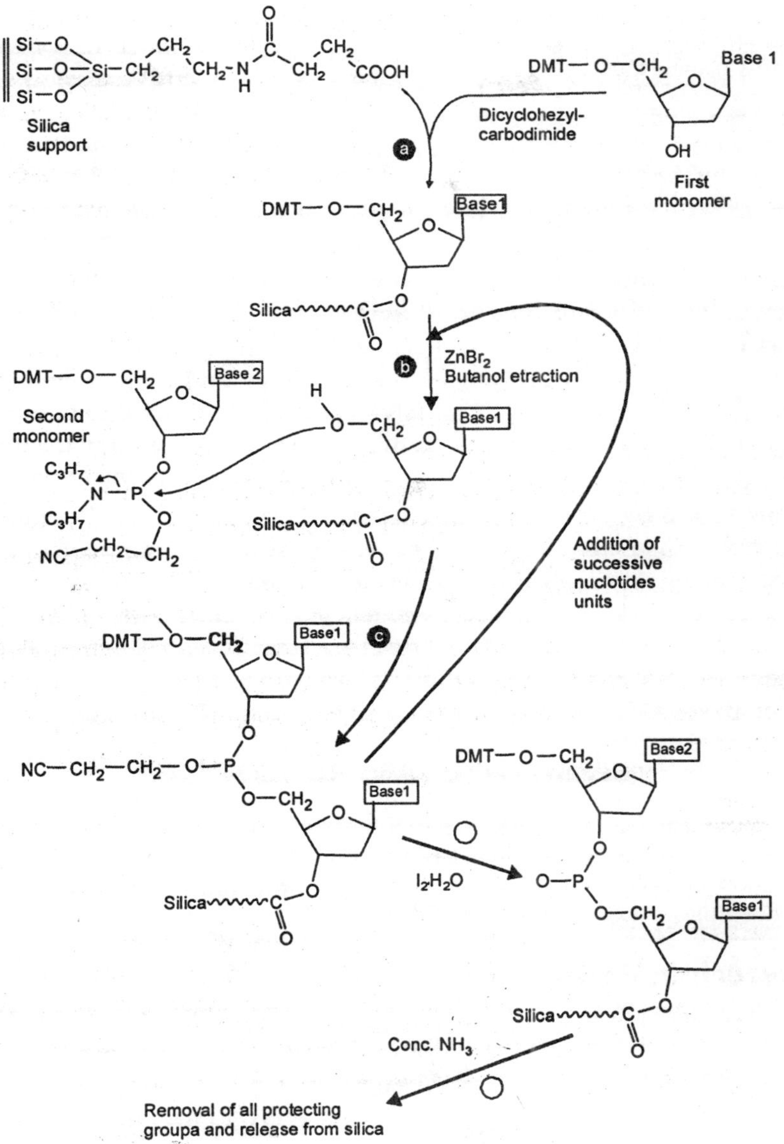

Silica support

Dicyclohexyl-carbodimide

First monomer

$ZnBr_2$
Butanol etraction

Second monomer

Addition of successive nuclotides units

I_2H_2O

Conc. NH_3

Removal of all protecting groupa and release from silica

...(1.25)

The Polymerase Chain Reaction (PCR)

This important technique was first described in 1971-1974 by Khorana and associates but was not used until it was rediscovered in 1983 by Mullis. It was quickly developed and has played a major role in biochemistry ever since. It continues to be applied in numerous ways. The PCR technique provides a way of "amplifying" a small number of DNA molecules, *i.e.*, to produce many copies. This is often done by cloning but PCR offers a quick and easy way to obtain millions of copies of a desired relatively short segment of DNA.

Standard PCR can be used for up to about 5000-nucleotide pieces. More recently modified procedures have allowed 35-kb segments to be amplified. The basic PCR procedure is initiated by hybridizing two oligonucleotide primers onto opposite strands of denatured DNA, one at each end of the section chosen for amplification. A DNA polymerase is then used to convert each of the separated strands into a duplex.

The mixture of products is then heated to denature the two new duplexes. After cooling, the primers, which are present in great excess, hybridize to all four strands. In a second cycle of polymerase action these are all converted to duplexes, etc. After 20 cycles millions of copies will be made. At first, copies with tails extending beyond the limits specified by the oligonucleotide primers will be formed. However, it is easy to see that after a few cycles, most molecules will be of just the desired length.

A heat-stable polymerase from *Thermus aquaticus (Taq* polymerase) is used so that the enzyme is not denatured by the repeated cycles of heating and cooling, which are conducted automatically by a simple apparatus. The polymerase chain reaction is being used to speed up prenatal diagnosis of genetic diseases, to detect viral infections, for tissue typing needed for organ transplantation, in forensic procedures, and in the study of the DNA of ancient tissues such as those of frozen wooly mammoths. If suitable restriction enzyme sites are present in the primers, the amplified DNA can be cloned readily. The 3.3×10^{-9} fmol of a DNA sequence found in a diploid chromosome pair in a single cell can be amplified in 50 cycles to 5-500 fmol, enough to study by hybridization with radioactive probes.

A large sample of a few pg of DNA can be amplified in 20 cycles to micrograms. By placing sequencing primers within the amplified segments, it is possible to generate DNA that can be sequenced directly using the dideoxy sequencing technique without cloning. The PCR technique has also been used to amplify cDNA molecules formed from RNA transcripts present in very low abundance. One of the problems with the PCR is that priming may occur by DNA fragments other than the added primers. Contamination must be scrupulously avoided. Another problem is that errors are introduced into DNA during amplification by PCR. Perhaps 1 in 200 of the copies will contain an incorrect base. If such a molecule is cloned the error will be perpetuated. Good practice requires that more than one clone is selected and sequenced to allow such errors to be avoided.

Sequence Determination

Satisfactory (but slow) methods for determining sequences of RNA molecules have been known for over 30 years. The procedures are somewhat parallel to those used in sequencing proteins. However, no similar method could be devised for DNA. Little progress was made until rather recently when new approaches led to extremely rapid procedures for sequencing DNA. As a consequence, it is now much easier to learn the sequences of genes than it is to sequence the proteins which they encode!

Preparing The DNA

The first step is to obtain a sample of enough identical DNA molecules to permit sequence analysis. This in itself may be a complex undertaking. Perhaps we want to know the sequence of one particular gene in the 3,500,000 kilobase pairs of DNA present in a single human cell. How can this gene be found and the DNA be obtained for analysis? Three techniques have been essential: cutting the DNA with restriction endonucleases, hybridization, and cloning. More recently PCR and related methods have simplified the sample preparation. DNA can often be amplified using primers that contain sequences that will later serve as *sequencing primers*.

Restriction Maps and Southern Blots

Although it doesn't require the synthesis of a primer, the Maxam-Gilbert procedure usually demands that a "restriction map" of the DNA be prepared to help keep track of the fragments being sequenced. The restriction map of the mitochondrial DNA gene, *oxi3* from yeast. This gene, which encodes one of the subunits of cytochrome oxidase, consists of 9979 base pairs. It was cloned in a suitable plasmid after which the restriction map was prepared by cutting with 18 different restriction enzymes.

The protein subunit contains 510 residues and therefore requires a coding capacity in the DNA of 1530 base pairs. This is only 16% of the total length of the gene, the majority of whose DNA is found in the three large introns. One way to select a desired segment of DNA from a digest of chromosomal DNA is to sort out the "restriction fragments" by gel electrophoresis. The DNA from the gel can be transferred to a nitrocellulose sheet while retaining the separation pattern using a method devised by Southern. In this *Southern blot* technique, solvent flows from a pool beneath the gel up through the gel and the nitrocellulose sheet into paper towels. The DNA is trapped on the nitrocellulose in the same pattern observed in the electropherogram.

A suitably labeled probe such as cDNA with incorporated ^{32}P is flowed repeatedly across the nitrocellulose sheet under conditions that favor formation of hybrids. Only the DNA complementary to the cDNA probe will retain the label. This DNA can then be located with the help of a autoradiogram. It is important that single-stranded DNA be used.

If double-stranded restriction fragments are separated on the electropherogram they must be denatured while in place in the gel before hybridization is attempted. Once the desired piece of DNA has been identified it is usually necessary to increase its amount. The conventional approach is to incorporate the DNA fragment into a plasmid and clone. The selected DNA can usually be cut cleanly from the plasmid used for cloning with the same restriction endonuclease originally used in fragmenting the DNA. An alternative procedure is to clone a mixture of DNA fragments and then sort colonies of bacteria containing the cloned fragments using DNA-RNA hybridization. Bacteria from the selected colonies are then propagated to produce large amounts of the plasmid DNA. Alternatively, PCR can be used directly on the selected DNA fragment. This is more often the preferred choice.

The Sanger dideoxy method

The rapid sequencing methods all depend upon the fact that single-stranded DNA fragments under denaturing conditions migrate on electrophoresis in polyacrylamide gels strictly according to their length. Thus, if a mixture contains all lengths of radiolabeled polynucleotides from very short oligonucleotides to fragments containing 200 or 300 bases, the polynucleotides will all appear, one above the other, as a series of bands that can be visualized by radioautography. The first of these methods was published by Sanger and Coulson in 1975 and was followed in 1977 by the method which is now used. A sample of double-stranded DNA is denatured.

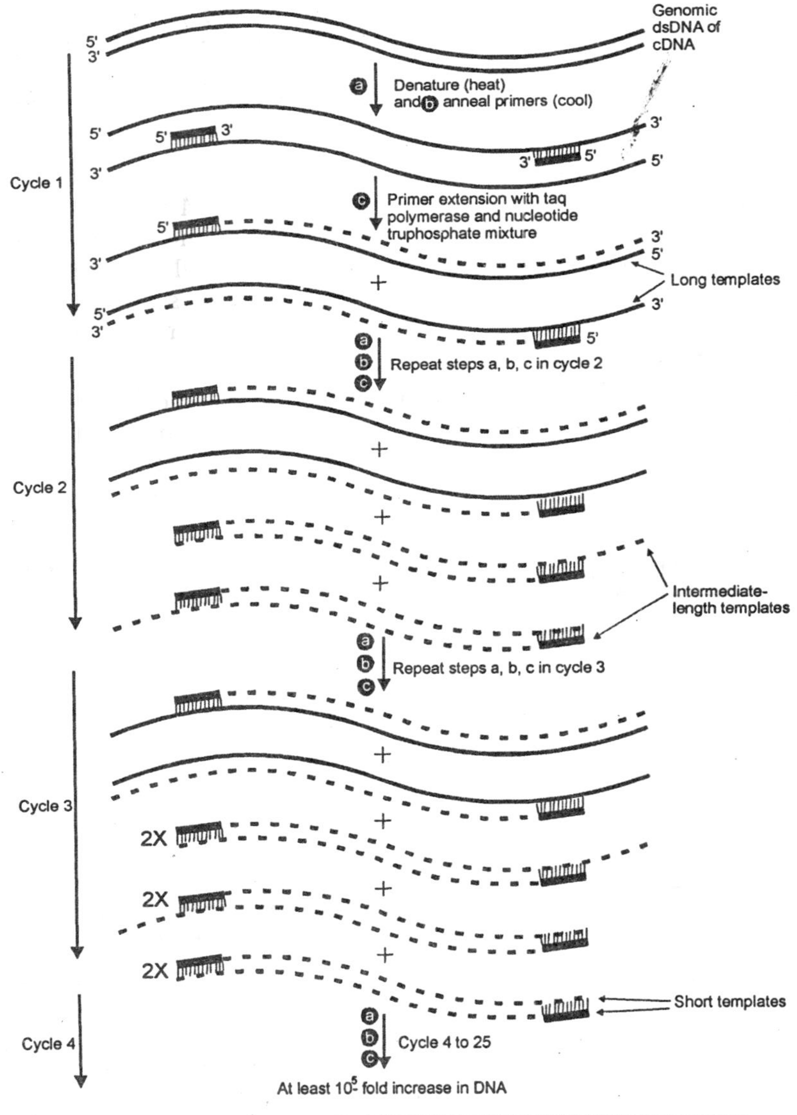

Fig. 1.43. Amplification of DNA using the polymerase chain reaction (PCR). Double-stranded DNA is denatured by heating to 90-99° C (step *a*) and oligonucleotide primers complementary to short 12-18 nucleotide sequences at the two ends of the piece of DNA to be amplified are annealed to the separated strands by cooling to 40-75° C (step *b*). The two DNA strands serve as templates for synthesis of new complementary strands using a heat-stable DNA polymerase and a mixture of the four nucleotide triphosphates. Nucleotide units are added to the 3′ ends of the primers, with the new chains growing in the 5′ → 3′ direction (step c). Steps *a*, *b*, and *c* are then repeated as many as 30 times using a thermal cycler device that periodically raises and lowers the temperature with a cycle time of a few minutes. The polymerase is unharmed by the heating and is reused in each cycle. An excess of the primer and of the nucleotide triphosphates sufficient for all of the cycles is present initially. In the early cycles new long and intermediate length templates are created. However, the number of short templates increases exponentially and the final product consists predominately of the short selected DNA segment (short templates).

One of the resulting single strands is used as a template to direct the synthesis of a complementary strand of radioactive DNA using a suitable DNA polymerase. The "Klenow fragment" of *E. coli,* DNA polymerase I, reverse transcriptase from a retrovirus, bacteriophage T7 DNA polymerase, *Taq* polymerase, and specially engineered enzymes produced from cloned genes have all been used. Before me sequencing begins it is necessary to prepare a short *primer* that is complementary to a sequence at one end of the DNA strand to be sequenced.

This may be prepared enzymatically, or by non-enzymatic synthesis. The short primer is annealed to the end of the DNA and the resulting molecule is incubated with a DNA polymerase and a mixture of the four mononucleotide triphosphates, one of which is radio-labeled in this position. Four reaction mixtures are prepared. Each mixture contains all four nucleoside triphosphates and also one of four different *chainterminating inhibitors,* the most popular of which are the 2′, 3′-dideoxyribonucleoside triphosphates:

Thymine
Cytosine
Guanine of
Adenine

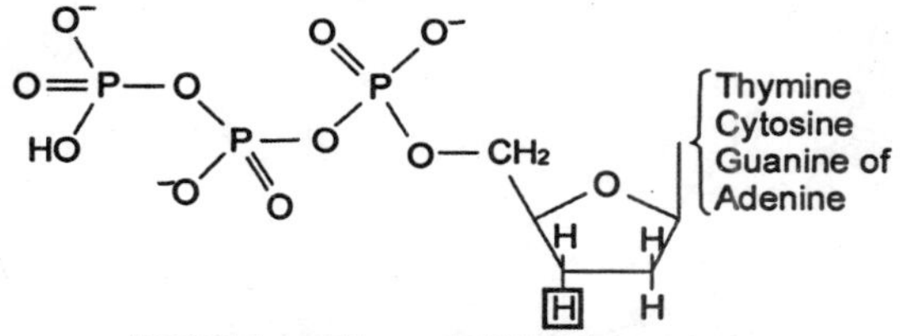

2′,3′-Dideoxyribonucleotide triphosphates

These inhibitors are added in a ratio of about 100:1 with the natural substrates. In this ratio they are incorporated into the growing DNA chain about once in 200 times on the average. However, in the various growing DNA chains they are incorporated at different points ranging from the very first nucleotide to the last. Since the incorporated dideoxy monomer lacks the 3-hydroxyl group needed for polymer formation, chain growth is terminated abruptly.

Synthesized polynucleotides are denatured and subjected to electrophoresis in four adjacent lanes. The resulting patterns contain bands corresponding to all of the successive oligonucleotides but not all in the same lane. A given lane will contain only the bands of the oligonucleotides terminated by the particular inhibitor used. The other bands will be found in the other three lanes. Each band will have been terminated by the inhibitor employed in that lane. The nucleotide sequence can be read directly from the banding pattern. Arabinosyl nucleotide triphosphates have also been used as chain-terminating inhibitors. A sequence determined by the Sanger method is usually checked by also sequencing the complementary strand.

A major factor in the success of the dideoxy sequencing method has been the development of cloning techniques that provide ssDNA in a form ready for use. In any cloning procedure the DNA that is to be sequenced has been covalently ligated to the end of a DNA strand of the *cloning vehicle,* a modified plasmid or virus. Sequencing is often done on DNA cloned in a modified ssDNA bacterial virus such as M13.

Although dsDNA is ligated to the ds Implicating form of viral DNA, the virus particles produced when the virus is propagated in *E. coli* cells are single stranded. Cleavage of the viral DNA with its incorporated *"passenger DNA"* with an appropriate restriction endonuclease releases the passenger DNA (to be sequenced) with a short piece of DNA from the cloning vehicle attached at the 3′ end. Since the sequence of this small piece of the cloning vehicle is known, a suitable primer of length ~12-18 nucleotides can be synthesized (or purchased) and annealed to the DNA. This serves as the primer and allows the sequence to be read from the 5′ end to the 3′ end of the synthesized complementary strand.

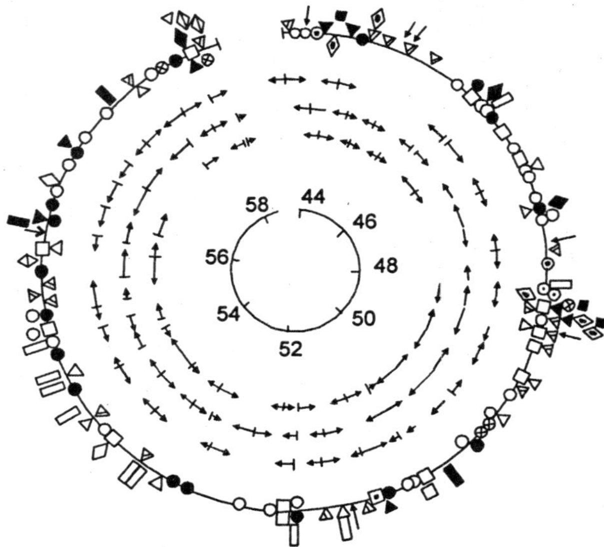

Fig. 1.44. A physical map of the *oxi3* locus of yeast mitochondrial DNA. The restriction fragments used for DNA sequencing are indicated by the *arrows*. The extent to which the sequences were read is represented by the lengths of the arrows. The map units are shown in the *inner circle*. The following symbols, together with the names of the restriction enzymes, are used for the restriction sites:

▲ *Hinf* I	○ *Alu* I	⊙ *Hha* I
△ *Hpa* II	⌽ *Pvu* II	⚠ *Rsa* I
▯ *Hae* III	◇ *Hinc* II	▮ *Hph* I
◆ *Taq* I	◇ *Hind* III	⊡ *Blg* II
● *Mbo* I	⊗ *Eco* RI	◇ *Bam* HI
□ *Mbo* II	↓ *Eco* RII	■ *Hpa* I

Double-stranded DNA attached to vehicles such as the pUC plasmids can also be sequenced directly if the DNA is denatured by alkali treatment. After neutralization and precipitation an appropriate primer is annealed to one or the other of the two strands. Since its introduction, many modifications and improvements have been developed. The sequencing gels have been improved.

The use of ^{35}S labeling has given sharper autoradiographs. Alternatively, a silver stain can be used with unlabeled primers. GC-rich DNA sequences are often difficult to sequence, probably because even in the denaturing polyacrylamide gels used for sequencing they tend to form hairpin loops, perhaps as a result of formation of Hoogsteen base pairs. Formation of these loops results in uneven spacing between the adjacent bands in the sequencing gel, so-called "compression artifacts." Use of a 7-deaza-dGTP in place of dGTP in the sequencing reaction ameliorates this problem.

Sensitivity can be improved by use of "cycle sequencing" in which a heat-stable polymerase such as *Taq* polymerase is used, and after heating the same template DNA is used repeatedly to give a higher yield of labeled fragments. About 200 – 400 bases can be successfully sequenced manually in a single run. By prolonging the time of electrophoresis in a second run, the sequence can be extended considerably.

By using the just obtained sequence information, it is possible to select a new start point 200 or more nucleotides further along the template chain and to synthesize an oligonucleotide primer to anneal to the template at this point. In this way it is possible to "walk" along the template adding additional sequences at each step. Another procedure is to delete by mutation various segments of the cloned DNA above the sequence that binds the primer. This permits analysis of the whole cloned fragment via an overlapping set of sequences from the deletion mutants. Many DNA sequences continue to be determined manually by the well developed long gel procedures as illustrated in Fig. 1.49.

However, sequencing whole genomes has dependedupon the development of high-speed automated procedures. Instead of radiolabeling, fluorescent dyes may be joined to the primer to allow detection of the chain fragments produced during sequencing. Automatic sequencers use four dyes that fluoresce with different colors. A different dye is used for each of the reaction mixtures. Then the four samples are mixed together and the DNA fragments are subjected to electrophoresis in a single lane. A laser beam excites the fluorescence, scanning several lanes with different samples as the electrophoresis progresses.

A photomultiplier tube records the fluorescence intensity of each band through a series of four filters in a rotating wheel.

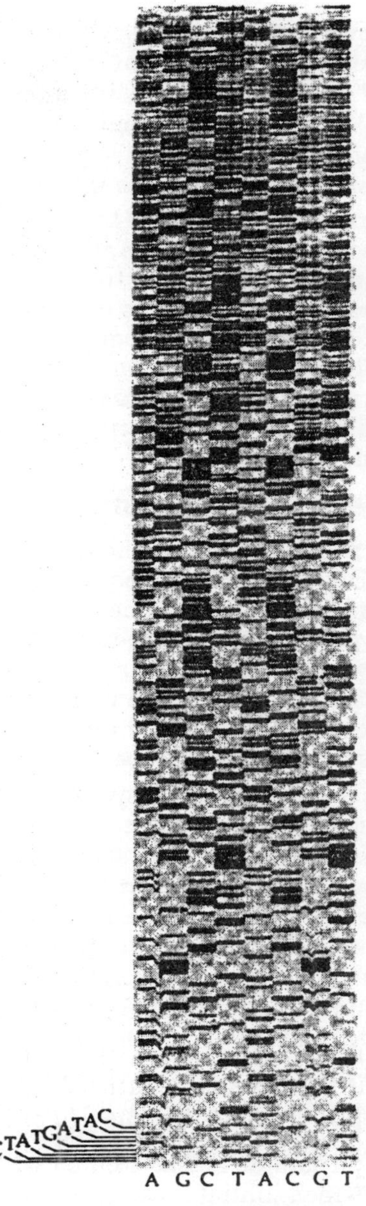

Fig. 1.45. A DNA sequencing gel obtained using a segment of DNA from salmon sperm selected by suitable oligonucleotide primers, amplified by PCR, and sequenced with a ^{35}S label in the primer. Four samples were used, one with each of the four dideoxy chain terminators (A, G, C, T, A, C, G, T from left to right). After electrophoresis the shorter fragments are at the lower end of the gel. The sequence of the strand complementary to the template strand whose sequence is being determined is read from the bottom of the gel. Here it starts CTATGATAC.

This allows automatic recognition of the four different colors of fluorescence and therefore of the nucleic acid base present in each position in the sequence. Improved strategies for "primer walking" and for "shotgun sequencing" have been devised. In the shotgun strategy, whole bacterial genomes have been cut by restriction enzymes into large numbers of overlapping fragments which have been separated and sequenced. A computer program is used to analyze and assemble the sequences into a complete genomic sequence.

An example is provided by the genome of the *Methanococcus jannaschii*. Its large circular chromosome contains 1,664,976 bp and there are two additional pieces to the genome, one containing 58,407 bp and the other 16,550 bp. The sequences were deduced from 36,718 individual sequencing runs on high-speed automatic sequencers. For each run, on average, 481 bp could be read. To sequence the human genome faster methods are needed. Capillary electrophoresis with a single laser beam scanning the output of 24 capillaries has been demonstrated. Extremely rapid sequencing of oligonucleotides up to 100 bp in length can be accomplished by mass spectrometry. This may be an important technique for diagnosis of genetic defects.

The Method of Maxam and Gilbert

The nonenzymatic method devised by Maxam and Gilbert can be used to sequence either ss or dsDNA. Before the sequencing is begun, a radioactive label is incorporated, usually at the 5′ end. This is often done by cleaving off any phosphate groups present on the 5′ end with alkaline phosphatase and then transferring a new radioactive phospho group with the assistance of the enzyme polynucleotide kinase and radioactive γ-^{32}p-labeled ATP. If dsDNA is used the strands are separated so that each has a label only at one end. The key step in sequencing by the Maxam-Gilbert procedure is to cleave chemically the DNA at random locations using reagents that have some specificity for particular bases.

The cleavage process involves three distinct steps: (1) chemical modification, as specific as is possible for the chosen base; (2) displacement of the modified base from the sugar; and (3) elimination and chain cleavage using amine catalysis. Two consecutive steps can often be combined. There are several versions of the method; one involves dimethyl sulfate as the specific reagent for guanine. It forms N^7-methylguanosine, which upon heating with the strong base piperidine at 90°C undergoes addition of hydroxyl ion with ring opening and displacement of the modified base according to Eq. 1.26. The product, a glycosylamine of piperidine, is in equilibrium with a Schiff base which can undergo chain cleavage as in Eq. 1.21. Asecond sample of DNA is treated with piperidine-formate buffer of pH 2 in the cold.

This promotes the acid depurination of both guanine and adenosine. A third sample is treated with hydrazine, with both cytidine and thymidine being cleaved to ribosylurea according to Eq. 1.13 again, this is followed by displacement and β elimination catalyzed by piperidine. The fourth sample is also treated with hydrazine but in the presence of a high concentration of NaCl which inhibits the reaction with thymidine, by lowering the pK_a of the thymine, and allows the cleavage to be more nearly specific for cytidine. Each of the reactions is conducted in such a way that on the average only one cleavage event occurs per molecule of DNA.

Since the cleavages occur at many different points, a family of nested radioactively labeled oligonucleotides, one from each original molecule, is produced. When these are sorted by polyacrylamide gel electrophoresis, the pattern of the oligonucleotides in the four adjacent channels allows the nucleotide sequence to be read directly from the autoradiogram. The Maxam-Gilbert method doesn't require synthesis of a primer and it sometimes works well for sequences that are difficult to obtain with the Sanger-Coulson procedure.

The two methods may both be used to provide additional certainty about a sequence. The Maxam-Gilbert method is very convenient for sequencing small oligonucleotides which often react poorly with the polymerase used for the chain termination method. The method usually requires that a restriction map be prepared.

Sequencing RNA

The first known RNA sequence, that of an aianine tRNA, was determined by Holley and associates in 1965. The RNA was subjected to partial hydrolysis with pancreatic ribonuclease and ribonuclease T_1. The small oligonucleotide fragments were separated by ion exchange chromatography under denaturing conditions (7 M urea) and were then characterized individually. The availability of additional enzymes such as ribonuclease U_2, the *B cereus* ribonuclease and ribonuclease Phy M of *Physarum* and the use of radiolabeling and of two-dimensional fingerprinting of digests have made the procedures more versatile.

A valine tRNA was sequenced independently by Bayev and Campbell. Since the development of the rapid methods for sequencing DNA, many mRNA sequences have been determined by using reverse transcriptase to make a cDNA strand complementary to the RNA. The cDNA is then sequenced. Rapid sequencing methods parallel to those used for DNA have also been devised.

Nearest Neighbor Analysis

A technique developed by Kornberg and associates before the availability of sequencing methods is the *nearest neighbor sequence analysis*. Using a single radioactive ^{32}P-containing nucleoside triphosphate together with the three other unlabeled nucleoside triphosphates, a primer chain of DNA is elongated from the 3′ end along a ssDNA template chain using a DNA polymerase. The incorporation of ^{32}P from the a position of the nucleotide triphosphate occurs in the bridge phosphates that connect the nucleotide originally carrying the ^{32}P to 3′ position of the neighboring nucleotide.

...(1.26)

Cleavage of the ^{32}P-containing product of the reaction with a mixture of micrococcal DNase and spleen phosphodiesterase, which catalyze *b*-type cleavage gives fragments in which the ^{32}P will now be attached to what was the 5′ nearest neighbor to the radioactive nucleotide in the DNA. Measurement of the radioactivity in each of the 3′ nucleotides of thymine, cytosine, adenine, and guanine gives the frequencies of the adjacent pairs, TA, CA, AA, and GA.

Using the other radioactive nucleoside triphosphates one at a time in separate experiments, all of the nearest neighbor frequencies can be obtained. From such an experiment it was possible to deduce that the strands in the double helix were oriented in an antiparallel fashion, as predicted by Watson and Crick. If the strands had been parallel, different nearest neighbor frequencies would have been observed.

Understanding Sequences

Sequences of over 20 million nucleotides from hundreds of organisms had been determined by 1988 and the number is doubling each 2 – 3 years. Sequencing the human genome has required rates of millions of bases per day. With the massive amount of data already available it has become of great interest to compare sequences of genes, whether they encode similar or dissimilar proteins, to make comparisons between species, and to search for sequences that bind specific proteins or that encode particular regulatory signals. Relationships of common evolutionary origin or homology as well as other sequence similarities are often sought. To handle the mass of existing data, powerful computer programs have been developed and various graphical procedures have also been developed to help the human mind comprehend the results.

One important problem is to define and locate what are called *consensus sequences*. The problem is best illustrated by examples. The cleavage site for the *Eco*RI restriction endonuclease is *GAATTC*. There is no ambiguity. In a DNA of random sequence this would be expected to occur by chance in about $(1/4)^6$ nucleotides (4 kb). On the other hand, the *HinII* restriction endonuclease cleaves within the consensus sequence *GTYRAC*. where Y = C or T and R = A or G. It would be expected to occur by chance in about $1/4^5$ nucleotides. Many binding sites for RNA polymerase, the so called *promoters* contain the consensus sequence *TAtAaT*, at position-10, ahead of the 5′ end of the sequence that is transcribed into mRNA. The lower case *t* and a used here imply that other nucleotides may often replace T or A at these positions.

There are many promoters and over 70% of those described have this consensus sequence. All have the less restricted sequence *TAxxxT*, where *x* may be any nucleotide. Our definition of consensus sequence is somewhat arbitrary. Now consider the problem of locating a –35 site whose consensus sequence is *TTGACA* but which may, for different genes, be shifted backward or forward by a nucleotide or two. This is a consensus sequence. Therefore, in many cases one or more substitutions in the sequence will have been made. The result is that the sequence of nucleotides in which the consensus sequence is to be found is likely to appear entirely random. Sophisticated computer programs are helpful in locating it.

Protein-DNA Interactions

The most detailed information about interactions of proteins with DNA is coming from X-ray crystallographic studies. Several other methods have also been very useful. Much has been learned from the effects of mutations in DNA-binding proteins or in regions of DNA to which a protein binds. Binding of proteins to DNA can also be recognized by its effects on the mobility of DNA during gel electrophoresis. Chemical or laser-induced crosslinking can show that within a complex a specific residue in a protein is adjacent to a certain sequence in the DNA. The technique of *protection mapping* or *"footprinting"* is widely used to determine which nucleotides in a sequence are covered by a bound protein.

A reagent which attacks and cleaves DNA nearly randomly is used. DNase I was first introduced for this purpose and has been used widely. An important finding is that certain sites that are readily cleaved (*hypersensitive sites*) are frequently located in chromatin undergoing transcription. Footprinting has also been accomplished with other nucleases, with dimethylsulfate (which acts on A and C), with carbodiimides (which act on U and G), and with the Maxam-Gilbert guanine-specific cleavage. One of the most popular methods employs cleavage by hydroxyl radicals. *Photofootprintinp* depends upon decreased or increased sensitivity to ultraviolet light at sites bound by proteins. In footprinting experiments the DNA to be studied is radioactively labeled at one end of one strand.

In the absence of the protecting protein, denaturation and electrophoresis of the cleaved fragments yields a nearly random "ladder" of DNA fragments. In the presence of the binding protein some cleavage products will be missing from the ladder. The bound protein leaves a "footprint". Related methods are being applied to the determination of the secondary structure of RNA molecules and to the study of interactions with proteins. For example, treatment with dimethyl sulfate under appropriate conditions methylates bases that are not paired, giving largely 1-methyladenosine and 3-methylcytidine.

Nuclear Magnetic Resonance

Much of the initial effort to study polynucleotides by NMR spectroscopy was directed toward transfer RNAs, only a few of which have been crystallized in a form suitable for X-ray diffraction. Study of the other tRNAs by NMR techniques has established that all of the tRNAs have a similar architecture and that the structures observed in the crystals are preserved in solution. Fig. 1.51 shows the low-field end of the NMR spectrum of a valine-specific tRNA from *E. coli.* The spectrum is run in H_2O rather than D_2O so that exchangeable hydrogens in the hydrogen bonds of the Watson-Crick base pairs can be observed.

The protons giving rise to the downfield resonances are primarily those attached to nitrogen atoms of the rings and in hydrogen-bonded positions. These protons are shielded by adjacent electron-donating groups and by their attachment to the semiaromatic rings of the bases. The NMR signals are further shifted downfield to varying degrees depending upon whether or not the proton being observed is attached to a base that is stacked with other bases. The size of the shift also depends upon which neighboring bases are present. The proton on N-3 of AU base pairs is deshielded more than the proton on the N-l of GC base pairs. Therefore, the AU protons appear further downfield than the GC protons.

The stronger ring current in A than in C enhances this separation. All of the resonances in Fig. 1.51 have been assigned to particular bases. This was done in part by varying the temperature, changing the magnesium ion concentration, and predicting shifts caused by ring currents in adjacent bases making use of the X-ray crystal structures. NMR spectra of hairpin helical fragments also provided essential information. From integration of the areas under the peaks it was concluded that the 20 resonances seen below –11 ppm represent 27 protons.

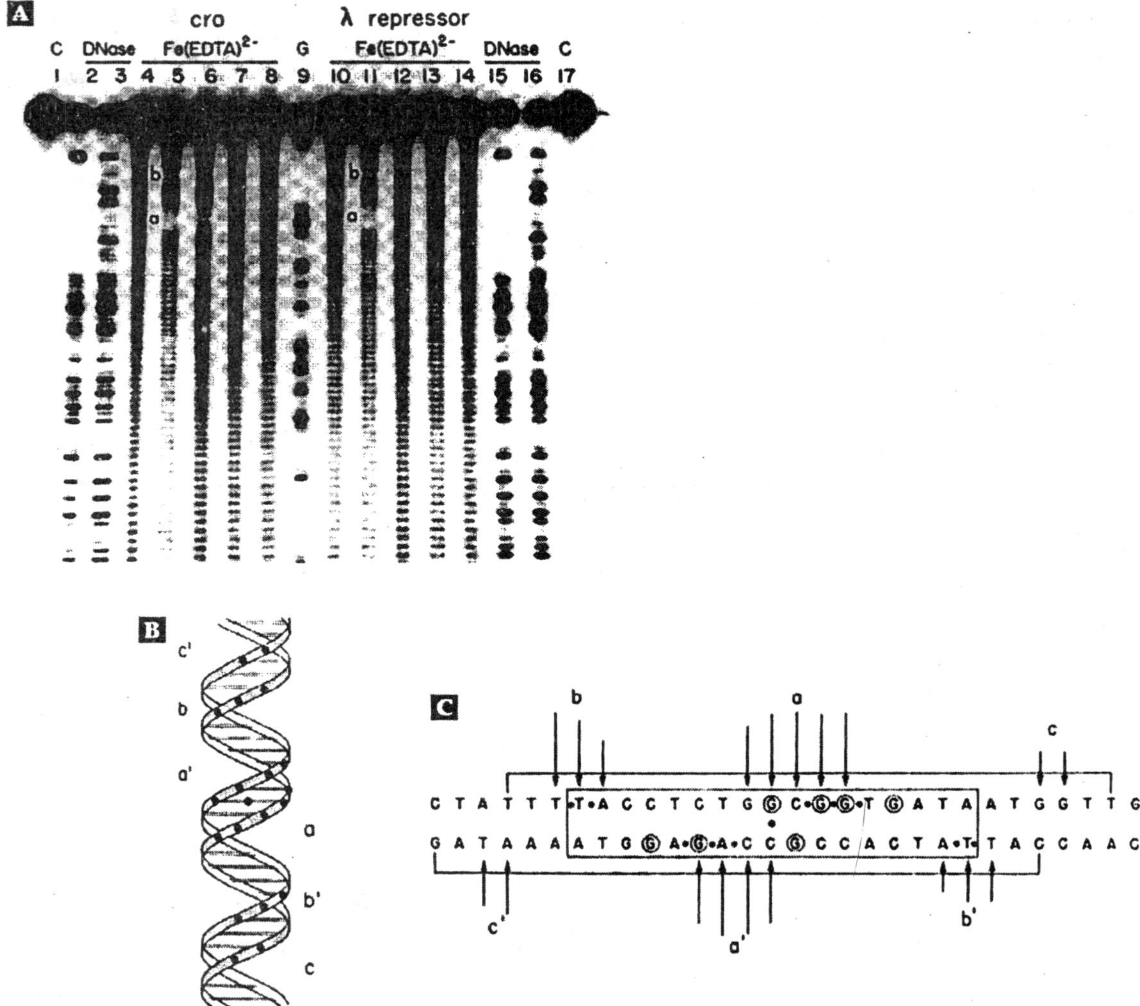

Fig. 1.46. (A) Hydroxyl radical footprints of phage X represser on both strands of the 120-bp restriction fragment of *E. coli* DNA carrying the O_R1 operator sequence. "Bottom" and "top" refer to the DNA strands as drawn in (C), and as diagrammed in (B). Labeling of the 5′ end of the *Bgl* II restriction fragment afforded data for the bottom strand and labeling of the 3′ *Bgl* II end afforded data for the top strand. Lanes 1 and 12, untreated DNA, labeled on the bottom and top strands, respectively; lanes 2 and 10, products of DNase I digestion of DNA labeled on the bottom and top strands, respectively, with no represser present; lanes 3 and 11, products of DNase I digestion of DNA labeled on the bottom and top strands, respectively, complexed with λ represser (675 nM); lanes 4 and 9, products of Maxam-Gilbert guanine-specific sequencing reactions performed on DNA labeled on the bottom and top strands, respectively; lanes 5 and 6, products of hydroxyl radical cutting of DNA labeled on the bottom strand, complexed with λ represser. Lane 5, 90-nM λ represser and lane 6, 675-nM λ represser. Lanes 7 and 8, products of hydroxyl radical cutting of DNA labeled on the top strand, complexed with λ represser. Lane 7, 675 nM λ represser. Lane 8, 90 nM λ represser. The labels a, b, c, a′, b′, and c′ mark the hydroxyl radical footprints.

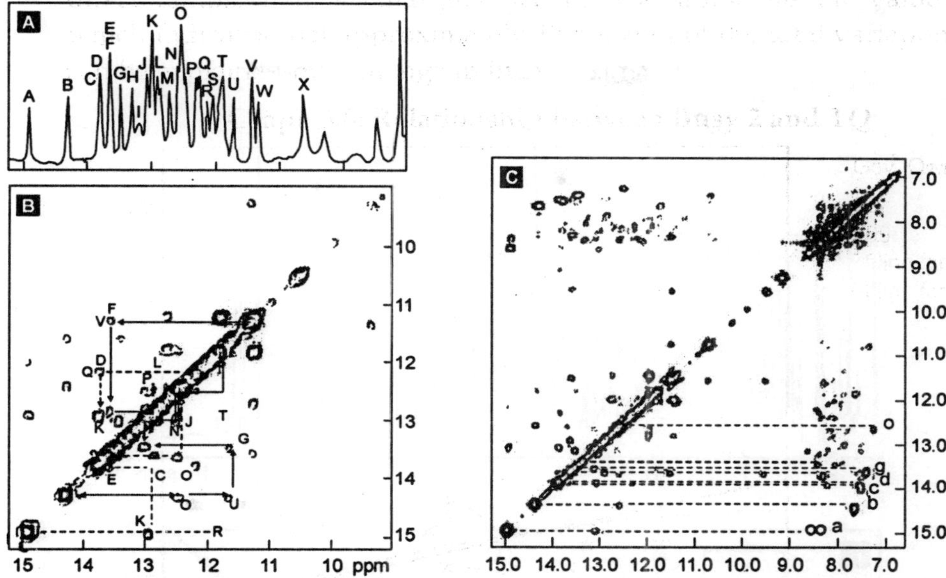

Fig. 1.47. (A) The low-field region of the one-dimensional ^{1}H NMR spectrum of *E. coli* tRNA$_1^{Val}$ at 27°C in H_2O. Resonances are identified by letters A - X. (B) NOESY spectrum of the same tRNA under similar conditions showing the imino-imino NOEs. In the lower right sector the connectivity traces of the acceptor helix and dihydrouridine helix are shown as solid and dotted lines, respectively. In the NOESY sample the two protons in peak EF are partially resolved whereas the two protons in peak T have coalesced. (C) NOESY spectrum of *E. coli* tRNA$_1^{Val}$ at 32°C showing the imino and aromatic proton regions. AU-type imino protons have been connected horizontally by a dotted line to the cross-peak of their proximal C2-H or C8-H in the 7 to 9 ppm region, which has been labeled with the corresponding lower-case letter.

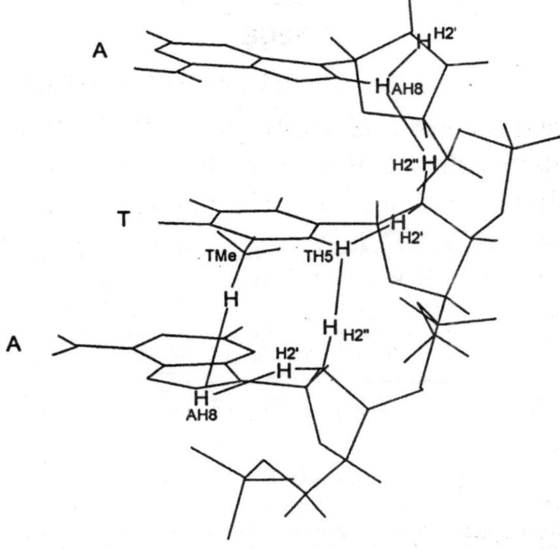

Fig. 1.48. Segment of a DNA chain in the B conformation illustrating some intrachain NOEs that may be observed. The close juxtaposition of proton pairs is provided by the H2'-*endo* (^{2}E) conformation of the sugar rings with *anti* base conformation.

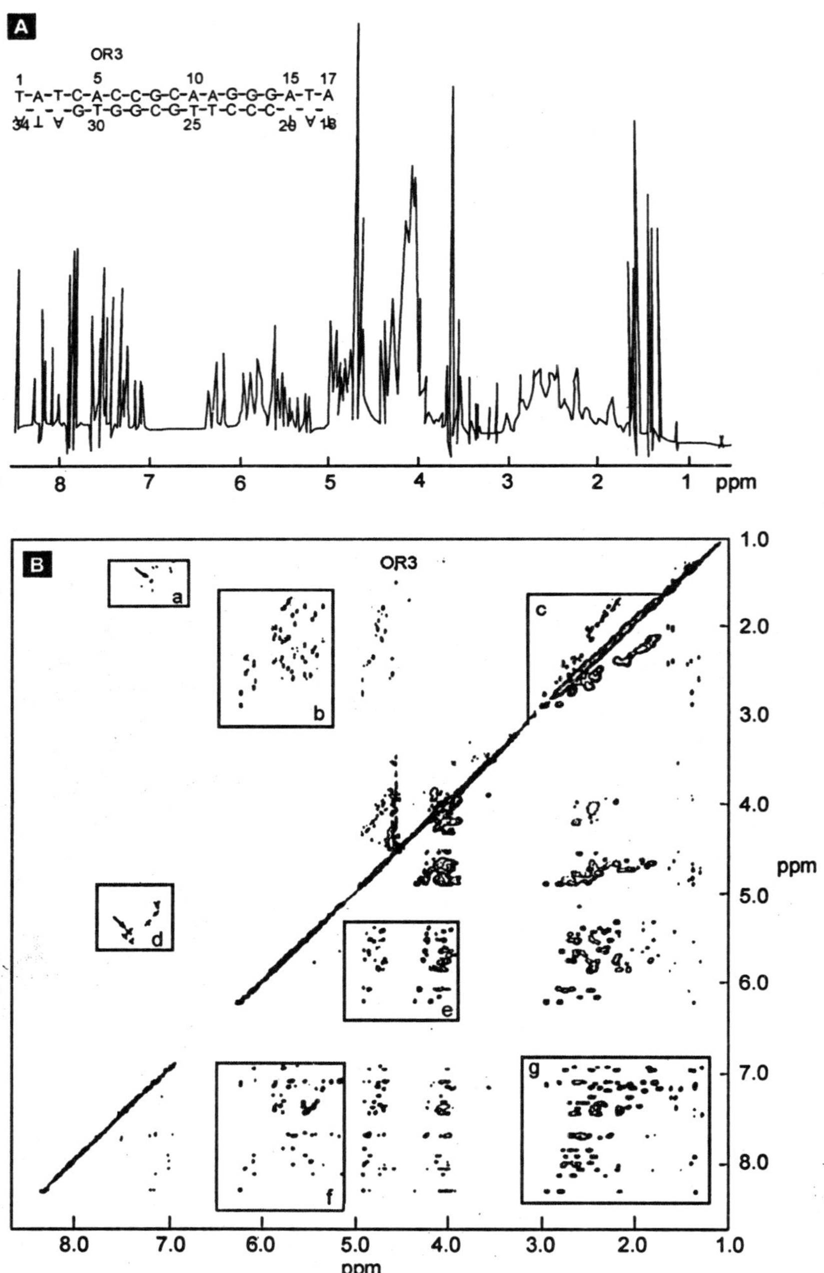

Fig. 1.49. (A) ^{1}H NMR spectrum of a 17 base-pair DNA segment from the operator sequence OR3 from bacteriophage λ in D_2O at 37°C. (B) Combined COSY above the diagonal and NOESY (below the diagonal) spectra. C5H and C6H J coupling is established from cross-peaks in box d for cytosines and in box a for thymines. Two unresolved cross-peaks give rise to the more intense spots marked by arrows. Box b contains cross-peaks from scalar coupling of the two H2′ protons to the H1′ protons of the deoxyribose rings.

Twenty of these are in Watson-Crick base pairs and correspond to those expected from the X-ray structure. Six more belong to protons involved in tertiary interactions, such as base pair triplets or non-Watson-Crick pairs. One of these, labeled "G" in the figure, is in the dihydrouridine stem and involves the ring proton of N-l of m^7G46 which is hydrogen bonded to N-7 of G22 in the major groove of the RNA. G22 is located at the beginning of the "extra loop." Measurements of the NOE of nearby protons in both small RNA molecules and DNA oligonucleotides provided much additional information.

Fig. 1.51B and C show NOESY spectra of the $tRNA^{Val}$ and the way in which weak cross-peaks between the H-bonded imino protons in adjacent base pairs can be used to establish connectivities. Beginning with resonance B, it is possible to establish the sequence of the NH groups giving rise to these resonances as OBUGJNT. Using other data as well, it was concluded that these represent the seven base pairs of the acceptor stem with resonance C representing the first GC pair, resonance B the second, etc. Resonance A was identified as coming from the base triplet containing a Hoogsteen base pair of 4-thiouracil at position 8 with A14. Its connectivity to the sequence KCEO is also outlined in Fig. 1.51B.

However, it could not be established without additional data which also helped to identify the sequence O-K as residues 10-13 of the dihydrouracil stem. Peak O is a multiproton peak representing not only GC 10 but also UA 7. In general, the GC protons are at the higher field side of the spectrum and the AU protons at the lower side. However, AU 7 is shifted to an anomolously high position. Cross-peaks between imino protons of uracil and the nearby C2 protons of adenine in Watson-Crick AU base pairs or C8 protons of Hoogsteen AU pairs can be observed in the 6.5 – 9 ppm region as shown in Fig. 1.35B. This region also contains information about other protons bound to the nuclei acid bases. Similar techniques are being used for the study of DNA.

The presence of a second hydrogen in the 2′ position of the deoxyribose rings of DNA adds several H-H distances, that can be measured in addition to those seen in RNAs. Characteristic differences are seen in the NOESY plots of A, B, and Z forms of DNA. Although detailed structural information has been obtained for short segments of DNA, spectra of larger oligonucleotides are impossible to analyze with two-dimensional methods because of extensive overlap of resonances. The difficulty is already apparent in the 17 base pair DNA segment for which a one-dimensional spectrum as well as COSY and NOESY spectra are shown in Fig. 1.53. Some help with the complexity can be obtained by incorporation of ^{13}C-enriched methyl groups into polynucleotides. One way to do this is to grow yeast on a medium containing methionine with ^{13}C in its methyl group.

The most intense peaks in ^{13}C NMR spectra of tRNA molecules from this yeast will represent methyl groups in modified bases in the tRNA. Incorporated ^{13}C also permits study of internal motion within tRNA or other oligonucleotides by NMR methods. However, as with protein NMR spectroscopy the major recent advances have come from systematic incorporation of both ^{13}C and ^{15}N into nucleic acids and the development of three- and four-dimensional NMR methods. Also important are methods for replacing some hydrogen atoms with deuterium to simplify spectra. The very sensitive ^{19}F nucleus can be introduced into tRNAs by incorporation of 5-fluorouracil in place of uracil. Phosphorus 31 NMR spectra, can provide information about conformations of the chain. A few recent NMR investigations of polynucleotides include studies of triple-helical DNA, Holliday junctions, double-stranded oligonucleotides containing adducts of carcinogens, of hairpin loops with sheared A•A and G•G pairs, and of proton exchange in both imino and amino groups.

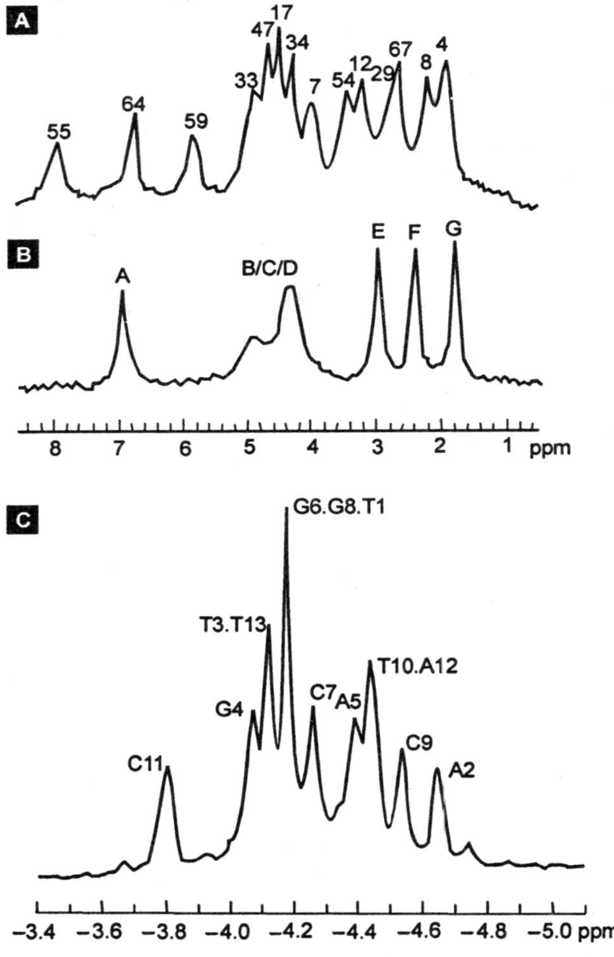

Fig. 1.50. (A) An ^{19}F NMR spectrum of the 76-residue *E. coli* tRNA^{Val} containing 5-fluorouracil in 14 positions. Recorded at 47°C. The numbers above the resonances indicate the position in the sequence. (B) A similar spectrum for a 35-residue "minihelix" that contains the acceptor stem of the tRNA^{Val} and seven fluorouracils. The broad peaks B, D, and E are shifted far upfield by reaction with bisulfite suggesting that they are not hydrogen bonded and are present in the loop of the stem-loop structure. Peaks A, E, F, and G correspond to resonances 64, 7, 67, and 4, respectively, in (A) and represent fluorouracil in the stem structure. (C) A ^{31}P NMR spectrum of a synthetic 14 base-pair DN A segment related to the *E. coli lac* operator. The palindromic sequence is TCTGAGCGCTCAGA. The numbers refer to the positions from the 5′ end.

CHAPTER

2 Mechanical Separation of Nucleic Acids

Of all the various nucleic acid molecules within the cell the three that are most often isolated are chromosomal DNA, mRNA from tissue or cells and plasmid DNA from bacteria. In all cases the principles for isolation are similar and can be divided into three stages.

1. The cells or tissue are broken open to release their contents.
2. The cell extract is treated to remove all components except the DNA or RNA.
3. The resulting DNA or RNA solution is concentrated.

ISOLATION OF TOTAL CELLULAR DNA

The tissue or cell sample is firstly homogenized in a buffer containing a detergent such as Triton X-100 and sodium deodecyl sulphate (SDS), which disrupts the cell and dissociates DNA-protein complexes. Protein and RNA are then removed by sequential incubations with a proteolytic enzyme (usually proteinase K) and ribonuclease. Finally the DNA is extracted into ethanol. Ethanol only precipitates long chain nucleic acids and so leaves the single nucleotides from RNA digestion in the aqueous layer.

Table 2.1. Isolation of Total Cellular DNA

Reagents

Lysis buffer,	Tris buffer (10 mmol l^{-1}) pH 7.6 containing 1% v/v Triton X-100 and 1% w/v SDS
Tris buffer	10 mmol l^{-1} pH 7.6
Proteinase K	10 mg ml^{-1}
Phenol-chloroform	1 : 1 v/v
Ribonuclease	10 mg l^{-1} in Tris buffer pH 7.6
Absolute ethanol	

Method

Homogenize tissue sample in 6 vols of lysis buffer.

Pellet cell debris by centrifugation at 2500 g for 20 mm at 4°C.

Incubate supernatant with proteninase K at 37°C for 2–4 h.

Remove the protein by extraction into phenol/chloroform.

Precipitate nucleic acids (both DNA and RNA) from the aqueous phase by addition of 2 vols of absolute ethanol. Centrifuge.

Resuspend nucleic acids in 1 ml Tris buffer and add 10 μl ribonuclease.

Incubate for 1 hour at 37°C.

Precipitate DNA with 2 vols of absolute ethanol. Centrifuge

Resuspend in buffer

Measure absorbance at 260 nm.

Calculation

A solution containing 50 μm ml^{-1} DNA has an absorbance of 1.0.

Concentration of DNA in the sample is therefore;

absorbance × 50 μg mol^{-1}

Caesium Chloride Density Gradient Centrifugation

An alternative method for DNA isolation is caesium chloride (CsCl) density gradient centrifugation, which separates protein, RNA and DNA according to their buoyant densities. A density gradient is produced by centrifuging a solution of CsCl at high speed. Macromolecules present in the CsCl solution during centrifugation will form bands at distinct points in the gradient depending on their buoyant density. DNA has a buoyant density of 1.7 g ml^{-1} and will therefore migrate to this point in the gradient. Protein has a much lower buoyant density and will float at the top of the tube, while RNA will pellet at the bottom. Density gradient centrifugation in the presence of ethidium bromide (EtBr) is more often used in the separation of plasmid DNA from chromosomal DNA. The binding of EtBr to DNA causes partial unwinding of the double helix.

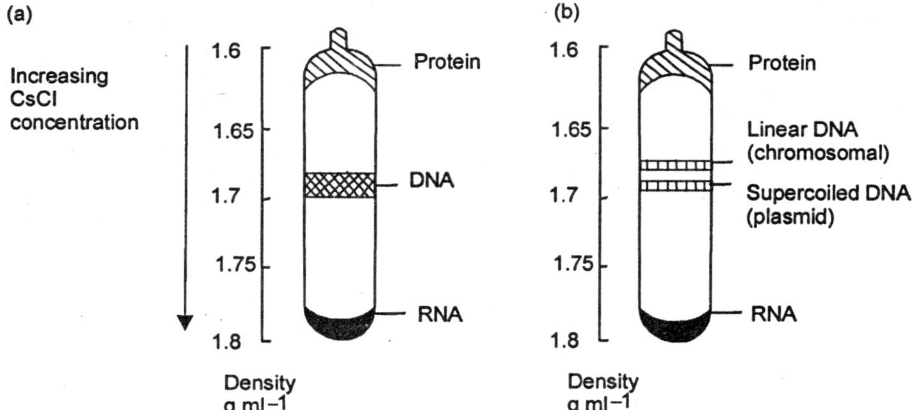

Fig. 2.1. Caesium chloride density gradient centrifugation for (a) the separation of DNA from RNA and protein and (b) the separation of linear DNA and supercoiled DNA.

This unwinding results in a decrease in buoyant density of about 0.125 g ml^{-1} for linear DNA. However, supercoiled DNA binds only a small amount of EtBr and the decrease in buoyant density is only about 0.085 g ml^{-1}. During extraction of DNA from bacteria the chromosomal DNA is broken into small fragments. This process causes the DNA to relax into a linear state. The much smaller plasmid DNA is not broken and remains in a supercoiled form. Therefore the supercoiled plasmid molecules will band in an EtBr-CsCl gradient at a different position to linear chromosomal DNA. The position of the DNA bands can be seen using ultraviolet light, which causes the bound EtBr to fluoresce.

ISOLATION OF RNA

The isolation of RNA is less straightforward than DNA. The sample is easily contaminated by ribonuclease causing breakdown of the RNA. Endogenous ribonuclease activity is prevented by addition of inhibitors in the early stages while exogenous activity is minimized by using pure chemicals and sterile glassware. A second problem is the tight association of RNA with protein as in polysomes. Harsh treatments are required to release the RNA from these complexes. Many methods have been described for the isolation of RNA including those shown below.

Proteinase K Method

In this method the cells are lysed by incubation in a hypotonic solution, which leaves the nuclei intact. The cell debris and nuclei are pelleted by centrifugation leaving the cytoplasmic RNA free from DNA in the supernatant. The RNA is released from the polysomes by incubation with proteinase K and the protein extracted into phenol/chloroform. The RNA is then precipitated from the aqueous phase using ethanol.

Guanidine Thiocyanate Method

The cells are lysed in a buffer containing strong chaotropic reagents such as guanidine thiocyanate and 2-mercaptoethanol, which completely denatures any ribonuclease present. The supernatant is then placed on a cushion of CsCl (5.7 mol l^{-1}) and centrifuged at 100000 g for 18 h. The RNA passes through the CsCl and is pelleted, while the DNA and protein remain in the aqueous solution. The RNA pellet is dissolved in buffer and concentrated by precipitation in cold ethanol. mRNA is then isolated from this total cellular extract by affinity chromatography using oligo-dT-cellulose or poly(U)-sepharose. The isolation of DNA and RNA from tissue has been simplified in recent years by the use of kits that enable the extraction to take place in a single step and therefore speed up the process. These reagents promote the formation of complexes of RNA with guanidine and water molecules and abolish hydrophilic interactions of DNA and protein.

In effect, DNA and proteins are effectively removed from the aqueous phase while RNA remains in this phase. After centrifugation to separate the two phases, the RNA is precipitated from the aqueous phase by ethanol or isopropanol. After the removal of the aqueous phase, DNA and protein in the organic phase can be recovered by sequential precipitation with ethanol and isopropanol. All types of RNA are isolated by this method. If necessary, mRNA can be isolated by affinity chromatography using oligo-dT-cellulose or poly(U)-sepharose.

ELECTROPHORETIC METHODS

The standard method used to separate, identify and purify DNA (and RNA) fragments is electrophoresis through agarose gels. The technique is simple, rapid to perform and capable of resolving mixtures of DNA fragments to a high degree. At neutral or alkaline pH the phosphate groups of DNA give rise to a uniform negative charge per unit length of DNA molecule, which therefore moves towards the anode in an electric field.

Molecules of different sizes can be separated by carrying out the electrophoresis through a supporting matrix. The size of DNA molecules depends on the qriain length and so DNA molecular sizes are quoted as numbers of base pairs. The matrix offers less resistance to small molecules, which therefore move faster. The rate of movement of a molecule of a particular length in the matrix depends on the pore size of the gel, which is determined by the concentration of agarose. Concentrations between 0.5 and 2% are commonly used covering a DNA size range from 200 to 50 000 base pairs. The separation can be done in vertical or horizontal apparatus in rods or slabs.

Because of the ease of use, horizontal slab electrophoresis systems are most commonly used in which the whole gel is submerged in buffer. The DNA within the gel can be directly located by including a low concentration of the dye ethidium bromide in the gel, which binds to the DNA, causing it to fluoresce. Adequate safety precautions must always be taken when using ethidium bromide owing to its hazardous nature. As little as 1 ng of DNA can be detected by direct examination of the gel in ultraviolet light. As ultraviolet light may cause eye damage, goggles must be worn when observing gels.

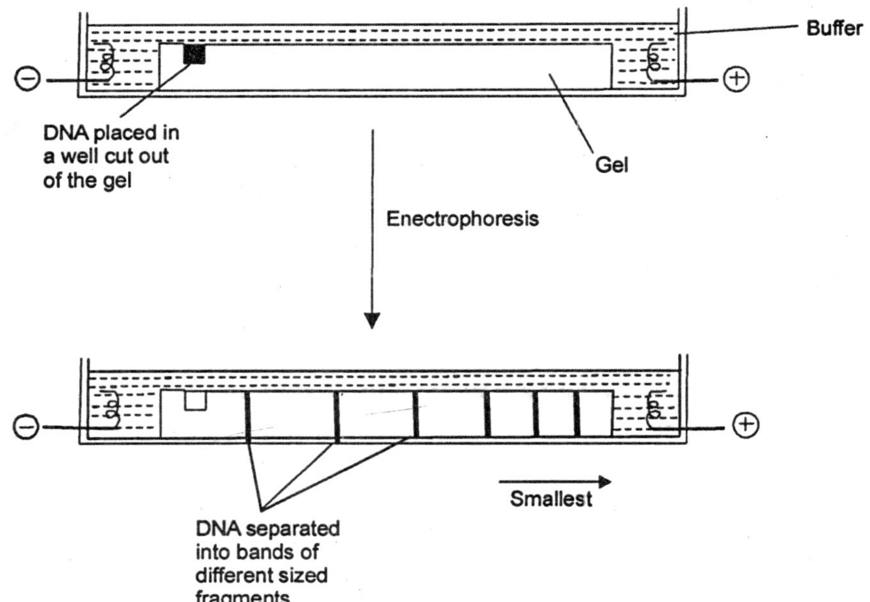

Fig. 2.2. The separation of DNA by agarose gel electrophoresis.

To avoid prolonged exposure to ultraviolet light, which can cause damage to the DNA as well as to the person, it is usual to photograph the gels for subsequent analysis. A transilluminator shines ultraviolet light of wavelength 300 nm on the gel from below and the fluorescent emission at 590 nm is photographed by a Polaroid camera. A red filter can be used to enhance the contrast between the fluorescent band and background. For the determination of DNA size suitable markers are included in one of the lanes. A straight line calibration graph can be obtained by plotting distance migrated against $\log_{10}$ base pairs of the markers.

Alternatives to Agarose Gel Electrophoresis

Agarose gel electrophoresis can only be used for DNA molecules that are greater than 200 base pairs in size. For molecules smaller than this polyacrylamide gel electrophoresis is required. This type of electrophoresis works on the same principle as agarose electrophoresis and separates molecules according to their size. Gels containing 12% polyacrylamide are used to resolve fragments between 20 and 100 base pairs and 3.5% polyacrylamide for fragments between 80 and 200 base pairs. Capillary gel electrophoresis is increasingly being used for the separation of DNA fragments.

Polyacrylamide gel-filled capillaries are usually employed because agarose gels are unable to withstand the heat generated by the high voltages used. This technique is used particularly to separate oligonucleotides and DNA sequence products, but double-stranded DNA, such as the products of restriction endonuclease digestion and PCR, can be separated with physical gels. These are relatively low viscosity polymer solutions, *e.g.* alkylcellulose.

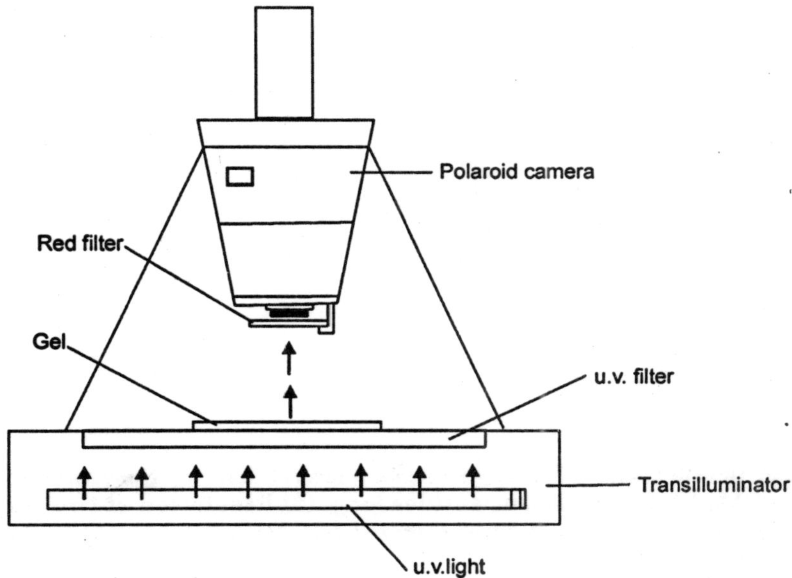

Fig. 2.3. Transilluminator and camera for the photography of DNA in gels.

Extraction of DNA from Gels

In a number of cases, *e.g.* for cloning and DNA sequencing, the DNA must be recovered from the gel in as high a yield as possible and without damage to the molecules. A number of methods have been tried, none entirely satisfactory, but electroelution is most often used.

The two major problems with this method are, firstly, that the agarose is often contaminated by sulphated polysaccharides, which are extracted from the gel together with the DNA, and are potent inhibitors of many enzymes and, secondly, that the yield of DNA, particularly for large molecules, is relatively low. The DNA band is cut from the gel Md placed in a dialysis bag containing a small volume of buffer.

The bag is placed in an electrophoretic tank and, when the current is switched on, the DNA passes out of the gel. The polarity of the current is reversed for a few moments to cause any DNA actually on the dialysis membrane to move back into the buffer within the bag. The DNA in the buffer is then precipitated with ethanol.

CHROMATOGRAPHIC METHODS

High Performance Liquid Chromatography

An alternative to nucleic acid isolation by electrophoresis followed by electrophoretic elution is HPLC. There are several advantages in the use of HPLC. It is simple and easy to usp and has a high degree of resolution, it is faster than electrophoresis and contamination of DNA fragments by impurities from the agarose gel is avoided. The two types of HPLC most commonly used are reverse phase and ion exchange.

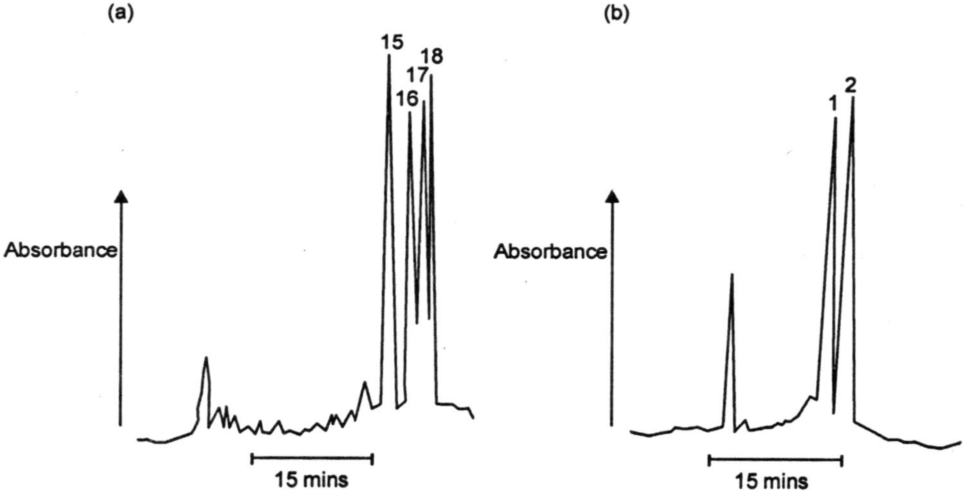

Fig. 2.3. Separation of oligonucleotides by reverse phase HPLC. (*a*) Resolution of a mixture of oligonucleotides of 15, 16, 17 and 18 nucleotides long, (*b*) Resolution of two oligonucleotides each of 18 nucleotides but differing in base composition. Peak 1 is TCACAGTCTGATCTCACC. Peak 2 is TCACAGTCTGATCTC-GAT.

Reverse-phase HPLC relies on the differences in hydrophobicity between nucleic acids of different lengths and base composition. Its major application is in the separation of small pieces of DNA (less than 50 base pairs) such as the products of oligonucleotide synthesis. Gradient elution with increasing amounts of acetonitrile is commonly used with the smaller, more polar nucleotides eluting first. The resolution is excellent and separation of oligonucleotides differing in length by only one base pair is easily achieved. Furthermore, the difference in hydrophobicity of the different bases means that oligonucleotides of the same length but with different base composition can also be separated. The order of polarity and therefore elution is G ≫ C > A > T.

In ion-exchange chromatography nucleic acids are separated on the basis of the difference in charge. Each nucleic acid has a different net charge based on the number of phosphate groups in the molecule (base length) and the respective charges on the bases (base composition). The latter effect is relatively minor and so separation depends almost exclusively on size. Separation is accomplished by slowly increasing the ionic strength of the mobile phase using NaCl, KC1 or KH_2PO_4.

The larger, more charged oligonucleotides elute later than the shorter ones. Ion-exchange HPLC can also be useful in the separation of larger nucleic acid molecules. One such application is as an alternative to CsCl density gradient centrifugation in the preparation of plasmids. Plasmid molecules typically consist of between 1000 and 10 000 base pairs. The plasmid is first isolated from the bacterial cell by alkaline lysis and pure plasmid obtained from this crude extract by a one-step chromatographic separation.

Affinity Chromatography for the Isolation of mRNA

Most eukaryotic mRNA molecules have up to 250 adenine bases at their 3′ end. These 'poly (A) tails' can be used in the affinity chromatographic purification of mRNA from a total cellular RNA extract. Under high salt conditions, poly (A) will hybridize to oligo-dT-cellulose or poly(U)-sepharose. These materials are polymers of 10 to 20 deoxythymidine or uridine nucleotides covalently bound to a carbohydrate support. They bind mRNA containing poly (A) tails as short as 20 residues. rRNA and tRNA do not possess poly (A) sequences and will not bind. After washing the mRNA can be eluted with a low salt buffer.

METHODS OF NUCLEIC ACID ANALYSIS

The method used to determine DNA concentration will depend on the nature of the sample. In recombinant DNA technology relatively pure samples of DNA or RNA are used and so simple spectrophotometric methods are adequate. However, when there are large amounts of interfering substances present, such as in a tissue or cell homogenate, a chemical method will be necessary.

SPECTROPHOTOMETRIC METHODS

The presence of conjugated double-bond systems in the purine and pyrimidine bases means that DNA and RNA absorb light in the ultraviolet region at 260 nm. For approximate determinations it can be assumed that a 50 μg ml^{-1} solution of double-stranded DNA (dsDNA) has an absorbance of 1 at 260 nm. More exact quantitation can be obtained by comparing the ratio of the absorbance of the sample at 260 and 280 nm. The term optical density (OD) is often used in place of absorbance.

Pure DNA preparations should have an OD 260/OD 280 of 1.8. Ratio less than this may indicate protein contamination while higher ratios may indicate the presence of RNA. Bases that are not hydrogen bonded absorb more strongly than when base paired. Therefore treatments that disrupt the hydrogen bonding between the base pairs, such as heat or alkali, will increase the absorbance of DNA. When a solution of double-stranded DNA (dsDNA) is heated slowly, initially there is little change in the absorption until the 'melting temperature' (T_m) is reached. Here the hydrogen bonds are broken, producing a rapid increase in absorbance to a higher value, which is not significantly changed on further heating. DNA is often present in amounts too small to be detected by direct spectroscopy. In this case, the fluorescent dye EtBr can be used to amplify the absorption. EtBr binds to the DNA molecule by intercalating between adjacent base pairs. It absorbs ultraviolet light at 300 nm and emits light at 590 nm in the red/orange region of the visible spectrum. The method can be used to determine the amount of DNA in a test-tube by comparing the EtBr-mediated fluorescence of the sample with that of standards of known amounts of DNA.

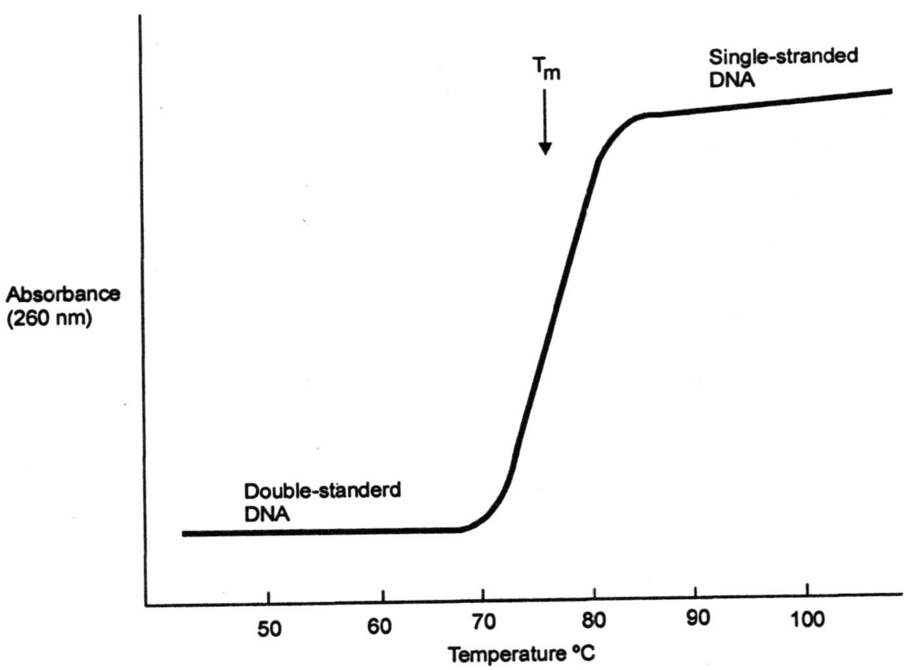

Fig. 2.4. Melting temperature for DNA. At this temperature the absorbance increases because the two strands of DNA come apart.

CHEMICAL METHODS

Burton Method

This is a spectrophotometric assay based on the reaction of diphenylamine with the deoxyribose moiety of DNA to produce a complex that absorbs at 600 nm. The reaction is specific for deoxyribose and RNA does not interfere. It can be used on relatively crude extracts where direct spectrophotometric determinations of DNA concentration are not possible.

Table 2.2. The Burton Method for the Quantitation of DNA.

Reagents

Tissue sample

Standard DNA samples ranging from 1 to 50 μg ml^{-1}

Perchloric acid 0.5 mol l^{-1}

Burton reagent

diphenylamine	15 g l^{-1}
sulphuric acid	0.25 mol l^{-1}
acetaldehyde	0.05 ml l^{-1}

Method

1. Add enough perchloric acid to cover the tissue sample Add the same volume of perchloric acid to the standard samples of DNA.
2. Heat both samples to 70-80°C for between 20 and 90 min.
3. Centrifuge at 300 g to remove the cellular debris.
4. Add 1 ml of sample or standard supernatant to 2 ml of reagent.
5. Mix and leave for 18 hours at 30°C.
6. Measure the absorbance at 600 nm.

Calculation

1. Prepare a calibration graph from the absorbance values of the standard DNA samples.
2. Read off the DNA content of the tissue samples.

Table 2.3. The DABA Fluorescence Method for the Quantitation of DNA.

Reagents

Tissue sample

Standard DNA samples ranging from 1 to 50 μg ml^{-1}

Hydrochloric acid	1.0 mol l^{-1}
Diaminobenzoic acid	320 g l^{-1}

Method

1. Spot 100 μl of homogenized tissue sample or DNA standard onto the bottom of a polypropylene tube and dry at 50°C.
2. Add 100 μl diaminobenzoic acid reagent and incubate at 55°C for 45 min.
3. Add 1.0 ml hydrochloric acid.
4. Measure fluorescence (excitation wavelength 410 nm: emission wavelength 510 nm).

Calculation

1. Prepare a calibration graph from the absorbance values of the standard DNA samples.
2. Read off the DNA content of the test samples.

DABA Fluorescence Assay

Diaminobenzoic acid (DABA) reacts with aldehydes of the form RCH CHO to produce a strongly fluorescent compound. Acid-catalysed removal of the purine base from the nucleic acid exposes the 1′ and 2′ carbons of deoxyribose and produces such an aldehyde group. Deoxyribose is the predominant aldehyde present in mammalian cells and essentially the only one present in acid precipitates of aqueous extracts. Hence, no purification is required and RNA does not interfere. The method can be used on very small samples (*e.g.* 100 μl).

ENZYMIC METHODS

Most DNA manipulations require the use of purified enzymes. Within cells these enzymes are used for DNA replication and transcription, breakdown of foreign DNA, repair of mutated DNA and recombination between different DNA molecules. Most of the purified enzymes are available commercially. They are usually supplied in Tris buffer containing 50% glycerol and are stored at –20°C.

The glycerol is included in the buffer to prevent freezing but must be removed prior to use as it alters the activity of some enzyme. For each enzyme, an optimized buffer solution is provided. Most of the buffers are based on Tris-HCl and have a pH range of 7.4 – 8.0. All contain magnesium ions as activators for the enzymes. The ionic strength of the buffers is important and different enzymes require different ionic strengths for optimal activity. In general, three strengths of buffer are used: low, medium and high, which contain 10, 50 and 100 mol l^{-1} sodium or potassium chloride respectively.

Restriction Endonucleases

Restriction endonucleases are enzymes isolated from bacteria. *In vivo* these enzymes are involved in recognising and cutting up foreign DNA, thus protecting the bacteria against phage and virus infection. They recognise specific base sequences in double-stranded DNA and break the phosphodiester bonds between two nucleotides within the sequence. These recognition sequences may be four, five or six nucleotides long. Over 350 different enzymes have been isolated so far and each recognises a different base sequence. Some of the more commonly used enzymes with their recognition sequences. Table 2.2. Some enzymes, *e.g. Eco* RI, cut the DNA at a different position in the two strands, producing a single stranded overhang or a 'sticky end'.

Other enzymes, *e.g. PvuII,* produce blunt-ended fragments with no sticky ends. The formation of sticky ends is useful because, by base pair matching of the sticky ends, any DNA fragment produced by the action of the same restriction enzyme can be joined together in a specific manner.

DNA ligases

Ligases are found in all cells but the two that are used in DNA analysis are either DNA ligase from *Escherichia coli,* which requires the nucleotide NAD as a cofactor, or T4 DNA ligase from the T4 phage, which requires adenosine triphosphate (ATP) as a cofactor. Both enzymes catalyse the synthesis of a phosphodiester bond between the 3′ hydroxyl group of one nucleotide and the 5′ phosphoryl group of the next nucleotide in a dsDNA molecule. Ligases are most often used to produce a covalent bond between nucleotides in DNA molecules joined by base pair matching of sticky ends.

Table 2.4. Recognition sequences of a selection of restriction endonucleases. The double-stranded DNA is cut at the positions arrowed.

Enzyme	*Organism*	*Recognition sequence*
Eco RI	*Escherichia coli*	↓ before G (top); GAATTC CTTAAG; ↑ before G (bottom)
Bam HI	*Bacillus amyloliquefaciens*	↓ before G (top); GGATCC CCTAGG; ↑ before G (bottom)
Pvu I	*Proteus vulgaris*	CGATCG GCTAGC
Pvu II	*Proteus vulgaris*	CAGCTG GTCGAC
Hin dIII	*Haemophilus influenzae* Rd	AAGCTT TTCGAA
Alu I	*Arthrobacter luteus*	AGCT TCGA
Hae III	*Haemophilus aegyptius*	GGCC CCGG
Hin fI	*Haemophilus influenzae*	GANTC CTNAG

DNA Polymerases

DNA polymerases synthesize a new strand of DNA complementary to an existing DNA or RNA strand. Short oligonucleotides complementary to the ends of the original DNA or RNA strand, known as primers, are added to the incubation mixture.

DNA Polymerase

For most applications DNA polymerase 1 from *E. coli* is used. The enzyme attaches to a short single-stranded region in a dsDNA molecule and then synthesizes a new strand of DNA, degrading the existing strand as it proceeds. When used *in vitro* this incubation is carried out at 12 – 15°C to prevent more than one round of replication occurring. It is used for *in vitro* labelling of DNA by the nick translation method (described below).

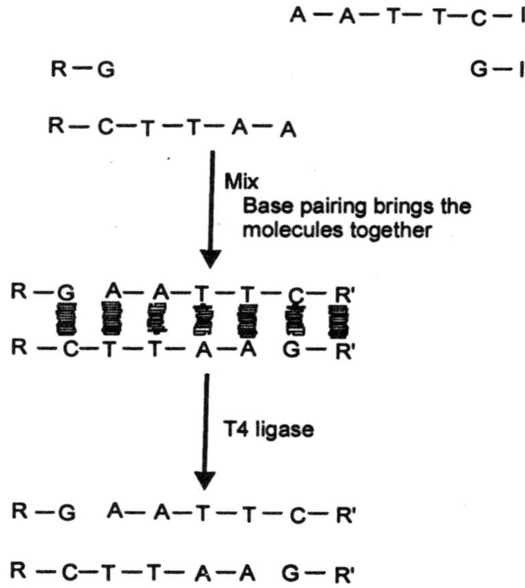

Fig. 2.5. The joining of two molecules of DNA by the enzyme T4 ligase. The enzyme produces covalent bonds between nucleotides in a strand of DNA. The DNA molecules are brought together by base pairing of their sticky ends.

The Klenow Fragment

The DNA polymerase 1 molecule contains its polymerase and nuclease activities on different parts of the enzyme molecule. These two parts can be separated by treatment with the enzyme *subtilisin*. The part which retains the polymerase function is known as the Klenow fragment. This enzyme sythesizes a new DNA strand complementary to the single strand of DNA (the template) only. It is used to create blunt ends in dsDNA and in the dideoxy method of DNA sequencing.

Reverse Transcriptase

This enzyme is involved in the replication of retro-viruses *in vivo*. It synthesizes a complementary DNA (cDNA) strand using RNA instead of DNA as its template. It is widely used to create a strand of cDNA from mRNA extracted from cells or tissue for cloning or for PCR analysis.

Taq Polymerase

Is a thermostable DNA polymerase which was originally isolated from the bacterium *Thermus aquaticus,* which lives in hot springs. This enzyme has a maximum activity at 70 – 80°C and remains active at temperatures up to 90°C. It is used extensively in the analysis of DNA using PCR.

DNA-Modifying Enzymes

Many other enzymes can be used to remove or add groups to the ends of the DNA molecule. The three major ones are alkaline phosphatase, which removes a phosphate group from the 5′ terminus; polynucleotide kinase, which adds a phosphate group to a free 5′ terminus; and terminal deoxynucleotidyl transferase, which adds one or more deoxynucleotides to the 3′ terminus.

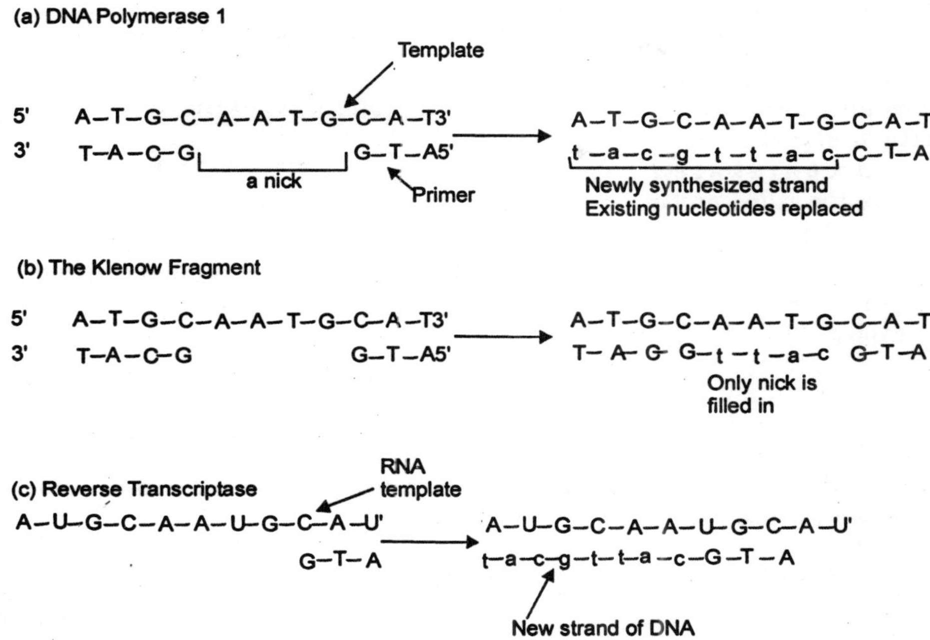

Fig. 2.6. The actions of the DNA polymerases used in recombinant DNA technology.

DETECTION OF SPECIFIC SEQUENCES OF DNA

If a single-stranded DNA molecule is placed with a complementary single DNA sequence the two molecules will hybridize. This hybridization forms the basis of a number of very powerful techniques for detecting and quantifying specific nucleic acid sequences. The hybridization may be carried out either in solution or more commonly with the DNA immobilized on nitrocellulose filters. The complementary DNA sequence is known as a cDNA probe. Probes for a large number of important nucleic acid sequences are now available.

Production of Labelled DNA Probes

Before using the probe for hybridization, it must be labelled to enable subsequent detection. One method of labelling is by the incorporation of ^{32}P-labelled nucleotides, a process that may be carried out in two ways.

Nick translation uses the enzyme DNA polymerase 1. Most pieces of DNA contain some nicked areas. DNA polymerase 1 can attach here and catalyse a strand replacement reaction. When the reaction is carried out in the presence of radioactively labelled deoxynucleotides the replaced DNA strand becomes labelled. Usually only one labelled deoxynucleotide is included.

End filling is a more gentle method but can only be used on DNA molecules that have sticky ends. It uses the Klenow fragment, which fills in the sticky ends by synthesizing the complementary strand. Again if the reaction is carried out in the presence of labelled deoxynucleotides the result is a strand of DNA which is radioactively labelled. Non-radioactive labelling systems are also used. These incorporate biotinylated or fluorescein-conjugated nucleotides into the cDNA probe in place of the ^{32}P-nucleotide.

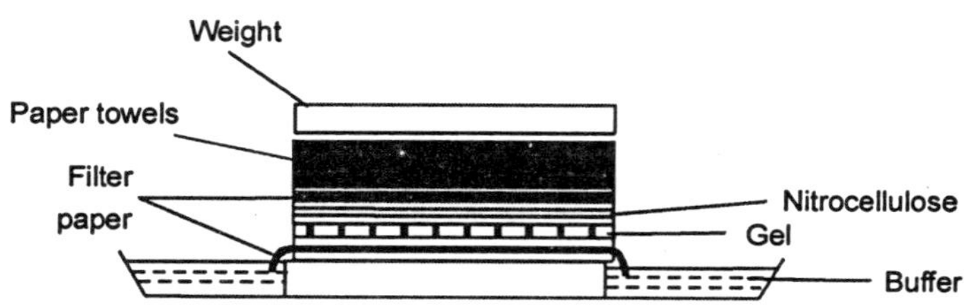

Fig. 2.7. The transfer of DNA from a gel to a nitrocellulose sheet Buffer soaks through the gel to the nitrocellulose and takes the DNA with it.

The radio-labelled probe can be detected directly by autoradiography, whereas the blotinylated or fluorescein-conjugated probes are detected using alkaline phosphatase or peroxidase-conjugated antibodies directed against the biotin or fluorescein. The binding of antibody to the nucleotide can then be visualized by incubation with the enzyme substrate and a colour reagent. Although non-radioactive labelling systems are advantageous with regards to safety and disposal, the visualization procedure requires several more incubation steps, which may be time consuming.

Southern Blotting

The method most often used in the detection of specific DNA sequences is Southern blotting. The DNA is fragmented by digestion with a restriction endonuclease and the fragments separated by agarose gel electrophoresis. The DNA bands are then transferred to a nitrocellulose membrane so that individual DNA molecules are retained during the hybridization process. The apparatus used for this transfer. The nitrocellulose is placed on the gel and buffer is allowed to soak through, carrying DNA from the gel to the membrane where the DNA is bound.

The result is that the nitrocellulose membrane carries a replica of DNA bands from the agarose gel. The DNA on the membrane is then incubated with labelled DNA probe, which will hybridize to any bands which contain sequences complementary to it. After washing, this hybridization can be visualized by autoradiography or by using an antibody detection systerm. The specificity of the hybridization between probe and DNA depends on the temperature and ionic strength of the buffer.

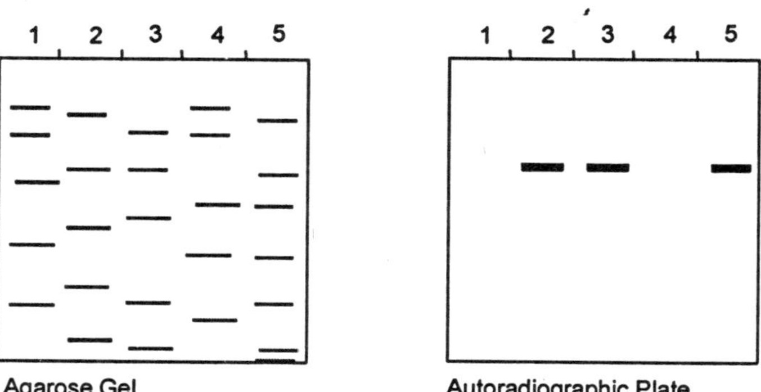

Fig. 2.8. Detection of a specific sequence of DNA by hybridization to a ^{32}P-labelled cDNA probe. DNA is transferred to nitrocellulose and incubated with the probe. After washing, specific binding is visualized by autoradiography. The DNA sequence detected by the probe is present in lanes 2, 3 and 5 but not 1 and 4.

At high temperatures and low salt concentration (high

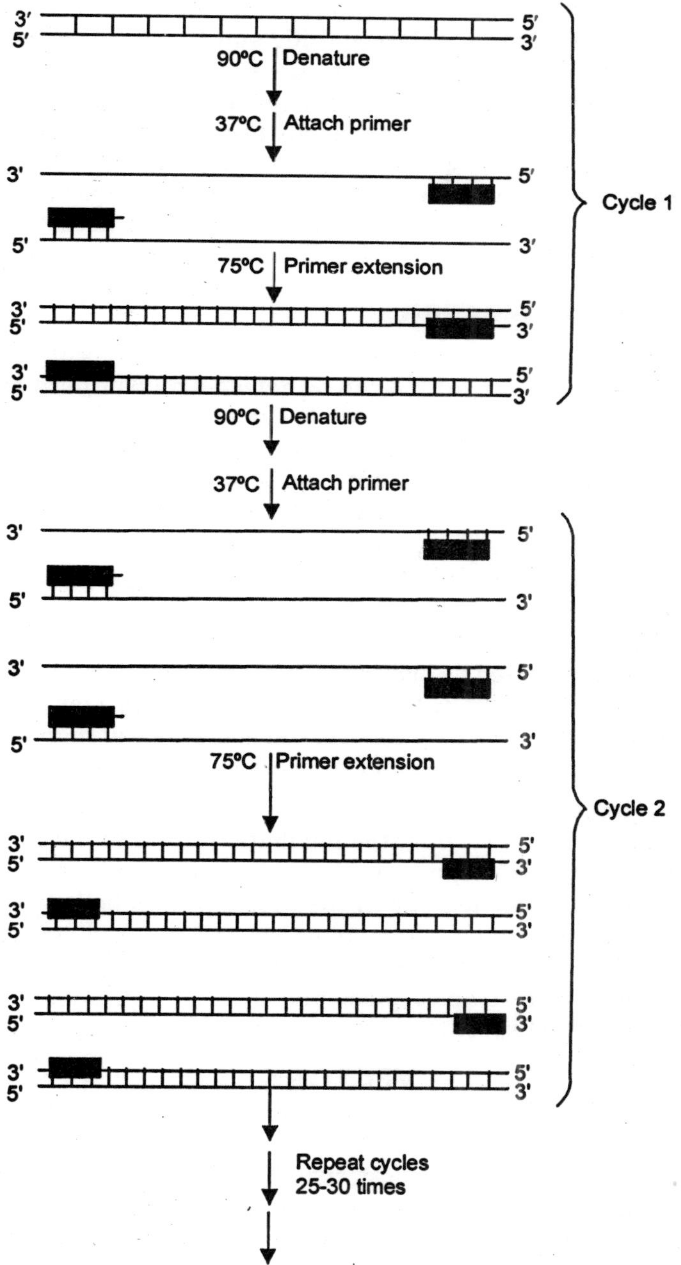

Fig. 2.9. The polymerase chain reaction for the amplification of NDA sequences. DNA is heated to separate the two strands. A primer is attached to the 5′ end of each strand and extended using NDA polymerase 1. The two new strands are separated as before and the cycle repeated up to 30 times.

stringency), hybridization is very specific. At lower stringencies (low temperatures and high salt concentration) the hybridization is less specific and the cDNA probe will bind to many sequences. In Southern blotting, incubation with the probe is carried out at low stringencies, which allows hybridization to a large number of fragments. The membrane is then washed at high stringency so that the probe remains bound only to highly homologous sequences. Another similar technique to Southern blotting is Northern blotting. Here, instead of DNA fragments, mRNA fragments are probed with a labelled cDNA probe after separation by electrophoresis and transfer to nitrocellulose membranes. Northern blotting is used to detect and quantify mRNA from tissue extracts.

The Polymerase Chain Reaction

Although with the use of cDNA probes containing many molecules of ^{32}P (high specific activity) as little as 20 pg of DNA or mRNA can be detected by Southern or Northern blotting, many DNA or RNA sequences within tissues are below this level of detection. The polymerase chain reaction allows massive replication of a target sequence of DNA and allows detection of DNA or RNA sequences that are present as only one or two copies per cell. The target sequence of DNA is extracted from the cell and denatured into its single strand by heating to 90°C.

Two oligonucleotide primers complementary to the opposing ends of the single strands of the target UNA are synthesized and annealed. Primers for PCR consist of a 20-30 nucleotide sequence specific for the target DNA and result in the amplification of that sequence of DNA only. The primed single strand is then extended using DNA polymerase 1. The newly synthesized double-stranded sequence is then denatured by heating to 90°C and new oligonucleotide primers added. This process is then repeated through 25 to 30 cycles and results in a large increase in DNA. The reaction depends upon the use of Taq polymerase, which is stable at high temperatures and can survive the repeated heating to 90°C without loss of activity.

The amplification process is carried out in a thermocycler, which allows the automatic control of the rate of heating and cooling, the incubation times at each temperature and the number of cycles. This information can be programmed into the thermocycler at the start of the reaction process, which is carried out in the presence of labelled nucleotides. The PCR product is detected using agarose gel electrophoresis and autoradiography. An extension of the PCR method is RT-PCR. Here mRNA is extracted from the cells or tissue, converted to cDNA using the enzyme reverse transcriptase and PCR is then carried out. This method is used to detect the expression of specific mRNA sequences in cels or tissues.

VECTORS

Pieces of DNA, for example genes or cDNA probes, are multiplied using a method based on *in vivo* replication in bacterial cells. The sequence of DNA is incorporated into a vehicle or vector which transports it into a cell host, usually *E. coli.* As the bacteria culture grows, the vector also replicates so producing more of the required sequence of DNA. Two of the naturally occurring types of DNA molecule which can be used as vectors are plasmids and viral chromosomes.

PLASMIDS

Plasmids are the circular double-stranded DNA molecules found in the cytoplasm of bacterial cells. To be a useful vector the plasmid must be small. The smaller the plasmid, the easier it is taken up into the bacteria. Most vector plasmids are about 3-4 kilobases (kb) in length. This allows incorporation of sequences of DNA up to 10 – 12 kb without impairing plasmid uptake. It is also helpful if the vector has a high copy number. The larger the copy number, the more molecules of DNA will be produced. The uptake of the plasmid into the bacterial cell is called transformation and in the laboratory can be induced in two ways. In one method the bacteria and DNA are placed in ice-cold $CaCl_2$ (50 m mol 1^{-1}). This induces the DNA to stick to the outside of the bacteria. The temperature is then increased to 42°C and the DNA enters the cell.

The second method is by electroporation, which involves the application of a brief high voltage pulse to a suspension of cells and DNA. This results in a transient membrane permeability and the subsequent uptake of DNA into the bacteria.

VIRAL CHROMOSOMES

The chromosomes of bacteriophages, which are viruses that specifically infect bacteria, can also be used as vectors. These are linear double-stranded DNA molecules which contain the genes required for viral replication within the host cell. During infection the DNA molecules are inserted into the bacterial host. The most widely used bacteriophage is lambda (λ) which infects *E. coli.* Within the *E. coli* lambda has two alternative modes of replication: the lytic and the lysogenic cycles. In the lytic cycle the DNA is incorporated into the *E. coli* chromosome.

Replication takes place, resulting in the production of a large number of mature phage particles, which are released from the cell by cell lysis. In the lysogenic mode the lambda DNA is incorporated into the chromosomal DNA in such a way that replication of the lambda DNA occurs but it is not packaged into mature particles and there is no cell lysis. Instead the lambda genome is inherited by each daughter bacterial cell at cell division. Lambda is maintained in this prophage state through the action of a represser protein, cl. If the repressor protein is inactivated the lambda DNA is excised from the host chromosome and a lytic cycle is activated. In one mutated form of lambda the repressor protein is inactive at temperatures above 37 °C but active at lower temperatures.

When this lambda is used as a vector the infected *E. coli* are grown first at 32°C to allow replication of the DNA in the lysogenic cycle. The temperature is then increased to 37°C to inactivate the repressor. This results in the excision of the lambda genome and release of lambda particles by lysis. Another viral chromosome that can beused as a vector is that of the filamentous phage Ml3. The M13 chromosome is a single-stranded DNA molecule which when inserted into the bacterial host replicates outside the bacterial chromosome in the cytoplasm. The virus is then reassembled and released from the bacterial cell without cell lysis.

SELECTION OF TRANSFORMED BACTERIA

The procedure of inserting DNA into a vector and its subsequent introduction into the bacterial host is extremely inefficient and will result in the majority of bacteria in the culture either containing none of the required vector or containing a vector in which no DNA is inserted. However, vectors have been produced that, by placing the bacteria in specific media, allow only the growth of those containing the vector and the inserted DNA. One such plasmid vector is pBR322. It contains two genes coding for resistance to the antibiotics ampicillin and tetracycline.

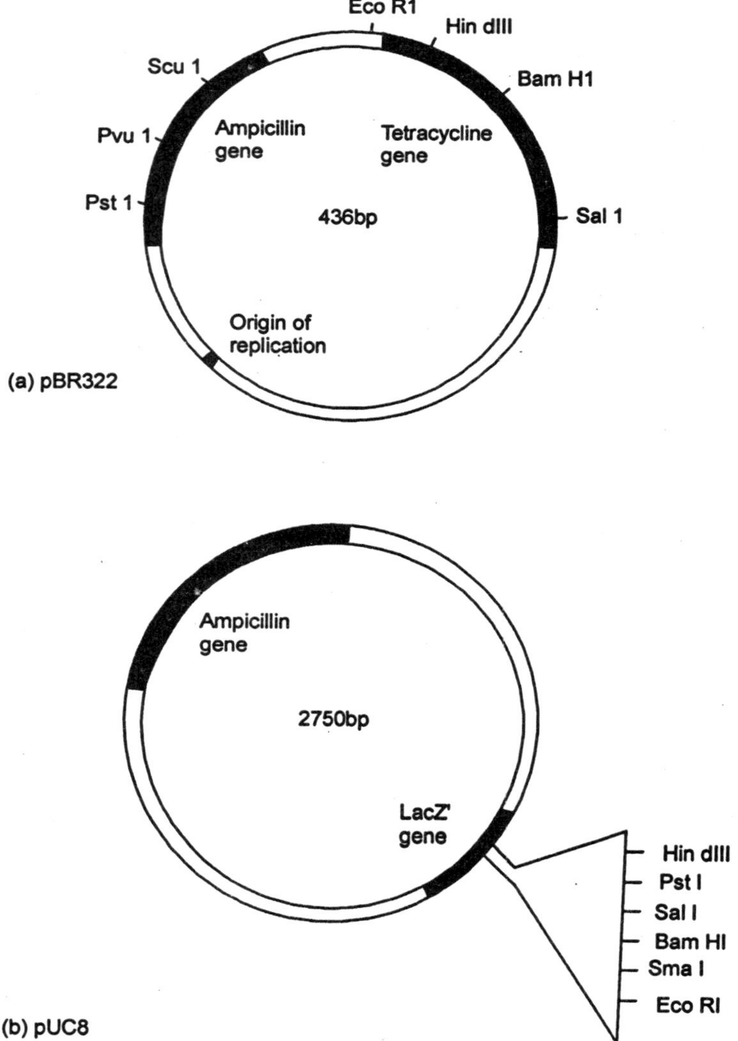

Fig. 2.10. Maps of the plasmids (*a*) pBR322 and (*b*) pUC8 showing restriction endonuclease sites in genes used to select recombinants.

Bacteria that take up the vector (transformed bacteria) will be resistant to the antibiotic and will grow in its presence, while non-transformed bacteria will not. The bacteria are therefore grown on media containing either ampicillin or tetracycline. The resultant colonies must contain bacteria that have taken up the vector. Within the genes coding for ampicillin and tetracyoline are unique recognition sites for many different restriction enzymes. DNA fragments are inserted into these sites using the appropriate enzyme. Insertion of the DNA fragment results in disruption of the gene and the resistance to the antibiotic is lost. Therefore any bacteria that have taken up the vector with the fragment of DNA inserted (recombinants) will not grow in the presence of that antibiotic. This process is called insertional inactivatipn.

To select tor a bacterium that contains arragment of DNA inserted into the tetracycline gene of pBR322, for example after using the *Bam* HI restriction enzyme, the bacteria are grown on a medium containing ampicillin.

Colonies that grow contain the plasmid but only a few contain the recombinant pBR322 molecules. The colonies are then replica plated onto media containing tetracycline. The colonies that grow are resistant to tetracycline and so have an intact tetracycline gene. Those that contain the recombinant molecule will not grow but can be taken from the original plate and grown up in bulk. Another widely used form of insertional inactivation is the *lacZ'* gene which was first used in the pUC8 vector.

The *lacZ* gene codes for the enzyme β-galactosidase, which breaks down lactose and is present in the *E. coli* chromosome. Some strains of *E. coli* have a modified *lacZ* gene *(lacZ')* which has a large segment missing. These mutants can only synthesize the whole enzyme when the missing *lacZ* segment is supplied by a plasmid such as pUC8. This *lacZ* segment can be inactivated by insertion of a DNA fragment using an appropriate restriction enzyme. pUC8 also contains an ampicillin resistance gene and so selection involves firstly growing the bacteria on ampicillin to select for transformants followed by transfer to media containing X-gal. X-gal is a synthetic substrate for the enzyme β-galactosidase and gives a blue product. Bacteria containing a recombinant pUC8 molecule will not supply the extra piece of enzyme, so X-gal is not broken down and the colonies are white.

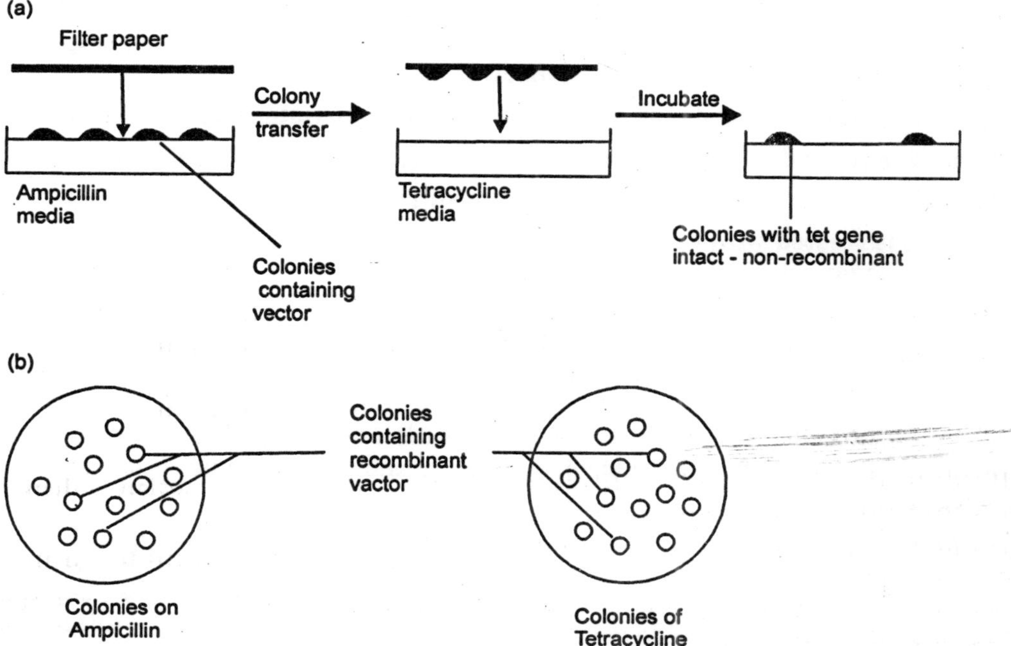

Fig. 2.11. Replica plating for the selection of bacteria containing a recombinant vector, (*a*) The method used to transfer colonies of bacteria between plates, (*b*) Comparison of the distribution of the colonies between the plates enables identification of the colonies containing the recombinant vector. The dotted circles represent colonies that do not grow on tetracycline.

Bacteria that contain the intact pUC8 molecule will supply the β-galactosidase fragment, so X-gal is broken down and blue colonies are formed. The white colonies are the ones required and can be taken directly from this plate.

DNA SEQUENCING

It is the sequence of the nucleotide bases in the DNA molecule that is fundamental to nucleic acid function. Hence the techniques that determine this sequence are extremely important in recombinant DNA technology. Two similar methods have been developed. Both depend on the production of a mixture of oligonucleotides labelled either radioactively or with fluorescein, with one common end and differing in length by a single nucleotide at the other end. This mixture of oligonucleotides is separated by high resolution electrophoresis on polyacrylamide gels and the position of the bands determined.

Known DNA sequences associated with specific genes are stored in computer libraries which are freely accessible. This allows the comparison of newly sequenced genes with those coding for known proteins.

THE MAXAM AND GILBERT METHOD

In this method the single-stranded DNA fragment to be sequenced is end-labelled by treatment with alkaline phosphatase to remove the 5′ phosphate, followed by reaction with ^{32}P-labelled ATP in the presence of polynucleotide kinase, which attaches ^{32}P to the 5′ terminal. The labelled DNA fragment is then divided into four aliquots, each of which is treated with a reagent which modifies a specific base as follows.

Aliquot 1 is treated with dimethyl sulphate, which methylates guanine residues.

Aliquot 2 is treated with formic acid, which modifies adenine and guanine residues.

Aliquot 3 is treated with hydrazine, which modifies thymine and cytosine residues.

Aliquot 4 is treated with hydrazine in the presence of NaCl (5 mol l^{-1}), which makes the reaction specific for cytosine.

After these reactions the four aliquots are incubated with piperidine, which cleaves the sugar-phosphate backbone of DNA next to the residue that has been modified. The modification reaction is carried out under conditions that allow only a limited number of bases in each DNA molecule to react. In perfect conditions this will be one base per DNA molecule. This means that cleavage with piperidine will result in the production of a mixture of oligonucleotides each having the same 5′ end but extending to the nucleotide next to the one that was modified.

Any base in the original DNA molecule is as likely to react as any other with the reagent which is specific for it. Therefore these reactions evenly distribute the radioactivity among the cleavage products. The oligonucleotides in each aliquot are placed in four different but adjacent wells of a polyacrylamide gel containing SDS and separated by electrophoresis. The separated fragments are detected by autoradiography. The resolution of these gels is such that.oligonucleotides differing by only one nucleotide can be distinguished.

The smallest fragments travel the furthest. Hence the band that has travelled the furthest contains only one nucleotide, while the next band up the gel contains two nucleotides, etc.

The lane in which the band is found indicates the next base in the sequence. The DNA sequence is therefore obtained by reading the gel from bottom to top.

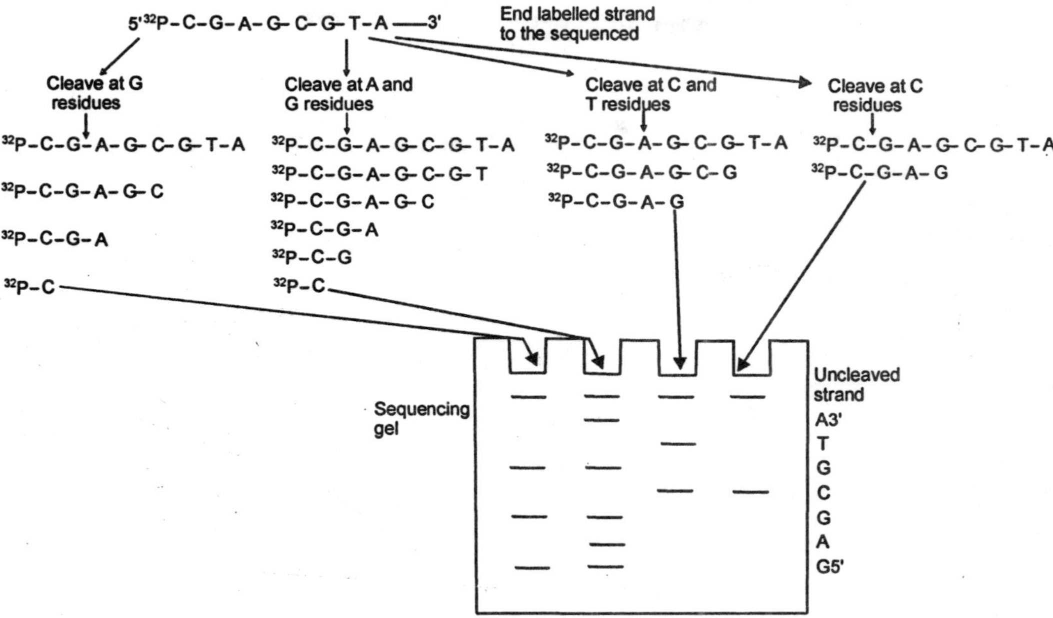

Fig. 2.12. The Maxam and Gilbert method of DNA sequencing. The labelled DNA strand is divided into four aliquots. Each is treated to cleave the strand next to a different base resulting in a mixture of different length nucleotides. These nucleotides are separated by polyacrylamide gel electrophoresis and the DNA sequence read from the gel.

THE DIDEOXY METHOD

This method was developed by Sanger and his colleagues and is also referred to as the chain termination method. The procedure requires a single-stranded DNA template and a short primer complementary to the 3′ end of the region of DNA to be sequenced. A complementary copy of the DNA strand is produced by the Klenow fragment of DNA polymerase 1. Synthesis is carried out in the presence of the four deoxyribonucleotide triphosphates, one of which is labelled with ^{32}P. In the presence of competing dideoxyribonucleotide triphosphates (ddNTPs), specific termination of DNA synthesis occurs where the dideoxy derivative is incorporated instead of the deoxyribonucleotide. Four incubations are carried out, each in the presence of a different dideoxy derivative.

Each incubation generates a heterogeneous population of labelled oligonucleotides terminating with the same nucleotide. Urea is added to each incubation to separate the two strands of DNA and the single strands are separated electrophoretically in adjacent lanes of an SDS polyacrylamide gel. Urea is also present in the gel to ensure that the strands of DNA stay separated. The gel is then autoradiographed. The DNA sequence can be deduced from the ascending order of the bands in the four adjacent lanes. The single-stranded template DNA required for this method of sequencing is usually provided by using the bacteriophage

Ml3 as a vector. The most widely used method of DNA sequencing using this system is the 'shotgun approach'.

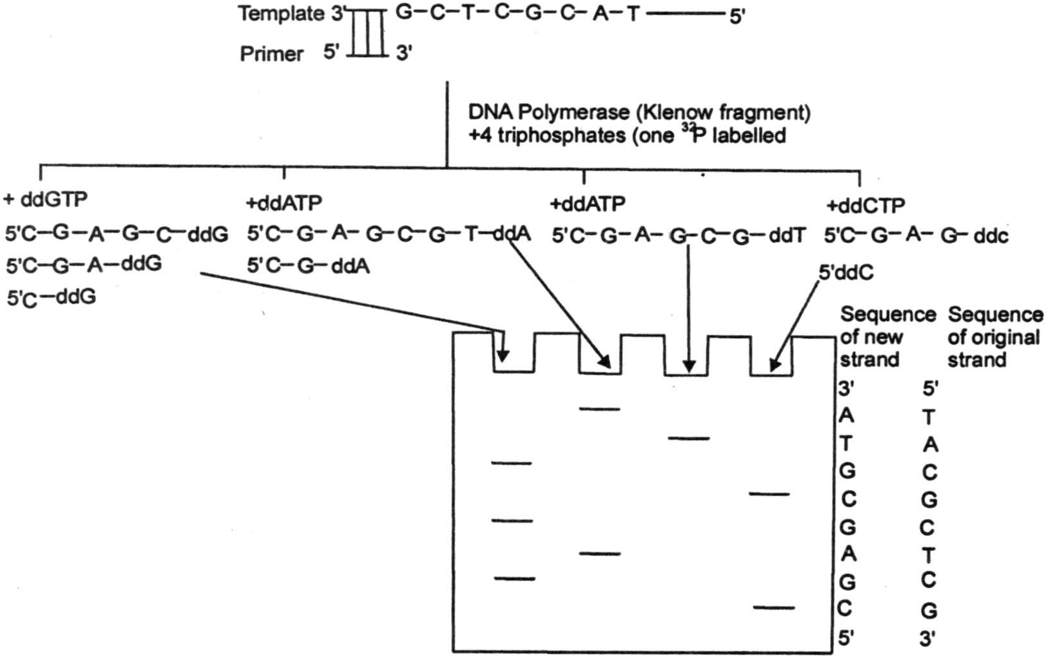

Fig. 2.13. The dideoxy (chain termination) method of DNA sequencing. A complementary copy of the DNA strand is synthesized in the presence of dideoxy analogues of each base. Chain termination occurs where the analogue is inserted in place of the true base. The newly synthesized strands of DNA are separated by polyacrylamide gel electrophoresis and the sequence deduced from thse pattern of bands.

A random series of fragments is produced from the DNA to be sequenced. This may be done by sonication or digestion with restriction enzymes. Each fragment is incorporated into the M13 vector and its sequence obtained. Comparison of the sequences of each fragment will show overlapping sections and allow deduction of the sequence of the original piece of DNA. Fluorescein-labelled deoxyribonucleotide triphosphates may be used in place of those labelled with ^{32}P.

Once the DNA sequences are separated by electrophoresis, the resulting DNA bands fluoresce and are analysed by a fluorogram imager, which produces a picture of the fluorescent bands similar to the autoradiography produced when using ^{32}P-labelled nucleotides. Kits containing all the reagents required for DNA sequencing including the DNA polymerase enzyme, M13, vector, primers, buffers and labelled and unlabelled nucleotides are available from a number of manufacturers. All the components of these kits have been optimized for use with each other.

CHAPTER 3

DNA Methylation Analysis

The explosion of genetic information available after the completion of the Human Genome Project has proved a gold mine for pharmacogenetic research to identify biomarkers. Early estimates predicted more than 1.42 million single nucleotide polymorphisms (SNPs) are present in the human genome. The 126 build of the public SNP repository, dbSNP (completed by May 2006, see http://www.ncbi.nlm.nih.gov/SNP/index.html), contained 27, 846, 394 submissions for the human genome, of which 5, 646, 244 were validated SNPs (20%). Polymorphisms in coding and control regions of genes can cause significant interindividual variation in the resulting protein function and activity, leading to important differences in disease susceptibility and drug metabolism. This expansion in evaluable SNPs has led to a number of detection methods.

Pyrosequencing is sequencing by synthesis, a simple-to-use technique for accurate and quantitative analysis of DNA sequences, and it produces specific sequence data in the form of peaks on a pyrogram where the light signal collected is transformed and presented based on the DNA base composition of interest. It does not require the presence of a restriction enzyme site, and polymerase chain reaction (PCR) product and internal primer sites can vary in size and position. In addition, it can also be used to identify triallelic, indel, and short-repeat polymorphisms, as well as to determine allele percentages for DNA methylation or pooled sample assessment. The DNA methylation assay by Pyrosequencing is to detect a special "C to T" polymorphism as after treatment with sodium bisulfate, unmethylated cytosine residues are converted to thymine, whereas methylated cytosine residues are retained as cytosine. Being an intrinsic feature of the Pyrosequencing method, the availability of sequence directly adjacent to the polymorphisms allows internal quality control checks to be made for each sample. Pyrosequencing is performed on a 96-well platform, and in an average day, more than 3000 individual genotypes can be measured. This method has been used to genotype a number of clinically relevant polymorphisms and DNA methylation status. Included in these methods are assay details for commonly analyzed clinically relevant polymorphisms in the cytochrome P450 and adenosine triphosphate (ATP)-binding cassette transporter genes CYP3A4 and ABCB1, and also for DNA methylation levels in the DNA mismatch repair and fluorouracil catalytic enzyme genes MLH1 and DPYD.

MATERIALS

DNA Template

DNA from any source can be used in Pyrosequencing assays, Commonly used kits for DNA extraction, including Centra (Minneapolis, Minnesota, U.S.A.) and Qiagen (Hilden, Germany), do not inhibit the assay. For use in methylation assays, the original genomic DNA must be converted through the widely used bisulfate conversion method. After treatment with sodium bisulfate, unmethylated cytosine residues are converted to thymine; whereas methylated cytosine residues are retained as cytosine.

Bisulfate Conversion of Genomic DNA and SssI Methylase Treatment

1. 4 M sodium bisulfate (Sigma, St. Louis, Missouri, U.S.A.) at pH 5 (freshly prepared).
2. 10 mM hydroquinone (Sigma) (freshly prepared).
3. 2M NaOH.
4. Mineral oil from Sigma.
5. Heating block for DNA incubation.
6. DNA clean-up kit from Qiagen.
7. SssI methylase from New England Biolab.

Polymerase Chain Reaction

1. Primer design software. (Any free or custom-purchased software is appropriate. Pyrosequencing also sells custom primer design software).
2. 1 to 5 ng DNA template.
3. PCR mastermix, for example: 30 mM Tris-HCl, 100 mM potassium chloride, pH 8.05, 400 μM dNTP, and 5 μM magnesium chloride.
4. Hot start Taq polymerase.
5. DNase- and RNase-free 18.2 mΩ water.
6. DNA oligonucleotides (primers).
7. Unskirted 96-well PCR trays.
8. 96-well sealing film or silicon mat for covering 96-well plates in a thermocycler.
9. Thermocycler with 96-well capacity and heated lid.

Agarose Gel Electrophoresis

1. Agarose.
2. 50X TAB buffer: for 1 L, add 242 g Tris base, 57.1 mL glacial acetic acid, and 18.6 g ethylenediamine tetraacetic acid to 18.2 mΩ water. Store at room temperature. Dilute to IX with water prior to use.
3. Microwave.
4. Ethidium bromide: 4 μL of 10 mg/mL ethidium bromide/100 mL agarose; add *after heating*.

5. 6X loading dye: for 100 mL: 30 mL glycerol, 70 mL water plus a pinch of bromophenol blue and a pinch of xylene cyanol FF (amount can be varied depending on the desired color). Store at room temperature.
6. Gel apparatus: casting tray, gel tank, lid, and power supply.
7. UV gel documentation system with thermal printer.

Processing PCR for Pyrosequencing

1. Centrifuge with rotor/buckets to handle 96-well plates.
2. 2X binding buffer: for 1 L, add 1.21 g Tris, 117 g NaCl, 0.292 g ethylenediamine tetraacetic acid to water, pH 7.6, with 1 M HC1. Sterile filter then add 1 mL Tween-20.
3. Bead mix: 240 μL streptavidin coated Sepharose beads, 4560 μL 2X binding buffer, and 3600 μL 18.2 mΩ water per 96-well plate (the magnetic bead processing protocol for a PSQ96 or PSQ96MA is described elsewhere). Excess Sepharose/binding buffer mix can be stored in a glass bottle at 4°C.
4. 96-well plate shaker (*e.g.*, Eppendorf thermomixer; Fisher Scientific, Hampton, NH).
5. Vacuum prep tool and troughs (Biotage, Uppsala, Sweden).
6. 70% ethanol in 18.2 mΩ water.
7. 0.2 M NaOH in 18.2 mΩ water.
8. Washing buffer: for 1 L, add 1.21 g Tris to water, pH 7.6 with 4 M acetic acid. Sterile filter.
9. Annealing buffer: for 1 L, add 2.42 g Tris, 0.43 g magnesium acetate tetrahydrate to water, pH 7.6 with 4 M acetic acid. Sterile filter.
10. Pyrosequencing primer mix: 12 μL of 0.3 μM Pyrosequencing primer in annealing buffer per well dispensed into a 96-well Pyrosequencing plate (Biotage).
11. Heating block capable of at least 80°C.
12. Pyrosequencing plate adaptor set (base and iron) (Biotage).
13. Adhesive sealing film for 96-well plates.

Pyrosequencing

1. PSQhs96 or PSQhs96A Pyrosequencer with Pyrosequencing 96A version 1.1 or 96MA software or higher. A detailed protocol for the PSQ96 or PSQ96MA has been described previously.
2. PSQ cartridge, capillary dispensing tips or nucleotide dispensing tips, and reagent dispensing tips for hsPSQ96 and hsPSQ96A (Biotage).
3. Pyrosequencing HS reagent kit (Biotage).
4. DNase and RNase free 18.2 mΩ water.
5. Microcentrifuge.

METHODS

Pyrosequencing is based on sequencing by synthesis. The assay takes advantage of the natural release of pyrophosphate whenever a nucleotide is incorporated onto an open 3′ DNA strand. The released pyrophosphate is used in a sulfurylase reaction releasing ATP. The released ATP can be used by luciferase in the conversion of luciferin to oxyluciferin. The reaction results in the emission of light, which is collected by a CCD camera and recorded in the form of peaks, known as programs. When a nucleotide is not incorporated into the reaction, no pyrophosphate is released, and the unused nucleotide is removed from the system by degradation through apyrase. This four-enzyme process is performed in a closed system in a single well.

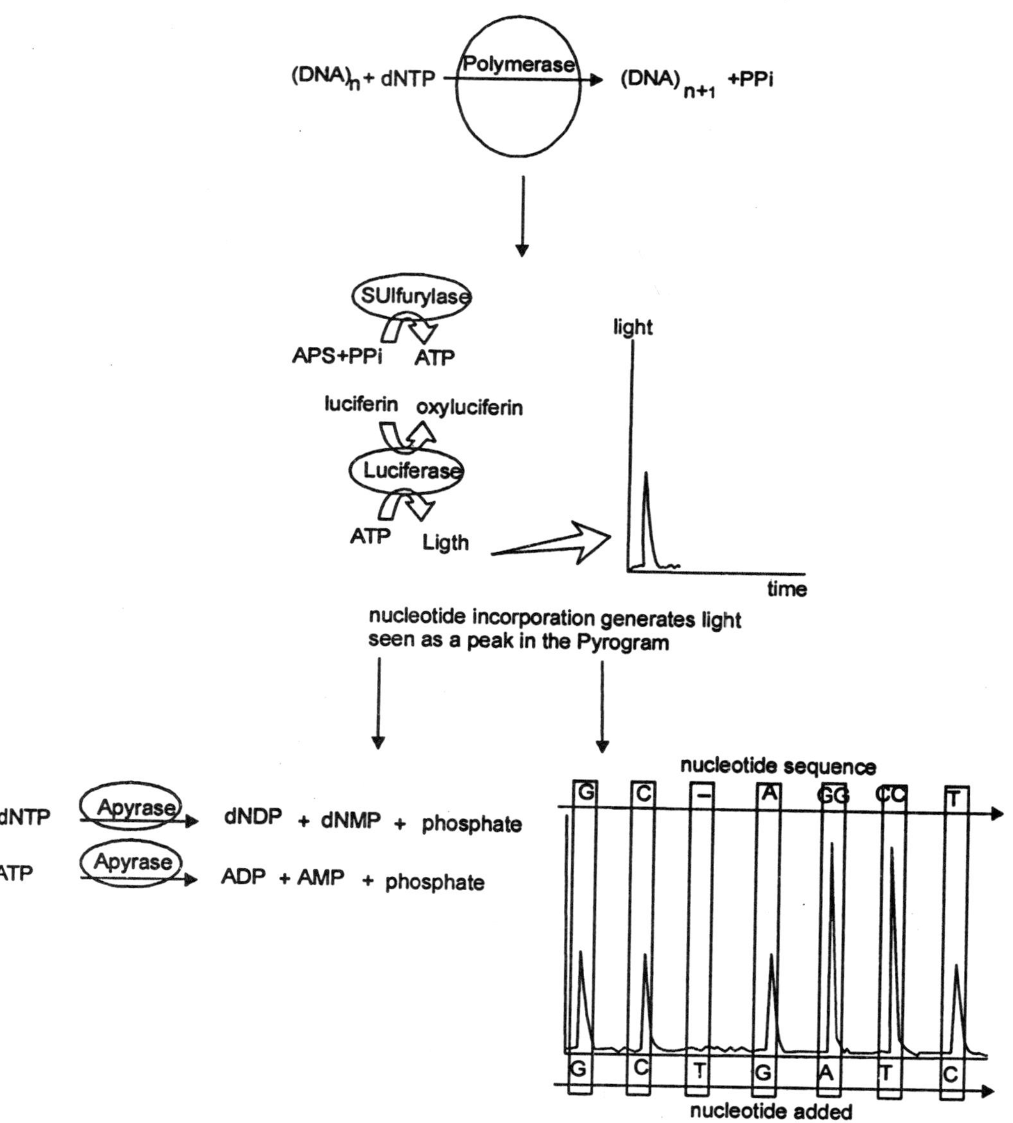

Fig. 3.1. The Pyrosequencing reaction. A modified ATP is used for the nucleotide dispensations to prevent its direct use by luciferase in the reaction.

Bisulfate Conversion of Original Genomic DNA

1. Before the bisulfate conversion, a normal genomic DNA or low-methylation sperm DNA can be treated with the *SssI* methylase to generate a methylation positive control DNA sample (*i.e.*, high-methylation DNA): 1 μg DNA is incubated with 1 unit of *SssI* methylase and 160 mM *S*-adenosylmethionine at 37°C overnight.
2. Treatment of genomic DNA with sodium bisulfate (NaHSO): Add 5 μL 2M NaOH (0.2 M final) into 45 μL DNA solution (0.5 to 1 μg) and incubate tube at 37°C for 15 min; and then add 550 μL methylation solution composed of 520 μL 4 M NaHSO (pH 5.0 M) and 30 μL 10 mM hydroquinone (final 0.5 mM), and invert tube to mix well. Slowly add 300 μL mineral oil to avoid evaporating and incubate at 55°C for 14 to 16 hours under dark light.
3. Clean-up the DNA with Qiagen kit: Place one GFX column (Amersham-Pharmacia Biotech, Uppsala, Sweden) in a collection tube and add 500 μL capture buffer to the GFX column. Remove the mineral oil from DNA sample tube and transfer the posttreatment DNA sample to the GFX column (dividing 600 μL of each sample into three columns). Mix thoroughly by pipetting the sample up and down 4 to 6 times and then centrifuge at 14,000 rpm for 30 s. Discard the flow-through by emptying the collection tube.

   Place the GFX column back inside the collection tube and add 500 μL wash buffer to the column. Centrifuge at 14,000 rpm for 30 to 60 s, discard the collection tube, and transfer the GFX column to a fresh 1.5-mL tube. Apply 50 μL elution buffer (IX Tris-EDTA buffer, pH 8.0) (or DNase- and RNase-free water) directly to the top of the glass fiber matrix in the GFX column and then incubate the sample at room temperature for 1 min. After centrifugation at 14,000 rpm for 1 to 2 min, the purified DNA is recovered.
4. Final modification of DNA with NaOH: Add 9 μL 2 M NaOH into 47 to 50 μL eluted DNA at room temperature for 5 min, and then add 400 uL 100% ethanol and 15 μL 3 M sodium acetate (pH 5.5), invert tube to mix well for rich precipitation of DNA. Keep the DNA sample on ice for 30 min, and then centrifuge at 14,000 rpm for 20 min at 4°C. After centrifugation, pour off solution, and resuspend DNA pellet in 50 μL water.
5. Store the purified bisulfate-converted DNA in –20°C for the methylation PCR assay.

PCR Primer Design

1. Any primer design software, freely available or custom purchased, may be used to design PCR primers for Pyrosequencing. The polymorphism may be in any position of the PCR amplicon except the PCR primer sequence itself, from only 1 base away from the 3′ end of the PCR primer sequence to the center of PCR amplicon. SNPs, indels, repeats, and so forth, do not require specific PCR primer design modifications.

   For the DNA methylation assay, the primer design strategy is just like designing a C/T SNP, but it can include as many CpG dinucleotide loci as the assay capability allows. Currently, the Pyrosequencing run can be carried out for up to 70-bp sequence of interest covering ~20 CpG loci.

2. Primers should be between 15 and 30 bases long, with an optimum size of 20 bases, ideally with a GC:AT ratio around 50% (although not essential, as you are at the mercy of the location of the polymorphism).
3. Amplicon sizes less than 300 bp are optimal, amplicon sizes 300 to 500 bp give acceptable results, amplicon sizes greater than 500 bp will require optimization during the processing of the PCR product for Pyrosequencing. Amplicon sizes of 100 to 200 bp are suitable for most template sources, including fragmented DNA.
4. Care should be taken to avoid any possible template loops from primers or the single-stranded amplicon doubling back on themselves, as these can lead to background problems during the Pyrosequencing assay.
5. Optimum primer melting temperature (T) is 60°C; however, again, the position of the polymorphism determines the ability to design optimum primers and 50°C to 69°C will work. The individual primers should ideally have T values within 2°C of each other to allow effective optimization of the PCR.
6. Primer specificity should be checked by screening the primers across available human genome sequence using the NCBI Blast program (http://www.ncbi. nlm.nih.gov/blast). Extra care should be taken when designing assays for gene family members (*e.g.*, cytochromes) or genes with known pseudogenes (*e.g.*, DHFR), as cross-hybridization of primers can lead to high background, reduced signal, and/or false-positive results.
7. One primer needs to be biotinylated at the 5' end. Which primer to biotinylate is dependent on the Pyrosequencing primer orientation.

Pyrosequencing Primer Design

1. The entire PCR amplicon sequence, including forward and reverse primer sequences, is required to generate the optimum Pyrosequencing primer. It is recommended that the users purchase the custom Pyrosequencing primer design software.
2. Unless multiplexing is required, the software should be defaulted to find both forward and reverse primers to improve the likelihood of obtaining the optimum primer sequence. The software will list all possible forward and reverse primers by score. A score of 100 is "perfect," scores between 90 and 100 are considered "high." Often, "medium" scores yield usable primers, as certain scoring parameters are more critical than others. Template loops likely cause background will be highlighted by the software.
3. The orientation of the Pyrosequencing primer will determine the PCR primer to be biotinylated. Forward Pyrosequencing primers require a biotinylated reverse PCR primer; and reverse Pyrosequencing primers require a biotinylated forward PCR primer.

PCR Optimization

1. Primer optimization of magnesium concentration and temperature should be carried out in advance for new assays. Ideally, a gradient PCR with different magnesium concentrations should be performed, if a thermocycler with a gradient block is available. If a pre-made PCR mix is used, only temperature optimization need be performed. An example gradient setup based on a 96-well PCR block with gradient function follows :

Mastermix

130 µL AmpliTaq Gold PCR mastermix (Applied Biosystems, Foster City, CA)

Forward primer (10 pM final concentration)

Table 3.1. Primer Details for MLH1 and DPYD Methylations.

Gene	*PCR stage*	*Orientation*	*PCR primers (5′– 3′)**	*Sequence to analyze*
MLH1	First round	Forward	tttTtTaaTtTtgtgggttgTtggg	
		Reverse	AAaAAccacaaAaAcaAAAccaa	
	Second round	Forward	5′-Biotin/TtgTTcgTtaTTtagaaggatatg	
		Reverse	tctActcctattAActAAatattc	
	Pyrosequencing	Reverse	TcgTtaTTtagaaggatatg	5'-CGGACAGCGATCTCT AACGCGCAAGCG-3′
DPYD	First round	Forward	tttagTagtttagagattaaaggTTagt	
		Reverse	AAAccatAAcaAtAcctacaAtc	
	Second round	Forward	ggTtgaaTtgggaagg	
		Reverse	5′-Biotin/aAtctAccaAtAacaaaccctc	
	Pyrosequencing	Forward	ggTtgaaTtgggaagg	5′-CGGGCTGCGGCTGC CGTGCCTGGCGCGGG AGCCG-3′

*Uppercase letters denote a transition of T or G to A.

Reverse primer (10 pM final concentration)

13 µL DNA

Up to 260 µL with 18.2 mΩ water

Add 20 µL mastermix to row of a 96-well plate or 12 0.2-mL tubes and place the gradient block (ensure samples cover a continuous row).

PCR Program [based on an MJ Research (Reno, NV) gradient block]

93 °C for 20 min (or appropriate temperature/time to activate Taq)

30 cycles of:

94°C for 30 s

55°C to 72°C for 30 s

72°C for 30 s

Then:

72°C for 5 min

Store at 4°C.

2. The gradient PCR should be visualized using a 1% or 2% agarose gel. The optimal temperature should give the brightest single band at the appropriate amplicon size. Care should be taken to avoid temperatures where a smeared or multiband product

can be seen, as these can increase Pyrosequencing background or reduce specificity if coamplifying a different DNA region. Where several temperatures of equal band intensity are available, the highest temperature should be picked to ensure specificity.

PCR for Pyrosequencing

1. Care should be taken to avoid contamination. The bench area should be swabbed with 70% ethanol or 5% bleach solution before each PCR set-up and barrier tips should be used for all pipetting steps (if a robot workstation with fixed tips is used, tips should be cleaned in 5% bleach solution between every DNA dispensation and between every 96-well plate mastermix dispensation).
2. 1 μL of (1 to 5 ng) DNA (depending on source); should be dispensed into an unskilled 96-well PCR tray. At least one well should not contain DNA to act as a negative control.
3. A 20-μL PCR is ideal for Pyrosequencing; however, if the PCR product is especially strong or wide-peak pyrograms occur, a 10-μL reaction will work well. For a 20-mL reaction based on ABI AmpliTaq Gold PCR mastermix.

   10 μL ABI AmpliTaq Gold PCR mix

   Forward PCR primer (10 pM final concentration)

   Reverse PCR primer (10 pM final concentration)

   Up to 19 μL with 18.2 mΩ water

   1 μL template
4. The PCR plate should be well sealed using a silicon mat or adhesive film. The following PCR program should be run.

   93°C for 10 min (or relevant temperature/time for Taq activation) 55 cycles of:

   95°C for 30 s

   X°C for 30 s (based on gradient-derived annealing temp)

   72°C for 30 s

   Then:

   72°C for 5 min

   Store at 4°C.
5. It is possible to directly use the PCR product for Pyrosequencing; however, it is advisable to check the product and the negative control on a 1% to 2% gel to ensure the reaction has been performed successfully and no contamination is present. Contamination is identifiable at the Pyrosequencing stage; however, it is cheaper and faster to run an agarose gel than to process and run a contaminated/failed Pyrosequencing plate. 96-Well plates should be briefly centrifuged and the lid removed with care to prevent sample aerosol and inadvertent cross-contamination.

   Typically, 5 μL of the negative control and 5 μL of 5 to 6 wells should give an idea of the success of the PCR. The Pyrosequencing will not be affected by the reduction in volume in these wells. Because of the unusually large number of PCR cycles, some smearing may be visible on a gel, even if the optimum annealing temperature has been used. At this stage, the smearing typically does not affect the Pyrosequencing reaction if the PCR primers are specific and the negative control does not contain product.

6. The PCR product can be stored at 4°C until needed. PCR trays should be briefly centrifuged as condensation may occur on the lid, which is a possible source of post-PCR contamination.
7. In the DNA methylation assay, the PCR products may be obtained by a two-round nested PCR because of the insensitivity of PCR primers to the DNA methylation status of starting DNAs. Specifically in the examples of MLH1 and DPYD methylation assays, the first-round external PCR reaction U was carried out for 40 cycles using AmpliTaq Gold PCR master mix 5 pmol of each primer, and 10 ng of the bisulfate-converted genomic DNA in a 20-μL reaction.

   The external reaction was run at 94°C, 55°C, and 72°C for 1 min each, and the second-round nested reaction was run for 55 cycles at 94°C, 60°C, and 72°C for 1 min each for *MLH1* and 30 s each for *DPYD*. The nested reaction was carried out with 5′-biotinylated reverse or forward primers in a 20-μL reaction containing 2 μL of 1:5 diluted first-round PCR products and 4 pmol of each primer.

PCR Processing for Pyrosequencing

This protocol assumes the use of a streptavidin/Sepharose bead setup for Pyrosequencing on an hsPSQ96 or an hsPSQ96A system. The magnetic bead processing method for the PSQ96 or PSQ96MA is described elsewhere.

1. A 96-well Pyrosequencing plate containing Pyrosequencing primer mix should be set up.
2. The small volume readily evaporates, if the setup time is longer than 10 to 15 min; cover the plate with adhesive film. Primer plates can be aliquoted in advance and stored at 4°C. It is advisable to allow them to reach ambient temperature and briefly centrifuge them before use after storage.
3. Add 70 μL of Sepharose bead mix to each well of the PCR product. Replace silicon lid/adhesive film securely.
4. Shake the 96-well plate for 5 min at room temperature. If using the Eppendorf thermomixer, 1400 rpm is the optimum speed. This allows the streptavidin-coated Sepharose beads to anneal to the biotin tag on the PCR primer. Use the plate immediately; if the plate is allowed to sit, the beads will settle to the bottom of the wells and will not be accessible to the vacuum tool. If settling has occurred, briefly return the plate to the shaker to disperse the beads.
5. Align reagent troughs, PCR product/bead mix tray and Pyrosequencing primer tray.
6. With the vacuum switched OFF, shake the vacuum tool tips into clean 18.2 mΩ water. Discard water, refill trough, and switch the vacuum on. Place filter tips into trough until all water has been removed (~30 s).
7. Place filter tips into the wells containing the PCR/bead mix. Ensure all liquid has been removed from the tray by slightly rocking the vacuum tool can prevent surface tension from causing liquid to remain in the wells. The beads attached to the biotin primer will prevent the PCR product from going through the filters.

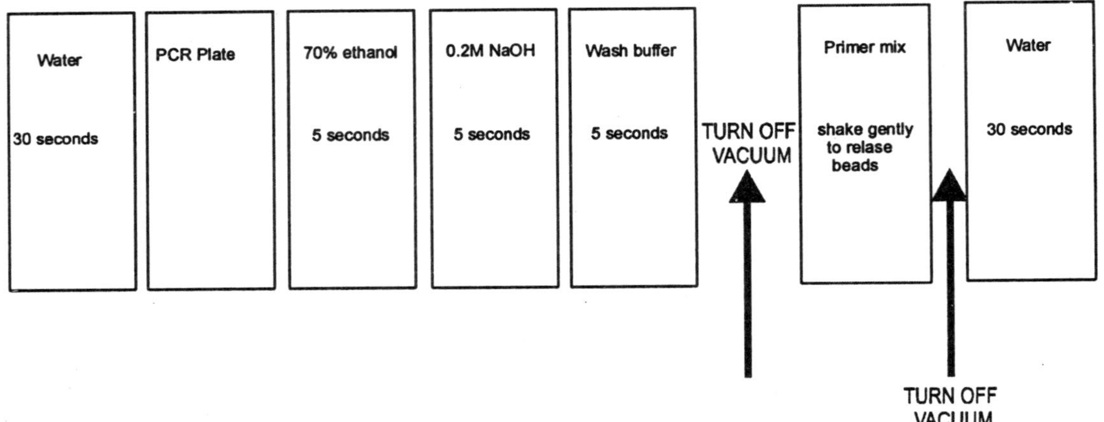

Fig. 3.2. Reagent layout for processing PCR plates for Pyrosequencing. The same orientation is important for the PCR and Pyrosequencing plates.

8. With the vacuum still on, place the filer tips in the 70% ethanol. Wait a few seconds until a good flow of liquid is seen through the tubing and allow the tips to suck up ethanol for 5 s. Repeat with 0.2 M NaOH and washing buffer. The NaOH denatures the DNA; therefore, only single-stranded PCR product remains adhered to the filter tips.
9. Switch the vacuum off or remove the vacuum hose from the vacuum tool and place the filter tips into the Pyrosequencing plate containing the Pyrosequencing primer/annealing buffer mix. Residual vacuum will cause the primer mix to be sucked up through the tips so ensure it is fully off. Gently rock the tips in the wells to disperse the PCR product.
10. Place the Pyrosequencing plate onto a heating block at 80°C for 2 min. Ensure the plate sits on the Pyrosequencing plate adaptor with the corresponding lid (or "iron") placed over the plate to prevent evaporation. After 2 min, remove from heating block and place on a bench surface to cool. Once the plate is cool to the touch, cover with an adhesive seal (unless it will be run within 10 to 15 min) to prevent evaporation. If evaporation has occurred, adding 12 μL of annealing buffer will rescue the plate. Covering the plate while it is too hot will cause condensation on the lid, which can lead to cross-contamination of the wells.
11. Processed plates can be stored at 4°C until needed.

Pyrosequencing

Entering Assay Details

1. Open the Pyrosequencing software. A user name and password is required. This is usually set up with instrument installation. Individual or group-wide passwords can be used.
2. If the assay is not already entered into the software, on the left of the screen click "simplex entry". In the menu tree to the right of the simplex entry icon, scroll to the top, right click over "simplex entry," and select "new entry."

3. The required fields are a unique name for the assay (usually gene/SNP name) and a sequence to analyze. Usually, 5 to 6 bases after the SNP position provides enough information for the assay. SNPs should be denoted as, for example, T/C (triallelic or tetra-allelic SNPs can also be entered, *e.g.,* G/A/T or G/A/T/C) and indels as, for example, [GATC]. Short repeats should be entered as a series of indels, for instance, [TA][TA][TA]. Clicking "dispensation order" will automatically generate the least amount of nucleotide dispensations required for optimum genotype information. The dispensation order can be edited manually by typing in the dispensation order field, which is useful for troubleshooting problem assays.
4. Select "show histograms" and the predicted pyrogram pattern will be displayed on the right. The default screens show both homozygous patterns and the heterozygous pattern. It is possible to scroll through histograms on the lower panel, useful if multiplex of multiple indels are to be analyzed, and so forth. Selecting individual or all predicted histograms on the box below the dispensation order and clicking "export" opens the histograms in a browser window where they can be printed or saved.
5. Click "save." At this stage, the parameters can no longer be altered, a duplicate setup with a unique name will need to be created for any alterations to the assay.

Entering an SNP Run

1. Select the "SNP run" icon on the far left of the screen.
2. On the menu tree, right-click over "SNP run" and select "new SNP run".
3. The essential parameters on the setup tab are a unique run name (*e.g.,* gene/SNP/sample set/date) and the active well map. The default plate map is for a full 96-well plate. Individual wells can be selected (hold down control for nonadjacent wells), clicking the "activate wells" button will gray-out unused wells.

   In addition, instrument parameters must be selected from the drop-down menu. Usually, "instrument parameters" is a default file; however, care should be taken to ensure the appropriate parameters are selected for nucleotide or capillary dispensing tips, as they are not interchangeable. Parameter set-up instructions can be found with the dispensing tip packaging.
4. The essential parameters on the setup tab are to select the SNP assay by clicking on the drop-down menu under "simplex" and selecting the assay name and to fill the plate map by clicking and dragging over the active (white) wells.
5. Once the run has been set up, click "save." This can be edited post-save, and changes can be re-saved.
6. If multiple plates of the same assay are to be run, on the menu tree right click over the SNP run you have just entered and select "duplicate SNP run." The only parameter necessary is a unique run name.

Individual Plate Run for PSQhs96 and PSQhs96A

1. On the SNP run setup page click the "view" tab and select "run." This will list the appropriate volumes of nucleotides, enzyme, and reagent needed for the individual run.

2. Set up the cartridge holder. It is essential that all nucleotide/capillary and reagent tips are clean before use. To check for blockages in the nucleotide and reagent tips, fill with 18.2 mΩ water and apply pressure over the top of the tip. Water should squirt from the bottom of the tip. If this does not occur, try filling/emptying the tip several times with water and re-try forcing liquid through. If the tip remains blocked, discard. For nucleotide dispensing tips, do *not* force water through them. The hydrophobic disks may dislodge and prevent the tip from functioning. Rather, ensure the tip has been rinsed several times in water and has been stored in a clean, lint-free environment.

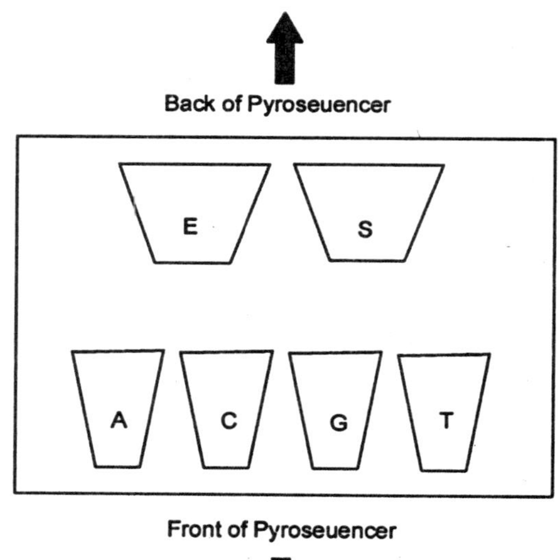

Fig. 3.3. Reagent and nucleotide cartridge orientation. E, enzyme; S, substrate; A, C, G, and T, nucleotides. A modification of dATP is used to prevent the nucleotide from being a direct source for the oxyluciferase.

3. Nucleotides, enzyme, and substrate are sold as a reagent kit. Each vial is clearly labeled. Nucleotides come as a solution, enzyme and reagent are lyophilized and should be resuspended with 18.2 mΩ water before use, and the volumes vary per kit and are clearly marked on the labels. The enzyme and substrate both dissolve rapidly and no mixing or shaking is required. Indeed, this should be avoided as air bubbles in the liquid could cause tip blockages or inconsistent dispensation. Unused resuspended enzyme and substrate can be stored at –20°C for future use.
4. If using the nucleotide dispensing tips, the nucleotides should be microfuged for 10 min and care should be taken to not aliquot from the bottom of the vial in case any precipitate is present, which could cause tip blockage. For all dispensing tips, it is recommended that non-barrier pipette tips are used as fibers can cause tip blockage.
5. If the capillary dispensing tips are used, the nucleotides should be diluted 1:1 with Tris-EDTA buffer (pH 8.0), and mixed well before use.
6. The nucleotide and reagent dispensing tips should be filled according to the volumes suggested by the software. Nucleotide dispensing tips should be filled by *doubling* the amount suggested by the software. Care should be taken not to pipette air bubbles and to gently angle the liquid down the sides of the tips.

   Capillary and reagent dispensing tips can allow minute air bubbles without affecting their performance. With nucleotide dispensing tips, it is extremely important to check all of the tips for air bubbles. These can usually be removed by gently tapping the sides of the tips until the air bubbles surface, or, if necessary, dislodging them with a clean pipette tip.

7. A test plate should be run after each cartridge refill. This is extremely important when using the nucleotide dispensing tips; and three or four test plates should be run in succession to ensure no blockages are present. The substrate reagent-dispensing tip is also prone to blockage if the substrate is allowed to sit in the tip at room temperature for any length of time. To run a test plate: place the cartridge in the pyrosequencer and the test plate in the 96-well plate platform.

   On the far left of the software screen, select the "instrument" tab, and then select "instrument" and "manage." Click "test." A warning will appear asking you to check that you have placed the test plate, into the instrument. Click "ok." The test takes approximately 30 s. Remove the plate. In the center there should be six wells with liquid: four nucleotides, a reagent, and a substrate. If there are fewer than six wells with liquid, a blockage has occurred.
8. Remove the adhesive film carefully from the Pyrosequencing plate and place it in the pyrosequencer. Close all levers and click "run" on the plate run setup. The pyrosequencer will now automatically dispense enzyme, substrate, and nucleotides in the predetermined dispensation order. The progress of each individual well can be monitored at any time by selecting the relevant well on the 96-well plate map on the screen.
9. To automatically analyze the data once the run has completed, select "analyze all."

Batch Runs Using the PSQhs96A

1. SNP runs should be set up, saved, and closed.
2. Select the "Batch run" icon on the far left of the Pyrosequencing software, on the menu tree right click over "batch runs" and select "new batch run." One to 10 plates can be run in each batch. A unique name for each batch must be provided, and the instrument parameters must be selected for each batch. If barcoded plates are not used, uncheck the "barcode" field.
3. On the far left of the software, click on the "SNP runs" icon. From the menu tree, click and drag your SNP runs into the 1 to 10 slots on the batch window.
4. On the top menu bar, select "batch" and "setup information." This will open a browser window (may take a few seconds) with the total amount of nucleotides (which should be *doubled* for the capillary dispensing tips), enzyme, and reagents needed for the entire batch.
5. The cartridge should be set up. The dispensing tips should be cleaned between every batch, and a test plate should be run before every batch.
6. Remove the adhesive film from the Pyrosequencing plates and stack them (check that the plates can be lifted free without sticking to the lower plates, occasional warping may occur, causing plates to stick together, which jams the robotic arm). Place plates in the robot stacker unit. The correct plate orientation is shown on the top of the stacker unit. Ensure the plates lie flat on the base of the stacker unit and are between the grooves. Plate 1 on the batch setup should be on the top, plate 10 (or the last plate in the batch setup) should be on the bottom.
7. Ensure the stacker unit is firmly pushed into place. The nucleotides will not dispense if the unit is only partially home.

MLH1 Sequence to analyze C/TGTTTGC/TGC/TGTTAGAGATC/TGTTGTTC/TG

Theoret cal outcomes for 50% methylation C/TC/T C/T C/T C/T

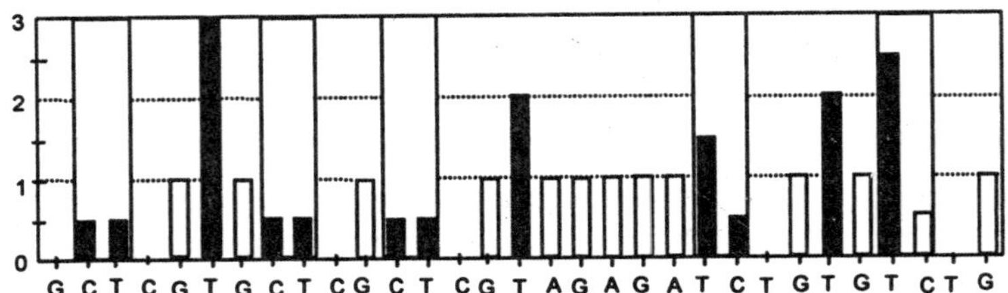

Experimental pyrogram for a methylated DNA sample

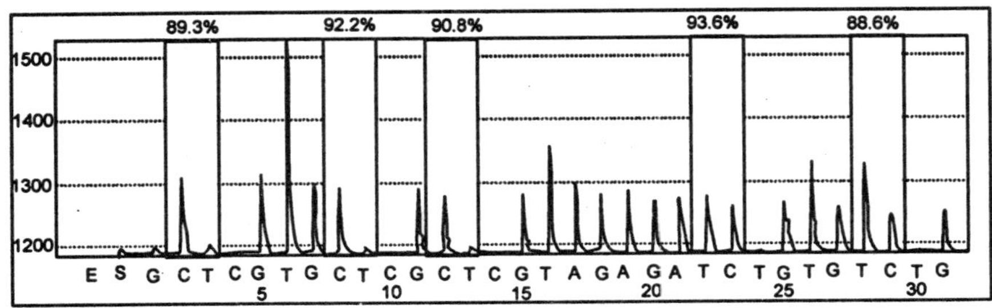

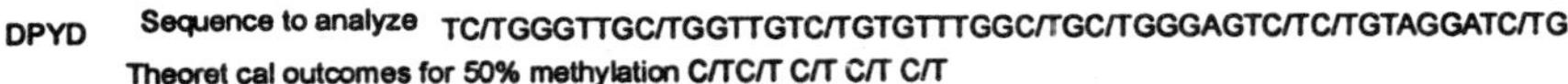

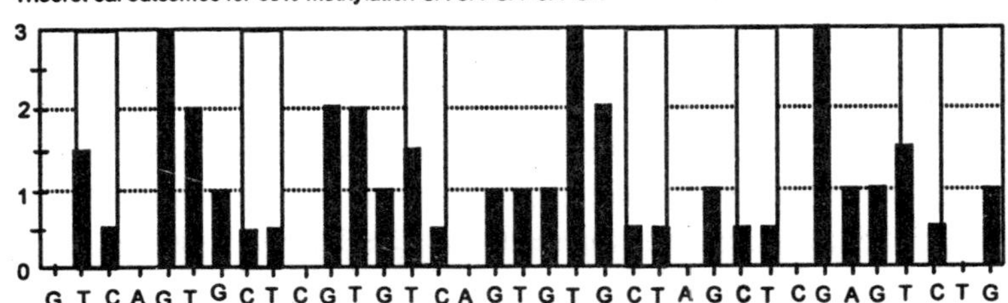

Experimental pyrogram for a methylated DNA sample:

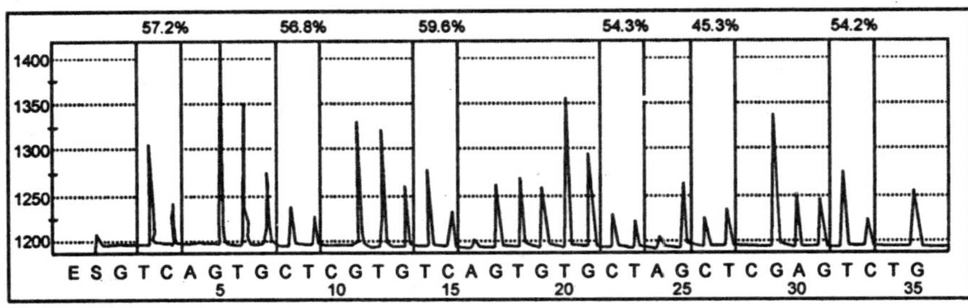

Fig. 3.4. Predicted histograms and typical programs for MLH1 and DPYD promoter methylation assays.

8. Click the "play" icon. Plates will automatically load and be discarded throughout the batch.
9. Plates will automatically be analyzed by the software when run in batch mode. They can be accessed from the batch setup window or from the individual SNP run files.

Analysis of Pyrosequencing Results

1. Once the Pyrosequencing run has been analyzed by the software, the 96-well plate map will be color-coded according to the result. Blue indicates a well in which the pyrogram matches one of the predicted histograms and a genotype can be accurately called. Orange indicates a possible match with a predicted histogram; however, human intervention is required to validate the call. Red indicates a failed well, where no match with a predicted histogram can be found.
2. The well(s) where no DNA was added in the PCR should automatically be scored failed. Nonspecific peaks in the negative control(s) may be evident. These are likely to be caused by looping of the internal primer and can aid troubleshooting assays by identifying whether the internal primer is the culprit for background peaks.
3. Samples checked (orange) for human intervention can be edited by clicking on the specific well and opening up the predicted histograms from the "histogram" tab on the right. If a genotype consensus is reached, the sample call can be manually edited by right-clicking over the genotype above the pyrogram. Genotypes can be selected, and pass/check/fail can be altered. The well on the plate map will show a dark circle, indicating that manual editing has taken place.
4. The data can be exported as a report, as a tab delimited file, or as an XML file. Custom export options are also available. The export function can be accessed by selecting "report" and then saved as the appropriate file type. Selected wells or the entire plate can be saved/exported. Pyrograms (all or selected) can also be saved or printed, up to six per page.
5. In the DNA methylation assay, the Pyrosequencing run should be analyzed by the Allele Quantification function within the software. The allele C and T peak percentage data can be exported along with the quality control marks (pass/check/fail) for each of the C/T loci. An average percentage of the C allele (*i.e.*, C/T ratio) from three experiments can be used for the DNA methylation status at each CpG locus.

For the overall DNA methylation status of a gene, the mean C/T ratio of all CpG loci tested for that gene can be calculated. When the mean C/T ratio is less than 0.05, it is defined as unmethylated (UM). Otherwise, 0.05 to 0.20 is denoted as low methylation (LM), 0.21 to 0.50 as medium methylation (MM), and greater than 0.50 as high methylation (HM). For example, in our study, overall there were 19 of 48 (39.6%) and 20 of 49 (40.8%) tumors that were defined as DNA-methylated (C/T ratio $\geqq$ 0.05) in DPYD and MLH1, respectively. Among these methylated tumors, MM and HM were observed in 2 and 4 (12.5%) for DPYD and 5 and 9 (28.6%) for MLH1, and the methylation levels across multiple loci for these two genes was similar ($p > 0.1$) (*i.e.*, so-called dichotomous trait).

Table 3.2. DNA Methylation Result of 52 Colorectal Cancer Patients

	CpG locus	*No. of methylated DNA samples*	*Mean methylation level of methylated DNA samples*	*UM*	*LM*	*MM*	*HM*
	1	13	0.37				
	2	18	0.29				
DPYD (n = 48*)	3	22	0.35	29	13	2	4
	4	24	0.24				
	5	13	0.29				
	6	13	0.35				
	1	15	0.56				
	2	23	0.44				
MLH1 (n = 49*)	3	15	0.60	29	6	5	9
	4	24	0.36				
	5	20	0.54				

UM, unmethylated; LM, low methylation; MM, medium methylation; HM, high methylation.
*Four and 3 samples failed in the DPYD and MLH1 assays, respectively.

NOTES

1. Pyrosequencing has been successfully performed on DNA from cell lines, blood, serum, plasma, paraffin-embedded tissue, frozen tissue, and whole-genome-amplified product. In addition, complementary DNA from various sources has also been successfully pyrosequenced.
2. The actual starting concentration of DNA depends on the quality of the template. DNA extracted from blood is highly accessible for PCR and consequently 0.5 to 1 ng can produce reliable, reproducible product. DNA from plasma, serum, frozen tissue, and whole-genome-amplified methods tend to be fragmented and more template may be necessary for optimum PCR. A test in advance of serial dilutions of the template DNA should be performed with the PCR primers to find the appropriate concentration that gives a clean, high-yield PCR product. In addition, for the quantitative methylation assay, the starting amount of the DNA should be larger than that for regular SNP assays. At least 10 ng of bisulfate-converted DNA should be used in the methylation PCR.
3. Pre-made mixes of buffer, magnesium, dNTPs, and Taq polymerase are recommended as they provide consistent results and minimize pipetting errors.
4. Non-hot-start Taq is also suitable; however, primer dimers are less of a problem with hot-start Taq and this is recommended.
5. All solutions should be made using 18.2 mΩ water. Solutions other than the NaOH and 70% ethanol should be sterile filtered prior to the addition of Tween-20. 10X washing buffer, annealing buffer, and NaOH can be made and stored at room temperature for dilution to the working concentrations. All solutions can be stored at room temperature.

6. Problem template loops will also be flagged in the Pyrosequencing primer design software.
7. This protocol is based on simplex assays; however, multiplexing with up to three internal primers can be performed, either from the same PCR product or from different PCR products. The primer design software can only determine one internal primer at a time; often the first-choice primers for each will not be useful in a multiplex assay where the combined sequence to analyze is best designed to generate unique SNP dispensations. In addition, the orientation of the primers is vital for multiplex assays as only one PCR primer can be biotinylated.
8. Critical scoring parameters on the Pyrosequencing primer report:

Misprinting

If the internal primer can anneal to multiple positions within the amplicon, the 3′ ends of the annealed region can incorporate nucleotides leading to incorrect genotype calls or unacceptable background.

Duplex Formation

If the internal primer can dimerize with itself, as for the mispriming, unacceptable background or reduced signal intensity owing to suboptimum primer annealing may result.

Hairpin Loop

If the primer forms secondary structures, the amount of primer available for the reaction is diminished, and reduced signal can result.

Template Loop

Loops of more than approximately four to five GC-rich regions will be flagged by the software, and these primers should be avoided. Loops less than four bases should also be avoided if possible to reduce the likelihood of background.

Noncritical scoring parameters on the Pyrosequencing primer report:

Repeated Base at SNP Sequence

This reduces the score if there are a string of identical bases around the SNP. This is not something that can be controlled or optimized for as the SNP position is not movable. Typically, the pyrograms can accommodate as many as three bases in a row with no problems. Four to six bases may be difficult to read manually as the scale will be affected. More than six repeated bases are not recommended because distinguishing the peak heights becomes very difficult.

Primer Length

Primer lengths longer than 15 bases give a reduced score based on the expense of longer primers. The length of the primer is not critical to the reaction.

9. If an appropriate Pyrosequencing primer cannot be found as the critical scoring parameters are flagged, it is possible to “trick” the software to improve the search.

As the software will only look five bases to either side of the SNP for a suitable primer, entering a fake SNP five bases before or after will extend the region searched. This may help to overcome mispriming and dimer problems. To eliminate template loops, adjusting the 5′ end of the PCR primer that would cause the loop will help (*e.g.*, shifting the primer two to three bases to the left or right or trying a PCR primer in a slightly different region). As only one primer is likely to cause the loop problem, if a primer in the opposite orientation is available (even if not the highest score), this is often the easiest solution.

10. Pre-made PCR mixes are usually a fixed magnesium chloride concentration. If primer conditions are not optimized through temperature alone, extra magnesium chloride may be added to the PCR mix. In addition, problem assays may be improved by the addition of 5% to 10% dimethyl sulfoxide or 1 M Betaine. This will not affect the Pyrosequencing.
11. The mix is for 13 samples, allowing one extra sample for pipetting discrepancies.
12. If a larger volume of DNA is necessary, adjustments can be made to the PCR mastermix (reducing the water volume), or DNA may be dispensed into the plate and allowed to dry down overnight at room temperature. The DNA is reconstituted once the PCR mastermix is added.
13. For multiple primer sets/plate, at least one negative control/primer set should be included.
14. Fifty-five cycles are run to ensure all primers and nucleotides are exhausted and not available to cause background during the Pyrosequencing. If wide peaks occur in the program, reducing the number of cycles to 40 may help to prevent these.
15. Multiple assays can be run in a 96-well plate; indeed, each well could contain a different internal primer. The wells corresponding with the negative controls from the PCR setup should contain internal primer, as this is a valuable troubleshooting method for program background issues.
16. A workstation platform is available from Pyrosequencing, which holds the reagent troughs and plates in specified positions. Any method to hold the reagent troughs stationary is appropriate, for example, rigid plastic tip box lids.
17. If a multiplex assay is to be set up, select the "multiplex entry" icon, right click over "multiplex entry" on the menu tree, and select "new entry." Type in the three separate dispensation orders for each internal primer. The computer-generated dispensation order will give a combined dispensation for the three SNPs. The field requirements here are the same as for the simplex entry except two or three sequences to analyze may be entered.
18. The menu tree for SNP runs can be organized into folders so multiple users can easily access their files. If this has been done, right click over the relevant folder and select "new run."
19. Each well can contain a different simplex/multiplex entry if desired, simply select the entry and click in the appropriate well until all active wells are filled.
20. Pyrosequencing provides specific storage boxes for the tips with the instrument, and more are available from the company if required.

21. To save on plate costs, attach adhesive film to the top of the test plate. The dispensation will occur on the film rather than in the wells, and this can be wiped off and the plate can be reused.
22. If multiple primer sets are used/plate, the negative controls for each primer set should be checked for contamination.
23. The report structure is available in forms readily transferable to most database/spreadsheet systems.
24. Pyrosequencing can generate quantitative data for DNA methylation level. However, the concept of the DNA methylation level can be confused in the literature, because the DNA methylation level may be interpreting either how many molecules with a gene sequence of interest in a DNA sample or how many CpG loci in a gene sequence for a certain DNA sample have been methylated. In fact, most DNA samples from tissue specimens have mixed genomes from a variety of cell types including cells not of interest to the study (*e.g.*, stromal cells), and different cells or tissues can be methylated differentially at a specific CpG locus. The DNA mixture from different types of cells can mask the true level of DNA methylation at a specific CpG site for a specific type of cells. This is particularly important for quantitative DNA methylation studies; and laser capture microdissection and fluorescence-activated cell sorting can provide highly pure cells of specific types.

CHAPTER 4

Manipulating DNA

A broad array of techniques has made it possible to investigate the fine structure of DNA as well as to redesign existing genes. The impact of these techniques cannot be overestimated. It is possible to map and isolate genes from complex eukaryotes—feats that could never have been accomplished by classical genetic methods. Genes can be redesigned to achieve a variety of practical goals. Newly designed genes may be inserted into the same species from which the original unmodified genes came or into other species.

In this chapter we first consider DNA sequencing. Then we explore the different approaches for amplifying and isolating specific genes or gene segments. Following this, we examine the methods currently available for restructuring existing DNA sequences. Finally, we look at some of the major advances in our understanding of gene structure and location that have resulted from the new technology. This part of the discussion focuses on two examples: The mapping of the human globin gene family and the mapping of the gene responsible for the genetically inherited disease, cystic fibrosis.

SEQUENCING DNA

If we are going to manipulate a segment of DNA, it is most useful to know something about its primary structure. The most complete information comes from total sequence analysisr. Initial efforts at sequencing nucleic acids were confined to RNA molecules that could be readily isolated in pure form. The first sequence to be determined was that for tyrosine tRNA from yeast. From roughly 100 pounds of yeast Robert Holley was able to isolate enough of the tyrosine tRNA to carry out a sequence analysis. This historically significant effort required a number of enzymes and chromatographic techniques. We shall not elaborate on this accomplishment because sequencing of RNA is no longer done directly.

In fact, the sequencing of RNA has been replaced by the sequencing of "cDNA," which results from the reverse transcription of RNA into DNA. Two quite different methods have been developed for sequencing DNA. One method, developed by Walter Gilbert and Alan Maxam and involving cleavage of preexisting DNA, uses a chemical approach. A second method, involving premature termination of newly synthesized DNA, uses an enzymatic approach that was developed by Fred Sanger. For pure sequencing, Sanger's is the method of choice. It employs chain-terminating dideoxynucleoside triphosphates to produce a continuous series of fragments in reactions catalyzed by DNA polymerase.

Dideoxynucleoside triphosphates (ddXTPs) resemble deoxynucleoside triphosphates except that they lack a 3′—OH group. They can add to a growing chain during polymerization, but they cannot be added onto, and as a result they act as chain terminators. The DNA being sequenced is mixed with a suitable primer, radioactive dXTPs, DNA polymerase I (PolI), and a small amount of one ddXTP. The primers determine where DNA synthesis starts, and the ddXTP determines the base type where elongation stops.

The products of four separate reaction mixtures, each differing only by the ddXTP it contains, are analyzed in Fig. 4.1. As depicted, reaction 1, using ddATP, contains all fragments with an A terminus; reaction 2, using ddCTP, contains all C terminations; and so on. After the newly synthesized oligonucleotides are separated from the template by denaturation they are fractionated by electrophoresis for a limited time on polyacrylamide gels. The positions of the fragments on the gel are detected by autoradiography. The sequence is read directly from the composite autoradiogram, starting with the fastest moving (smallest) band at the bottom of the gel and moving up. If the first band is in reaction 3, it is a G residue; the next band up, appearing from reaction 4, would be T; and so on. Up to 800 residues can be read from a single gel.

METHODS FOR AMPLIFICATION

Cellular genomes are very large; even *E. coli* contains more than a million base pairs, and eukaryotic genomes frequently contain a billion or more base pairs in one complete genome. Because of their large size it is impractical to fractionate the cellular genome and expect to obtain enough of a particular DNA segment for sequencing or other investigations. Two methods of amplifying defined segments of DNA have been developed. The first of these takes the desired segment and amplifies it *in vitro* with DNA PolI using DNA primers that bind to the ends of the region of interest. This method involves repeated cycles of synthesis and is appropriately named the *polymerase chain reaction (PCR)* method. The second method inserts the DNA segment of interest into a plasmid or virus that can be amplified *in vivo*. We discuss both of these methods because they are both useful in many ways.

AMPLIFICATION BY THE POLYMERASE CHAIN REACTION

PCR entails enzymatic amplification of specific DNA sequences using two oligonucleotide primers that flank the DNA segment to be amplified. The primers must complement opposite strands so that after annealing, their 3′ ends in effect face each other.

The PCR procedure has three steps, which are usually repeated many times in a cyclical manner:

1. Denaturation of the original double-strandsd DNA sample at high temperature.
2. Annealing of the oligonucleotide primers to the DNA template at low temperature (37°C).
3. Extension of the primers using DNA polymerase.

These steps are illustrated in Fig. 4.2. Each set of three steps comprises a cycle. The extension products of one primer provide a template for the other primer in a subsequent cycle so that each successive cycle essentially doubles the amount of DNA. This results in the exponential accumulation of the specific target fragment by approximately 2^n, where n is the number of cycles. The specific target fragment is also referred to as the "short product" and is defined as the region between the 5′ ends of the extension primers.

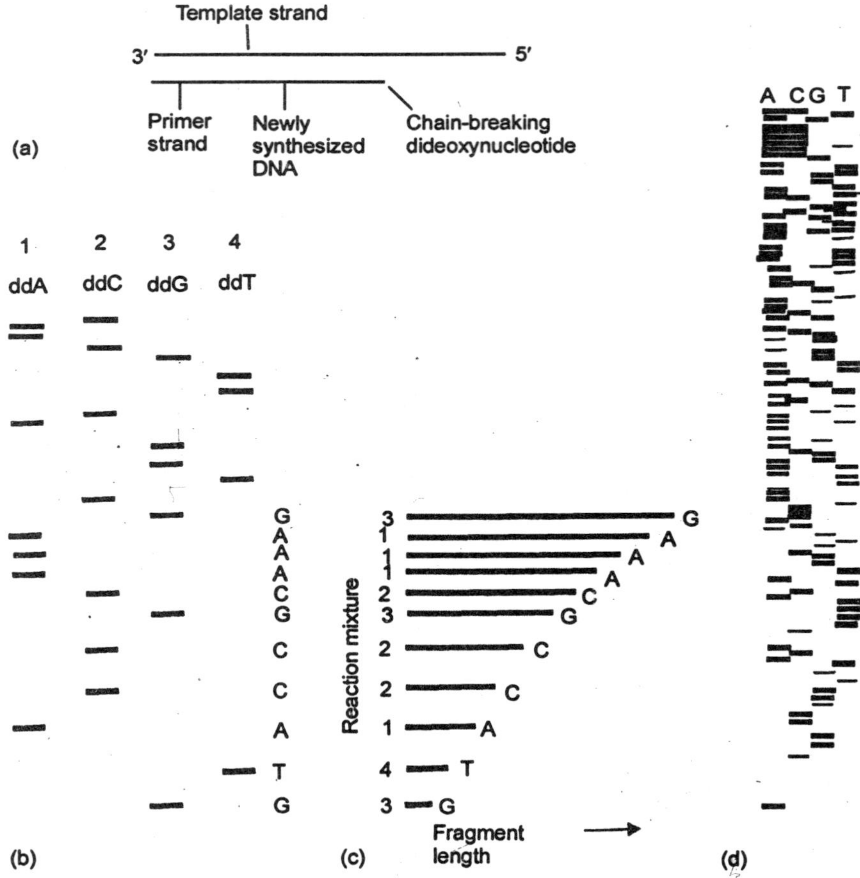

Fig. 4.1. The Sanger dideoxynucleoside method of sequencing DNA. *(a)* A suitable template is chosen, and the primer is chosen so that DNA synthesis begins at the point of interest. The primer is radioactively labeled. In addition to the template-primer complex the reaction mixture contains all four radioactive deoxyribonucleoside triphosphates and small amounts of a single dideoxynucleoside triphosphate. The dideoxy compound serves as a chain terminator. (*b*) After synthesis in the presence of DNA polymerase I, the products of the reaction mixture are separated by gel electrophoresis and analyzed by autoradiography. The gel is run under denaturing conditions in warm urea so that single-stranded fragments separate strictly according to size. For a given dideoxy compound all fragments terminating with that particular base should give rise to bands on the gel. The interpretation of the gel pattern is given in (*c*). The smallest labeled fragment moves the fastest and appears at the bottom of the gel. (*d*) A typical sequencing film. The sequence begins CAAAAAACGG Coutesy of GIBCO-BRL.

Each primer is physically incorporated into one strand of the short product. Other products are also synthesized during the succession of cycles, such as the "long product" of indefinite length, which is derived from the template molecules.

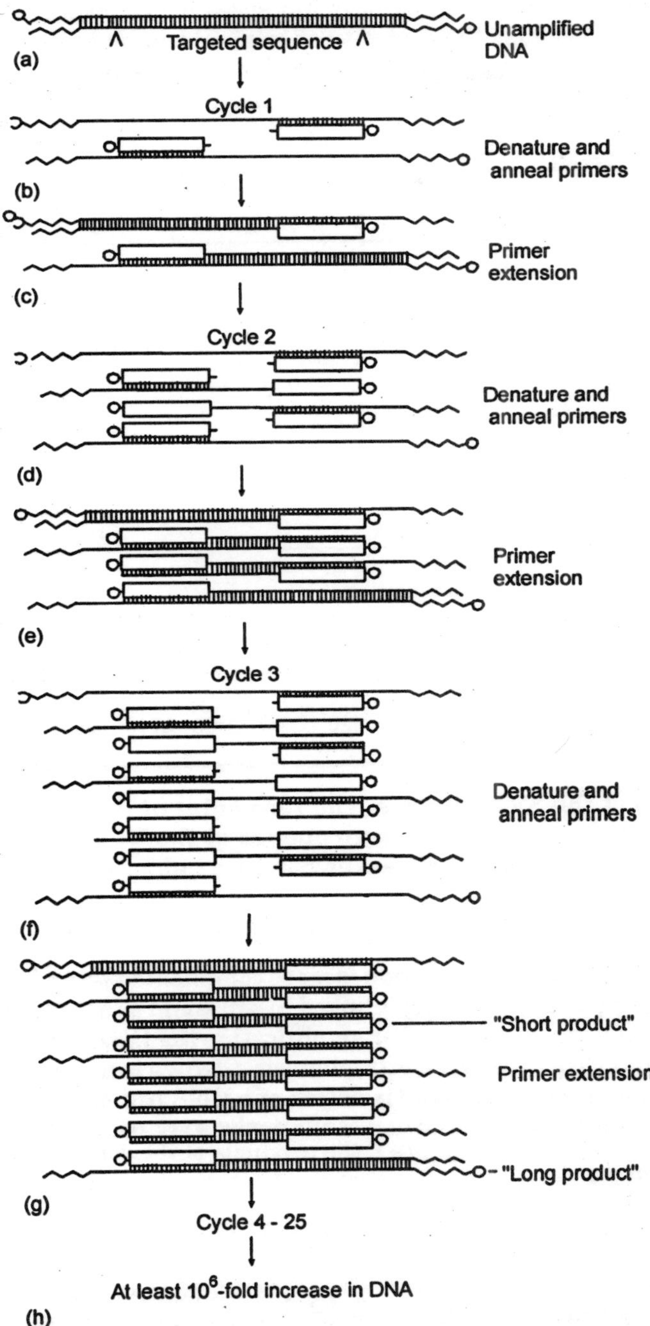

Fig. 4.2. Steps in the polymerase chain reaction (PCR). The DNA to be amplified is denatured and annealed with two oligonucleotides that flank the region of interest. These oligonucleotides (or primers) are extended. Extension continues to the ends of the DNA strands. The products are again denatured and annealed to primers for a second round of extension. This process of denaturation, annealing, and primer extension is repeated many times. The primary product of the reaction is duplex DNA, bounded by the sequences of the priemrs.

However, the amount of long product only increases arithmetically during each cycle of the amplification process because the quantity of original template remains constant. At the end of the PCR process the short product is so overwhelmingly abundant compared with the long product that its purification is not required for most purposes.

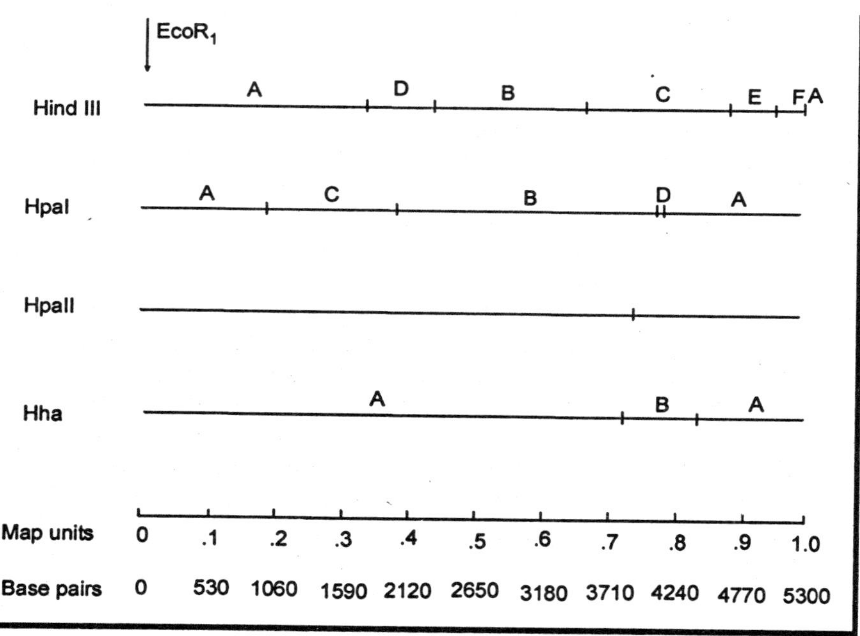

Fig. 4.3. Cleavage map of the SV40 genome. The zero point of the map is the unique *EcoR*1 site. For clarity, the circular genome is shown opened at the R1 site, and the cleavage sites (and resulting fragments) for each restriction enzyme are indicated on a separate line.

DNA CLONING

The second method for DNA amplification is more complicated than PCR, but it has several advantages. DNA to be amplified by cloning is linked to a plasmid or a virus that can be replicated indefinitely in the appropriate host cell. After amplification the DNA of interest can be cut from the plasmid or virus and reisolated by gel electrophoresis. Cloning is not only useful for amplifying a segment of DNA, it can be adapted to the isolation of a DNA segment of interest from a large mixture such as is obtained from the isolation of the entire genome.

Restriction Enzymes

Most of the enzymes that are absolutely essential for cloning were discussed in the previous chapter. The most important enzymes that have not been discussed yet are the *restriction enzymes*. Systematic cleavages of duplex DNA at specific sites requires restriction enzymes. Each species of bacteria harbors a unique restriction enzyme, and hundreds of restriction enzymes with different specificities have been isolated, giving researchers a great deal of choice as to how and where DNA is cut. Some of the most commonly used restriction-enzymes and their recognition sites are indicated in Table 4.1.

Most of these enzymes recognize a sequence of either four or six contiguous base pairs. The cleavage sites are situated so that a blunt-ended or staggered-ended DNA results from the cleavage reaction. As a rule the recognition sites are located on an axis of symmetry so that the freshly cleaved segments have identical structures at their ends. A viral genome cleaved exhaustively with a particular restriction enzyme usually yields several fragments.

Table 4.1. Recognition Sequences and Cutting Sites of Selected Restriction

Enzyme	*Recognition Sequences*	*Enayme*	*Recognition Sequences*
AluI	↓ AGCT TCGA ↑	HpaII	↓ CCGG GGCC ↑
BamHI	↓ GGATCC CCTAGG ↑	KpaI	↓ GGTACC CCATGG ↑
BglII	↓ AGATCT TCTAGA ↑	MboI	↓ GATC CTAG ↑
ClaI	↓ ATCGAT TAGCTA ↑	PstI	↓ CTGCAG GACGTC ↑
EcoRI	GAATTC CTTAAG ↑	PvuI	CGATCG GCTAGC ↑
HaeII	↓ GGCC CCGG ↑	SalI	↓ GTCGAC CAGCTG ↑
HindII	↓ GTPyPuAC CAPuPyTG ↑	SmaI	↓ CCCGGG GGGCCC ↑
HindIII	↓ AAGCTT TTCGAA ↑	XmaI	↓ CCCGGG GGGCCC ↑

Some restriction enzyme cleavage sites for the 5,300 bp (5.3 kb) SV40 virus genome are shown in Fig. 4.3. The duplex fragments obtained after cleavage can be separated according to size by gel electrophoresis. Nondenaturing conditions are used so that the duplex strands stay together. The larger a fragment is, the slower it migrates on the gel. After electrophoresis for a time sufficient to separate the fragments, the gel is stained with a fluorescent dye such as ethidium bromide and viewed under long-wavelength ultraviolet light (long-wavelength UV is used because it does not damage the DNA). Individual fragments may be extracted from the gel for sequencing, PCR amplification, or cloning (described later on). The problem of determining how a set of restriction fragments are normally connected is resolved by determining the sequences by a second set of fragments cut with a different restriction enzyme.

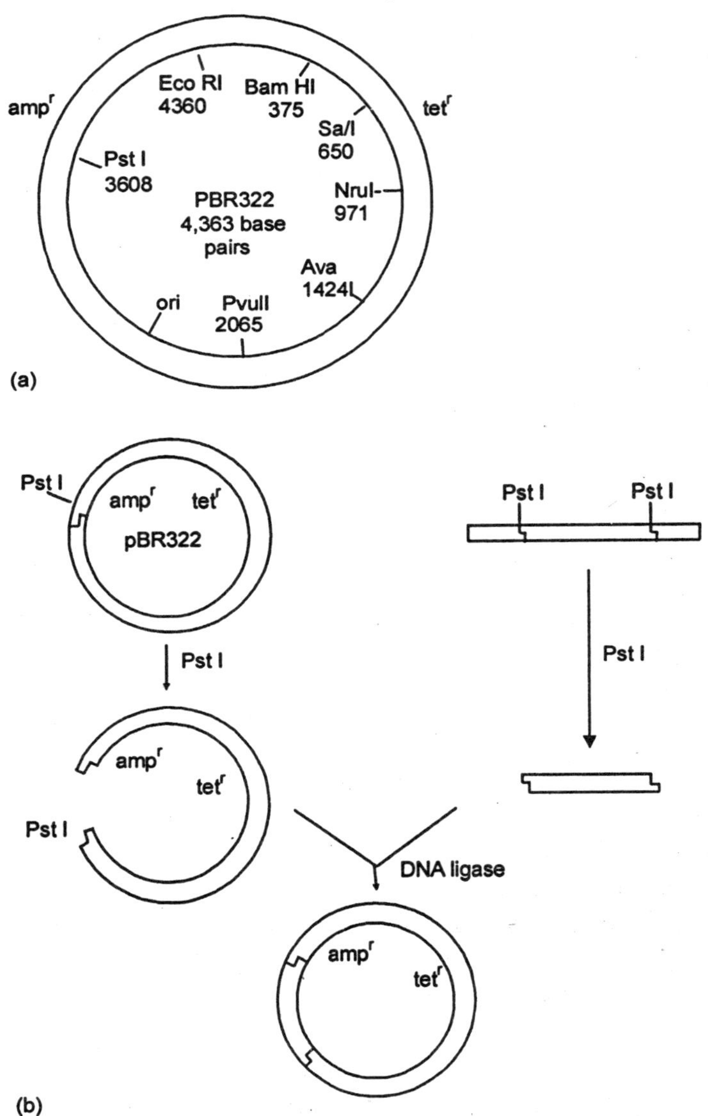

Fig. 4.4. Structure of the pBR322 plasmid *(a)* and construction of a hybrid plasmid containing the pBR322 vector and a segment of foreign DNA *(b)*. For pBR322 the unique sites for various restriction enzymes are indicated. Also indicated are the locations of the tetracycline (*tet*^{r}) and the ampicillin (*amp*^{r}) resistance genes and the origin for DNA replication. The hybrid plasmid is constructed by treating the plasmid and the foreign DNA with the *Pst*I restriction enzyme and mixing the two DNAs together in the presence of DNA ligase.

The overlapping information obtained from the two sets of fragments permits a determination of the complete sequence of the intact genome. The strategy of sequencing overlapping fragments is identical to that used in primary structure determination of proteins.

Cloning of Small Pieces of DNA

In a simple procedure for "DNA cloning," an autonomously replicating plasmid and insert DNA are cut with a restriction enzyme and then the pieces are annealed and covalently joined by the action of DNA ligase. The resulting recombinant molecules are then transfected into *E. coli,* where they replicate. When plasmid vectors are used, a population of permeabilized cells is bathed in the plasmid DNA containing the inserted DNA. Because only a small number of cells become transfected by this procedure, a way to select cells that carry the desired hybrid plasmids is needed. A particularly useful plasmid vector for selectmg transfected cells called pBR322 is itself a hybrid plasmid.

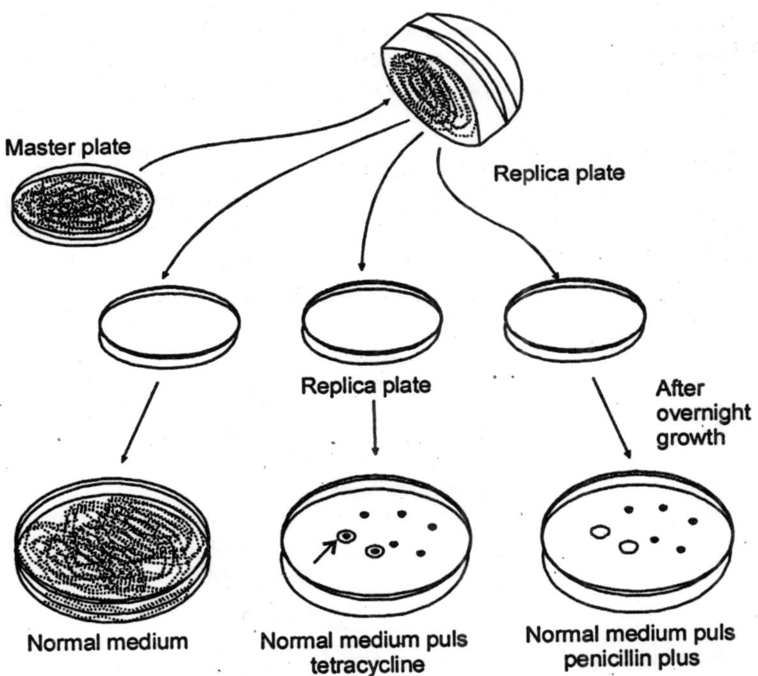

Fig. 4.5. Application of the replica plating technique to the detection of hybrid plasmid-containing cells. About 10^7 bacteria are spread on a plate. After overnight growth the plate (master plate) appears as a uniform "lawn" of bacteria, but in reality it consists of very small colonies that have merged to give a uniform appearance. A piece of velvet is lightly pressed against the surface of this lawn, and some cells stick to the velvet. Several essentially identical impressions of the lawn are transferred to fresh plates, which contain normal medium or normal medium supplemented with antibiotics. The results on the replica plates after overnight growth are indicated. The plate in normal medium again gives rise to a lawn of cells as virtually all of the cells transferred grow into colonies. The plates containing antibiotics only give rise to a few colonies, each of which is derived from a single cell that carries the plasmid-conferred drug resistance(s).

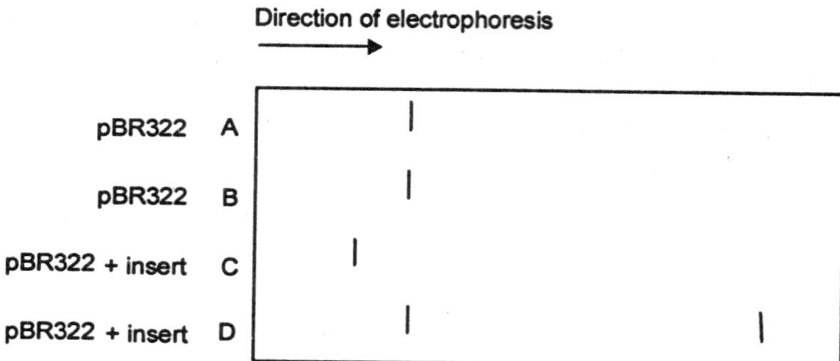

Fig. 4.6. Electrophoretogram of restriction enzyme digests of pBR322 and pBR322 with a DNA insert at the *Pst*I site. The insert is assumed to have no internal *BamHI* restriction sites. In channels A and B the pBR322 is predigested with *Pst*I and *BamHI*, respectively. The resulting DNA migrates with the same mobility because the plasmid has one site for each of these enzymes and therefore has the same molecular weight. In C and D the hybrid plasmid containing a DNA insert is treated with *BamHI* and *Pst*I, respectively. In C the hybrid plasmid has been linearized by one cut at the *BamHI* site in the *tet*ʳ gene. It runs more slowly than the pBR322 because it is larger due to the DNA insert. In D Ihe plasmid cuts at two *Pst*I sites located between the pBR322 sequences and the insert sequences. Consequently, one segment migrates at the rale of a linearized pBR322 plasmid. The other segment, also linearized, migrates at a rate characteristic of the size of the DNA insert. The electrophoresis is run from left to right; fragments are stained with ethidium bromide and photographed with UV light.

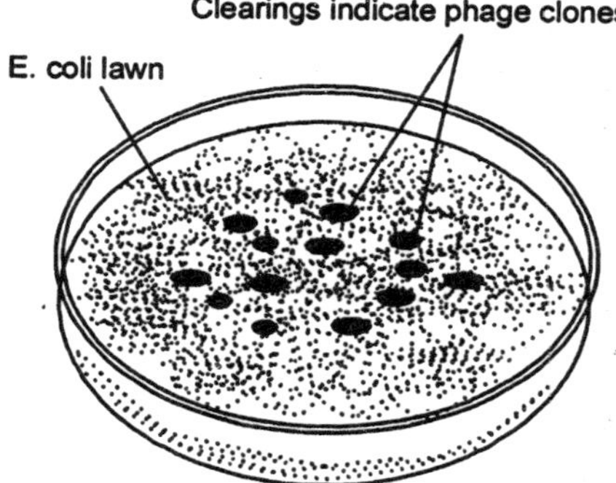

Fig. 4.7. The nutrient agar plate contains a continuous lawn of *E. coli* bacteria except for circular clearings that represent phage plaques. Each plaque was originally derived from a single phage infecting a single *E. coli* bacterium. After infection the phage multiplies, ultimately producing about 100 mature viruses. The phages also produce an enzyme that causes the harboring cell to lyse. When this happens the phages are released, and each of them infects a neighboring cell and goes through the same infectious cycle. The process continues. Each cycle takes about 30 min. Eventually a visible clearing can be seen on the plate.

This plasmid contains two genes, amp^r and tet^r, which confer resistance to penicillin and tetracycline, respectively. *Pst*I restriction fragments of foreign DNA may be inserted into the unique *Pst*I restriction site on pBR322. This is done by digesting pBR322 with *Pst*I, mixing with the restriction fragments to be cloned at low temperatures to permit annealing to take place between the two DNAs, and finally ligating the annealed fragments with DNA ligase.

The product contains some of the original pBR322 and some pBR322 with inserted foreign DNA. When this mixture is used in transfection, most cells are not transfected, some are transfected with pBR322, and some are transfected with the desired hybrid plasmid.

The three types of cells may be readily distinguished by their drug-resistant properties. Normal cells do not grow in the presence of tetracycline or penicillin. Transfected cells with the DNA inserted in the plasmid are tetracycline-resistant but penicillin-sensitive, because the insert has disrupted the amp^r gene. Cells containing the desired plasmids can be distinguished from those containing pBR322 by "replica plating". The first step when using this approach is to spread a large population of treated bacteria on an agarose plate containing growth medium. Within 12 h a seemingly homogenous "lawn" of cells develops on the surface of the agarose. Actually, the lawn results from the growth of many microcolonies to the point of confluency. At this point a piece of velvet is lightly pressed against the surface of the plate, and this impression is transferred to other agarose plates containing growth medium with tetracycline or penicillin plus tetracycline.

Only the transfected cells produce colonies on the plates containing the antibiotics, and because of their small number, each of these gives rise to readily detectable clones. The clones present on the tetracycline-containing plates, which are missing on the penicillin plus tetracycline plates, most likely contain the desired hybrid plasmids. These clones are usually plucked from the tetracycline plates and retested to eliminate any uncertainty about the original drug testing. Once this is confirmed, the appropriate plasmid-containing cells are grown in liquid culture.

After a moderate density of growth is achieved, the plasmid DNA is selectively amplified by overnight growth. Plasmid DNA replication continues for several hours, until each cell contains 1,000 to 2,000 copies of the small circular plasmid DNA. This DNA is readily separable from the host DNA and can be characterized by its rate of migration on gel electrophoresis, or other more specific tests to determine if it contains the inserted DNA sequence. If desired, the inserted sequence may be removed from the plasmid vector by digestion with *Pst*I, the restriction enzyme used in the initial construction of the hybrid plasmid. The cleaved fragments can be separated readily by gel electrophoresis. Many plasmid vectors, other than pBR322, have specific advantages for other purposes.

Cloning Larger DNA Segments

Bacteriophage λ possesses a number of advantages as a cloning vector. DNA fragments as large as 24 kb can be propagated using such vectors. The primary pool of clones can be amplified by limited phage growth as plaques, and the entire collection of phage clones (recognized as clear plaques) can be stored for long periods in a small volume.

Because it does not accommodate molecules of DNA that are much longer than the viral genome, the use of A as a vector for cloning substantial DNA fragments requires the removal of a significant portion of the viral DNA before-hand.

Fortunately, the central third of the genome contains genes that are not essential for phage production and can therefore be deleted. Bacteriophage λ vectors that accommodate foreign DNA fragments generated by a variety of restriction endonucleases have been constructed. The recombinant DNA molecules that incorporate some of these vectors can be introduced directly into *E. coli* by transfection. Alternatively, recombinant DNA molecules can be packaged into phage particles and subsequently infected into suitable host cells.

Cosmids

Although plasmids and bacteriophage λ are both highly useful vectors, the size of the DNA fragments that can be cloned in them is limited. With plasmids, the larger the fragment of foreign DNA inserted, the lower the efficiency of ligation and transfection, making the cloning of DNA fragments larger than 15 kp experimentally difficult. In λ vectors, the length of the nonessential region of λ DNA limits fragment size to 24 kb or less.

Also, the original λ vectors do not allow propagation of viable bacterial cells that carry the inserted DNA fragment; the inert is propagated as part of a virus that lyses the cell. Cosmids were developed as vectors for cloning large DNA fragments. The first part of their name, "cos," comes from the fact that cosmids contain the cohesive ends, or *cos,* sites of normal λ. These ends are essential for packaging the DNA into λ phage heads. The last part of their name, "mid," indicates that cosmids carry a plasmid origin of replication like the one found in the pBR322 plasmid. Such cosmids can be used for cloning in the same way as any other plasmid vector.

However, because cosmids also contain the *cos* sites, cosmid DNA along with an inserted DNA fragment can be packaged as a λ phage. The result after packaging is a defective but nevertheless infectious phage particle. Once the cosmid and the inserted DNA fragment are introduced by infection into a λ-sensitive cell, the plasmid replicates. Since cosmids lack the entire bacteriophage genome except for the region adjacent to the *cos* sites, these vectors can propagate exogenously derived DNA fragments of up to 40-50 kb in length.

Shuttle Vectors

Vectors that include replication systems derived from more than one host species are known as *shuttle vectors*. Such vectors commonly include a replication system able to function in *E. coli* and one that works in a second host, which may be bacterial or eukaryotic. Initial cloning and amplification of the DNA segment to be studied is often carried out in *E. coli* because it is easier to make large quantities when culturing in *E. coli*. The recombinant DNA molecule, consisting of the "bifunctional vector" plus the cloned segment of DNA, is then introduced into the second host, where the purpose is usually to measure the expression of the genes carried by the vector. Shuttle vectors that can replicate in both *E. coli* and yeast are the most common.

CONSTRUCTING A "LIBRARY"

Cloning can involve a single vector-linked DNA fragment or a collection of independently isolated vector-linked DNA fragments derived from a single organism. Such a collection is termed a *"Library"* and may serve as the source of well-defined sequences from a given organism. Each clone of a library harbors a particular DNA segment from the desired organism. Within the entire library a sequence may be repeated, but other sequences may be missing.

The ideal library, which can only be approached, represents all of the sequences with the smallest possible number of clones. A library from the same cell or organism can be prepared in two ways. The genome may be fragmented and ligated to the appropriate vector to produce a *genomic DNA library*. An alternative approach is to construct a *cDNA library* in which the DNA fragments to be cloned are obtained by reverse transcription from the cellular RNA. Each of these libraries has advantages and disadvantages, and for a specific purpose, one library is usually preferred over the other. The vast majority of DNAs within a library are uncharacterized. As a rule, the task of finding the desired genes or sequences within a library greatly exceeds the task of constructing the library.

A Genomic DNA Library

A major concern in constructing a genomic DNA library is to maximize the probability that all segments of the genome are represented. If the genomic DNA is prepared by cutting with a restriction enzyme, an added concern is that the enzyme cleaves genes of interest at one or more sites. To increase the likelihood of isolating desired genes in one piece, different restriction enzymes can be used on different parallel preparations. But even if the genes of interest are not cut by the enzyme(s) chosen, the DNA fragments produced may be inconveniently small to work with.

Table 4.2. Theoretical Number of Clones Required to Fully Represent the Entire Genome of Various Organisms

Size of Cloned DNA Fragment (bp)	*2×10^6 (e.g., bacteria)*	*Genome Size (bp) 2×10^7 (e.g., fungi)*	*3×10^9 (e.g., mammals)*
5×10^3	400	4,000	600,000
10×10^3	200	2,000	300,000
20×10^3	100	1,000	15,000
40×10^3	50	500	75,000

An enzyme that recognizes a sequence of six bases (a six-cutter) gives an average fragment size of 4,096 bp[b], which is a reasonable size for making a plasmid library but much smaller than the size desirable for cloning in λ or a cosmid vector. Therefore, when large, randomly generated fragments are desired, the method of choice usually entails making an incomplete digest with a four-cutter restriction enzyme, which produces overlapping ends that can be readily cloned into the chosen vector *as* described *earlier. The* extent of digestion is controlled so that cleavage occurs at only some of the restriction enzyme recognition sites and the average size of the fragments produced is in the desired *range*.

The conditions used thus depend on whether the product is going to be cloned in a plasmid, a λ phage, or a cosmid vector. Table 4.2 gives the minimum number of clones (*i.e.*, the size of the library) required to fully represent the entire genome in a genomic DNA library, as a function of the average size of the cloned fragments and the size of the genome. Since DNA fragments in a population are cloned on a random basis, the chance of finding a given single-copy gene in a library of the indicated size is 50%. A clone bank should be 3 to 10 times the minimum size to giye a high probability that a particular segment is represented.

A cDNA Library

A cDNA library consists of a collection of clones that contain DNA copies of the cellular or organismic RNA. If the RNA is obtained from a differentiated multicellular organism, then the library varies in composition according to the type of cell used as the RNA source and to the physiological state of the cell. This variation is a reflection of the relative abundances of particular mRNAs made by different cell types. If a cDNA species corresponding to a particular gene product is desired, it is often possible to select a cell type suspected to synthesize a large amount of the corresponding mRNA or mRNA-related protein.

Thus, pituitary cells can be used if cDNA encoding growth hormone is desired, whereas liver cells can be used if a serum albumin cDNA is the goal. The mRNAs present in low amounts clearly require the screening of a larger library than the mRNAs present in medium or high abundance. Once the crude mRNA fraction has been isolated from the chosen cells or tissue, it is converted to duplex DNA molecules with the help of reverse transcriptase. This duplex DNA does not have "sticky ends" for insertion into a vector. For this purpose *DNA linkers* are attached to the ends.

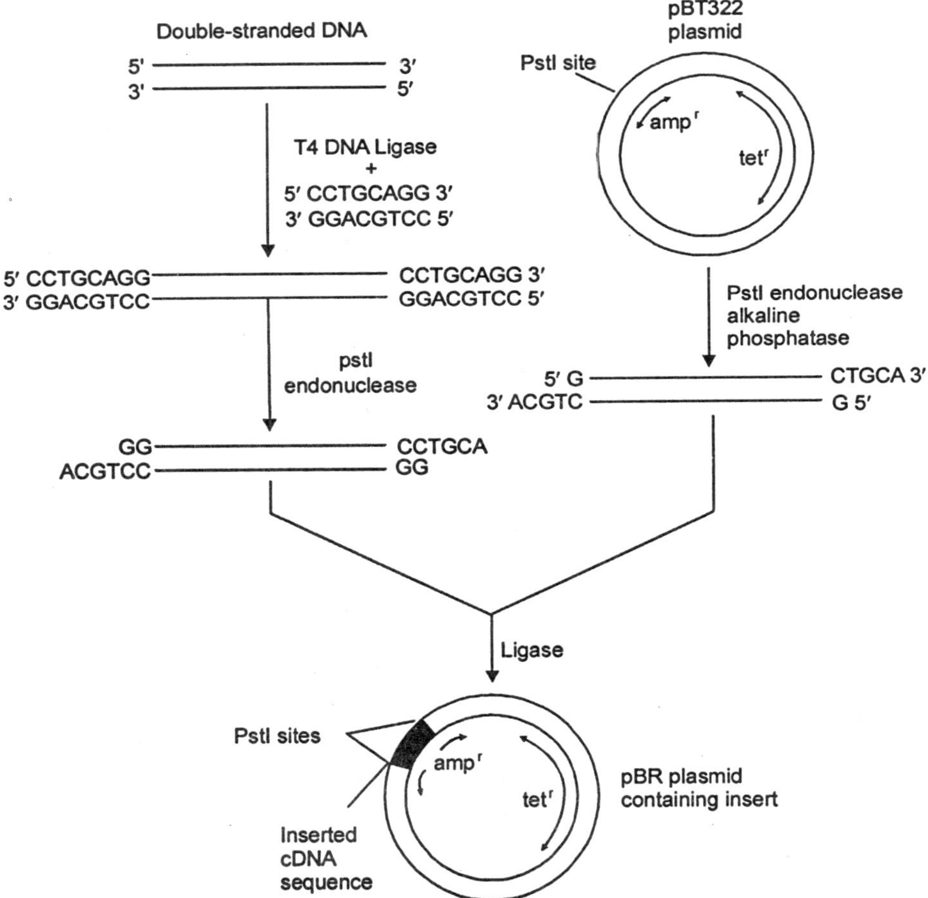

Fig. 4.8. Insertion of cDNA into pBR322 plasmid by the linker method. The strategy here is to open up the plasmid with a restriction enzyme that makes staggered cuts and to attach linkers that contain the same recognition site to the cDNA. After the linkers are attached to the cDNA, the duplex is treated with the same restriction enyzme *(PstI)* to expose the overhangs. The two DNAs are mixed together and ligated. After transfection, cells containing the hybrid plasmids are recognized by tetracycline resistance and ampicillin sensitivity. Identification of the insert is discussed in the text.

Linkers are synthetic single-stranded oligonucleotide segments (6, 8, 10, or 12 bases in length) that self-associate to form symmetrical, blunt-ended, double-stranded molecules containing the recognition sequence for a particular restriction enzyme. Fig. 4.8 shows an eight-base linker (CCTGCAGG) containing a *Pst*I recognition site. This linker self-associates to produce an eight-base, blunt-ended, duplex structure that adds to the double-stranded cDNA in the presence of T4 ligase. The resulting product is treated with *Pst*I to produce the characteristic 3′ overhang. The plasmid, linearized with *Pst*I, and the two DNAs are mixed and reacted with ligase to produce plasmid with the insert.

Picking the Correct Clone from a Library

A library can contain thousands or even tens of thousands of different kinds of clones, making it a challenging endeavor to isolate a clone with the DNA of interest. Most currently used procedures for screening large numbers of colonies for plasmids or phage that contain specific DNA inserts are variants of the *colony hybridization method* developed by Grunstein and Hogness. This procedure makes use of a specific radioactive probe that contains some sequences complementary to those in the DNA of interest.

The colonies to be screened are first grown on agar petri plates. A replica of each plate is made on another agar plate, which is stored for reference. A replica is also made on a nitrocellulose filter. The colonies formed on the filter are lysed and the contents denatured simultaneously by treatment with sodium hydroxide. After heating, the denatured DNA is fixed on the filter at each site where a colony was located. The DNA on the filter is then hybridized with a radioactively labeled nucleic acid probe complementary to the specific DNA sequence to be selected.

The presence of hybridized probe at sites occupied by DNA derived from colonies that include the DNA fragment of interest is detected by autoradiography. The colony whose DNA hybridizes with the nucleic acid probe can then be picked from the reference plate, which contains a viable bacterial colony at a corresponding location.

Escherichia Coli

Despite the success and broad applications of *E. coli* cloning systems, instances occur in which gene products cannot be made in this bacterium. Either they are not synthesized in their entirety, or they are rapidly broken down after synthesis. In addition, the study of certain processes indigenous to other species (*e.g.*, photosynthesis, antibiotic production) often requires the use of a host bacterial species that naturally carries out the process, which may exclude *E. coli*. Effective cloning systems are available for a variety of bacterial hosts, including *Bacillus subtilis, Streptomyces spp.*, and *Agrobacter tumefaciens*. Cloning systems have also been developed for eukaryotic hosts such as the yeast *Saccharomyces cerevisiae*, mammalian cells in tissue culture, and plant cells.

SITE-DIRECTED MUTAGENESIS

By combining different procedures of manipulation. it is now possible to make discrete changes in genes. This technique, called *site-directed mutagenesis*, is one of the most important in modern genetics and biochemistry. The first site-directed mutagenesis studies were carried out by David Shortle and Daniel Nathans in 1978 with the help of the mutagen sodium bisulfite, which deaminates C residues so that they become converted into U residues. Directed mutagenesis as it is practiced today is based on the chemical synthesis of a deoxyoligonuclcotide that contains discrete changes in its sequence from that normally observed in the genome under investigation.

These changes may be single-base or multibase; they may involve base changes, base deletions, or base additions. Many variations of site-directed mutagenesis exist. One can start out with a circular, single-stranded DNA and anneal it to a synthetic primer DNA carrying the desired changes. This primer can be extended, and the resulting product can be transfected. Finally, one selects clones of cells containing the plasmid with the desired changes. The polymerase chain reaction (PCR) will probably be the method of choice in the future for carrying out site-directed mutagenesis.

In its most general form the use of PCR for this purpose requires a piece of duplex starting DNA, two outside flanking primers that are perfect complements to segments of opposing strands, and two complementary primers with the degired changes in their sequence. The steps followed parallel the steps of PCR amplification. PCR amplification can be coupled with classical cloning methods using cloning vectors. For this purpose the PCR product should contain restriction sites at its ends that are suitable for cloning. Thus, PCR amplification and cloning need not be thought of as alternatives for particular purposes but as complementing each other to give a greater variety of approaches.

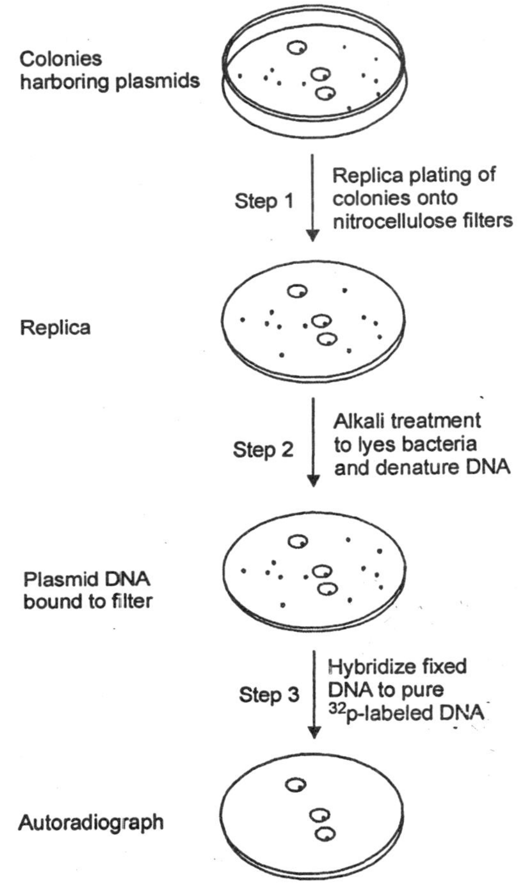

Fig. 4.9. Colony hybridization procedure used to identify bacterial clones harboring a plasmid containing a specific DNA. Step 1: Replica-plate the colonies containing plasmids onto nitrocellulose paper. Step 2: Lyse cells with NaOH and fix denatured DNA to paper. Step 3: Hybridize to ^{32}P-labeled DNA carrying the desired sequence and autoradiograph the product. Locations of desired DNA should be emphasized in the autoradiograph. Clones carrying desired plasmids (circled) may then be isolated from a corresponding agar replica plate carrying untreated colonies.

RECOMBINANT DNA TECHNIQUES

The human globin family is a paradigm for studying differential gene activity during development and the molecular basis of genetic disorders in gene expression. Hemoglobin is a tetramer containing two α-like and two β-like subunits. These proteins are encoded by a small number of genes that are expressed sequentially during development. The information summarized in Fig. 4.12 indicates that, the α-like and β-like globin gene families have coordinated programs for expression: Two switches exist for the β-like genes, whereas a single switch results in activation of adult α-globin production early in fetal life. A combination of classical and recombinant DNA techniques has been used to show that the α-like genes are located in a single cluster on chromosome 16, and the β-genes are located in a single cluster on chromosome. We focus on the contributions to our understanding of the globin genes that have resulted from investigations using the recombinant DNA approach.

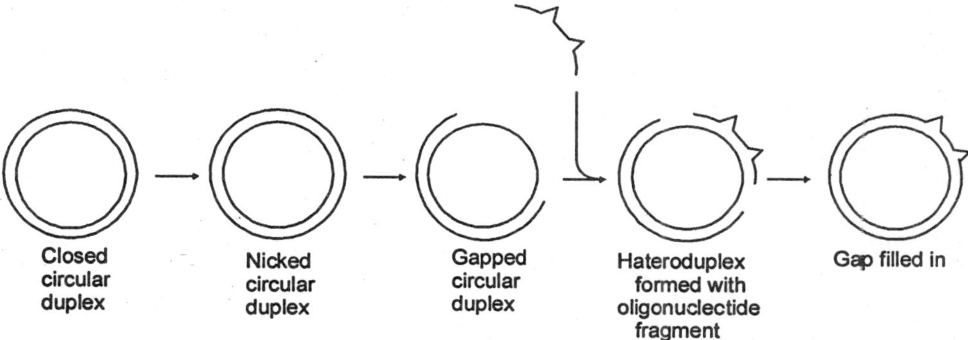

Fig. 4.10. Scheme for oligonucleotide-directed mutagenesis of double-stranded circular plasmid DNA. Supercoiled plasmid circles are nicked in one strand and rendered partially single-stranded by treatment with exonuclease. The gapped circles are hybridized with a homologous oligodeoxynucleotide carrying, by design, some mismatches. *In vitro* DNA synthesis, primed in part by the oligodeoxynucleotide, leads to heteroduplex plasmid circles.

Detecting Defective Hemoglobin Genes

All of the hemoglobin genes are represented by two or more, alleles within the human population; the genes are said to be *polymorphic*. This polymorphism frequently shows up in readily detectable phenotypes when the differences occur in vital areas of the polypeptide chains. Polymorphisms show up in the DNA even more frequently because the DNA contains sequence differences in both the coding and the noncoding regions of a gene, and these differences can be detected even when no visible effect is apparent in the organism. Because of their frequency and ease of detection by recombinant DNA methods, *DNA polymorphisms* have become extremely useful in mapping the human genome.

Whereas DNA polymorphisms should be recognizable by sequence differences, it is usually more convenient to detect these polymorphisms by differences in the size of DNA fragments obtained with restriction enzymes. Differences observed in this way are called *restriction fragment length polymorphisms* (*RFLPs*). Kan and Dozy were the first to discover an allelelinked DNA polymorphism in the globin genes. With it they predicted which fetuses carried normal and which carried abnormal sickle-cell genes for β-globin.

To analyze the DNA for these differences they used a technique called *Southern blotting*. The steps involved in Southern blotting are illustrated in Fig. 4.14. First the genomic DNA from the test subject is digested with a restriction enzyme to yield specific DNA fragments.

These fragments are separated according to size by agarose gel electrophoresis. Next the DNA is denatured and transferred from the agarose gel to a cellulose nitrate sheet. The DNA firmly bound to the sheet is hybridized with a radioactively labeled DNA probe, which carries some of the sequences of interest. The radiolabel, which hybridizes to specific regions of the sheet, is detected by autoradiography.

By comparing the results obtained from the DNA of different individuals one can see if the labeled DNAs move with the same or a different mobility. If they move differently, there must be a RFLP difference between the individuals. Detection of an RFLP by this means usually depends on the restriction enzyme used in the initial digestion. Some enzymes show a difference; others do not. To apply this technique to their hemoglobin studies Kan and Dozy first had to prepare a DNA probe that carried specific sequences in the β-globin gene. This was a relatively simple task because reticulocytes, which contain vast quantities of hemoglobin, are also greatly enriched in the β-globin mRNA. From the purified β-globin mRNA isolated from reticulocytes, a cDNA probe was made with the help of reverse transcriptase.

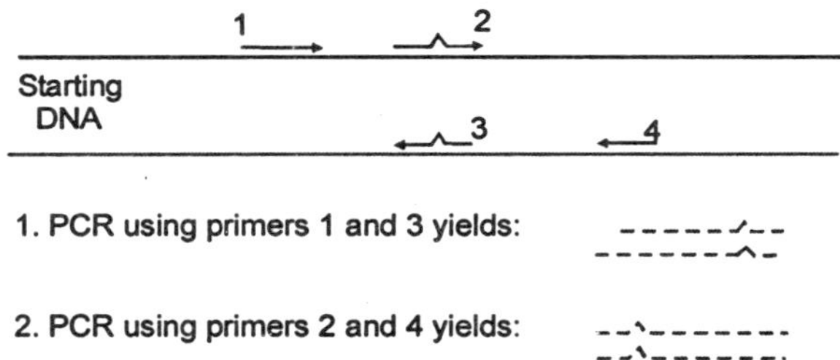

3. Mix products from steps 1 and 2 above, denature, and anneal. This should result in a mixture of four products, A, B, C, and D.

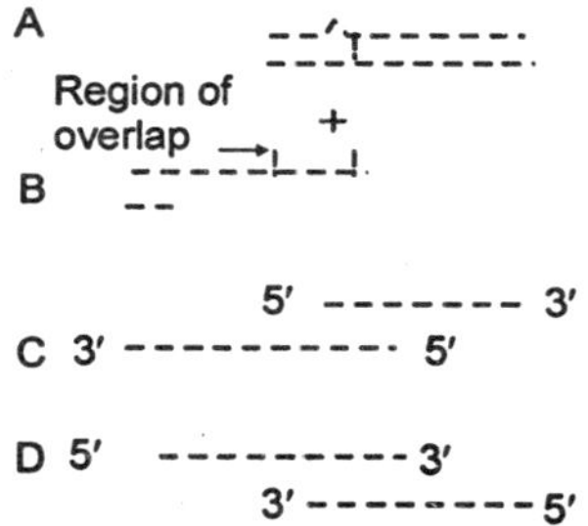

4. Extend D with DNA polymerase and amplify with primers. 1 and 4 to give final mutated DNA segment.

Fig. 4.11. Illustration of a general method of mutagenesis using PCR. Primers are represented as short lines with arrowheads pointing in the 3′ direction. The bump in primers 2 and 3 and their products represent a mismatched base, a deliberate alteration in base sequence from that present in the starting DNA. Of the four major products resulting from step 3 only D is extendable by DNA polymerase.

Using a *Hpa*I restriction enzyme digest of the total genomic DNA in conjunction with Southern blotting, Kan and Dozy showed that the normal β-globin gene was contained within a 7.6-kb *Hpa*I restriction fragment, whereas the β-globin gene of sickle-cell anemia, Hb^s was contained within a 13-kb fragment. Further analysis showed that the RFLP resulted from a *Hpa*I restriction site 5 kb to the 3′ side of the β-globin gene, that was present in the normal case and absent in Hb^s. Subsequent analysis showed sequence differences within the coding regions. One may wonder why the RFLP outside the coding region was so commonly associated with the abnormal gene. A possible explanation for this is that in the distant past a mutation occurred which resulted in the 7.6-kb type and a few 1 3-kb types before introduction of the sickle-cell gene mutation. After the Hb^s mutation was introduced into the 13-kb type, it became greatly expanded because of the selective advantages of this gene in heterozy.

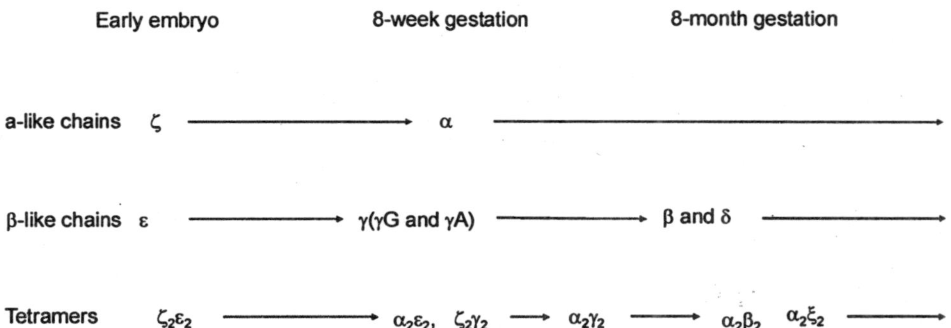

Fig. 4.12. Changes in types of hemoglobin observed in early development. A single switch in gene expression is observed for α-like chains. Two switches in gene expression are observed for β-Iike chains. The corresponding tetrameric hemoglobin molecules observed at different stages in development are also indicated.

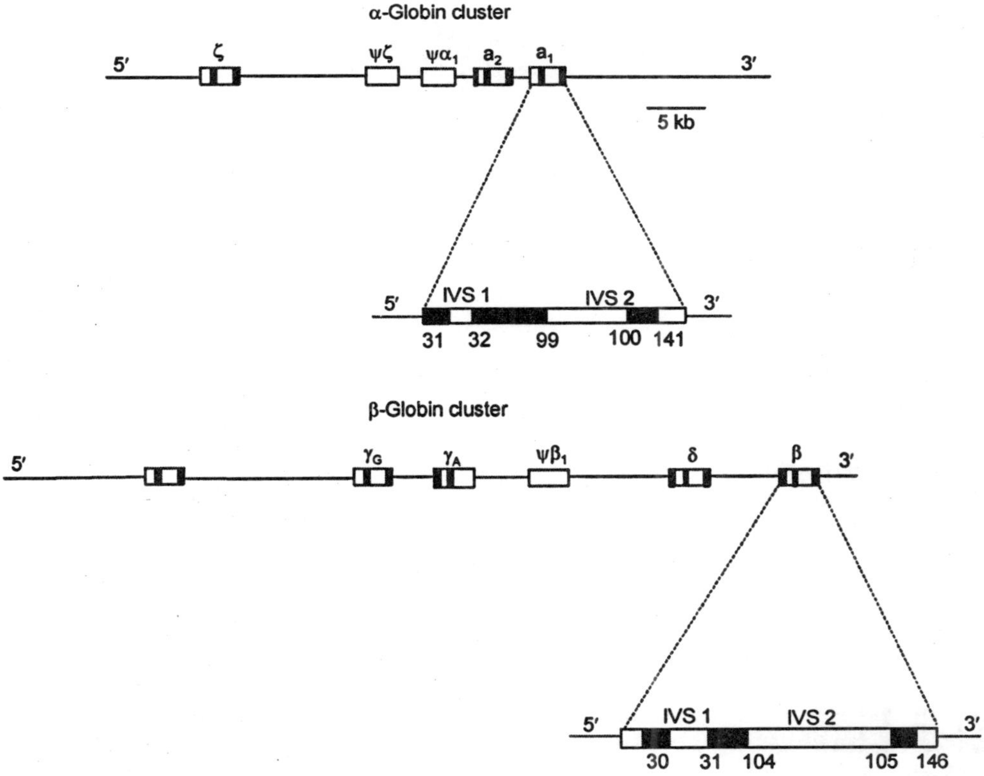

α-Globin cluster

5′ ζ ψζ ψα_1 a_2 a_1 3′

5 kb

5′ IVS 1 IVS 2 3′

31 32 99 100 141

β-Globin cluster

5′ γ_G γ_A ψβ_1 δ β 3′

5′ IVS 1 IVS 2 3′

30 31 104 105 146

Fig. 4.13. The chromosomal localization and genomic organization of the human globin genes. The α- and β-globin gene complexes are positioned on chromosomes 16 and 11, respectively. For each complex, the arrangement of genes on the chromosome is depicted above, and the general structure of the major gene is shown below, together with the location of the intervening sequences, or introns (IVS), and codon numbers. Coding regions are shown by solid boxes and IVS regions by open boxes. Genes with the ψ symbol in front are called pseudogenes because they are sequence-related but not expressed.

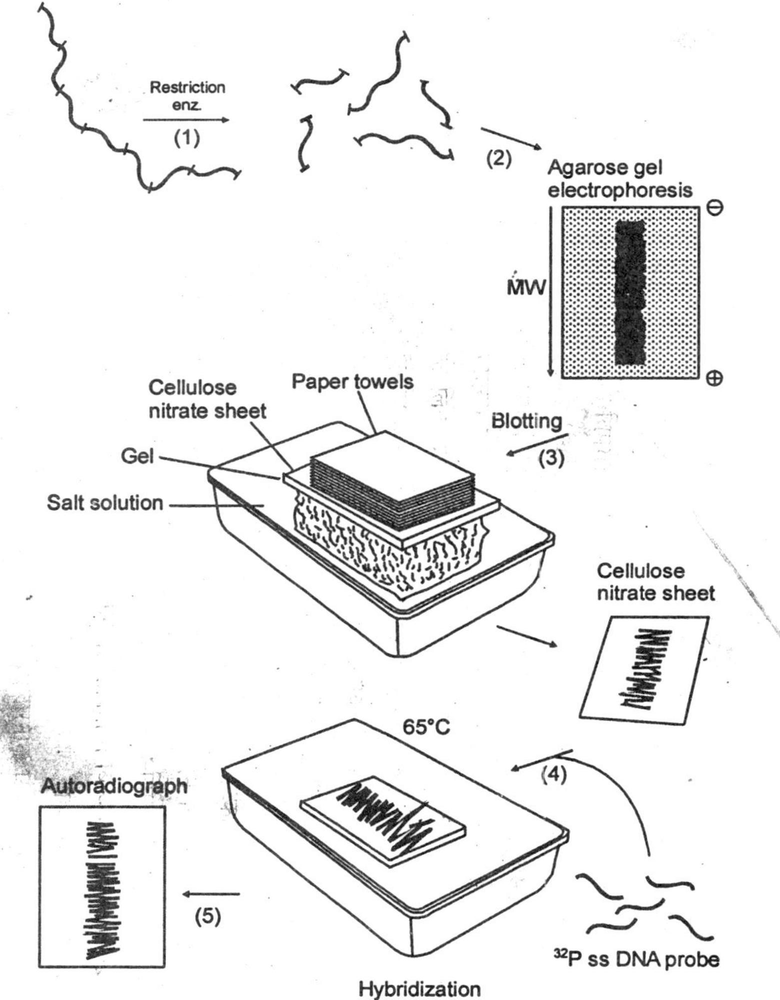

Fig. 4.14. The steps involved in assaying by Southern blotting. The DNA to be analyzed is digested with a restriction enzyme (1). The resulting fragments are electrophoresed on an agarose gel (2). The DNA fragments on the gel are transferred to a cellulose nitrate sheet by placing the cellulose nitrate sheet next to the gel and passing solvent through the gel into the sheet. Flow of the solvent is maintained by blotting the far side of the cellulose nitrate sheet with paper towels. The DNA, first denatured with alkali, flows with the solvent but gets stuck in the sheet (3). The sheet is hybridized to radioactively labeled DNA containing the gene sequence of interest (4). The hybridized sheet is autoradiographed to determine the location of the labeled restriction fragment on the gel (5).

In this connection it should be noted that heterozygotes carrying one normal gene and one sickle-cell gene fare far better when infected by malaria. As a result in areas where malaria is prevalent the heterozygote has a selective advantage over the normal homozygote. Knowledge of this and other polymorphisms has been used for pre- or postnatal diagnosis of

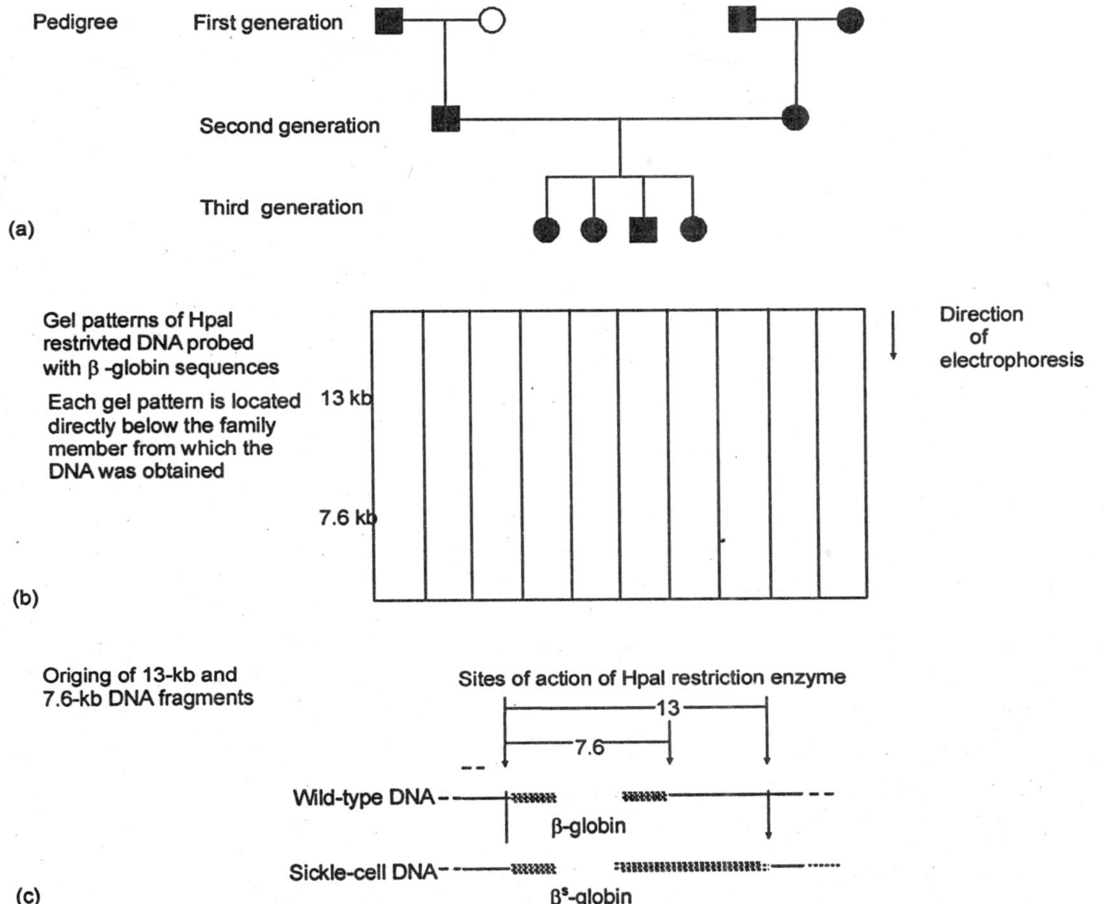

Fig. 4.15. Inheritance pattern of an RFLP associated with sickle-cell disease. Humans carry two alleles for the same gene, and each offspring inherits one allele from each of its parents in an entirely random fashion. Normal individuals are homozygous for normal *Hb* alleles; individuals with sickle-cell trait are heterozygous, with the one normal *Hb* allele and one *Hb*^{s} allele; and individuals with sickle-cell disease are homozygous for the *Hb*^{s} allele. At the top (a) we see a three-generation pedigree analysis for a family that carries both the normal and the sickle-cell gene for β-globin. Males are represented by squares and females by circles. A purple circle or square indicates an individual who is homozygous normal. A half-filled circle or square (red/purple) indicates a heterozygous individual with sickle-cell trait. A filled circle or square (red) indicates a homozygous individual with sickle-cell disease. In (*a*) both sets of grandparents produce a heterozygous individual with sickle-cell trait. Because one of the grandparents is homozygous normal and the other is heterozygous, there is a 50% chance that the grandparent mating will give rise to a heterozygous offspring as shown and also a 50% chance that they will have normal offspring (not shown). The two heterozygous parents have an increased chance of having abnormal offspring because in this mating each parent carries one abnormal gene or allele. There is a 25% chance of a homozygous sickle-cell anemic offspring, a 50% chance of an abnormal offspring with sickle-cell trait, and a 25% chance of a normal offspring. Below the pedigree chart is the electrophoretic pattern of a *Hpa*I digest probed with β-globin cDNA by the Southern blotting technique (*b*). At the bottom we see an interpretation of the normal and abnormal DNAs (*c*).

the sickle-cell gene. Such information can be of great practical value in genetic counseling. Incidentally, it is now possible to diagnose sickle-cell disease (which occurs in individuals that are homozygous for the sickle-cell gene) with greater certainty because the point mutation leading to the defect in the coding region itself produces a recognizable RFLP.

The β-Globin cDNA

Detailed mapping with DNA probes was first successfully executed on the human β-globin gene. All members of a human genomic library that annealed to the radioactive cDNA probe for the β-globin gene were isolated, and each of these was sequenced. This analysis resulted in a complete description of the β-globin gene. The β-globin gene is appreciably longer than the β-globin mRNA; in addition to containing regions that are present in the final mRNA the gene contains two "intervening" regions not represented in the mRNA sequences. We have more to say about the significance of these intervening sequences (IVS, also called *introns*) in the next chapter.

Chromosome Walking

The original cDNA probe carrying the β-globin mRNA sequences could only detect members in the genomic library that contained sequences homologous to those present in the probe. To explore the region flanking the β-globin gene the genomic library was probed further. For this purpose members of the genomic library that hybridized with the original cDNA were themselves converted into radioactive probes, and these were used to locate additional members of the library that contained sequences flanking the β-globin gene.

A cyclic repetition of this process resulted in a gradual extension of sequence information in and around the β-globin gene. Using the library in this manner to extend the map is known as *chromosome walking*. A parallel approach was used to extend the map around the adult α-globin gene. It can be seen that in both cases several genes occur in a cluster for each of the protein types. Most of these can be correlated with the genes that are expressed at different times during development. In addition, genes, called pseudogenes, occur that have strong sequence similarities to known genes but are never expressed. It is not clear if these pseudogenes have a function or simply represent evolutionary "junk," which has not been removed.

Walking and Jumping

By combining the linkage information obtained from RFLP mapping with other DNA manipulation techniques, it has been possible to locate genes causing serious genetic disorders even when these genes are only known from their inheritance patterns. The list of serious disorders that can be linked to single genetic loci is growing. It includes Huntington's disease, Duchenne's muscular dystrophy, polycystic kidney disease, cystic fibrosis, chronic granulomatous disease, peripheral neurofibromatosis, central neurofibromatosis, familial polyposis coli, and multiple endocrine neoplasia.

One of the most spectacular achievements has been the determination of the gene causing cystic fibrosis (CF). *Cystic fibrosis* is the most common serious genetic disorder in caucasian populations. The major clinical symptoms include chronic pulmonary disease, pancreatic exocrine insufficiency, and an increase in the concentration of sweat electrolyte. Bearers of this disease are readily diagnosed and often die of congestive lung complications before age 30. Pedigree analysis shows that, as in the case of sickle-cell anemia, a single gene inherited in autosomal recessive fashion results in the disease syndrome. The frequency of the disease is 1 in 2,000, from which it may be calculated that the carrier frequency is about 1 in 20 (the frequency of the heterozygote for a rare allele is twice the square root of the frequency of the homozygote).

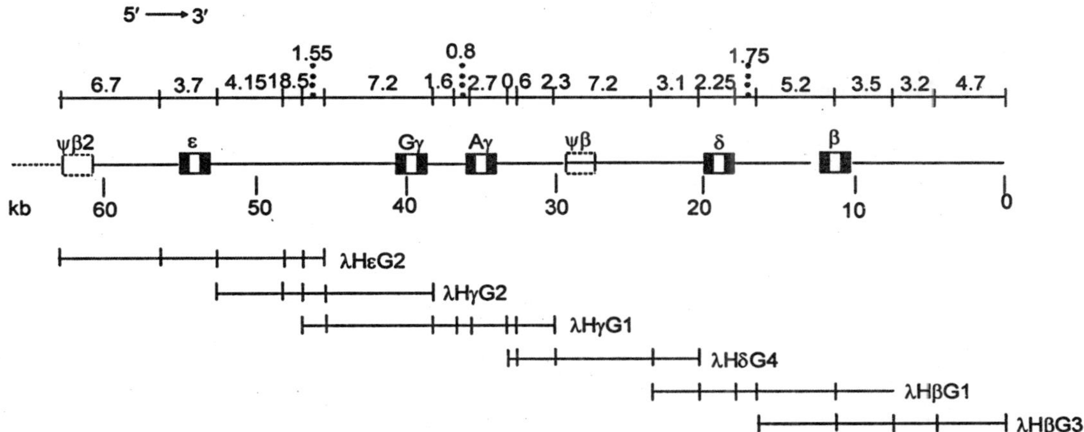

Fig. 4.16. The linkage map of the human β-globin gene locus as shown by the structural analysis of overlapping λ genomic clones. Both λHβG1 and λHβG3 clones contained the entire β-globin gene. Other clones detected by "walking" led to the discovery of other β-globinlike genes. These included four genes that are expressed and two pseudogenes that are not expressed. The genomic segments of the clones isolated are shown together with the cleavage sites for the enzyme *Eco*R1. The numbers on the top line indicate the size of the fragments in kilobase pairs.

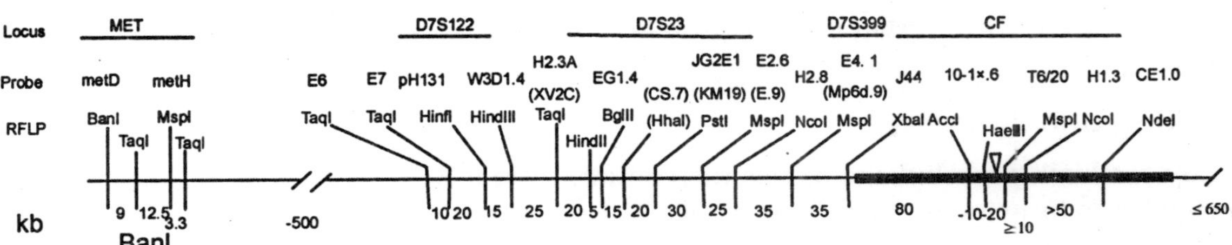

Fig. 4.17. Map of restriction fragment length polymorphisms (RFLPs) closely linked to the cystic fibrosis (CF) gene. The inverted triangle near the right-hand end indicates the location of the ΔF_{508} mutation characteristic of most persons with cystic fibrosis disease.

By classical genetic analysis the CF locus has been assigned to the long arm of chromosome 7 near the *met* locus. In many genetic disorders, cystic fibrosis included, the analysis begins before the responsible gene and its protein product are known. Thus, one cannot locate the gene directly, as in the case of the β-globin gene and then determine its approximate location in subsequent analysis. For genes such as those responsible for cystic fibrosis the approximate location is determined by conventional genetic analysis, and then one attempts to close in on the gene by a process called *"reverse genetics."*

Conventional genetic mapping by recombination frequency is not practical with human genes below map distances of 1 centimorgan (cM) because of the small number of test recombinant crosses that are ordinarily available for observation. Unfortunately, a centimorgan on the human genome is equivalent to a physical distance of about 1,000 kbp, so this presents the problem of "walking along" a chromosome for a million base pairs. Chromosome walking procedures are not suited to such long distances because of the size limitation of probes. Even cosmids, which provide the largest probes, cannot harbor probes larger than 40 kbp'. It would take 25 cosmids end-to-end to span 1,000 kbp. A walk involving this many probes would take a very long time even if it were possible. Another hindrance to such a long walk is that the human genome is sprinkled with segments of repetitious DNA.

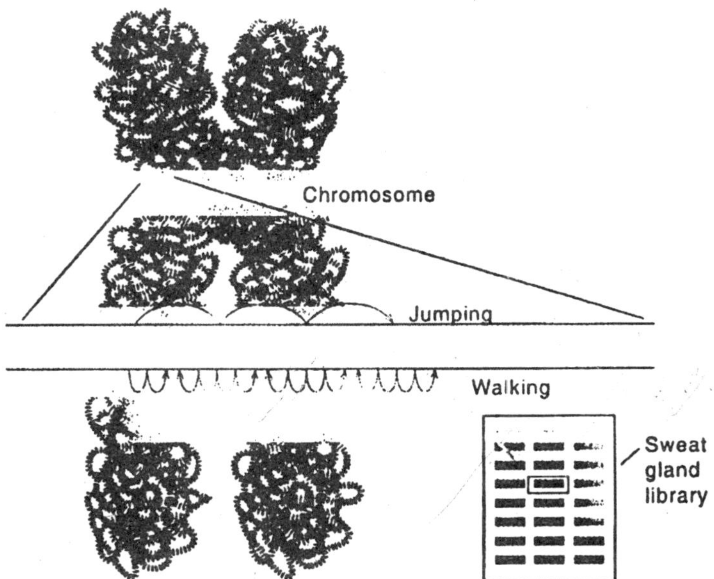

Fig. 4.18. Jumping and walking to find the cystic fibrosis gene. Following each jump, the locus defined by the jump was used as a starting point for a chromosome walk. Each DNA segment so found was hybridized to a sweat gland cDNA library until a match was found. It seemed likely that the sweat gland cDNA library would have a good representation of the cystic fibrosis gene transcript because the disease involves the sweat glands.

Those regions "interrupt" a walk because a probe that encounters such a region anneals to many members of the library that could be situated almost anywhere in the genome. Walking breaks down under such conditions and would benefit by an alternative procedure. To traverse long distances of the genome and to skip troublesome regions of repetitious sequences, a procedure called *chromosome "jumping"* was devised. Probes for jumping carry small segments from the same chromosome, which are ordinarily about 500 kb apart. The precise distance is variable and a function of how the jumping probes are made. We do not go into the complexities here of constructing such a library except to say that there are several ways in which this can be done.

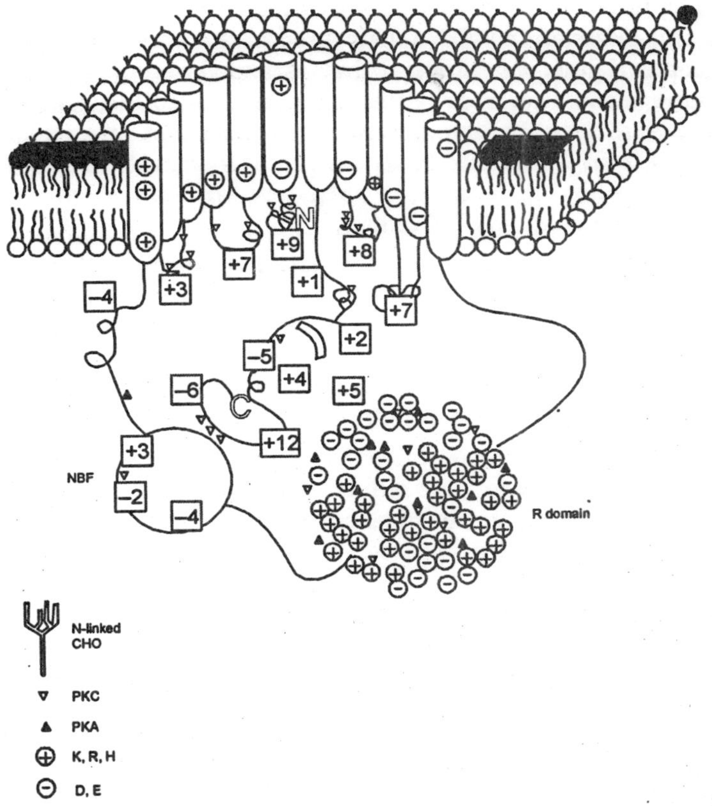

Fig. 4.19. Predicted structure of the CFTR protein. Cylinders represent membrane-spanning helical segments. The cytoplasmically oriented NBFs are shown as blue spheres with slots to indicate the points of entry of nucleotides. R represents the large polar domain, which is linked to two halves of the protein molecule. Charged amino acids are shown as small circles with the charge sign. Net charges on the internal and external loops joining the membrane cylinders and on regions of the NBFs are contained in open squares. Potential sites for phosphorylation by protein kinases A or C (PKA or PKC) and N-glycosylation (N-linked CHO) are indicated. (K = Lys; R = Arg; H = His; D = Asp; E = Glu.)

Usually a jumping library and a walking library are prepared from the same genome and used in conjunction with each other. The region around each locus recognized by the jumping library is scrutinized by the walking library. This complex analysis would be in vain if one did not have some criterion for knowing when the goal of finding the cystic fibrosis gene was reached. This is where clues from the physiological nature of the condition became useful. In a brilliant strategy a cDNA library was prepared from the mRNA fraction of sweat gland tissue. Recall that in cystic fibrosis sweat glands malfunction; therefore, the mRNA for the CF gene might be well represented in the mRNA fraction of the sweat gland cells. While the walking and jumping process was in progress each new segment mapping in the general region of interest was tested against the sweat gland cDNA library.

Finally, a member of the walking library was found that annealed with a member of the sweat gland cDNA library. Was this match fortuitous or did it mean the CF gene had been found? To answer this question the cDNA discovered in this way was used to probe the genomic library. By this means a gene was mapped that extended over a region of about 250 kb with 23 introns. A unique transcript, approximately 6,500 nucleotides in length was detected in extracts of sweat gland tissue which matched the transcript size expected from this gene. The protein predicted from a sequence analysis of this transcript consists of two similar motifs, each with (1) a domain having properties consistent with membrane association and (2) a domain believed to be involved in ATP binding.

Finally it was discovered that many CF patients carry a three base deletion in this transcript which should result in the loss of a phenylalanine residue from the protein. This defect correlates with the notion that CF patients have a faulty membrane protein that leads to the secretion problems characteristic of the disease. The fact that the abnormal gene is located as close as one can tell by classical genetics to the CF locus adds additional support to the notion that the CF gene has been found.

Finding the disease gene does not, of course, mean that a cure is in the offing. However, the characterization of the disease gene will be a tremendous aid in diagnosing carriers and fetuses that are homozygous for the disease gene. It also should be a help in focusing approaches to finding a cure for the disease.

CHAPTER

5

Single Nucleotide Polymorphisms

The full integration of genetic and phenotype data will only be achieved when a complete catalog of all genetic and epigenetic features of the human genome is available. A crucial step in this huge enterprise is the complete characterization of the human transcriptome. This includes (*i*) the identification of all human genes, (*ii*) the identification of all transcripts for these genes, including variants, and (*iii*) the characterization of their spatialtemporal pattern of expression. Various strategies have been developed to explore the transcriptome. With the advent of molecular biology in the 1970s, Northern blot started to be used routinely to identify genes expressed in several conditions, both normal and pathologic.

More recently, the development of reverse transcription-polymerase chain reaction (RT-PCR) represented a significant advance over the Northern blot, especially in the past few years when automated, robot-dependent protocols have appeared. In the early 1990s, the automated DNA sequencers allowed the development of a strategy to evaluate expressed sequences in a large scale. Expressed sequence tags (ESTs) were developed by Craig Venter's group with the aim of generating an expression fingerprint for a given tissue or cell. The strategy is a huge success with the resulting database, dbEST (maintained by the National Center for Biotechnology Information; NCBI), containing today more than 40 million sequences from a variety of organisms. However, in terms of throughput, none of these technologies are comparable with the ones more widely used today including serial analysis of gene expression (SAGE).

Massively Parallel Signature Sequencing (MPSS) and microarray. Furthermore, these last methods are intrinsically quantitative, a feature normally missed by the EST data set in which most of the libraries are normalized. Although these methods are primed primarily toward the transcriptome, they are also useful in the identification of single nucleotide polymorphisms (SNP).

Specific software using different statistical approaches has been developed with the aim to identify SNPs in the EST data set. Furthermore, SNPs that affect the tag-to-gene assignment in SAGE and MPSS experiments can be easily genotyped by scanning the respective libraries. In this review, we intend to describe the SAGE methodology and the bioinformatics approaches developed to explore the data derived from SAGE. Furthermore, we will discuss the use of SAGE data in the identification of biologically meaningful SNPs.

SAGE

SAGE uses automated DNA sequencing to count a large number of mRNA transcripts from basically any type of biological sample. The deeper coverage of the transcriptome achieved by SAGE, when compared with EST sequencing for example, comes from the fact that SAGE increases the number of genes to be counted by minimizing the portion of the transcript sequenced. This is achieved because SAGE works by cloning and sequencing a small (10 to 17 bp) tag of the cDNA at a defined position near the 3′ end of the transcript. This defined position corresponds mostly with the fragment immediately downstream of the 3′ most NlaIII site. Tags are ligated, cloned end-to-end, and sequenced allowing the serial analysis of multiple transcripts.

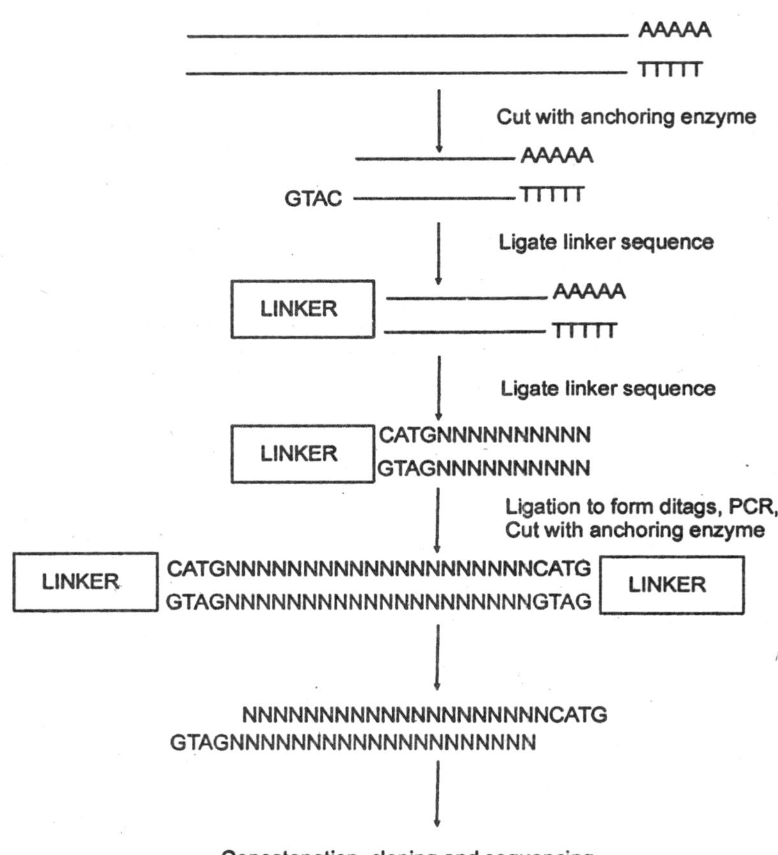

Fig. 5.1. Schematic view of the SAGE methodology.

The number of times a particular tag is observed in a given library is directly correlated with transcript abundance. To generate the tag data, polyA+ RNA is extracted and used as a template for cDNA synthesis using biotinylated polyT oligos. This allows the recovery of 3′ most cDNA fragments after digestion with an anchoring enzyme (mostly NlaIII) and affinity purification using streptavidin. These fragments are divided in half and ligated to two different linkers. Both linkers contain a site for the tagging enzyme (BsmFl) and the anchoring enzyme overhang. Each linker contains a specific prime site for PCR amplification. Fragments containing the linker plus a transcript sequence (tag) are obtained by digestion with BsmFl.

The isolated linker tags are blunt ended, and the two different sets of fragments, each with a different linker, are ligated. Linker-ditags-linker constructs are amplified using primers specific to each linker, and the ditags are released by digestion with the anchoring enzyme. These ditags are purified, concatenated, cloned, and sequenced. There are two major methodological versions of SAGE: the conventional one, described above, which generates 10-bp-long tags, and the Long SAGE, a very similar strategy that generates 17-bp tags. The longer length of these latter tags makes them less ambiguous. Because of that, the genome sequence, when available for the corresponding species, can be used for mapping purposes. Several different methods have appeared based on the same principles that govern the SAGE technology. Table 5.1 lists the major methods with differences between them.

As mentioned before, there is MPSS, a proprietary technology that goes deeper in the transcriptome coverage through the parallel sequencing on beads of transcripts' tags. This approach can generate up to a million tags in a single experiment, which represents roughly a 5X coverage of a cell transcriptome. The length of the tag (16 to 18 bp) is longer than the conventional SAGE but shorter than Long SAGE. Methods that target the 5′ end of transcripts have emerged in the past few years.

Cap analysis of gene expression (CAGE) is based on the construction of high-quality, full-length cDNA libraries. Instead of using the polyA tail as target, CAGE targets the 5′ end of cDNA using the cap-trapper technology. Tags are then generated, concatenated, and sequenced. CAGE has been extensively used for a variety of purposes.

Kodzius et. al., Wei et. al., developed the 5′and 3′ long serial analysis of gene expression (LS). The major aspect of both 5′ and 3′ LS is the extraction, concatenation, sequencing, and mapping of terminal transcripts' tags onto the genome sequence to define the transcription initiation site and the polyadenylation site. There is a wide range of applications for SAGE. The most powerful and attractive feature of SAGE is its ability to measure the expression of thousands of genes in a quantitative way and without *a priori* knowledge of the gene sequence. One is then able to perform expression profiling of normal and tumor tissues and identify messages that are differentially expressed in cancer. The use of Long SAGE has also allowed the identification of new transcribed loci in the human genome.

Table 5.1. Comparison of Different Methods Using the Same Principle as SAGE

Method	*Target*	*Anchoring enzyme*	*Tagging enzyme*	*Tag length (bp)*
SAGE	3′	NlaIII	BsmF1	14
Long SAGE	3′	NlaIII	MmeI	21
MPSS	3′	DpnII	BbuI	16-20
CAGE	5′	Not used	GsuI/MmeI	16-20
Super SAGE	3′	NlaIII	EcoP1SI	26
LS	3′/5′	Not used	MmeI	20

BIOINFORMATICS OF SAGE

Bioinformatics is fundamental for the processing and interpretation of any SAGE experiment. First, specific software is needed to process the sequence reads extracting the correct tags from the concatenated sequence and calculating their frequency in a given library. This complete catalog of tags and their frequencies is not fully informative due to the short length of the tags. Therefore, much effort has been devoted to the development of protocols that aim to assign tags to genes. Because tags are, in principle, located immediately downstream of the 3′ most NlaIII site, it is possible to deduce a "virtual" tag from the cDNA sequence of all known genes.

A database of these virtual tags can then be searched using the "real" tags as queries. The first attempt to build a computational protocol for the annotation of SAGE tags came from NCBI with their SAGEMap. SAGEMap constructs a "reliable" mapping by selecting the best connection between a tag and a gene above a specific threshold. The score of this connection is given by an equation that takes into account the diversity of the nucleotide sequence types (linked to a given tag) and the number of sequences agreeing on the connection. SAGE Genie is an additional protocol for the mapping of tags to genes and vice versa.

The basis of SAGE Genie is a differential scoring of databases according to the representation of the 3' most SAGE tag. For instance, a tag matching a more reliable database, like the one represented by sequences from the Mammalian Gene Collection (MGC) initiative, gets a higher score than a tag matching the EST database. SAGE Genie takes into account as well the effect of several phenomena on the assignment. Alternative polyadenylation can generate an alternative tag through the deletion of the original 3′ NlaIII site (in the case of a shorter transcript) or the creation of a new 3′ most site (in the case of a longer transcript). An artifact commonly found in cDNA libraries is internal priming, in which an artifactual cDNA is synthesized through the annealing of a polyT primer to an internal stretch of As (a common feature in the UnTranslated Regions (UTRs) of mammalian genes). Both features were evaluated in the original publication of SAGE Genie.

More recently, da Silva et. al., evaluated the effect of SNPs on the tag-to-gene assignment. This will be further discussed later. All the data from SAGE Genie is available through a portal on the World Wide Web (http://cgap.nci.nih.gov/SAGE/). In the same portal, several computational tools are available for the analysis of the data. A serious shortcoming from both SAGE Genie and SAGEMap is the absence of a service that allows the user to submit a batch of sequences for processing.

Very recently, Galante et. al., developed Automatic Correspondence of Tags and Genes (ACTG), a Web-based tool that allows a large-scale tag-to-gene mapping using several reference databases. ACTG can map data from SAGE, MPSS, and Sequencing By Synthesis (SBS)-based technologies. Long SAGE and MPSS data are also suitable, due to their length, for genome mapping. This is important to place the tag into a genomic context. Nowadays, it is relatively common for a given species to have its genome sequenced without, however, many corresponding cDNAs. In this case, with a longer tag, one can map the tag to the genome, a process that helps the tag-to-gene assignment.

SAGE AND MPSS EXPERIMENTS

In principle, alternative tags can be associated with SNPs present within the tag sequence or within the restriction site used for SAGE and MPSS library construction. Using a computational strategy based on the comapping of tags, SNPs, and genes onto the genome sequence, da Silva et. al., performed a genome-wide analysis on this association. The identification of SNP-associated alternative tags was categorized in three different types. The first category encompass all cases in which a SNP generated a new restriction enzyme site downstream to the original tag, producing in this way a new 3′ most tag.

The second type corresponds with those cases in which the presence of a SNP disrupted the 3′ most restriction site associated with the original tag making the immediate upstream site to become the 3′ most site. Finally, a SNP could be located within the tag sequence itself producing a SNP-associated tag with one single mismatch when compared with the original tag. We found more than 2000 SNP-associated tags using publicly available SAGE and MPSS data. A fraction of these tags were validated using the Generation of Long cDNA fragments for Gene Identification-Massively Parallel Signature Sequencing (GLGI-MPSS) methodology. Furthermore, we were able to genotype the HB4a cell line for the presence of the SNP associated with the alternative tag.

PROTOCOL FOR MICROSAGE

Here we provide a day-by-day procedure for the construction of MicroSage libraries. The length of the procedure can be longer if stopping points (usually precipitation steps) are used.

Day 1

Total RNA Isolation

1. Start with at least 100,000 cells in order to obtain 1 µg total RNA, the absolute minimum amount of starting material. Optimally, 10 µg total RNA would be used for MicroSAGE.
2. Use RNA isolation kit/protocol of your choice.
3. Check quality of RNA by gel.

Kinase Linkers

Perform the following protocol in PCR-free conditions.

1. Dilute Linker IB, 2B, 1A, and 2A to 350 ng/µL
2. Make the following mixes to phosphorylate Linkers 1B and 2B:

Contents	*Tube 1*	*Tube 2*
Linker 1B (350 ng/µL)	9 µL	–
Linker 2B (350 ng/µL)	–	9 µL
LoTE	6 µL	6 µL
10X kinase buffer (NEB)	2 µL	2 µL
10 mM ATP	2 µL	2 µL
T4 polynucleotide kinase (10 U/µL)	1 µL	1 µL

3. Incubate at 37°C for 30 min.
4. Heat inactivate at 65°C for 10 min.
5. In one tube (labeled "Linker 1AB"), mix 9 μL Linker 1A (350 ng/μL) to the 20 μL phosphorylated Linker 1B prepared above (final concentration 200 ng/μL); in a second tube (labeled "Linker 2AB"), mix 9 μL Linker 2A (350 ng/μL) to the 20 μL phosphorylated Linker 2B prepared above (final concentration 200 ng/μL).
6. In order to anneal the linkers, perform the following incubations for both tubes:

   95°C, 2 min

   65°C, 10 min

   37°C, 10 min

   Room temperature, 20 min

   Store at –80°C
7. Phosphorylation reaction should be tested by self ligation as follows (both Linker 1AB and Linker 2AB should be tested):

Anneal mix	2 μL (from **step 6,** above)
5X ligase buffer	2 μL
T4 ligase (5U)	1 μL
ddH_2O	5 μL
Total	10 μL

8. Incubate at 37°C for 30 min.
9. Analyze on a 1.5-mm, 12% PolyAcrilamide Gel Electrophoresis- Tris-AcEtate (PAGE-TAE). Analyze the size of the bands by comparing to 1-kb and 100-bp DNA markers. Run the gel until the purple bromophenol band gets near the bottom of the gel. The ligated dimers should run at 80 bp and the monomers at 40 bp.
10. Load all of the self-ligated linkers and 1 μL of the annealed mixes (as negative controls). Phosphorylated linkers should allow linker-linker dimers (80 to 100 bp) to form after ligation, whereas dephosphorylated linkers will prevent self-ligation. Only linker pairs that self-ligate >70% (with <30% of reaction unligated) should be used in subsequent steps.

Day 2

Isolation of Poly + RNA and cDNA Synthesis

Thoroughly resuspend Dynabeads oligo (dT)25 (from mRNA direct kit; Dynal). Transfer 100 to 120 μL to an RNase-free, 1.7-mL Eppendorf tube.

Place on magnetic Eppendorf tube holder (Dynal).

After ~30 s, remove the supernatant. Always place the pipette tip at the opposite side of the tube, push the pipette tip to the bottom, and pipette very slowly, so you won't disturb, and thus lose, the beads. (This is a washing step.)

Resuspend the beads in 500 µL lysis/binding buffer (from kit) to preserve the beads until ready to add the RNA. In all the washing steps, add solution to the tube while it is still on the magnetic stand in order to minimize "drying out" the beads, then close the cap and remove it from the magnet. Resuspend the beads by "flicking" the tube or by gentle pipetting. Leave the beads in this buffer until ready to add them to the RNA.

Mix 1 to 10 µg of total RNA with 1 µL lysis/binding buffer—the appropriate volume of RNA depends on the RNA concentration, so be sure that RNA volume does not exceed 100 µL.

Remove the supernatant from the Dynabeads.

Add the RNA/buffer mixture to these prewashed Dynabeads.

Incubate at room temperature for 5 min with constant agitation.

Place the tube on the magnet for 2 min or until solution is completely clear, then remove supernatant (the supernatant can be saved and used for DNA prep).

Wash 2 × 1 mL Buffer A/washing buffer containing Lithium Dodecil Sulfate (LIDS) + 20 µg/mL glycogen (add 1 µL 20 mg/mL stock glycogen for 1 mL of buffer)—remember to wash by pipetting up and down, placing the tube in the magnet, and removing the supernatant once the solution is clear (removes non-polyA$^+$ RNA and genomic DNA).

Wash 1 × 1 mL Buffer B + 20 µg/mL glycogen. This buffer does not contain LiDS, so the beads will not conglomerate on the magnet as quickly and cleanly as the previous washes—be careful not to remove any beads!

Wash 4 × 100 µL 1X first strand buffer (dilute 5X first strand buffer from cDNA synthesis kit [Invitrogen] 1:5 with Di Ethyl Pyro Carbonate (DEPC) dH_2O).

Remove the supernatant and resuspend beads in the following first strand synthesis mix (components from the cDNA synthesis kit) in the following order:

DEPC dH_2O	54 µL
5X First strand buffer	18 µL
0.1 M DTT	9 µL
10 mM DNTP	4.5 µL
Total	85.5 µL

Place the tube at 42°C for 2 min.

Add 3 µL Superscript RTII.

Incubate at 42°C for 1 h and mix the beads every 10 min by gently tapping the tube.

After incubation, place the tube on *ice* to terminate the reaction.

Immediately, with the tube still on ice, add the following components for the second strand synthesis (components from the cDNA synthesis kit), in the order shown below:

DEPC dH_2O (prechilled)	302 µL
10X second strand buffer	75 µL
10 mM dNTP	15 µL

E. coli DNA ligase*	3 µL
E. coli DNA pol I	12 µL
E. coli Rnase H	3 µL
Total	500 µL

Incubate at 16°C for 2 h, mix beads every 10 min by flicking the tube gently. During this incubation period, prepare the solutions for the rest of the protocol.

After incubation, place the tubes on ice and terminate the reaction by adding 40 µL 0.5 M EDTA, pH 8.0.

Place the beads on the magnet and remove the supernatant.

Wash the beads 1 × 0.5 mL 1X BW + 1% SDS, made as follows (if necessary, heat the buffer at 75°C for 2 min to dissolve any precipitated SDS or BSA):

2X BW buffer	500 µL
20% SDS	50 µL
ddH_2O	450 µL
Total	1 mL

Resuspend the beads in 200 µL 1X BW + 1% SDS and heat at 75°C for 20 min.

Wash 4 × 200 µL 1X BW + 2X BSA, made as follows:

2X BW buffer	500 µL
100X BSA	20 µL
ddH_2O	480 µL
Total	1 mL

Wash 2 × 200 µL 1X Buffer 4 (NEB Biolabs) + 2X BSA (transfer beads to a new tube after the first wash) made as follows:

Buffer 4	100 µL
100XBSA	20 µL
ddH_2O	880 µL
Total	1 mL

We use 2X BSA in all the washing buffers because more BSA seems to reduce bead stickiness and improve the efficiency of the washes and the quality of the library. After the SDS washing/heating step, the beads are sticky until the first BSA wash, but then they are okay.

Cleavage of cDNA with Anchoring Enzyme (NlaII)

Resuspend the beads in the following mix and incubate at 37°C for 1 h (mix every 15 min):

LoTE	171 μL
BSA (100X, NEB)	4 μL
10X Buffer 4 (NEB)	20 μL
NlaIII (NEB, no. 125S, in –80°C in lab)	5 μL

Ligating Linkers to cDNA

1. After incubation, remove the supernatant and wash the beads 2 × 200 μL 1X BW + 0.1% SDS + BSA. Prepare as follows:

2X BW	500 μL
20% SDS	5 μL
100X BSA	20 μL
ddH_2O	475 μL
Total	1 mL

   Place on a hot block because SDS and BSA precipitate. (The supernantant precipitates to a certain degree in the SDS buffer, but it is worth using, otherwise the beads stick too much leading to bead loss and inefficient washes.)

2. Wash 4 × 200 μL 1X BW + 2X BSA (freshly made):

2X BW buffer	500 μL
100X BSA	20 μL
ddH_2O	475 μL
Total	1 mL

3. Wash 2 × 200 μL 1X T4 ligase buffer, which is prepared right before use as follows:

5X T4 ligase buffer (thawed just before use)	100 μL
Water	400 μL
Total	500 μL

4. Divide the beads into two tubes and place them on a magnet.
5. Remove the last wash and resuspend the beads as follows:

	Tube 1 (linker 1)	Tube 2 (linker 2)
LoTE	6 μL	6 μL
5X T4 ligase buffer	2 μL	2 μL
Linker 1AB (diluted 1:10 in ddH_2O)	1 μL	0
Linker 2AB (diluted 1:10 in ddH_2O)	0	1 μL

6. Heat tubes at 50°C for 2 min then let sit at RT for 15 min.
7. Add 1 μL 5 U/μL T4 ligase (high concentration; Invitrogen) to each tube (10 μL reaction).
8. Incubate at 16°C for 2 h. Mix beads every 10 to 15 min by gently tapping the tube.

Release Tags Using Tagging Enzyme (BsmF1) of cDNA

1. After ligation, wash each sample 2 × 200 μL 1X BW + 0.1% + 2X BSA (freshly made as follows). (*Note:* pool tube 1 and tube 2 together after first wash in order to minimize loss in subsequent steps.)

2X BW	500 μL
20% SDS	5 μL
100X BSA	20 μL
ddH_2O	475 μL
Total	1 mL

2. Wash 4 × 200 μL 1X BW containing 2X BSA (freshly made as follows):

2X BW	500 μL
100X BSA	20 μL
ddH_2O	480 μL
Total	1 mL

3. Wash 2 × 200 μL 1X Buffer 4 + 2X BSA (freshly made as follows) and transfer to new tubes after the first wash:

Buffer 4	100 μL
100X BSA	20 μL
ddH_2O	880 μL
Total	1 mL

4. Resuspend the beads in the following mix:

LoTE	170 μL
10X Buffer 4	20 μL
100X BSA	4 μL
BsmF1 (NEB, no. 572S)	2 μL
Total	196 μL

BsmF1 produces a 5′ sticky end of 4 bases (10 bp + 4 bp into unknown sequence).

5. Incubate at 65°C for 1 h, mix intermittently.

6. Centrifuge at > 20,000 × *g* for 2 min at 4°C.
7. Transfer supernatant to new tube (not siliconized!). *Keep the supernatant—this contains your actual tags!*
8. Wash beads once with 40 μL LoTE.
9. Pool the 200 μL supernatant and the 40 μL wash together (240 μL final volume).
10. Add 240 μL PC8.
11. Vortex, centrifuge for 5 min at > 11,000 × *g* at 4°C.
12. Transfer the upper aqueous phase to a new tube.
13. High-concentration EtOH precipitate by adding the following components, in order, to the aqueous phase: *Save sample in* –80°C freezer if stopping here, otherwise precipitate by placing in a dry ice/methanol bath for at least 15 min.

7.5 M ammonium acetate	133 μL
20 g/L glycogen	5 μL
100% EtOH	1000 μL

14. Centrifuge for 30 min at >20,000 × *g* at 4°C.
15. Remove the supernatant and wash pellet twice with 200 μL 75% EtOH.
16. Centrifuge again after the last wash, carefully remove the residual liquid with pipette tip, and let dry until all the residual liquid has evaporated.
17. Resuspend pellet in 10 μL LoTE.

Blunt Ending Released Tags

1. Now we have the tags with a protruding end that needs to be filled in to be able to blunt end ligate to form the ditags. Add, in order, the following mix to the resuspended pellet from the previous step:

ddH_2O	25 μL
5X Second strand buffer	10 μL
25 mM dNTPs	1 μL
100X BSA	1 μL
Klenow (2 U/μL)	1 μL

2. Incubate at 37°C for 30 min.
3. Add 190 μL LoTE (240 μL final volume).
4. Add 240 μL PCS (now have 480 μL total).
5. Vortex and then centrifuge for 5 min at >20,000 × *g* at 4°C.
6. Remove 200 μL of the upper aqueous phase containing the nucleic acids and aliquot into a tube labeled "ligase+" and aliquot the rest of the aqueous upper phase into a tube labeled "ligase–" (you can pool several "ligase–" samples together, but make sure

to equalize the volume of "ligase+" and "ligase–" samples by adding 160 μL LoTE to the "ligase–" sample).

7. High-concentration EtOH precipitate both the "ligase+" and "ligase–" tubes by adding, in order, to each tube the following:

7.5 M ammonium acetate	133 μL
Glycogen	5 μL
100% EtOH	900 μL

8. Centrifuge for 30 min at >1 1,000 × g at 4°C.
9. Remove the supernatant but keep it just in case it contains unprecipitated sample.
10. Wash 2 × 200 μL 75% EtOH.
11. Carefully remove residual liquid with pipette tip and air-dry for 5 min (do not overdry).
12. Resuspend the "ligase+" reaction in 5 μL LoTE.
13. Resuspend the "ligase–" reaction in 3 μL LoTE.

Ligating Tags to Form Ditags (Part 1)

1. Prepare 2X mixes are follows :

	2X "ligase+" mix	*Double "ligase+" mix*	*2X "ligase–" mix*
3 mM Tris pH 7.5	1.25 μL	2.5 μL	2.25 μL
5X ligase buffer	1.5 μL	3.0 μL	1.5 μL
T4 ligase (5 U/μL)	1 μL	2.0 μL	0 μL
Total		7.5 μL	3.75 μL

Prepare 3 mM Tris by adding 30 μL 100 mM Tris, pH 7.5, to 1 mL ddH_2O.

2. Add 5 μL 2X "ligase+" mix to the "ligase+" tube. Add 3 μL 2X "ligase–" mix to the "ligase–" tube. Therefore, the total volume of "ligase+" is 10 μL, and the "ligase–" is 6 μL.
3. Incubate overnight at 16°C.

Day 3

Ligating Tags to Form Ditags (Part 2)

1. Add 10 μL LoTE to the "ligase+" tube (20 μL total volume).
2. Add 14 μL LoTE to the "ligase–" tube (20 (μL total volume).
3. Aliquot the 20 μL into four aliquots of 5 μL each so that it isn't necessary to take the entire amount out of the freezer at once in order to reduce degradation.

PCR Amplification of Ditags

1. In order to identify the correct dilution of ditags to use for large-scale PCR amplification, a test PCR must be performed. Dilute 1 μL of each of the "ligase+" and "ligase–" reactions as follows:

1/50 "ligase+"	1/100 "ligase+"	1/200 "ligase+"
1 μL (+)cDNA →	10 μL 1/50 "ligase+" →	10 μL 1/100 " ligase+"
+ 49 μL ddH_2O	+ 10 μL ddH_2O	+ 10 μL ddH_2O

Reaction mix only.

1/50 "ligase–"	1/100 "ligase–"	1/200 "ligase–"
1 μL (–)cDNA →	10 μL 1/50 "ligase–" →	10 μL 1/100 "ligase–"
+ 49 μL ddH_2O	+ 10 μL ddH_2O	+ 10 μL ddH_2O

Therefore, we have 8 tubes total—use the small 8 well connected PCR tubes for the test PCR.

2. **Perform PCR using the following mix:**

	1 reaction	*10 reactions*
(TEST)		
10X PCR buffer (Appendix A)	5 μL	50 μL
25 mM dNTP (GibcoBRL)	3 μL	30 μL
DMSO (Sigma, no. D-2650	3 μL	30 μL
ddH_2O	35 μL	350 μL
Primer 1 (350 ng/ μL)	1 μL	10 μL
Primer 2 (350 ng/ μL)	1 μL	10 μL
Taq polymerase (5 U/ μL)	1 μL	10 μL
Template (various dilutions)	1 μL	10 μL
	50 μL	500 μL

3. **For the Hybaid MBS thermocycler, use the following optimized reaction scheme:**

Cycles	*Temperature / time*
1 cycle	94°C 1 min
26 cycles	94°C, 30 s; 55°C, 1 min; 70°C, 1 min
1 cycle	70°C, 5 min at end

Use 27 cycles for the "ligase+" reactions and 35 for the "ligase–" reactions. The appropriate cycle number is critical for isolating an adequate amount of ditag DNA for SAGE. Too few cycles will result in a low yield and may cause problems with subsequent steps. Too many cycles will give erratic results and can also result in low yields. Therefore, we recommend trying various cycle numbers (*e.g.*, 26, 28, 30) to determine the optimal number.

4. Remove 10 μL from each reaction and mix with 1 μL loading dye.
5. Electrophorese the samples on a 1-mm 12% polyacrylamide gel. Use 5 μL of a 10-bp ladder and 5 μL of a 100-bp ladder as markers. Electrophorese the gel at 160 V until the purple bromophenol blue band is at the bottom of the gel. Amplified ditags should be 102 bp in size. A background band of equal or lower intensity occurs around 80 bp. All other background bands should be of substantially lower intensity. The "ligase–" samples should not contain any amplified product of the size of the ditags even at 35 cycles.
6. After the PCR test to determine the appropriate dilution, perform large-scale PCR of 300 × 50 μL PCR reactions of the optimal dilution to pool the products. For instance, if the 1/50 dilution works the best and you need 300 reactions, then you will need 6 μL of the "ligase+" dilution + 294 μL LoTE = 1 μL per reaction × 300 reactions = 300 μL. We usually use a 300 reaction PCR premix that we aliquot into three 96-well plates with 50 μL PCR reaction/well.
7. Therefore, you will perform large-scale PCR as follows:

	1 Reaction	*300 Reactions (large scale)*
10X PCR buffer	5 μL	1500 L
25 mM dNTP	3 μL	900 μL
DMSO	3 μL	900 μL
ddH$_2$O	35 μL	10,500 μL
Primer 1 (350 ng/μL)	1 μL	300 μL
Primer 2 (350 ng/μL)	1 μL	300 μL
Regular Taq polymerase	1 μL	300 μL
Template (dilution)	1 μL	300 μL
	50 μL	15 mL (50 mL Falcon tube)

The PCR conditions for the large-scale amplification are as follows:

Cycles	*Temperature / time*
1 cycle	94°C 1 min
26 cycles	94°C, 30 s; 55°C, 1 min; 70°C, 1 min
1 cycle	70°C, 5 min at end

(Use one less cycle for the large scale)

Isolation of Ditags

1. After PCR is complete, centrifuge the 96-well microplates in the swinging bucket centrifuge at 2500 rpm for 10 min.

2. If possible, use a multichannel pipette to pool all the samples in a reagent reservoir to collect all of the PCR products (otherwise use a regular pipette) and then transfer the pooled PCR reactions into a 50-mL Falcon tube.
3. Add an equal volume of PCS and centrifuge in a swinging bucket rotor for 10 min at 2500 rpm.
4. Transfer the upper aqueous phase to a new 50-mL conical tube and EtOH precipitate as follows:

7.5 M ammonium acetate	5.1 mL
Glycogen	191 μL
Pooled PCR products	11.5 mL (use all of the PCR products)
100% EtOH	38.3 mL

5. Invert the tube several times. (At this point, you can keep the PCR products as a precipitate at –80°C and proceed to step 7 the next day or incubate in a dry ice/methanol bath for 15 min and continue).
6. While PCR products are precipitating, make four 1.5-mm-thick 12% acrylamide TAB gels.
7. Divide the sample into two polypropylene/Teflon tubes and balance up to two decimals with 100% EtOH.
8. Centrifuge in a fixed-angle rotor at 11,952 × *g* at *4°C* for 30 min.
9. Wash with 5 mL 75% EtOH and centrifuge for 5 min in the fixed-angle rotor at 11,952 × *g* at 4°C.
10. Air-dry the pellet.
11. Resuspend each pellet in 200 μL LoTE, pool the two tubes into one 400-μL sample, and place it on ice.
12. Add 40 μL 10X loading dye to the purified PCR products.
13. Apply a total of 110 μL to each of the four 12% acrylamide TAE gels.
14. Mix 5 μL of 10-bp DNA ladder with 5 μL 100-bp DNA ladder and load this mixture into a single well.
15. Electrophorese the gel for approximately 1.5 h at 160 V until the blue xylene band is near the bottom of the gel and the purple bromophenol blue band has run off the gel.
16. Stain the gel using SYBR Green I at a 1:10,000 dilution (add 10 μL SYBR Green I to 40 mL IX TAE buffer) in a foil-wrapped container (two gels per box); let the gel soak in the stain for 15 min on the shaker.
17. Visualize the gel on a UV box using the yellow SYBR Green filter. Protect the DNA by putting an EtOH-cleaned glass plate on the UV box, and then put the gel on the glass plate.
18. Cut out only the amplified ditag (102 bp) band from the gel. Be sure to not take the markers and 80-bp background band.
19. Place one third of each band in an 0.5-mL microcentrifuge tube (12 tubes total) processed.

20. Place the 0.5-mL microcentrifuge tubes in 2-mL round-bottom microcentrifuge tubes and centrifuge in microfuge at > 20,000 × *g* for 5 min at RT (this serves to break up the cut-out-bands into small fragments at the bottom of the 2.0-mL microcentrifuge tubes). If most of the gel is not contained in the larger microfuge tube, then centrifuge for 5 more min.
21. Discard 0.5-mL tubes and add 250 μL LoTE and 50 μL 7.5 M ammonium acetate to each 2.0-mL tube.
22. Tubes can remain at this point at 4°C overnight (do not leave for more than one night or the acrylamide can break down too much, and do not store at –80°C).
23. Vortex each tube, place at 65°C? for 15 min, and centrifuge briefly to collect the condensation at the bottom of the tube.
24. Place 5 μL LoTE on the center of the membrane of each of 12 Spin-X tubes.
25. Transfer contents of each tube to a Spin-X tube by mixing the contents well and using a 1000 tip that has been trimmed to have an opening of approximately 2 mm to transfer to Spin-X tubes. Use a spatula to scoop out and transfer any residual gel pieces.
26. Centrifuge each Spin-X tube for 5 min at >20,000 × *g* at RT and throw away Spin-X filter.
27. EtOH precipitate the eluates in new 1.7-mL tubes as follows:
    300 μL sample
    133 μL 7.5M NH_4OAc
    5 μL glycogen
    1000 μL 100% EtOH.
28. Centrifuge in a microcentrifuge at >20,000 × *g* for 30 min at 4°C.
29. Wash 2 × 200 μL 75% EtOH; for each wash, centrifuge for 5 min at 4°C.
30. Resuspend DNA in 10 μL LoTE/tube.
31. Pool samples into one tube (120 μL total).
32. Remove 1 μL for cDNA quantitation. The total amount of DNA at this stage should be 10 to 20 μg. If the concentration is not this high, redo the large-scale PCR prep/precipitation and then combine both large-scale 102-bp band preps. If the DNA concentration is too high, then the sample can be split in half and processed as two samples on day 4. If the sample is split in half, then the reaction should be brought to volume with LoTE.

Day 4

Purification of Ditags

1. Digest the 102-bp DNA with NlaIII by adding the following to the same sample tube:

PCR products in LoTE	120 μL
LoTE	196 μL
10X NEB Buffer 4	40 μL
100X BSA	4 μL
NlaIII (10 U/μL)	40 μL
Total	40 μL

Throw away the last of the NlaIII because it is necessary that the enzyme works perfectly, and if the enzyme is frozen and dethawed again, it will loose some activity.

2. Incubate 1 h at 37°C.
3. Extract with equal volume PCS (400 μL).
4. Vortex, centrifuge for 5 min at >20,000 × *g* at 4°C. (*Note:* be very careful in the following steps because now we are working with small 24- to 26-bp DNA segments. Keep reactions on ice).
5. Transfer the upper aqueous phase into two tubes (200 μL each) and then EtOH precipitate as follows:

   200 μL sample

   66 μL 7.5 M NH_4OAC

   5 μL glycogen

   825 μL 100% EtOH

   (*Note:* Can stop here to precipitate at –80°C overnight if desired. If doing so, the dry ice/methanol bath is not needed for the quick DNA precipitation. Continue the next day with centrifugation.)
6. Place in dry ice/methanol bath for 15 min.
7. Warm the tubes at RT for 2 min until solution has melted.
8. Centrifuge the tubes at 4°C for 30 min at >20,000 × *g*.
9. Remove the supernatant.
10. Wash 1 × 200 μL cold 75% EtOH. The EtOH is cold to protect the 26-bp ditags from denaturation. The melting point of 26-bp DNA is below room temperature.
11. Remove all EtOH by pipetting and briefly air-drying the pellet.
12. Resuspend the pellet in each tube in 10 μL of cold TE (not LoTE). This high-concentration Tris buffer is needed to protect the 26-bp DNA ditags.
13. Pool the resuspended DNA into one tube (20 μL total) and place on ice.
14. Add 2 μL 10X loading dye/sample buffer (22 μL total volume).
15. Load 5 μL of 10-bp and 100-bp ladders in separate wells on both sides of the four lanes of sample. Leave an open lane between the sample lanes and the marker lanes. Load 5.5 μL of this sample into four lanes of a 1.5-mm, 12% polyacrylamide TAE gel (10-well) and electrophorese at 100 V to 140 V. *Note:* do not run the voltage too high or your 26-bp product could melt. Do not run both loading dyes off the gel: the bromophenol blue front (runs about 45 bp) should be approximately 1.5 to 2 cm from the bottom of the gel.
16. Stain the gel using SYBR Green I stain, at a 1:10,000 dilution.
17. Cut out the 24- to 26-bp band from the four lanes (the band is between 25 bp and 30 bp). Remember to put an EtOH-cleaned glass plate on top on the UV box and then place the gel on the plate.
18. Place two excised bands in an 0.5-mL microcentrifuge tube (two tubes total). More 0.5-mL microcentrifuge tubes can be used if more lanes are loaded with 26-bp ditags or if the bands are large. Centrifuge in the microfuge at >20,000 × *g* at RT for 5 min or until the excised band is completely collected in the 2-mL tubes.

19. Discard the 0.5-mL tubes, and add 50 μL 7.5 M ammonium acetate and 250 μL LoTE (in this order) to the 2-mL tubes.
20. Vortex the tubes, and place at 37°C (*not* 65°C) for 20 min.
21. Use four Spin-X tubes to isolate the eluate. Spin for 5 min at >20,000 × *g* at RT.
22. EtOH precipitate in three tubes (200 μL each) as follows:
    200 μL sample
    66 μL 7.5M NH_4OAc
    5 μL glycogen
    825 μL 100% EtOH.
23. You can stop here and incubate overnight at –80°C to increase the amount of DNA precipitated. If you do this, the next day start at the centrifugation step and skip the dry ice/methanol bath step.
24. Place in a dry ice/methanol bath for 15 min.
25. Centrifuge at 4°C at >20,000 × *g* for 30 min.
26. Wash 2 × 200 μL of cold 75% EtOH.
27. Air-dry to remove all residual EtOH.
28. Resuspend each DNA sample in 2.5 μL cold LoTE (be sure the total volume is 8 μL after resuspending the pellet).
29. Remove 1 μL of the purified ditags for quantitating while ligating.

Day 5

Ligation of Ditags to Form Concatamers

Length of ligation time depends on quantity and purity of ditags. Typically, several hundred nanograms of ditags are isolated and produce large concatamers when the ligation reaction is carried for 1 to 3 h at 16°C (lower quantities or less pure ditags will require longer ligations).

1. Mix the following:

Pooled purified ditags	7 μL
5X ligation buffer (BRL)	2 μL
T4 ligase (5 U/μL)	1 μL

   Thaw the ligation buffer slowly because it contains ATP, which degrades easily. If the volume of pooled purified ditags is high, the reaction volume can be increased (*e.g.*, 10 μL ditags + 3 μL 5X ligase buffer + 1 μL ligase + 1 μL LoTE).
2. Incubate for 1 h 10 min if you started with 10 μg of total RNA at 16°C. (Fewer ditags require a shorter incubation.)
3. Heat the sample 65°C for 5 min.
4. Place it on ice for 10 min.
5. Add 1.5 μL 10X sample buffer/loading dye (5 μL 10X dye and 5 μL water) to ligation reaction.
6. Electrophorese a 1-mm, 8% polyacrylamide TAE gel electrophoresis (with a 10-well comb) to separate the concatamers. Have this gel ready by the end of the ligation. In the first lane of the 8% polyacrylamide, load 10 μL of a 1-kb ladder (25 ng/ μL) as a marker. Load entire concatenated sample into the fifth well (*i.e.*, one lane). Load as follows:

(1) empty
(2) 100-bp ladder
(3) 1-kb ladder
(4) empty
(5) entire sample
(6) empty
(7) 1-kb ladder.

7. Samples are electrophoresed at 130 V until the purple bromophenol blue dye reaches the bottom of the gel. Keep an eye on the gel because the running speed depends on the amount of buffer added to the buffer chamber.
8. Stain the gel with SYBR Green I 1:10,000 dilution. (This gel is thin and prone to rip. Be careful when taking it apart—wet the gel with buffer to take it off the plastic plate).
9. Visualize on UV box using the yellow SYBR Green filter. Concatamers will form a smear on gel with a range from about 100 bases to several kb. We usually isolate region 800-1500 bases and 1.5-3.0 kb.
10. Place each of these gel pieces into an 0.5-mL microcentrifuge tube with a needle-pierced bottom (two tubes total).
11. Place the tubes in a 2.0-mL round-bottom microcentrifuge tube and centrifuge at >20,000 × *g* for 5 min at RT.
12. Discard the 0.5-mL tubes and add 300 μL LoTE to the 2-mL tubes.
13. Vortex the tubes, and place at 65°C for 20 min.
14. Transfer the contents of each tube into one Spin-X microcentrifuge tubes (two Spin-X tubes total).
15. Centrifuge the Spin-X tubes in microcentrifuge for 5 min at >20,000 × *g* at 4°C.
16. EtOH precipitate both the high and low weight concatamers as follows:
    300 μL eluate
    133 μL 7.5 M NH_4OAc
    3 μL glycogen
    1000 μL 100% EtOH.
17. Can stop here and precipitate at –80°C overnight, or precipitate in an EtOH/dry ice bath for 15 min and continue.

Cloning Concatamers

Prepare the pZero cloning vector and test on a gel. pZero is aliquoted in 2.5-μL aliquots and stored at –20°C. (If the pZero preparation is done in the morning, then the ligation of the concatamers into the pZero can be done in the afternoon for an overnight incubation.)

Digest as follows:

pZero (1 μL)	1 μL
NEB buffer 2	1 μL
SphI	1 μL
Sterile H_2O	7 μL
Total	10 μL

Incubate for 20 min at 37°C.

Remove 1 μL of the digestion to electrophorese on a 1% agarose TAE gel. Electrophorese 1 μL of the digestion on a 1% agarose TAE gel beside a 1-kb ladder, and pZero vacant. pZero should migrate in the gel at approximately 2.5 kbp.

Add 30 ML LoTE to the rest of the sample and heat it at 70°C for 10 min to inactivate the enzyme. It is now 25 ng/μL, and 1 μL can be used per ligation.

Precipitate the concatamers in a methanol/dry ice bath for 15 min.

Centrifuge the concatamers and pZero at >20,000 × *g* for 30 min at 4°C.

Wash the concatamers and pZero with 2 × 200 μL 75% EtOH, centrifuge, and remove EtOH.

Resuspend each tube of the purified concatamer DNA in 6 μL LoTE.

Resuspend the pZero in 30 μL LoTE (approx. 30 ng/μL).

Ligate concatamers into pZero as follows [the following volumes are in oligo (dT)25]:

Fraction	*High fraction*	*Low fraction*	*Vector only*	*No ligase added*
Vector	1	1	1	1
5X ligase buffer	2	2	2	2
T4 ligase (5 U/μL)	1	1	1	0
Dwater	0	0	6	7
Concatamers	6	6	0	0

Incubate overnight at 16°C in 1.7-mL Eppendorf tubes. (*Note:* Before setting up the ligation, make sure that you can continue the next day.)

Day 6

Electroporation of Ligation Products

1. Add 190 μL LoTE to the ligation mix to bring the sample volume to 200 μL.
2. Add 200 μL PC8.
3. Briefly vortex and centrifuge for 5 min at >20,000 × *g* at RT.
4. Transfer the upper aqueous upper layer to a new tube.
5. EtOH precipitate the aqueous phase in 1.7-mL Eppendorf tubes as follows:

   200 μL sample

   133 μL 7.5 M ammonium acetate

   5 μL glycogen

   777 μL 100% EtOH.
6. Place EtOH precipitating DNA in the –80°C freezer until ready to electroporate.
7. Make the LB media and the LB/Zeocin plates.

8. Centrifuge at >20,000 × g at 4°C for 30 min.
9. Wash 4 × 200 µL 70% EtOH by centrifuging for 5 min at >20,000 × g at 4°C.
10. After the last wash, remove EtOH and air-dry pellets.
11. Resuspend each pellet in 10 µL LoTE—these are your ligation mixtures—and place on ice.
12. For electroporation, place the required number of 0.1 cm electroporation cuvettes and 1.7 mL microcentrifuge tubes on ice (four for one SAGE library).
13. Thaw one vial (100 µL) of ElectroMAX DH10B cells on ice. Before use, mix by pipetting the cells up and down several times and pipet 25 µL of the competent cells into each 1.7-mL microcentrifuge tube.
14. Pipet 1 µL of the ligation mixture into the competent cells of the appropriate 1.7-mL microcentrifuge tube. Keep the rest of ligation mixtures at –80°C in case a second electroporation is required.
15. Pipette the ligation mixture-competent cells into a chilled disposable cuvette. Keep the cuvettes on ice.
16. Set the *E. coli* electroporator pulser apparatus to 1.8 kV when using the cold 0.1-mm cuvettes (Do not touch the metal portion of the cuvette that touches the power supply.)
17. Pulse once. Hold the current for 4 s. Check that the electroporation worked for the required amount of time by pressing the left two buttons at the same time. The display should read 5.0 if the electroporation took place for 4 s. After electroporation, do not place the sample back on ice.
18. Remove the cuvette from the chamber and *immediately* add 1 mL LB medium and quickly but gently resuspend the cells.
19. Transfer the cells to a 15-mL conical tube and incubate while shaking at 100 rpm at 37 °C for 30 min to 1 h. Only close the Falcon tubes to the first closure point so the culture can "breathe" while shaking.
20. Plate the transformation mixture onto fresh LB-Zeocin plates (100 µg/mL Zeocin). Various dilutions of both the high and low concatamer fractions are pipetted onto the LB-Zeocin plates as follows:

    100 µL undiluted electroporated competent cell culture (two plates)

    10 µL electroporated competent cell culture diluted with 100 µL LB media (two plates)

    100 µL of the two negative controls (vector only, no ligase) (two plates)
21. The cells are spread by using glass beads. Use standard sterile technique, including flaming all bottles when opening and closing.
22. Incubate overnight at 37°C. Make sure to incubate the plates upside down. Keep rest of transformation mixtures at 4°C. Insert-containing plates should have hundreds to thousands of colonies while control plates should have 0 to <20 colonies.

Day 7

Colony PCR

1. Check insert sizes of the clones by colony PCR. Check at least 20 colonies from each ligation dilution plate to calculate the average insert size. Make the appropriate amount of PCR master mix in the PCR free conditions as follows:

	Per reaction
10X PCR buffer	2.5 μL
DMSO	1.25 μL
25 mM dNTPs	0.5 μL
M13F (350 ng/μL)	0.5 μL
M13R (350 ng/μL)	0.5 μL
ddH_2O	19.25 μL
Taq (5 U/μL)	0.5 μL
Total	25 μL

2. Aliquot the PCR reactions a 96-well plate (25 μL/well) in PCR free conditions.
3. For each colony, use a sterile pipet tip or a toothpick to gently touch one colony, and then dip the tip into the PCR mix.
4. Perform PCR as follows (optimized for HYBAID OMNE machine) :

1 cycle	95°C, 2 min
30 cycles	95°C, 30 s; 56°C, 1 min; 70°C, 1 min
1 cycle	70°C, 5 min

5. Load a 1-kb and 100-b ladder to determine the insert size. Electrophorese 5 μL of each reaction on a 1.5% agarose TAB gel at 100 V until the purple bromophenol blue band is near the bottom of the gel. A library is good if >90% of the colonies have inserts of at least 700 bp after subtracting the 226 bp (represents no insert). When estimating the average insert size, include the blank lanes.

CONCLUSION

The identification of alleles associated with specific traits will enter a revolutionary period with the new sequencing technologies. Genome and cDNA sequencing will become even more common. The acquisition of both a personal genome and phenome will represent the effective pillar for personalized medicine. We envisage that the strategy developed by us will be particularly informative as it allows the association of expression and genotype data. Furthermore, it can quantify the expression of the different alleles in a heterozygous subject by simply counting the number of the respective tags.

CHAPTER

6

Biochemical Events During Replication

Each daughter cell produced on cell division contains an identical copy of the genetic material. Since DNA carries the genetic blue-print of the cell encoded in the sequence of nucleotides, the question as to how DNA is reproduced in the cell has attracted a great deal of attention. Studies with intact cells have shown that replication is semiconservative and that it occurs in a discontinuous manner and work with purified enzymes has indicated the detailed mechanism of the polymerization reaction. However, it was the realization of the complexity of the reaction which is required to replicate DNA faithfully and rapidly that led to the development of several *in vitro* systems and finally to the reconstitution from purified components of complexes capable of limited replication. Modern techniques of gene cloning, *in vitro* mutagenesis and DNA sequencing have facilitated the dissection of several origins of replication and we are now beginning to gain some insight into the controls exerted over the event of initiation of replication which is an important, and possibly the major, site at which cell multiplication is controlled.

SEMICONSERVATIVE REPLICATION

At the end of their paper suggesting the double-stranded structure for DNA with its complementary base pairs, Watson and Crick wrote: 'It has not escaped our notice that the specific pairing we have postulated suggests a possible copying mechanism for the genetic material'.

Let us suppose that a short length of a DNA double helix has the nucleotide sequence shown in Fig. 6.1 and further that the two strands can be untwisted and separated another to form two single chains (middle), and that each base in the single strands can attach to itself the complementary deoxyribonucleotide by the same hydrogen-bonding

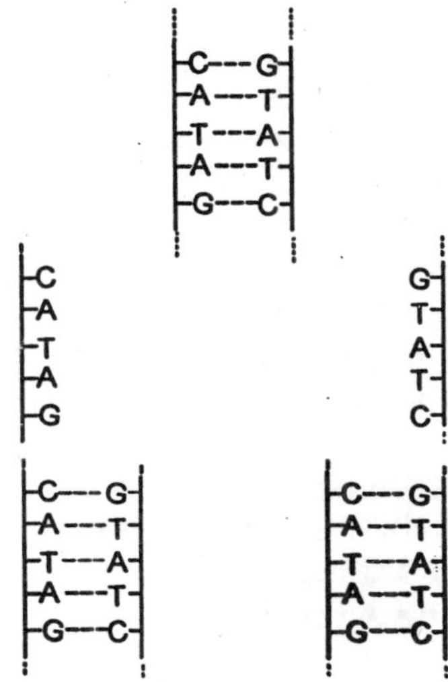

Fig. 6.1. Schematic illustration of the separation of the two strands of aportion of DNA with the formation of a new strand on each.

which exists in the intact DNA double helix. Finally, if these attached mononucleotides are polymerized to form a polynucleotide chain, the end result will be the formation of two complete DNA double helices identical with each other and with the original molecule. One strand of each daughter molecule will be derived from the original DNA molecule; the other will be the product of the new synthesis.

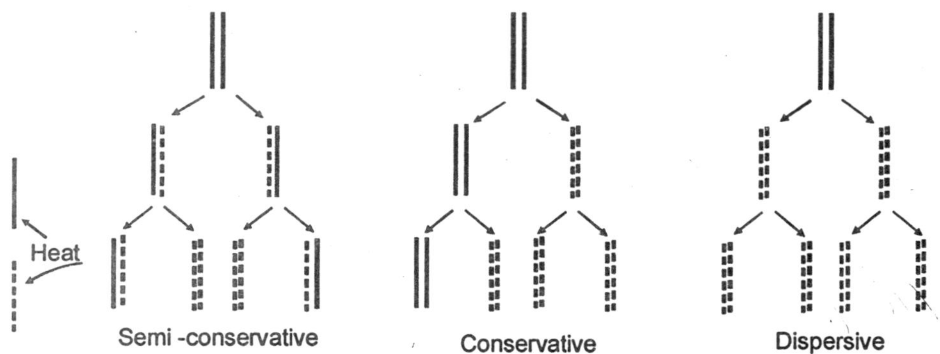

Fig. 6.2. Possible mechanisms of replication. The dotted lines represent filial strands. For details see text.

This mechanism is described as *semicon servative* to distinguish it from the other possible mechanisms. In the *conservative* mechanism the two strands do not come apart but act together as a template to form a completely new double-helical molecule; in this case one daughter molecule would be wholly new and the other totally derived from the parent. In the *dispersive* mechanism the parental molecule is partly degraded and the fragments are incorporated into two new daughter double helices. Convincing evidence for semiconservative replication of DNA was obtained by Meselson and Stahl who grew *E. coli* in a medium containing $^{15}NH_4Cl$ of 96.5% isotopic purity for fourteen generations so as to label the DNA very heavily with ^{15}N. The cells were then transferred to a medium containing $^{14}NH_4Cl$ and samples of bacteria were withdrawn at intervals for several generations. Each sample was lysed by means of sodium dodecyl sulphate and was centrifuged in a concentrated solution of caesium chloride at 140 000g for 20 hours to enable the DNA to attain sedimentation equilibrium.

The bands of DNA were found in the CsCl gradient in the region of density 1.71 g cm^{-3} and were well isolated from all other macromolecular components of the bacterial lysate. Ultraviolet absorption photographs taken during the course of the run revealed the position of the DNA bands. At the start of the experiment the DNA appeared as one single band corresponding to the heavy ^{15}N-labelled nucleic acid. Macromolecules containing half this level of ^{15}N then began to appear, and one generation time after the addition of ^{14}N these hybrid (light-heavy, ^{14}N-^{15}N) molecules alone were present. Subsequently a mixture of light-heavy (^{14}N-^{15}N) DNA and unlabelled (light-light, ^{14}N only) DNA was found. When two

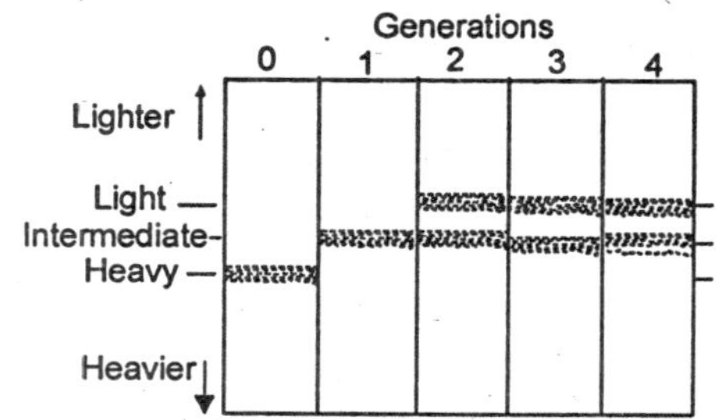

Fig. 6.3. The pattern of results from the Meselson and Stahl experiment.

generation times had elapsed after the addition of ^{14}N, the half-labelled and unlabelled DNA molecules were present in equal amounts. During subsequent generations the unlabelled DNA accumulated. Moreover, when the hybrid ^{14}N–^{15}N molecules were heated they separated to give a ^{14}N strand and a ^{15}N strand. Such experiments indicate that in DNA synthesis each existing DNA molecule is split into two subunits, each subunit going to a different daughter molecule.

The other subunit of each daughter molecule is the product of new synthesis. These subunits do not undergo any fragmentation but remain intact for many generations. Experiments at the chromosomal level also support the view that DNA replication is semiconservative. When the replication of the *E. coli* chromosome is followed by radioautography after labelling of the DNA with tritium, the amount of tritium per unit length of newly synthesized DNA is consistent with the presence of only one newly synthesized strand in a daughter chromosome. In plants to which tritiated thymidine has been given as a specific precursor of DNA during the period of DNA synthesis, both the daughter chromatids are found to be labelled at the time of cell division. However, at the next round of duplication after withdrawal of the [^{3}H]thy-midine, these two chromosomes each produce one labelled and one unlabelled chromatid as would be expected.

THE REPLICATION FORK

Discontinuous Synthesis

In the above section we have shown the two parental DNA strands separating from one another before the daughter strands are synthesized. However, in a long DNA molecule replication only occurs over one short stretch at a time and the two parental strands separate only at the point of replication to produce a Y-shaped molecule as the replication fork passes along the DNA. Replication forks can be seen in electron micrographs when they appear as bubbles in replicating DNA molecules. One daughter DNA chain grows in the $5' \rightarrow 3'$ direction and this occurs by the addition of an incoming—deoxyribonucleotide to the 3′-hydroxyl at the end of the growing *primer* chain. This is the reaction catalysed by DNA polymerase. As the two DNA chains are antiparallel (*i.e.,* one is running $5' \rightarrow 3'$ and other $3' \rightarrow 5'$) this poses the problem as to whether the two daughter strands are synthesized by two different mechanisms: one by adding nucleotides to a growing 3′-end and the other to a growing 5′-end.

Addition to the 5′-end may involve deoxyribonucleoside 3′-triphosphates or, alternatively, may result in DNA molecules having terminal 5′-triphosphates. Such termini have not been found and, moreover, addition of nucleotides to a 5′-triphosphate-terminated primer is incompatible with the proof-reading function of DNA polymerase. This function requires the removal of a mismatched terminal residue. If this residue held the energy for the polymerization, *i.e.* if it were a triphosphate, then this energy would be dissipated on proof-reading and polymerization would halt.

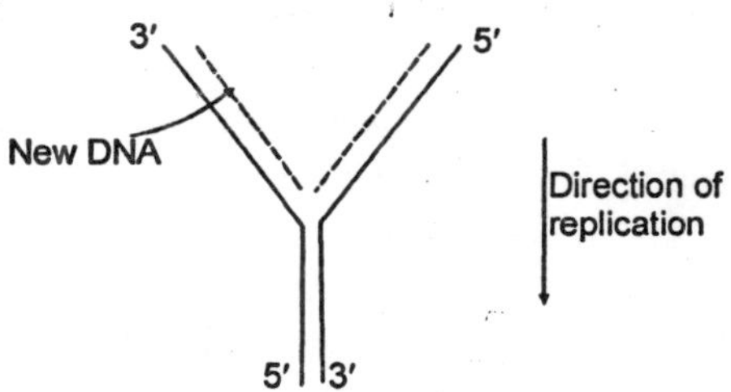

Fig. 6.4. A replication fork.

Other evidence argues against the extension of DNA chains at the 5′-ends.

(1) Many DNA **polymerases have** been investigated **and they all** add **deoxyribonucleoside 5′-phosphates** on to a 3′-OH **group on the growing** DNA chain, *i.e.* the **primer.**

(2) No enzymes **have been found** which will generate or polymerize deoxyribonucleoside 3′-triphosptiates, and neither have these putative substrates been found in cells.

(3) Both daughter strands have been shown to grow in the 5′ → 3′ direction.

The alternative explanation put forward by Okazaki is that both chains are synthesized by the same mechanism but that one is made 'backwards' in short pieces which are subsequently joined together by DNA ligase. Careful investigation of electron micrographs (which give a resolution of about 100 nucleotides) shows that on one side of a replication fork the DNA is single-stranded and this supports the '*discontinuous*' mechanism. In this model the strand of DNA which is made continuously in the direction of fork movement is called the *leading strand* while that made discontinuously in the reverse mode is called the *lagging strand*.

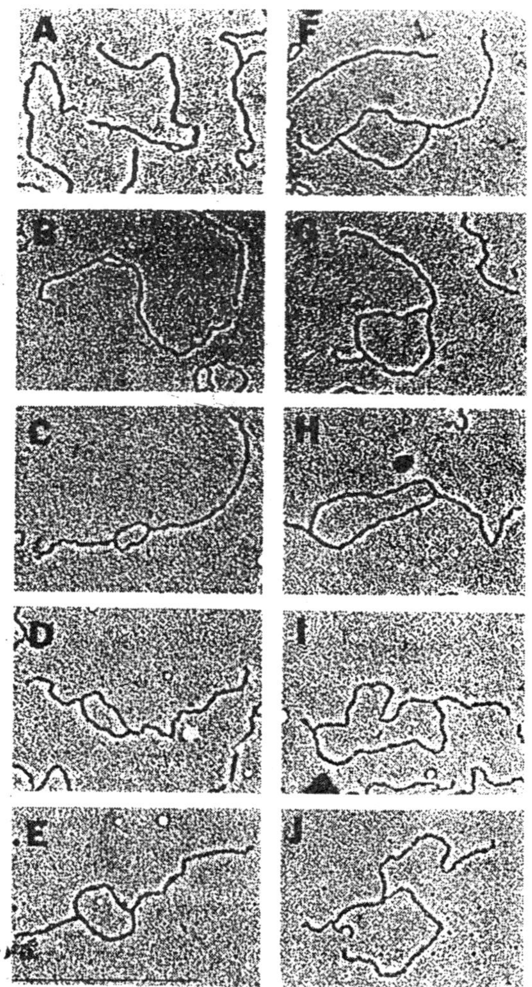

Fig. 6.5. Replicating SV40 DNA molecules. The molecules have been cut at a unique site with the restriction endonuclease EcoRI and have been arranged in increasing degree of replication (A through J) and oriented with the short branch at the left.

Okazaki Pieces

The major support for discontinuous synthesis stems from the work of Okazaki with *E. coli* and T4-infected *E. coli.* He showed that the most recently synthesized DNA, *i.e.* that labelled with a brief pulse of tritiated thymidine, can be isolated *after denaturation* as short pieces now known as Okazaki pieces.

In this system the short pieces sediment at 8-10 S on a gradient of alkaline sucrose, representing chain lengths of 1000-2000 nucleotides. In animal cells the fragments are much shorter, being only about 100-200 residues long, *i.e.* 4-5 S. In order to detect Okazaki pieces it is essential to lower the temperature to slow down the rate of reaction and to give a *very brief* pulse of tritiated thymidine. Under these conditions considerably more than half the radioactivity is found in small pieces which led to the proposal that Okazaki pieces are made on both sides of the replication fork. (There is, however, some controversy over this; see below.) An alternative explanation is that discontinuities are introduced into DNA during repair of regions of DNA containing uracil in place of thymine. It has been calculated that in *E. coli* during DNA replication one uracil will be accidentally incorporated into DNA for every 300 thymines which may lead to a cut in the DNA every 1200 bases; a size not too different from that of the Okazaki pieces. However, the frequency of uracil incorporation in *B. subtilis* is very low and in those mutants which fail to excise it (*ung*⁻) uracil is present less than once per 5000 bases.

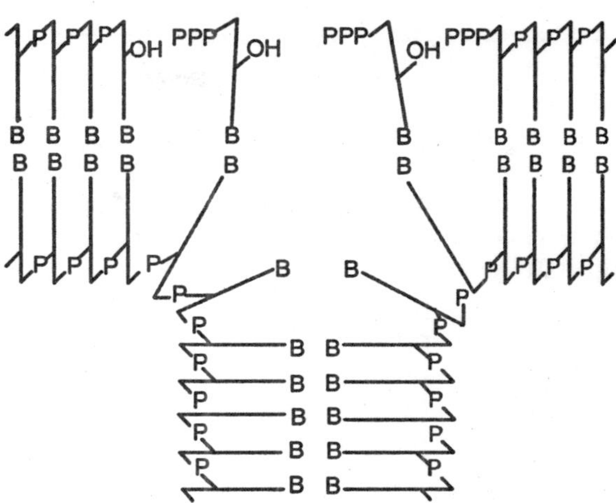

Fig. 6.6. A possible mechanism of DNA chain growth.

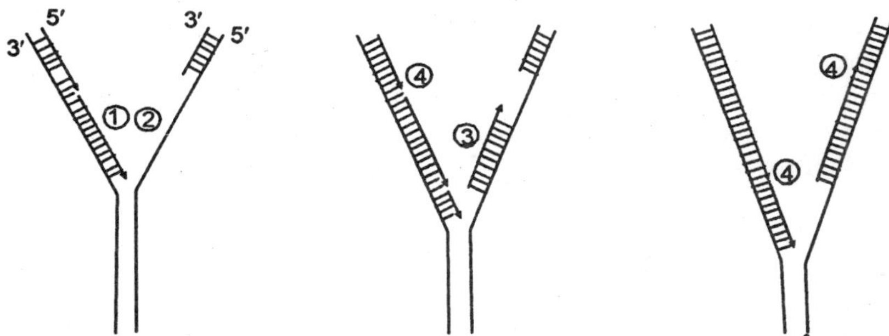

Fig. 6.7. A working model of DNA synthesis. Synthesis occurs in the 5′ → 3′ direction down the left-hand strand of the parental DNA molecule (1). This synthesis may be continuous or discontinuous and it exposes the right-hand strand of the parental DNA molecule. This single-stranded region (2) is probably stabilized by association with DNA-binding proteins. At some point DNA synthesis starts on the right-hand strand (3) which is copied backwards (*i.e.* 5′ → 3′) until the growing strand has filled the gap. DNA ligase then joins the newly synthesized pieces (4).

Incorporation of uracil is accentuated in mutants lacking an active dUTPase (*dut*⁻). Such cells have elevated levels of dUTP and this competes with and is incorporated into DNA in place of dTTP. A uracil N-glycosidase removes the uracil and the damaged DNA is repaired by an excision repair mechanism which involves cutting the DNA chain on the 5′-side of the damage *dut* mutants produce short Okazaki fragments (*sof*) only 100-200 nucleotides long. The discovery of such *sof* mutants was initially misinterpreted as indicating that Okazaki fragments were formed by joining of several much shorter pieces.

In double mutants lacking the uracil N-glycosidase (*ung*⁻) as well as the dUTPase the DNA contains significant amounts of uracil which is not excised. Normal-sized Okazaki fragments are formed on the lagging strand of *dut ung* double mutants showing that the majority of these fragments are not normally produced as a result of repair but are apparently true intermediates in DNA replication. Moreover Okazaki pieces accumulate in mutant *E. coli* cells deficient in DNA ligase or DNA polymerase I whether or not uracil excision occurs and this implies that they are true intermediates in DNA synthesis and that these two enzymes are involved in their subsequent incorporation into high-molecular-weight DNA.

Fig. 6.8. A partial denaturation map of the lambda replicating chromosome. Broken lines represent single-stranded regions of DNA and the arrows indicate such regions occurring at the replication fork.

Direction of Chain Growth

When T4, growing at 8°C to reduce the rate, of DNA synthesis, is incubated with [^{14}C]thymidine for 150 seconds, and for the final 6 seconds with [^{3}H]thymidine, Okazaki pieces can be isolated which apparently contain tritium at only the 3′-end. This was shown by degrading the isolated Okazaki pieces with exonuclease I of *E. coli* (which degrades single-stranded DNA from the 3′-end; when the ^{3}H label is released before the ^{14}C label. The complementary experiment using a nuclease from *B. subtilis* which acts from the 5′-end causes release of much of the ^{14}C before the ^{3}H is rendered acid-soluble. These experiments demonstrate that Okazaki pieces are made in the 5′ → 3′ direction and support the discontinuous mechanism.

Initiation of Okazaki Pieces

In wild-type *E. coli* rifampicin (an inhibitor of *E. coli* RNA polymerase) inhibits replication of phage M13 and certain plasmids. Replication is normal in mutants with a rifampicin-resistant RNA polymerase. Moreover, some RNA synthesis has been shown to be essential for the *in vitro* conversion of M13 single strands into the duplex form. Such evidence, considered with the fact that all DNA polymerases require a 3′-hydroxyl priming end, suggested that this end may be provided by an oligoribonucleotide. That rifampicin does not inhibit continued replication of *E. coli* DNA (it does block the initiation of new rounds or the replication of other phages such as ϕX174 was interpreted to mean that in these cases a second RNA polymerase (resistant to the drug) was involved in the synthesis of the priming oligoribonucleotides. Indeed, the product of the *dnaG* gene is believed to be such an enzyme and has been shown *in vitro* to synthesize short lengths of RNA on single-stranded DNA of phage G4.

Although not required for the initial stages of M13 replication, the *dnaG* gene product is essential for duplication of the replicative form. This and similar enzymes are called primases. The *dnaG* protein has been purified and shown to be a rifampicin-resistant RNA polymerase of molecular weight 64 000. As well as polymerizing ribonucleotides it can also use deoxyribonucleotides, and make a mixed primer though the first nucleotide is always rA. Okazaki fragments containing oligoribonucleotides at their 5′-end have now been isolated from both prokaryotic and eukaryotic systems. However, the experimental proof of their existence was initially subject to much controversy because non-covalent RNA-DNA interactions gave some spurious results. However, methods to detect RNA covalently linked to the 5′-end of DNA molecules have shown that, although not all Okazaki pieces have ribonucleotides at their 5′-end, the shortest ones do.

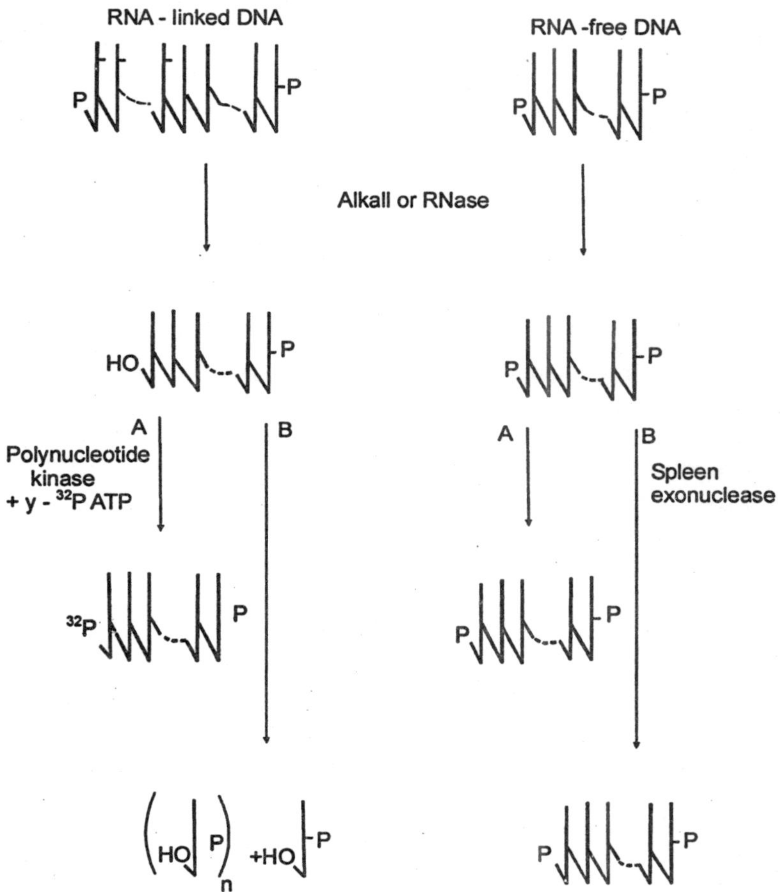

Fig. 6.9. Methods to detect RNA-linked Okazaki pieces produced *in vivo*. The isolated Okazaki pieces are first treated with polynucleotide kinase and non-radioactive ATP to ensure that all 5′-ends are phosphorylated. Subsequent treatment with alkali or ribonuclease removes any RNA from the 5′ end and leaves a 5′-hydroxyl group. This group can be detected using polynucleotide kinase and [γ-^{32}P]-ATP (Method A). However, this method is not selective for nascent DNA. Method B overcomes this disadvantage by starting with tritiated Okazaki pieces resulting from a pulse labelling experiment. Spleen exonuclease is used under conditions where only 5′-OH terminated DNA (*i.e.* that initially linked to RNA) is degraded. These methods have been criticized on the grounds that treatment with alkali of DNA cleaved during repair may also yield 5′-OH groups.

Moreover, RNA-linked Okazaki pieces accumulate in mutant *E. coli* cells deficient in either the polymerase or the 5′ → 3′ exonuclease of DNA polymerase I. Even so, using such a mutant, primers were detected on only about 30% of the Okazaki pieces and in a wild-type *E. coli* only three to six of the total of 20-40 chains in the size range 3000-9000 bases chase into high-molecular-weight DNA. When precautions are taken to eliminate all but covalent attachment of RNA to Okazaki pieces the size of the RNA segments obtained is only one to three nucleotides long and of no particular sequence which contrasts strongly with the 50-100 nucleotides erroneously attributed to the RNA primer.

In animal cells about 40% of Okazaki pieces have RNA primers and the RNA primers of animal viruses are about 10 nucleotides long. They do not have a specific base sequence. The use of toluenized *E. coli* enables α-[^{32}P]phosphate-labelled (*deoxyribo*nucleoside triphosphates to penetrate the cell membrane when transfer of radioactivity to a *ribo*-nucleotide occurs. This is definite evidence for a covalent attachment of RNA to DNA. When DNA synthesis is studied in nuclei isolated from polyoma-infected cells, the nascent DNA chains are found with a length of about ten ribonucleotides at their 5′-ends. When α-^{32}P-labelled deoxyribonucleoside triphosphatases are injected into the slime mould *Physarum polycephalum* there is transfer of the ^{32}P to ribonucleotides, which is indicative, as described above, of a covalent attachment of RNA and DNA. Failure to detect RNA primers may simply reflect their transient nature.

Indeed Okazaki's group have shown that many Okazaki pieces do not have RNA primers, and RNA is not found in mature DNA. An exception to this is the DNA of the plasmid Col E1 grown in the presence of chloramphenicol. In this case it is inferred that the drug prevents the normal excision of the RNA primers. Mitochondrial DNA also frequently has a number of ribonucleotides left in the mature duplex. Studies with the primase indicate that it shows a very strong preference to initiate with adenosine followed by guanosine and this suggests that initiation of Okazaki pieces may occur at particular sites on the lagging strand.

This idea is supported by evidence from *E. coli* suggesting that the sites of initiation are at or very close to sites of adenine methylation. However, the small phage P4 which requires only about 20 Okazaki pieces per round of replication shows no preferential initiation sites.

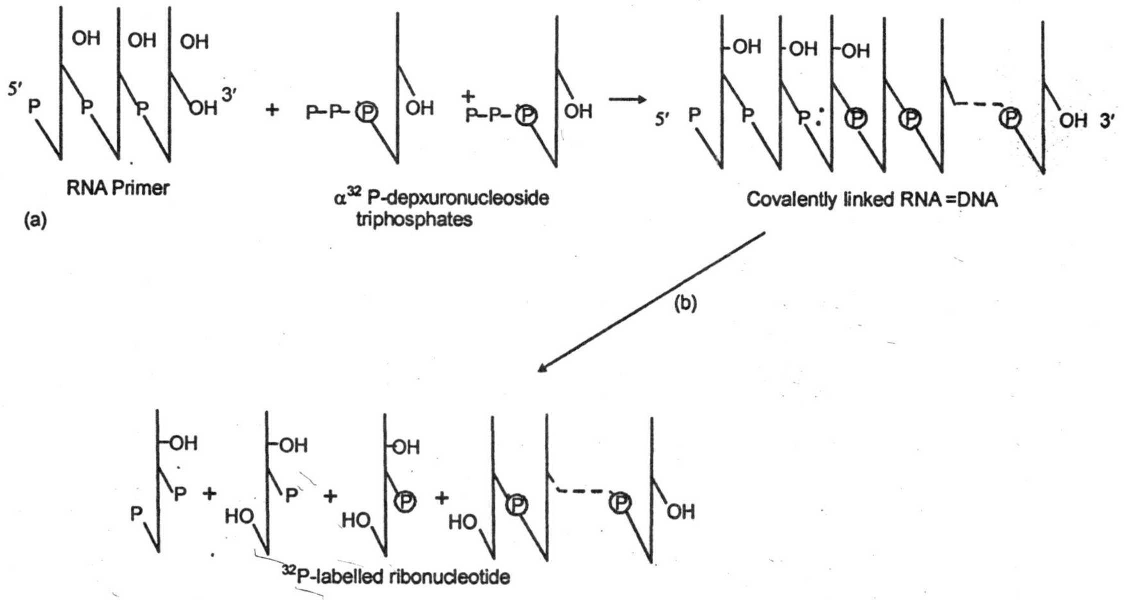

Fig. 6.10. The formation of the ^{32}P-labelled ribonucleotide is indicative of the occurrence of a covalently linked RNA-DNA molecule where the radioactive phosphate forms the bridge between the two nucleic acids, (a) Extension of RNA primer; (b) Alkaline hydrolysis of product; (ρ) = radioactive phosphate.

Continuous Synthesis

Discontinuous synthesis of one strand of DNA was postulated in order to overcome certain problems, but no problems appear to exist with the chain growing in the 5′ → 3′ direction, and so there is no *a priori* reason why this chain should not be made continuously. Indeed, this may well be the case when DNA synthesis is proceeding rapidly. Under these conditions nascent DNA is found in both large and small pieces. However, when deoxyribonucleotides are limiting and the rate of chain elongation slows, both daughter strands appear to be made discontinuously.

One possible explanation is that there is competition between the propagation of the growing chain and the initiation of new chains. Any initiation of an Okazaki piece on the leading strand is unlikely to be far in advance of the growing strand and so ligation will occur very rapidly, most probably before the Okazaki piece has grown to a detectable length. This is in contrast to the synthesis of an Okazaki piece on the lagging strand which (*a*) cannot be ligated until its synthesis is completed and (*b*) cannot be initiated until a region of single-stranded DNA of comparable length has been exposed on the lagging strand.

Under conditions where primer removal and ligation is restricted (*i.e.*, in the absence of a functional DNA polymerase I) Okazaki pieces are formed on both lagging and leading strands of the phage P2. However, under non-restrictive conditions discontinuities are not detected on the leading strand. *In vitro* studies using the cellophane disc assay have led to the conclusion that discontinuities on the leading strand are most likely produced as a result of uracil excision and that the normal continuous synthesis of the leading strand can be masked by uracil excision. A similar system for eliminaton of uracil residues from DNA exists in eukaryotes.

ENZYMES OF DNA SYNTHESIS

As we have said DNA synthesis occurs at a replication fork which progresses along the DNA molecule. A variety of proteins is required to bring about this process in an efficient manner. The major proteins are as follows:

(*a*) A DNA polymerase is required to add dNTPs to the growing leading strand and to growing Okazaki pieces on the lagging strand. In general DNA polymerases act alongside other helper proteins.

(*b*) A primase is required to help initiate Okazaki pieces.

(*c*) An exonuclease is required to remove the primers.

(*d*) A ligase is required to join the Okazaki pieces together.

(*e*) An unwinding protein is required to help unwind the duplex DNA at the replication fork.

(*f*) A topoisomerase is required to relax the tension engendered by unwinding duplex DNA.

(*g*) A binding protein is required to stabilize single-stranded DNA exposed by progression of the leading strand prior to initiation of Okazaki pieces.

(*h*) A gyrase may be required to help unwind the double helix ahead of the replication fork or for initiation of replication.

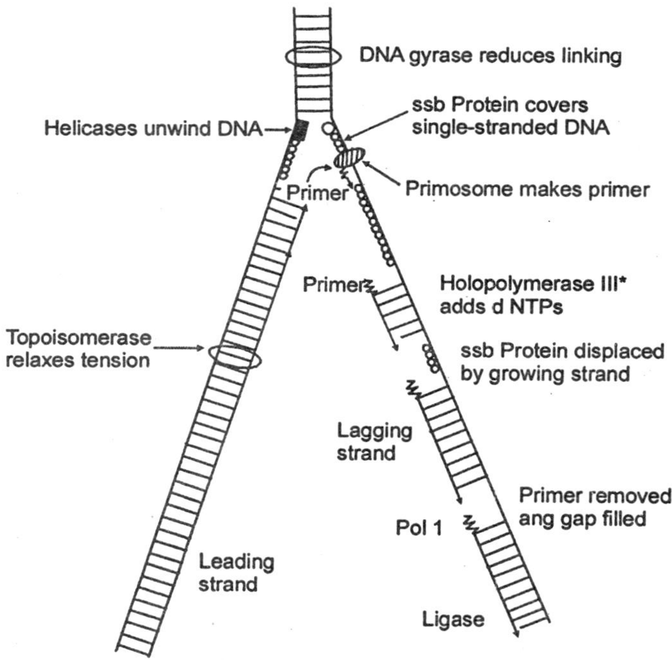

Fig. 6.11. Site of action of enzymes involved in DNA replication at the growing fork in *E. coli*.

DNA Polymerases

(a) Mechanism of Action

DNA polymerase is the name given to an enzyme which catalyses the synthesis of DNA from its deoxyribonucleotide precursors. Such enzymes have been purified from various sources (bacterial, plant and animal) but, apart from some details discussed below, the mechanism of the polymerization reaction is common to all preparations. DNA polymerase catalyses the formation of a phosphodiester bond between the 3′-OH group at the growing end of a DNA chain (the *primer*) and the 5′-phosphate group of the incoming deoxyribonucleoside triphosphate.

Growth is in the 5′ → 3′ direction, and the order in which the deoxyribonucleotides are added is dictated by base-pairing to a *template* DNA chain. Thus, as well as the four triphosphates and Mg^{2+} ions, the enzyme requires both primer

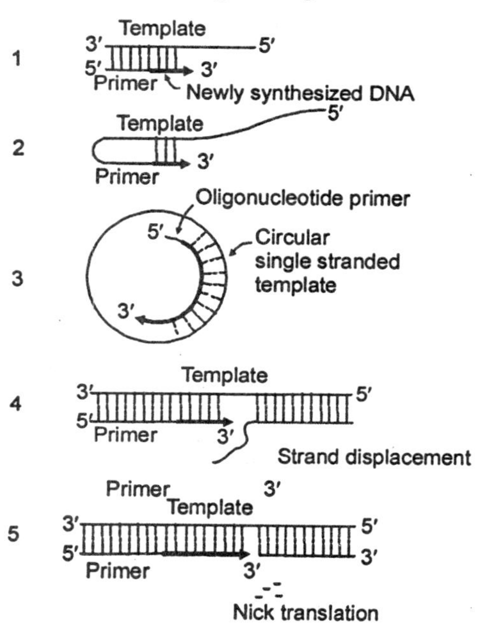

Fig. 6.12. Primer/templates for the replication of DNA.

and template DNA. No DNA polymerase has been found which is able to initiate DNA chains.

The simplest case to consider is that in which a single-stranded template has bound to it a growing strand of primer terminating at the growing point in a 3′-hydroxyl group. Such a situation is also illustrated diagrammatically for a piece of double-stranded DNA partially degraded by exonuclease III. The polymerase binds to the single-stranded template in the region of the 3′-hydroxyl end of the primer.

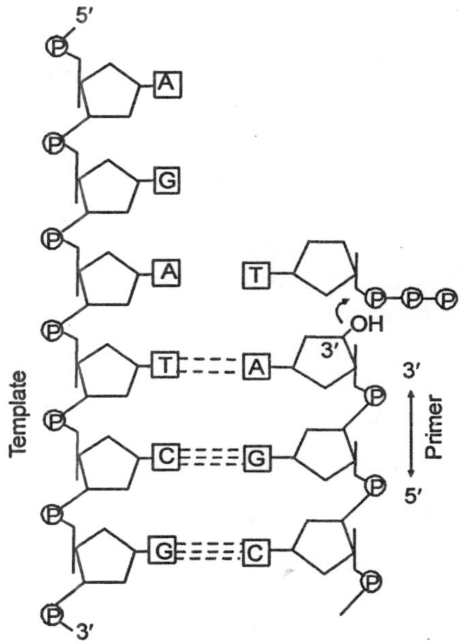

An incoming deoxyribonucleoside triphosphate containing a base which can pair with the corresponding base on the template becomes attached to the triphosphate binding site. The polymerase then catalyses a nucleophilic attack by the 3′-hydroxyl group of the primer on the α-phosphate of the deoxyribonucleoside 5′-triphosphate. Inorganic pyrophosphate is released, a phosphodiester bond is formed, and the chain is lengthened by one unit. The enzyme moves along the template by the distance of one unit and the newly added nucleotide with its 3′-hydroxyl group now occupies the primer terminus site. The process is then repeated until the enzyme reaches the end of the template strand. *In vivo* the pyrophosphate is hydrolysed by a pyrophosphatase which drives the equilibrium in favour of DNA synthesis.

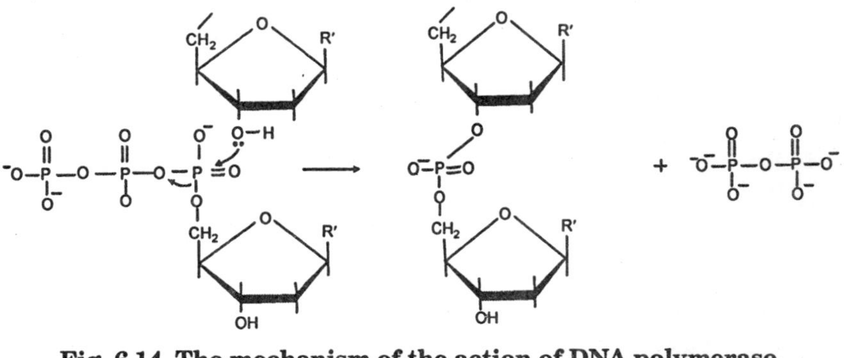

Fig. 6.13. Greater detail of the mechanism.

Fig. 6.14. The mechanism of the action of DNA polymerase.

In the replication of single-stranded DNA the 3′-hydroxy 1 terminus may loop back on itself and serve as a priming strand, or short lengths of oligonucleotide may act as primers by becoming hydrogen-bonded to the template. When a cyclic single-stranded DNA is used as template, replication can occur only in the presence of short oligonucleotides which become attached by base-pairing to the template and serve as primers. DNA polymerase cannot use double-stranded DNA as a primer/template. However, if a nick or gap is introduced into one strand the exposed 3′-OH group acts as a primer and the new growing strand will fill the gap and then may continue displacing the complementary strand from the 5′-end. In a strand displacement reaction, this can lead to complications since at some point the enzyme may leave the original template strand and begin to copy the complementary strand so that *in vitro* a branched structure is formed. In the presence of a 5′ → 3′ exonuclease the 3′-OH group at the nick can act as a primer for polymerase action while the exonuclease digests the DNA from the 5′-phosphate at the nick.

The nick thus moves along the duplex DNA in a so-called *nick translation* reaction. When native DNA is treated with exonuclease III, a partially single-stranded molecule is produced. It can be repaired and restored to the native double-stranded form by the polymerase. *E. coli DNA polymerase I,* Since a great amount of effort has been put into the study of NDA polymerase I from *E. coli* (the Kornberg enzyme) this archetypaol polymerase will be considered in some detail and then comparisons will be made with other enzymes.

Historical. In 1956 Kornberg and his collaborators used cell-free extracts from exponentially growing cultures of *E. coli* to demonstrate that, in the presence of ATP and Mg^{2+}, [^{14}C]thymidine, a specific precursor for DNA, was incorporated into an acid-precipitable material which was judged to be DNA from the fact that it was no longer acid-precipitable after treatment with deoxyribonuclease. Using this incorporation method for assay purposes, the bacterial extract was fractionated and it soon became evident that crude extracts contained a kinase system which converted thymidine into thymidine triphosphate (dTTP).

In subsequent work labelled dTTP was used as substrate and the enzyme responsible for its incorporation into DNA has been purified to homogeneity as judged by sedimentation, chromatography and electrophoresis. With a partially purified enzyme preparation it was shown that, for good incorporation of dTTP (labelled either with ^{14}C in C-2 of the thymine or with ^{32}P in the innermost phosphate group), the 5′-triphosphates of all four deoxyribonucleosides which normally occur in DNA had to be present together with Mg^{2+} ions and a denatured DNA template (calf thymus DNA was used in early experiments). Under these conditions excellent incorporation was observed

Table. 6.1. Incorporation of [^{32}P]dCTP into DNA.

Additions	*[^{32}P]DNA (pmol)*
dCTP (DNA omitted)	0.0
dCTP	2.5
dCTP + dGTP	5.1
dCTP + dGTP + dTTP	15.7
dCTP + dGTP + dTTP + dATP	3300

The incubation mixture (0.3 ml) contained 5 nmol of α[^{32}P]dCTP (7.2 × 10^7 cpm mol^{-1}) and 5 nmol of each of the other deoxynucleoside triphosphates as indicated, together with 1.9 μ mol of $MgCl_2$, 20 μ mol of glycine buffer (pH 9.2), 10 μg of DNa, and 3 μg of purified 'polymerse' from *E. coli*. The incubation was carried out at 37°C for 30 minutes.

Omission of any one triphosphate or Mg^{2+} or of DNA template, or pretreatment of the template with deoxyribonuclease, reduced the incorporation to a very low level.

Deoxyribonucleoside diphosphates could not replace the triphosphates. With the purified enzyme it was also possible by chemical analysis to demonstrate net synthesis of DNA, and increases of DNA by a factor of 10-20 could be obtained indicating that 90-95% of the isolated DNA was derived from the added deoxyribonucleoside triphosphates. As well as DNA polymerase activity, the purified enzyme shows several other activities each of which is associated with the same enzyme molecule. Thus exonuclease activity (both $3' \rightarrow 5'$ and $5' \rightarrow 3'$) is purified along with the polymerase activity in a ratio which is not altered by fractionation.

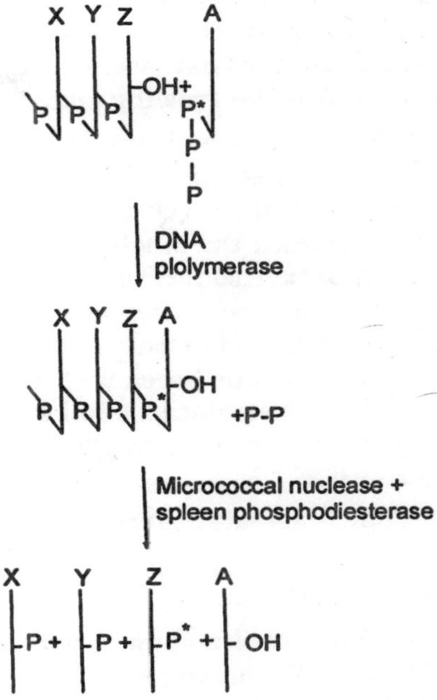

Fig. 6.15. Illustration of the method of nearest-neighbour sequence analysis.

Evidence for the Copying of the Template

The DNA polymerase is an enzyme with the unusual property of taking directions from a template and faithfully reproducing the sequence of nucleotides in the product. The evidence for this rests on several experimental observations.

(1) The most significant fact is that the enzyme can faithfully copy the nucleotide sequence of the single-stranded cyclic DNA bacteriophage ϕX174. The product can be converted into biologically active material by subsequent cyclization by the enzyme polynucleotide ligase. This experiment was done in 1967 by Kornberg and his colleagues who produced, *in vitro,* infective DNA identical with the natural material. These findings show that DNA polymerase I is an enzyme which has all the requirements necessary for the replication of DNA.

(2) When *E. coli* polymerase is used to prepare DNA under conditions such that only 5% of the sample produced comes from the template, the product has many of the same physical and chemical properties as DNA isolated from natural sources. The product appears to have a hydrogen-bonded structure similar to that of natural DNA and undergoes molecular melting in the same way.

It shows the same equivalence of adenine to thymine and guanine to cytosine that characterizes natural DNA. Moreover the characteristic ratio of A-T pairs to G-C pairs of a given DNA primer is imposed on the product whether the net DNA increase is 1% or 1000%. The base ratios in the product are not distorted when widely differing molar proportions of substrate are used. This is best illustrated by the use of the synthetic template poly(dA-dT) when, from a mixture of all four deoxyribonucleoside triphosphates, only dAMP and dTMP are incorporated into the product.

(3) The nucleotide sequence in the template DNA is reproduced in the product. This has been established by the Kornberg group by using the technique of *nearest-neighbour sequence* analysis. The partially purified *E. coli* enzyme is incubated with a particular template DNA

and all four deoxyribonucleoside triphosphates, one of which, say dATP, is labelled with ^{32}P in the innermost phosphate. During the synthetic reaction this ^{32}P becomes the bridge between the nucleoside of that labelled triphosphate (A) and the nearest-neighbour nucleotide containing the base Z at the growing end of the polynucleotide chain.

After the synthetic reaction is complete the DNA is isolated and degraded with micrococcal DNase and spleen phosphodiesterase to yield the deoxyribonucleoside 3′-monophosphates. The ^{32}P is thus transferred to the 3′-carbon of the neighbouring nucleotide in the chain *i.e.,* the one with which the labelled triphosphate has reacted (Z might be any one of the four bases). The four deoxyribonucleoside 3′-monophosphates are isolated by paper electrophoresis or HPLC, and their radioactivities measured to give the relative frequency with which the originally labelled nucleotide locates itself next to other nucleotides in the new chain. This procedure is carried out four times with a different labelled triphosphate each time so as to determine the relative frequencies of all sixteen possible nearest-neighbour (or dinucleotide) sequences. The results of such an experiment with DNA from *Mycobacterium phlei* as primer are shown in Table 6.2.

They illustrate several points:

(1) All sixteen possible nearest-neighbour sequences are present and they occur with widely varying frequencies.

(2) The results show a very striking deviation from the nearest-neighbour frequencies predicted if the arrangement of mononucleotides were completely random. Thus the frequency of TpA in the first row is quite different from that of ApT in the second row whereas these two frequencies would have to be identical in a random assembly. The nucleotides have therefore been assembled in accordance with a definite pattern.

Table 6.2. Nearest-neighbour frequencies of *Mycobacterium phlei* DNA.

Labeled triphosphate	*Deoxyribonucleoside 3′-phosphate-isolated*			
	Tp	Ap	Cp	Cp
dATP	TpA 0.012	ApA 0.024	CpA 0.063	GpA 0.065
dTTP	TpT 0.026	ApT 0.031	CpT 0.045	GpT 0.060
dGTP	TpG 0.063	ApG 0.045	CpG 0.139	GpG 0.090
dCTP	TpC 0.061	ApC 0.064	CpC 0.090	GpC 0.122
Sums	0.162	0.164	0.337	0.337

(3) The sums of the four columns show the equivalence of A to T and of G to C in the product and indicate both the validity of the analytical method and the replication of the overall composition of the primer DNA.

(4) The results indicate that base-pairing occurs in the newly synthesized DNA and that its two strands are of opposite polarity. According to the Watson-Crick model the two strands of the double helix are of opposite polarity, and it is presumed that each can act as a template for the formation of a new chain so as to give precise replication with the formation of two daughter helices identical with each other and with the parent helix. The results of nearest-neighbour sequence analysis support this mechanism. For example, the frequencies of ApA

and TpT sequences are equivalent, and so are the frequencies of GpG and CpC. The matching of the other sequences depends upon whether the strands of the double helix are of similar or opposite polarity.

If the strands are of opposite polarity the following matching sequences can be predicted:

CpA and TpG
GpA and TpC
CpT and ApG
GpT and ApC

whereas if the strands are of the same polarity the matching sequences would be:

TpA and ApT
CpA and GpT
GpA and CpT
TpG and ApC
ApG and TpC
CpG and GpC

The results in Table 6.2. favour the helix with strands of opposite polarity. The ApA and TpT, and the CpC and GpG sequences match in both models.

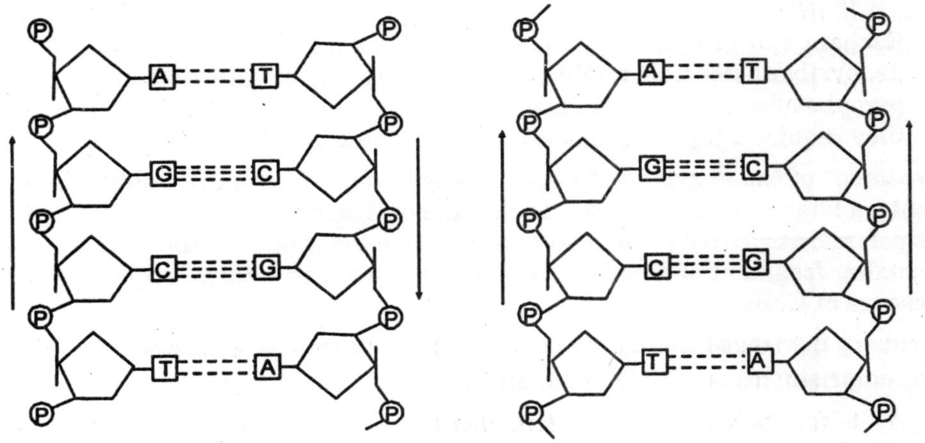

Fig. 6.16. Possible structures of the DNA molecule showing opposite polarity of strands (left) and the same polarity of strands (right).

(5) The nearest-neighbour frequencies measured by the method described above are those of the newly synthesized DNA. To verify that they are an accurate reflection of those in the original DNA template, an enzymically synthesized sample of calf thymus DNA in which only 5% of the total DNA consisted of the original template was itself used as template in a sequence analysis. The results showed good agreement between the sequence frequencies of the products primed by native DNA and by enzymically produced DNA whereas DNAs from other sources gave quite different results. It can be concluded therefore that the polymerase yields a DNA product with strands of opposite polarity and that the sequence of bases is faithfully reproduced.

(iii) The chemical nature of DNA polymerase I. The purified Kornberg enzyme is a protein of molecular weight 109 000 in the form of a single polypeptide chain. This chain can be unfolded in guanidine-HCl-mercaptoethanol so as to denature the protein. When the reagent is diluted out renaturation occurs with restoration of activity.

The protein migrates as a single band on SDS/acrylamide gel electrophoresis; it contains only one sulphydryl group and one disulphide group; the residue at the *N*-terminus is methionine. One *E. coli* cell contains about 400 molecules of enzyme, each spherical and of diameter 6.5 nm. It can form dimeric forms which can be visualized in electron micrographs.

(iv) The active centre of the enzyme. On the basis of binding experiments Kornberg has concluded that the enzyme is able to catalyse the following operations:

(1) Extension of a DNA chain in the $5' \rightarrow 3'$ direction by the addition to the 3′-hydroxyl terminus of mononucleotides from deoxyribonucleoside triphosphates at the rate of 1000 nucleotides per minute.

(2) Hydrolysis of a DNA chain from the 3′-hydroxyl end in the $3' \rightarrow 5'$ direction to yield 5′-monophosphates.

(3) Hydrolysis of DNA chain from the 5′-phosphate (or 5′-hydroxyl) terminus in the $5' \rightarrow 3'$ direction to yield mainly 5′-monophosphates.

(4) Pyrophosphorolysis of a DNA chain from the 3′-end; this is essentially the reversal of the polymerization reaction.

(5) Exchange of inorganic pyrophosphate with the terminal pyrophosphate group of a deoxyribonucleoside triphosphate as a result of alternating polymerization and pyrophosphorolysis. These last two reactions are of doubtful *in vivo* significance since they require a high concentration of inorganic pyrophosphate.

Treatment of the enzyme with the protease *subtilisin* from *B. subtilis* breaks it into two fragments, a larger fragment known as the Klenow fragment of molecular weight 76 000 which retains polymerase activity and $3' \rightarrow 5'$ nuclease activity but not the $5' \rightarrow 3'$ nuclease activity, and a smaller fragment of molecular weight 34 000 which retains nuclease $5' \rightarrow 3'$ activity in the presence of DNA.

Kornberg envisages the active centre of the, enzyme as a specially adapted polypeptide surface comprising at least five major sites.

(1) A site for the binding of the template chain in the region where base pairs are formed and for a few nucleotides on each side of this.

(2) A site for the growing primer chain which is, of course, base-paired to the template.

(3) A site for the special recognition of the terminal 3′-hydroxyl group of the primer. This point is the start of the $3' \rightarrow 5'$ hydrolytic cleavage.

(4) A triphosphate binding site for which all four triphosphates compete.

(5) A site which allows for the $5' \rightarrow 3'$ cleavage of a 5′-phosphoryl-terminated chain. It is presumably this area that is broken off by subtilisin.

These sites determine the nature of the DNA which can bind to the enzyme. For example, linear single-stranded DNA binds readily on site (1) whereas an intact linear duplex such as the DNA of bacteriophage T7 does not bind if it has been prepared with great care so as to avoid internal breaks. An intact circular duplex such as plasmid DNA or ϕX174 replicative form NDA does not bind to the enzyme until a 'nick' has been introduced in one of the strands by an appropriate nuclease yielding 3′-hydroxyl and 5′-phosphate termini. Such nicks are active points for replication whereas nicks introduced by micrococcal nuclease with 5′-hydroxyl and 3′-phosphoryl termini are not replication points although they bind the enzyme.

One molecule of enzyme is bound at each nick in either case. Recently evidence has been obtained that there are two distinct sites which recognize the 3′-OH group of the primer. When the terminal nucleotide is correctly matched to the template the 3′-OH group resides in the polymerase site. However, if the terminal nucleotide is not correctly matched to the template the 3′-OH group is displaced into a site where it is susceptible to the 3′ → 5′ exonuclease. The gene coding for *E. coli* DNA polymerase I has been sequenced and the sequence has been used to predict the primary and secondary structure of the protein, X-ray crystallography has shown the enzyme to have a deep crevice capable of binding duplex DNA.

Poly d(A-T) And Poly(dG) Poly (dC)

When DNA polymerase I is incubated without primer in the presence of dATP, dTTP and Mg^{2+} an interesting polymer is formed containing adenine and thymine nucleotides Polymer formation occurs only after a lag period of several hours and then takes place rapidly until 60-80% of the triphosphates have been utilized. The product of the reaction contains equal amounts of adenine and thymine. Nearest-neighbour frequency analysis shows that the frequencies of ApT and TpA are each 0.500 whereas the sequences ApA and TpT are undetectable. The polymer therefore contains alternating residues of A and T. The molecular weight calculated from the sedimentation valuce and reduced viscosity is between 2×10^6 and 8×10^6.

The polymer melts sharply at 71°C with an increase of 37% in the absorbance at 260 nm. The process is completely reversible on cooling. Such physical data, including the X-ray diffraction pattern, suggest that the molecule is a long fibrous double-stranded structure with the strands joined by hydrogen bonds between adenine and thymine bases. A somewhat similar polymer is formed when *E. coli* DNA polymerase I is incubated with high concentration of dGTP and dCTP in the presence of Mg^{2+}. Again, a lag period of several hours is found. This product contains guanine and cytosine, not necessarily in equal amounts, and nearest-neighbour sequence analysis shows that the frequencies of GpG and CpC are each 0.500.

Mild acid hydrolysis releases all the dGMP but none of the dCMP. Sedimentation and viscosity measurements yield values similar to those found with the poly[d(A-T)] but the T_m value is much higher (83°C). These observations are consistent with the view that the molecule consists of two homopolymers, one containing only guanine and the other only cytosine, hydrogen-bonded throughout their lengths.

The 3′ → 5′ Exonuclease and Proof-Reading

Before it was realized that this was an integral function of the polymerase, the 3′ → 5′ exonuclease activity was ascribed to an exonuclease II. It appears that this exonuclease functions to recognize and eliminate a non-base-paired terminus on the primer DNA. When

E. coli DNA polymerase I is provided with four triphosphates and a primer/template with a mismatched end, the non-matching terminus is removed by the 3′ → 5′ exonuclease before polymerization begins. (In the absence of the triphosphates the exonuclease continues to remove nucleotides from the frayed ends of the I , DNA molecules. Such a nuclease by correcting errors occurring on polymerization may be expected to increase dramatically the fidelity of the base-pairing mechanism. This proof-reading mechanism provides a justification. for involving 5′-deoxyribonucleoside triphosphates in the 5′ → 3′ extension of a 3′-OH on the growing DNA chain.

Addition of nucleotides to a growing chain terminating in a triphosphate would result, during proof-reading, in the removal of the energy required for further extension. However, there are arguments for and against the significance of this proposed proof-reading mechanism. Certain *E. coli* mutants show an increased or decreased rate of mutation which can be correlated with a change in the relative activities of the polymerase and 3′ → 5′ activities of NDA polymerase I. However, as we shall see below, DNA polymerase I is believed to have a role not as the major polymerase involved in replication but as an enzyme involved in the repair of DNA.

Furthermore, the rate of generation of dNMP by exonuclease action could account for only a minor increase in fidelity. The *error rates* in eukaryote polymerases which do not have associated 3′ → 5′ exonucleases are comparable with that for *E. coli* DNA polymerases I. On the other hand in *E. coli* infected with phage T4 (which codes for its own NDA polymerase) there are *mutator* and *antimutator polymerases* which show (*a*) decreased or increased rates of the pyrophosphate exchange reaction and (*b*) increased or decreased rates of incorporation of non-complementary nucleotides, both of which have been correlated with changes in 3′ → 5′ exonuclease activities.

The 5′ → 3′ Exonuclease

This activity, which is present in the smaller fragment released from DNA polymerase I by *subtilisin* treatment, cleaves base-paired regions of DNA, releasing oligonucleotides from 5′-ends. Because of its ability to jump several bases at a time, this nuclease can act on DNA molecules containing mismatched bases or distortions which render them unsuitable as substrates for polymerase. It may thus serve a function, for instance, in the elimination of thymine dimers from DNA exposed to ultraviolet radiation. When *E. coli* polymerase I binds to a nick on double-stranded DNA two reactions occur simultaneously. Polymerization extends the 3′-OH end and 5′ → 3′ exonuclease degrades the 5′-phosphate terminus.

This results in *nick translation*. If DNA polymerase I is incubated with a nicked duplex DNA molecule and radioactive deoxyribonucleoside triphosphates then the nick translation reaction results in replacement of long sections of the DNA molecule with a radioactive stretch of identical sequence, a reaction of great importance to modern molecular biology. Nick translation may only end when the enzyme reaches the end of the DNA molecule. Alternatively nick translation may end if a long 5′-terminus is released from the nick thereby rendering it no longer susceptible to the exonuclease. The branched product typical of reaction with native DNA primer/templates would then be formed. Normally DNA polymerase I is highly processive.

For each DNA/enzyme association event hundreds of nucleotides are polymerized before dissociation occurs. However, a mutant of DNA polymerase I shows all the normal properties of the wild-type enzyme but has a reduced rate of polymerization and nick translation caused by a lowered affinity of the enzyme for the DNA primer/template.

The Role of the Kornberg Enzyme in Vivo

Although the DNA polymerase I is exceedingly effective in the copying of a single-stranded DNA template when provided with a primer, it is much less effective with double-stranded DNA. This observation and other considerations led to doubts as to the role of the Kornberg enzyme (DNA polymerase I) in the replication of DNA *in vivo* and to the suggestion that it is concerned merely in maintenance and repair of DNA. These considerations may be summarized as follows:

(1) The purified enzyme cannot replicate intact double-stranded DNA semiconservatively to yield a biologically active product.

(2) The purified enzyme catalyses the incorporation of 1000 nucleotides per minute per molecule of enzyme whereas the estimated rate of incorporation *in vivo* is 100 times faster.

(3) Mutants of *E. coli* have been isolated which contain apparently normal Kornberg enzyme but are defective in DNA duplication. This demonstrates that other enzymes (which *may* include other polymerases) are required for *in vivo* DNA synthesis, and this may help to explain the deficiencies enumerated under (1) and (2) where only the purified Kornberg polymerase was present.

The evidence against the Kornberg enzyme being *essential* for replication *in vivo* is based on the properties of a mutant of *E. coli* (PolA1 or PolA$^-$) isolated by de Lucia and Cairns. This mutant and several others discovered later multiply normally but contain 1% or less of the Kornberg enzyme activity present in wild-type cells. Such mutants, however, show a reduced ability to join Okazaki fragments and an increased sensitivity to ultraviolet and ionizing radiation and to alkylating reagents. This implies a role for the enzyme in 'gap' filling, and perhaps also in excision of RNA primers and mismatched base pairs.

However, these functions are not completely lacking in mutants lacking DNA polymerase I and it is suggested that they may also be performed to a limited extent by other enzymes, *e.g.,* polymerases II and III and the RecB and C nuclease. Double mutants of PolA and RecB are non-viable, suggesting an essential function which can be carried out by more than one enzyme.

E. coli DNA Polymerase II

The discovery of the PolA$^-$ mutant of *E. coli* which grows well and replicates its DNA in the usual manner in spite of the absence of the Kornberg enzyme gave an impetus to the search for another enzyme apparatus which can synthesize DNA. Two further polymerases, designated DNA polymerase II and III, were found in extracts of PolA$^-$ cells. These enzymes had not been detected previously because, in extracts of wild-type cells, they show little activity relative to DNA polymerase I when single-stranded or nicked DNA is used as template. DNA polymerase II and III show significant activity only with a 'gapped' DNA template.

The enzymes can be separated from one another by chromatography on phosphocellulose, DEAE-cellulose, or DNA-agarose. Purified DNA polymerase II has a molecular weight of 90 000-120 000 and is homogeneous as judged by SDS/polyacrylamide gel electrophoresis. It synthesizes DNA in the 5′ → 3′ direction and for maximal activity it requires all four triphosphates, Mg^{2+}, NH^{4+}, and a native DNA template containing single-stranded gaps 50-200 bases long. The rate of reaction falls off with longer gaps, but may be restored by addition of *E. coli* DNA-binding protein.

Table 6.3. DNA polymerases of *E. coli*.

Polymerase (gene)	*Molecular weight*	*Molecules per cell*	*Nucleotides polymerized/s (a) per enzyme molecule (b) per bacterial cell*	*Direction of (a) polymerization (b) exonuclease action*	*Template (all require 3′-OH primer)*
I (*polA*)	109 000	400	(*a*) 16-20 (*b*) 8000	(*a*) 5′ → 3′ (*b*) 3′ → 5′ and 5′ → 3′	Denatured Nicked Gapped
II (*polB*)	120 000	17-100	(*a*) 2-5 (*b*) 500	(*a*) 5′ → 3′ (*b*) 3′ → 5′	Gapped
III (*polC*)	180 000	10	(*a*) 250-1000 (*b*) 10 000	(*a*) 5′ → 3′ (*b*) 3′ → 5′ and 5′ → 3′	Gapped

The enzyme also requires a 3′-OH primer. The purified enzyme like polymerase I, only synthesizes DNA at rates a fraction of those found *in vivo*. The enzyme also possesses 3′ → 5′ exonuclease activity, but no 5′ → 3′ exonuclease activity.

The function of *E. coli* DNA polymerase II *in vivo* is unknown and mutants lacking the enzyme appear normal in all respects. However, double mutants lacking both polymerase I and polymerase II join Okazaki fragments even more slowly than do mutants lacking polymerase I alone.

E. coli DNA polymerase III

E. coli with a temperature-sensitive mutation in the gene for DNA polymerase III (*dnaE* or *polC*) are not viable at the restrictive temperature and lysates prepared from them are defective in DNA synthesis. Complementation of such lysates with NDA polymerase III purified from normal cells restores their DNA synthetic ability. This is strong evidence that, unlike DNA polymerases I and II, polymerase III is *essential* for DNA synthesis. DNA polymerase III has been purified about 20 000 fold and shown to be highly sensitive to salt in dilute solutions (*i.e.*, normal assay conditions) but not when assayed by complementation (*i.e.*, at high protein concentration).

Although there appear to be only about 10 molecules per bacterial cell, its high rate of polymerization of nucleotides shows it to be capable of its proposed role in DNA synthesis. The best template for DNA polymerase III is double-stranded DNA with many small gaps containing 3′-OH priming ends. DNA polymerase III exists in a complex holoenzyme form in the cell and it proved difficult to determine the molecular weight and subunit composition of the enzyme. The α subunit of the holoenzyme (that coded for by the *dnaE* or *polC* gene) has a molecular weight of 140 000 and in DNA polymerase III core enzyme it is associated with two small subunits: ∊ (molecular weight 25 000) and θ (molecular weight 10 000). The core enzyme has both 3′ → 5′exonuclease (which could be involved in proof-reading) and 5′ → 3′ exonuclease activities though the latter is only manifest *in vitro* on duplex DNA with a single-stranded 5′-tail. For this reason the core enzyme cannot use a nicked primer/template but requires a gap. The 3′ → 5′ exonuclease activity resides in the ∊ subunit.

Table 6.4. Subunit Composition of *E. coli* DNA Polymerase III

Subunit	*Gene*	*Molecular weight*	*Name*				
α	*polC*	140 000	Polymerase III				
∊	*dnaQ*	25 000	3′ → 5′ exounclease	Polymerase III core enzyme			
θ		10 000			pol III′		
τ		83 000				III*	Polymerase III holoenzyme
γ	*dnaZ*	52 000	EFII				
δ	*dnaX*	32 000	EFIII				
β		440 000	EFI, copol III*				

DNA polymerase III core enzyme cannot use long single-stranded DNA molecules as a template even when provided with a primer. In the presence of the tau (τ) subunit it dimerizes to form polymerase III′ – $(\alpha\epsilon\theta)\tau_2(\alpha\epsilon\theta)$. Polymerase III′ associates with subunits γ and δ to form polymerase III*. In the presence of the β subunit (also known as copolymerase III* or factor I) the holoenzyme is formed which, with lipids and ATP, is able to extend an RNA or DNA primer hydrogen-bonded to circular, single-stranded M13, G4, or ϕX174 DNA. In the presence of ATP (which interacts with the α, τ, γ and δ subunits) the holopolymerase III forms an initiation complex with a primed template which results in hydrolysis of two ATP molecules and a rearrangement of the holopolymerase such that the β-subunit becomes inaccessible to antibody and the enzyme is unaffected by a concentration of KC1 (150 mM) which inhibits the non-activated enzyme.

The ATP-aetivated holoenzyme shows a much enhanced processivity, the one complex being capable of complete replication of a molecule of primed circular phage DNA. On completing replication the holoenzyme dissociates and, in the presence of ATP, will re-form on a new primed template. *B. subtilis,* a bacterium widely divergent from *E. coli,* also has three DNA polymerases which closely resemble those of *E. coli.* They differ in that polymerases I and II and also possibly III are devoid of nuclease activity and, in addition, DNA polymerase III is sensitive to 6-(*p*-hydroxyphenylazo)uracil (HPUra), an antibiotic active against Gram-positive bacteria. DNA synthesis in lysates of *B: subtilis* is also inhibited by HPUra.

DNA Polymerases in Eukaryotes

While the DNA polymerases of micro-organisms have been studied more intensively, similar enzymes are present in eukaryotic cells. In general they resemble the bacterial enzymes in their requirement for a template, a 3′-OH primer and four deoxyribonucleoside triphosphates. In higher eukaryotes there are three DNA polymerases known as α, β and γ. In Yeast, DNA polymerase I corresponds to DNA polymerase α. On cell fractionation the bulk of DNA polymerase β is found in the nuclear fraction and there is considerable evidence that the *in vivo* location of DNA polymerase α is intranuclear despite its recovery in high-speed supernatant fractions on cell disruption. In those cells making DNA there is evidence for association of DNA polymerase α with the nuclear matrix.

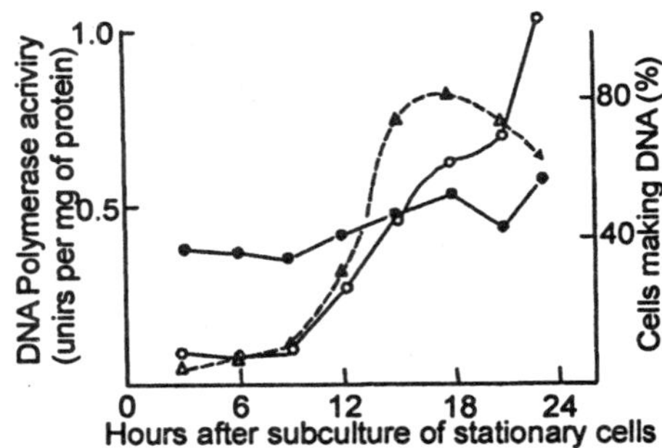

Fig. 6.17. Variation in the activities of nuclear (β) and supernatant (α) DNA polymerase with growth phase of cultured mouse L cells: •—•, nuclear; o—o supernatant; Δ – Δ % cells making DNA.

Indeed DNA polymerase α activity varies dramatically with the rate of cell division, being undetectable in non-growing cells and tissues. DNA polymerase α has a pH optimum in the range 6.5 – 8.0 and is highly susceptible to inhibition by thiol-active reagents (*i.e.* N-ethylmaleimide and *p*-chloromercuribenzoate). It is also inhibited strongly by araCTP and by (a tetracycline diterpene tetraol). In both instances inhibition is competitive with dCTP. Strangely it is resistant to inhibition by 2′, 3′-dideoxythymidine triphosphate (ddTTP). Use of these inhibitors together with consideration of the high enzymic activity found in growing cells has led to the conclusion that this enzyme plays the major role in DNA replication in eukaryotes.

A temperature-sensitive mutant cell line which is defective in DNA synthesis at the non-permissive temperature has a heat-labile DNA polymerase α. DNA polymerase α shows optimal activity with a gapped DNA template but shows a remarkable ability to use single-stranded DNA by forming transient hairpins. It will use synthetic single-stranded DNA templates provided with either oligoribo- or deoxyribonucleotide primers, *e.g.* poly(dT)•oligo(dA) or poly(dT)•oligo(A). It will not bind to duplex DNA. As normally isolated, polymerase a is only poorly processive, *i.e.* it adds about 11 nucleotides before it dissociates from the DNA.

The processivity can be dramatically increased in the presence of other proteins and ATP. In contrast to the enzymes from *E. coli* no nuclease activity normally is associated with the eukaryotic enzymes. Exceptions include a high molecular weight form, DNA polymerase δ, which has 3′ → 5′ exonuclease activity, and an enzyme from herpes-virus-infected cells. Because of its extreme lability to proteolysis during purification and its apparent heterogeneity, DNA polymerase α has only recently been purified to the same extent as *E. coli* polymerase I despite the fact that a 200-feld purification of the calf thymus enzyme was achieved in 1965 when it was separated from a terminal transferase. DNA polymerase *a* sediments in sucrose gradients as a broad peak in the 6-1 IS region and on gel filtration activity is found in a number of regions.

It is probable, but by no means certain, that the native, undegraded enzyme consists of two subunits: a 185 000-molecular-weight polymerase together with a 60-70 000-molecular-weight primase. However, on isolation the bulk of the polymerase is recovered as an active 155 000-molecular-weight protein together with several smaller active proteins which all share some primary sequence homology. Of particular interest is the presence of the associated *primase* activity which allows the enzyme to initiate replication on unprimed single-stranded cyclic DNAs. The primase is a novel RNA polymerase (*i.e.* not RNA polymerase I, II or III which synthesizes a short (8 – 15) oligoribo- nucleotide which, in the presence of deoxyribonucleotides, is extended by the DNA polymerase α activity. Primase does not act at random but primers are nearly always initiated with a GTP two to ten nucleotides 5′ to a hexanucleotide resembling the sequence C_2 A_2C_2. The association of primase with DNA polymerase α is restricted to the DNA-synthetic phase of the cell cycle and the complex forms part of the nuclear DNA replitase or replicase. DNA polymerase β is recovered in the nuclear fraction but shows little correlation between activity and the growth rate of the cells. This enzyme, which has been purified to homogeneity, has a pH optimum of around 8.6 – 9.0 and a molecular weight of 43 000.

However, in the final purification step used for the enzyme from Novikoff hepatoma or guineapig liver a polypeptide is removed from the enzyme whose molecular weight falls to 32 000. Re-addition of the polypeptide greatly stimulates enzymic activity. DNA polymerase β shows optimal activity with native DNA activated by limited treatment with DNase I (to produce single-stranded nicks and short gaps bearing 3′-OH priming termini) and shows negligible activity with denatured DNA. DNA polymerase β is relatively insensitive to thiol-active reagents, araCTP and aphidicolin but is strongly inhibited by ddTTP. The fact that DNA polymerase β never processes, *i.e.*, it leaves the DNA primer after addition of a single nucleotide, may account for its insensitivity to araGTP. It resembles DNA polymerase α in being devoid of nuclease activities.

Thus neither enzyme has the 'proof-reading' capacity of the bacterial enzymes, and in fact both will add on to a mismatched primer terminus. DNA polymerase β cannot use an oligoribonucleotide primer but, unlike polymerase α, given an oligodeoxyribonucleotide primer it is able to copy a ribonucleotide template [*e.g.* poly(A)•oligo(dT)]. DNA polymerase β is believed to play a role in repair of DNA and studies *in vitro* using nuclei from irradiated neuronal cells or from non-dividing mouse L929 cells treated with nuclease show the repair synthesis to be sensitive to ddTTP. However, damage by ultraviolet light or alkylating agents, especially when it is severe, involves DNA polymerase α in the repair reaction as well as DNA polymerase β. The use of an antibody to DNA polymerase β from calf thymus has demonstrated cross reacting molecules, of similar size in eukaryotes from protozoa to man indicating a lasting and essential role for this enzyme. Animal cells also have a small amount of DNA polymerase γ. This enzyme, probably identical to the previously identified mitochondrial DNA polymerase, shows greatest activity using an oligodeoxyribonucleotide primer and a ribonucleotide template (*e.g.* poly(A)•oligo(dT)).

Other than a presumed role in replication of mitochondrial DNA, polymerase γ has been implicated in the replication of adenovirus DNA/DNA polymerase γ from chick embryos is a tetramer having four identical subunits each of molecular weight 47 000. It is highly processive in contrast to DNA polymerases α and β and this may be relevant to its function. Chloroplasts also possess the enzymic mechanisms for synthesizing their own DNA and new enzymes are induced following viral infection. The DNA polymerase coded for by herpes simplex virus is very sensitive to inhibition by acyclovir and phosphonoacetic acid which makes these drugs suitable for use in the treatment of herpes virus infections. In addition to replicative DNA polymerase activities, which require the presence of all four deoxyribonucleoside triphosphates, animal cell extracts contain a separable enzyme responsible for the addition of nucleotidyl units to the *ends* of polynucleotide chains.

This second activity, which was originally described by Krakow in 1962, does not require a template strand but catalyses the incorporation of nucleotide units from single triphosphates into terminal positions in the DNA primer molecule. It is not further stimulated by the addition of the other three triphosphates but it is stimulated by cysteine. It has been called *'terminal transferase enzyme'* and may be used in the biosynthesis of homopolymers of deoxyribonucleotides. It is possible to distinguish between the two types of enzyme by the use of actinomycin D.Unlike the replicative enzymes, which are stimultated by low levels of EDTA, the terminal enzyme is completely inhibited by micromolar concentrations of EDTA which is believed to exert its effect by binding Zn^{2+}. Terminal transferase has a molecular weight of 58 000-62 000 but the purified enzyme shows bands at 26500 and 8000 molecular weight on electrophoriesis in sodium dodecyl suophate. This is a result of proteolysis during purification.

The activity of the terminal transferase is very low in all tissues except thymus and acute leukaemic lymphoblasts, *i.e.*, lymphoid progenitor cells. To what extent this reflects the peculiar physiological functions of such cells is unclear but the generation of short insertions at recombinational junctions formed during the rearrangements of immunoglobulin heavy chain genes is positively correlated with the presence of the enzyme.

Reverse Transcriptase or RNA-dependent DNA Polymerase

The enzymes so far discussed copy DNA strands in the synthesis of DNA, but great excitement was created in 1970 by the announcement of the existence in certain RNA viruses of RNA-dependent DNA polymerases which use RNA as a template for the synthesis of DNA. These enzymes were discovered simultaneously by Temin in the virus particles (virions) of Rous sarcoma virus (RSV) and by Baltimore in Rauscher mouse leukaemia virus (R-MLV). The observation was confirmed for more than half a dozen RNA viruses by Spiegelman, and was seen as an important break-through in cancer research since the RNA viruses involved were oncogenic, *i.e.* capable of bringing about malignant change. On treating the viral 70S RNA genome with agents which disrupt hydrogen bonds it dissociates into two genetically identical RNA molecules, and to several molecules of tRNA.

The unique RNA-directed DNA polymerase is also present in the virion. The enzyme uses one of the tRNA molecules ($tRNA^{Trp}$) as a primer to synthesize DNA on the template 35S single-stranded RNA, and this distinguishes it from any cellular-enzyme. Ribonucleoside triphosphates are without effect as substrates and the process is not sensitive to actinomycin D. The immediate product of the reaction is a double-stranded RNA-DN A hybrid which is the result of the synthesis of a complementary strand of DNA on the single-stranded viral RNA as template. The enzyme also shows ribonuclease H and DNA-dependent DNA polymerase activities which convert the hybrid into a duplex DNA.

The polymerase then proceeds to replicate the duplex DNA so as to provide more copies which may be integrated into the genome of the host cell. Reverse transcriptase may be used to make DNA copies of eukaryotic messenger RNA molecules. A thymidylate oligomer will anneal to the 3′-poly (A) tail of the mRNA and serve as a primer in a reaction which is of great importance in the production of complementary DNA (cDNA) molecules for cloning. Reverse transcriptase is the product of the *pol* gene of retroviruses. It is synthesized as a gag-pol polyprotein as the result of differential splicing of the transcript from the integrated proviral DNA or by translational suppression.

The polyprotein is processed by a protease, present in the *N*-terminal region of the polymerase, to give the β subunit (92 000 molecular weight) and then the α subunit (58 000 molecular weight) which in the leukosis virus (ALV) combine to give the most active (αβ) form of the enzyme. The enzyme from other retroviruses is a single polypeptide. The enzyme shows only low fidelity and lacks associated exonuclease activity though it is capable of specifically hydrolyzing the ribo strand of an RNA-DNA duplex. The 3′-end of the pol gene also encodes an endonuclease important for replication and integration which altogether makes this a remarkable enzyme.

DNA Ligases

The ligase or joining enzymes catalyse the repair of a single-stranded phosphodiester bond cleavage of the type introduced by endonuclease. They were first described in *E. coli* and have since been described in both animal and plant cells.

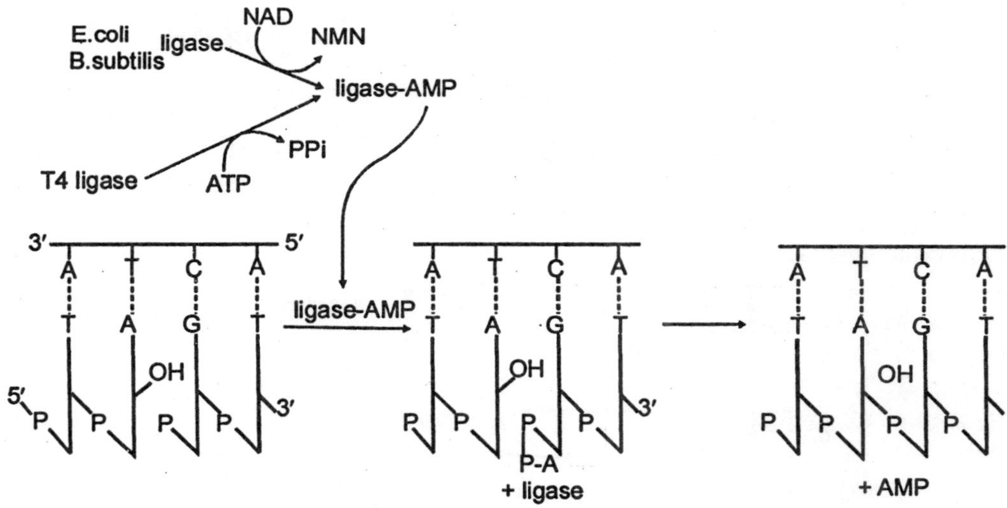

Fig. 6.18. Action of NDA ligases.

DNA ligases catalyse the formation of a phosphodiester bond between the free 5′-phosphate end of an oligo- or polynucleotide and the 3′-OH group of a second oligo- or polynucleotide positioned next to it. A ligase-AMP complex seems to be an obligatory intermediate and is formed by reaction with NAD in the case of *E. coli* and *B. subtilis* and with ATP in mammalian and phage-infected cells. The adenyl group is then transferred from the enzyme to the 5′-phosphoryl terminus of the DNA.

The activated phosphoryl group is then attacked by the 3′-hydroxyl terminus of the DNA to form a phosphodiester bond. DNA ligase will close single-strand breaks in double-stranded DNA or in either strand of a polyribonucleotide-polydeoxyribonucleotide hybrid polymer; double-stranded ribopolymers are not substrates. Breakage of a single phosphodiester bond without removal of a nucleotide to give 5′-phosphoryl and 3′-OH termini is essential for repair by ligase activity. Reaction is independent of the base composition around the cleavage point.

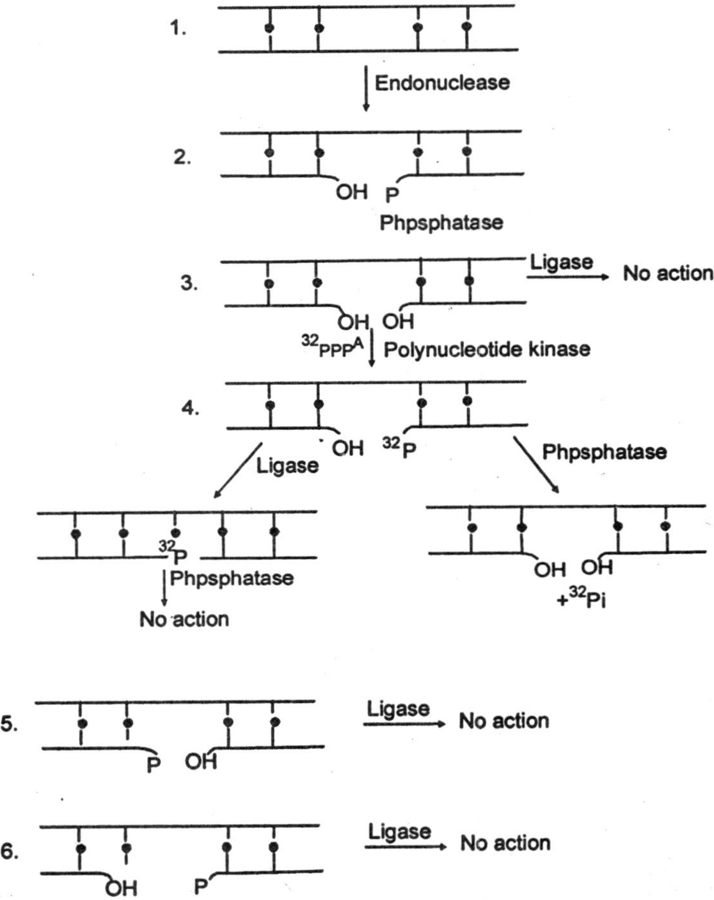

Fig. 6.19. An assay for DNA ligase. The substrate (4) is formed by treating double-stranded DNA (1) with DNase I to give single-stranded breaks (2), removing the 5′-phosphate residues with phosphatase (3) and replacing them with [^{32}P] phosphate groups with the aid of polynucleotide kinase and γ-labelled [^{32}P] ATP. Forms (3), (5) and (6) are inactive as substrates for the ligase.

The enzymes from *E. coli* or bacteriophage T4 will join short oligodeoxynucleotides in the presence of a long complementary strand, *e.g.* d(T-G)$_3$, d(T-G)$_4$ or d(T-G)$_5$ can be joined in the presence of poly(dC-dA) and blunt end ligation (*i.e.*, the joining of two duplex DNA molecules in the absence of any complementary overlapping sequence) can occur in the presence of excess ATP.

Assay of DNA Ligase

DNA ligase has been assayed in a variety of systems, including the formation of covalently closed circles of double-stranded DNA, restoration of transforming activity of nicked DNA and the formation of phosphatase-resistant radioactive phosphate.

Role of DNA Ligase

The nature of the reaction catalysed by DNA ligase makes it an important enzyme involved in DNA synthesis, DNA repair and genetic recombination. In all of these cases, its postulated role is in re-establishing continuity by joining a stretch of newly synthesized DNA to pre-existing DNA by the formation of a phosphodiester bond. It has, however, proved difficult to define a situation in which DNA ligase could be shown to be absolutely necessary for any one of these functions. Nevertheless, since *E. coli* ligase-deficient mutants selected in different ways exhibit abnormal uv sensitivity, it appears that ligase has some function in the repair process.

Temperature-sensitive ligase mutants also accumulate Okazaki pieces at the restrictive temperature. That these pieces are eventually joined is probably a reflection of the presence of a few molecules of ligase still remaining in the mutant cells. There are normally about 300 molecules of ligase per cell, and it has been calculated that these are capable of sealing 7500 breaks per minute at 30°C. Since only about 200 breaks per cell are formed on replication each minute there is a vast excess of enzyme. Infection of *E. coli* by T4 leads to the synthesis of a T4-specified ligase which, in addition to joining two DNA chains, will also join RNA to DNA in an RNA-DNA hybrid.

Some ligase is essential for T4 development, and phage with a temperature-sensitive ligase show increased susceptibility to uv radiation. There are two DNA ligases present in cells of higher eukaryotes. DNA ligase I (molecular weight 165 000-200 000) resembles DNA polymerase α in being recovered partly in supernatant fractions on cell homogenization and in being present in much greater amounts in dividing than in resting tissues. The smaller, nuclear DNA ligase II does not change in activity with the growth rate of the cells.

Helix-destabilizing Proteins (HD) or Single-stranded DNA-binding Proteins (ssb)

These are a group of proteins brought together under this heading because of a common property. They have been isolated from many sources and they do not nexessarily all perform the same function *in vivo*. The first of these to be characterized was that coded for by gene 32 of bacteriophage T4. This and similar proteins isolated from T7 and uninfected *E. coli* have the ability to convert double-stranded DNA into single-stranded from at a temperature 40°C below the normal melting temperature (T_m). The gene 32 protein has a molecular weight of 35 000 and it binds cooperatively to single-stranded regions of DNA, each protein covering some 10 nucleotides. Thus if a limiting amount of gene DNA some DNA molecules will become covered with protein and others remain protein free.

There are from 300 to 800 molecules of binding protein per uninfected *E. coli* cell, which is enough to cover about 1600 nucleotides at each replication fork. A similar length of T4 DNA can be covered by the 10 000 molecules of gene 32 protein present in an infected cell. (There are about 60 T4 replication forks per infected cell. Purified polymerases are stimulated in a very specific manner by the presence of their complementary DNA-binding proteins. Thus the T4 DNA-binding protein will stimulate only the T4 polymerase and the T7 DNA-binding protein will stimulate the T7 polymerase, the *E. coli* DNA-binding protein stimulates *E. coli* polymerase II and holdpolymerase III but does not stimulate polymerases I and III. It is suggested that the polymerase and DNA-binding protein form a specific complex that interacts with DNA and other proteins during replication. The HD proteins from *E. coli* and T4-infected cells are believed to play a role in replication, recombination and repair, and an *E. coli* mutant (ssb-1) has been isolated showing a defective DNA-binding protein and extracts fail to convert phage G4 single strands into duplex form.

In contrast an HD protein coded for by gene 5 of phage M13 plays an essential role in preventing further replication by stabilizing the single-stranded progeny viral DNA. DNA-binding proteins have also been isolated from eukaryotic cells. Thus helix-destabilizing proteins serve a passive role. Although *in vitro* they can bring about denaturation of DNA their role *in vivo* may simply be to stabilize single-stranded DNA.

DNA Unwinding Proteins or DNA Helicases

DNA helicases have the property of being able to unwind a DNA duplex obtaining the necessary energy from the hydrolysis of ATP. In *E. coli* there are several helicases which have been partially characterized by Hoffman Berling's group. Helicase I is present only in F^+ strains of *E. coli* (*i.e.* strains carrying the F sex factor and is the product of one of the transfer genes (*traI*) of the F factor. Thus a role for helicase I in bacterial conjugation is implied and it plays no role in cell DNA replication. It is a fibrous protein of molecular weight 180 000, similar to myosin.

About 80 molecules bind to a single-stranded region of DNA near a replication fork and travel down the single strand in a $5' \rightarrow 3'$ direction towards the fork unwinding the DNA. There are about 500 molecules per cell. There are about 3000 molecules of helicase II (molecular weight 75 000) per cell. It also moves towards a replication fork in a $5' \rightarrow 3'$ direction but remains bound to the single-stranded DNA generated. It interacts specifically with holopolymerase III to promote unwinding of DNA at the replication fork in the presence of *E. coli* ssb. It is the product of the *E. coli uvrD* gene, mutants in which show increased rate of mutation and recombination.

Immuhological studies show that helicase II is involved in the replication of *E. coli,* phage lambda and the plasmid Col E1. Rep is so called because it is essential for the replication of small single-stranded phages (*e.g.,* phage G4 and ϕX174) as well as *E. coli.* It differs from helicase I, II and III as it binds to a single-stranded region and translocates in a $3' \rightarrow 5'$ direction until it reaches the replication fork (*i.e.* along the leading strand). For its unwinding action it requires a single-stranded binding protein to bind to the two arms of single-stranded DNA produced. It is clear that rep acting on the leading side of the fork and helicase II acting on the lagging side would bring about efficient unwinding of the DNA duplex.

Two molecules of ATP are hydrolysed per nucleotide unwound. This is a high price to pay for unwinding DNA as it involves the release of 60 kJ of energy to break H-bonds whose energy varies from 5 kJ (A=T) to 21 kJ (G=C). Helicase III (molecular weight 2 × 20 000) is similar to rep but moves along the bound strand of DNA in a $5' \rightarrow 3'$ direction. It is clear that these helicases move along the DNA in combination with a DNA-polymerizing complex and similar enzymes are induced by T7 and T4 phage.

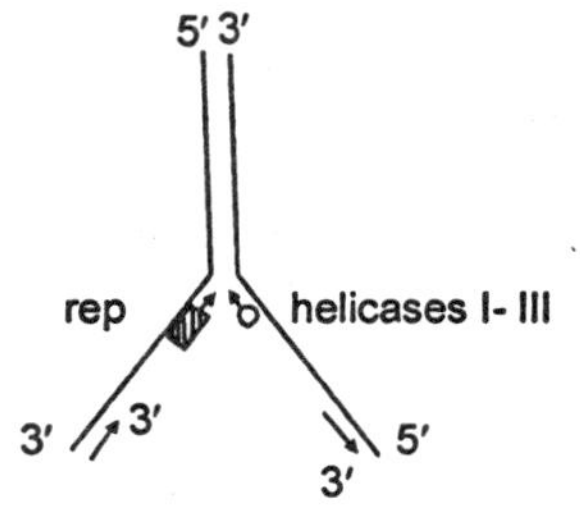

Fig. 6.20. DNA helicases acting at the replication fork.

The gene 4 product of bacteriophage T7 is similar to rep except that it forms a complex with the T7-coded DNA polymerase (the gene 5 product). Together they are able to extend a 3′-OH at a gap in the duplex, unwinding the DNA ahead of the complex and hydrolysing two ATP molecules for every dNMP incorporated. The gene 4 protein, in addition to acting as an unwindase, is also able to synthesize a specific primer for DNA synthesis (see below).

A similar function is catalysed by the T4 gene 41/61 complex and a second T4-coded helicase – the product of the non-essential *dda* gene-greatly speeds up the rate of replication fork movement in *in vitro* systems. The *dda* protein overcomes the inhibitory action exerted by RNA polymerase molecules which, when bound to the DNA template, can completely block fork movement. Unwinding enzymes are also found in eukaryotes. Other DNA-dependent ATPases are present in cells but in these cases the ATP hydrolysis is probably required to translocate an enzyme along the DNA chain (this may also be required for the unwinding enzymes).

Topoisomerases

When two DNA strands unwind as the replication fork moves forwards either the whole DNA molecule has to rotate at 10 000 rev./min (an impossibility for closed cyclic DNA molecules) or some sort of swivel must be present. Cairns first proposed the idea of a swivel, which might act by alternating action of an endonuclease and a ligase. However, the two functions are present in a single enzyme, examples of which have been isolated from both prokaryotes and eukaryotes. The activity was initially called a nicking closing enzyme but is now known as a topoisomerase as it catalyses the interconversion of different topological isomers of DNA.

The difference between a topoisomerase and the earlier suggested mechanism is that with the topoisomerase the cut ends are not released from the enzyme although rotation of the two strands about one another is effectively achieved. There are two types of topoisomerase. Topoisomerase I cuts only one strand of duplex DNA whereas both strands are cut by topoisomerase II. The cuts usually produce a 5′-OH group (which can be phosphorylated by polynucleotide kinase) and a 3′-phosphate which is covalently linked to a tyrosine in the enzyme. *In vitro,* topoisomerases are studied by following the result of their action on closed cyclic supercoiled DNA molecules such as SV40 DNA or the plasmid Col E1.

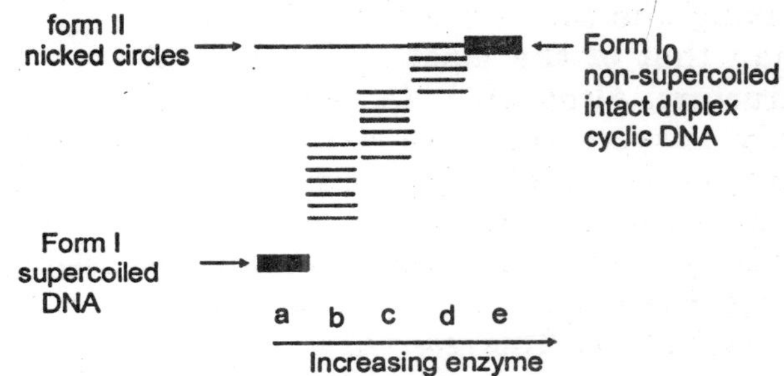

Fig. 6.21. Topoisomerase action. Diagram of agarose gel electrophoresis of SV40 DNA incubated with increasing amounts of topoisomerase I. Whereas an endonuclease would convert form I to form II directly, the topoisomerase produces DNA molecules of intermediate linking number.

The topoisomerases will bring about a step-by-step relaxation of such supercoiled molecules by changing the linking number. Type I topoisomerases change the linking number in single steps indicating that they produce an enzyme linked cut in one strand only of the DNA. The other strand is then passed through the cut which is resealed. This is equivalent to free rotation of the bond opposite to the nick. Type II topoisomerases produce an enzyme-bridged break in both strands of DNA. In an ATP-requiring action, another region of duplex DNA passes through the gap thus changing the linking number two steps at a time. In most instances both types of topoisomerase will act until all the supercoils are removed and the DNA is released as a fully relaxed closed cyclic DNA molecule. If the two regions of duplex DNA involved in topoisomerase II action are part of two different DNA molecules then these two molecules will become interlocked. Because of this the enzyme is capable of producing and resolving catenated and knotted molecules.

In a similar manner topoisomerase I can act opposite to a nick in a cyclic molecule and bring about the catenation and knotting of several supercoiled molecules to the nicked cyclic molecule. Both types of topoisomerase are found in prokaryotes and eukaryotes. The activity of topoisomerase II is higher in rapidly dividing eukaryotic cells and it is thought to be involved in DNA replication and the segregation of intertwined duplex DNA molecules.

In replicating cells it exists as a complex in association with other enzymes involved in DNA synthesis BIT]. Bacterial mutants defective in topoisomerase II are not viable unless compensatory mutations in topoisomerase I are also present. There are at least two types of topoisomerase II. The eukaryotic enzymes and the enzyme from T4 are only capable of relaxing supercoiled DNA. That is, they catalyse the conversion of supercoiled DNA into relaxed cyclic duplex. They are homodimers having two subunits each of approximate molecular weight 170 000. Bacterial type II topoisomerases are known as DNA gyrases. *E. coli* DNA gyrase is a 400 000-mol.wt. tetramer, the subunits of which are the products of two genes *gyrA* and *gyrB* (also known as *nal* and *cou).* The *gyrB* product is a DNA-dependent ATPase (molecular weight 95 000) which is sensitive to the antibiotics coumermycin and novobiocin. The *gyrA* product is a 100 000-110 000 molecular-weight 'nicking-closing' enzyme sensitive to nalidixic acid and oxolinic acid. In the presence of oxolinic acid or strong detergent these subunits remain bound to the 5′-ends of the cut and hence the tetrameric enzyme holds the cut ends together.

DNA gyrase, as well as relaxing supercoiled molecules can also reduce the linking number of cyclic duplex DNA molecules. Thus when incubated with ATP and DNA gyrase a relaxed duplex ring can be converted to a negatively supercoiled molecule with a linking number as much as 10% lower than that of the most stable B-DNA structure. Although DNA gyrase is the only enzyme that can catalyse this reaction *in vitro* it has been suggested that it is the normal reaction catalysed *in vivo* by all type II topoisomerases. A third type of topoisomerase II has been reported in archaebacteria. This *reverse gyrase* is able to convert a relaxed cyclic DNA duplex to a *positively* supercoiled molecule.

DNA gyrase cleaves duplex DNA to give a four-base-pair stagger, but there is controversy over whether or not it shows sequence specificity. Topoisomerase I may also show site selectivity.

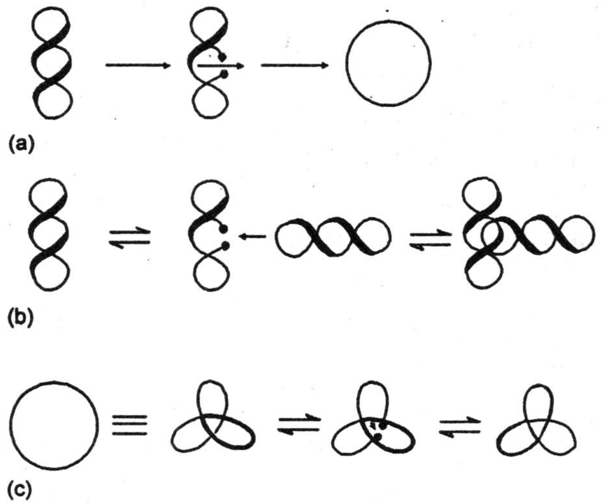

(a)

(b)

(c)

Fig. 6.22. Reactions catalysed by type II topoisomerases: (*a*) relaxation; (*b*) catenation and decatenation; (*c*) knotting and unknotting.

A model for the action of DNA gyrase has been constructed. The DNA first wraps itself around the protein to form a structure similar to a nucleosome. ATP then binds and causes a conformational change which drives the translocation of the DNA relative to the enzyme, leading to the formation of a positive superhelical loop on the DNA. A double-stranded breaking rejoining event removes the positive superhelical turn and on hydrolysis of ATP the enzyme returns to its original conformation leaving the DNA with a negative superhelical twist, *i.e.* partly unwound. The wrapping of the DNA around the gyrase produces overwound and underwound regions of duplex DNA such that when the DNA is left with an underwound region the overwound regions are selectively relaxed. DNA gyrase may play a role either by removing positive superhelical turns induced into DNA during replication or alternatively or additionally by putting negative supertwists into the DNA ahead of the replication fork thus making unwinding easier.

As treatment of *E. coli* with either coumermycin or nalidixic acid immediately stops DNA synthesis, the contribution of DNA gyrase is essential. DNA gyrase is also required for replication of phage TV DNA and phage T4 codes for a corresponding activity. The products of T4 genes 39, 52 and 60 form a complex with topoisomerase activity which is essential for initiation of replication of T4 DNA.

Decatenation (also catalysed by gyrase) may be an essential step in the physical separation of cyclic DNA molecules after replication. Single-stranded regions of DNA are required not only for initiation of DNA synthesis but also for recombination and transcription and DNA gyrase is believed to be required in these reactions.

It is possible that topoisomerase activity is markedly affected by enzyme phosphorylation or poly(ADP)ribosylation.

Fidelity Of Replication

To produce a sequence of bases in the DNA of a daughter cell which is *identical* to that in the parent the process of DNA replication would need to be completely faithful. This is obviously not the case as mutants do arise spontaneously. This is not altogether a bad thing as it is on this varied counterpane that natural selection is able to act.

There are also mechanisms which come into play to repair damage to DNA post-replicatively but even these mechanisms rely on the faithful production of DNA complementary to the template strand.

Fidelity may depend on a number of steps:

(*a*) The polymerase must select the correct deoxyribonucleoside triphosphate.

(*b*) A checking mechanism may exist which would allow rejection of incorrect triphosphates. Such enhanced discrimination may result from template-induced changes in enzyme conformation dictating the selection of the correct substrate or increased binding of the enzyme to the template in the presence of the complementary nucleotide.

(*c*) A kinetic proof-reading model proposed by Hopfield predicts that a high-energy [enzyme:template:dNMP] complex is produced with release of pyrophosphate. The polymerase can check the complex to see whether or not the correct base has been incorporated. The nucleotide (dNMP) may be released at this time but the chance of an incorrect nucleotide being released is much higher than for the correct nucleotide.

(*d*) Following incorporation, an exonucleolytic proof-reading mechanism could remove incorrect 3′-terminal nucleotides which are mismatched with the template strand prior to elongation. These nucleotides are also released as monophosphates.

Both the proof-reading mechanisms rely on the reduced rate of reaction with a mismatched nucleotide. In the kinetic mechanism it is the joining of the bound nucleotide which is delayed whereas with exonucleolytic proof-reading it is the addition of the next nucleotide which is delayed allowing a greater opportunity for exonuclease action. It is only if all these mechanisms fail that post-replicative repair is required but as explained above this is also subject to the demands that it produce a faithful complementary DNA strand. One of the reasons for increased interest in fidelity of replication arises from the fact that, unlike their prokaryote counterparts, eukaryote DNA polymerases lack the 3′ → 5′ proof-reading exonuclease activity yet still exhibit error rates much less than those predicted by a consideration of base-pairing alone.

An assay that has proved particularly useful in assessing the fidelity of *in vitro* replication involves the copying of the single-stranded DNA of the phage ϕX174. An amber mutant (ϕX174 am16 - see Table 6.5) is unable to grow on wild-type *E. coli* (but can grow on the suppressor strain *E. coli* C62). Faithful copying by polymerase of the mutant ϕX174 DNA will reproduce mutant progeny, but a single base substitution at any one of the three bases of the amber codon will result in progeny viral DNA which can grow on wild-type *E. coli.* Using purified DNA polymerase α from calf thymus Grosse *et. al.*, showed a reversion frequency of about 1×10^{-4}. Of course all the base changes would not result in substitution with the original wild-type base and, as this region of the ϕX174 genome codes for two different proteins, a variety of progeny phage can result.

For instance, insertion of a guanine opposite to the guanine of the amber codon (TAG) results in a change of an arginine into a threonine (AGO → ACG) in gene A and insertion of a tyrosine into gene B. This produces small phage plaques which show reduced growth at temperatures above 35°C.

Table 6.5. Nucleotide Substitutions Possible at the am 16 Codon of the ϕX174 Genome.

Gene A	Phe Arg		
Mutant DNA sequence	5′-A G A T T T A G G C T G G 3′		
Gene B	Ile am16 Ala		
Position at amber codon	Mispaired nucleotide	Gene A	Gene B
T	G	Phe-Arg	Glu
	T	Leu-Arg	Lys
	C (wt)	Leu-Arg	Gln
A	G	Phe-Arg	Ser
	A	Phe-Trp	Leu
	C	Phe-Gly	Trp
G	G	Phe-Thr	Tyr
	T	Phe-Lys	Ochre
	A	Phe-Met	Tyr

From a study of the plaque morphology and related sequence data it can be estimated that the frequency of mispairing dATP with a guanine in the template (1 in 20 700) is an order of magnitude more common than the reciprocal mispairing of dGTP with an adenine in the template. Nonetheless, the major mutations are purine:purine mismatches (transversions) and GT transitions. An alternative assay system involves comparing the competition between the correct and incorrect nucleotide for incorporation in place of a dideoxynucleotide in the DNA sequencing reaction. This system has the advantage that the effect of the environment of the substituted nucleotide can be studied. Having established an assay system, the effect on mispairing of altering the relative concentrations of precursors can be studied.

For instance, when the concentration of dATP is kept low and that of dCTP increased the chances of mispairing with the first base of the amber triplet is increased. However, other examples of an effect of pool bias were not found using polymerase α. The fidelity of the various eukaryote DNA polymerases is very variable and it is only the high-molecular-weight 9S form of polymerase α which shows a fidelity comparable to that of the prokaryote polymerases, both making only 1 error in 10^5 nucleotides incorporated.

The eukaryotic enzymes are working in the absence of a proof-reading exonuclease and the results suggest that a domain of polymerase a which is lost on proteolysis is vital for maintaining fidelity in a parallel but perhaps a different way from that in which the proof-reading exonuclease of *E. coli* polymerases is used. In the absence of exonucleolytic removal of a mismatched primer terminus, DNA polymerase α shows an elongation rate of only 5% that with a fully matched terminus. It is not only the first base after the mismatch which is added more slowly, for the effect extends to additions five bases following the mismatch. In contrast, the prokaryotic DNA polymerases have associated proof-reading exonucleases and in the ϕX174 am16 assay pool effects are more marked. Thus the frequency of a TAG → TGG transversion depends not only on the ratio of incorrect to correct triphosphate but also on the concentration of the next triphosphate, *i.e.* $[dGTP]^2/[dATP]$.

By using deoxycytidine [1-thio]triphosphates which, once incorporated, cannot be removed by exonuclease action, Kunkel calculated that proof-reading can increase the fidelity of *E. coli* DNA polymerase I or phage T4 polymerase by up to 20-fold. The efficiency of proof-reading depends on the adjacent nucleotide which is stacked together with the mismatched base and the fidelity of these enzymes and *E. coli* DNA polymerase III holoenzyme is dependent on the divalent ions and accessory proteins present. In an assay of fidelity which measures the production of dNMP, Loeb's group have failed to find significant evidence for proof-reading of either the kinetic or exonuclease variety. They favour a direct action of the polymerase in enhancing the specificity of base-pairing and show that the affinity of *E. coli* DNA polymerase I for a complementary nucleotide exceeds that for non-complementary nucleotides by one to three orders of magnitude.

Similar results have been obtained for the T4 DNA polymerase. It is difficult to reconcile the failure to find evidence for the importance of proof-reading by 3′ → 5′ exonuclease with the occurrence of mutator and antimutator mutations in prokaryotes. A way out is to envisage excision repair reactions occurring following an endonucleolytic cut in DNA on the 3′-side of a region of damaged DNA. The 3′ → 5′ exonuclease would then act to remove the defective region in a way that the 5′ → 3′ exonuclease can act if the endonucleolytic cut is 5′ to the lesion. A defective 3′ → 5′ exonuclease would fail to excise the mismatch leading to an increase in mutation frequency. It is the post-replicative mismatch correction mechanisms which contribute to the increase in fidelity (up to one error per 10^9-10^{11} nucleotides *in vivo*. The RNA-dependent DNA polymerases of retroviruses (reverse transcriptase) are very error-prone showing a fidelity similar to that of RNA polymerases, *i.e.*, one error per 10^3-10^4 nucleotides incorporated.

In Vitro Systems for Studying DNA Replication

DNA Mutants

Increased understanding of DNA replication has come about as a result of the intensive use of bacterial mutants, especially temperature-sensitive mutants, *i.e.* bacteria which grow normally at low temperatures but cease to grow at a higher, restrictive temperature. The study of such conditional lethal mutants, defective in DNA synthesis, has demonsrated that a number of proteins are essential for DNA replication in *E. coli.* Some of these proteins together with the genes which code for them are listed in Table 6.6.

Most of these proteins have now been purified and used with some success in attempts to reconstruct a system which will replicate DNA *in vitro*.The mutant cells, when transferred to the restrictive temperature, may cease immediately to make DNA or they may continue to grow for some time. The former mutants are defective at the stage of the elongation of the replicating DNA molecule whereas the latter are defective in the process of initiation of new rounds of DNA replication.The genetic approach thus indicates the involvement of a number of gene products in DNA replication but is able to say little about the detailed function of these proteins. The biochemist seeks to purify the proteins involved and to characterize their individual functions. A first step in this process involves the circumvention of the barrier of the cell membrane. This has been done in a number of ways.

Permeable Cells

Treatment of *E. coli* lipid solvents (ether or toluene whilst inhibiting cell division appears to leave the cells and constituent enzymes intact yet renders the membrane permeable to small molecules including nucleotide precurosors. Such cells when incubated with the four deoxyribonucleoside triphosphates, ATP and Mg^{2+} continue semi-conservative replication. The products of such replication are the 10S Okazaki pieces which are only joined into high-molecular-weight DNA in the presence of NAD, the cofactor for *E. coli* DNA ligase. It was with toluenized cells and α-^{32}P-labelled deoxyribonucleotides that Okazaki was able to show the covalent linkage of RNA and DNA.

Cell Lysates

The lysate formed on disintegration of the cell membrane is capable of maintaining DNA synthesis under certain conditions. It is very important not to dilute the macromolecular constituents of the cell but to maintain them near their *in vivo* concentration. One way in which this has been achieved is by lysis of cells on a cellophane disc held on an agar surface.

Small molecules are then allowed to diffuse through the disc into the concentrated lysate. Semiconservative DNA synthesis occurs in this and similar systems if the lysate is prepared from wild-type *E. coli,* but when it is prepared from *E. coli* cells defective in DNA synthesis then the lysate reflects the *in vivo* capabilities of the cells. Thus lysates prepared from mutants temperature-sensitive in the genes *dnaB, dnaD* or *dnaG* (fail to extend growing DNA chains at the non-permissive temperature; only synthesize DNA at low temperatures.

In lysates prepared from mutants temperature-sensitive in the genes *dnaA* or *dnaC* (fail to initiate DNA synthesis at non-permissive temperatures) DNA synthesis fails to occur only if the cells are maintained at the non-permissive temperature for one generation time prior to lysis.

Table 6.6. *E. coli* Genes whose Products Play a role in DNA Synthesis.

Gene	*Enzyme*	*Effect of mutation*
dnaA		No initiation
dnaB		No synthesis
dnaC		No initiation
dnaE (*polC*)	DNA polymerase III (α)	No synthesis
dnaF (*nrd*)	Ribonucleotide reductase	No synthesis
dnaG	Primase	No synthesis
dnaQ (*mutD*)	DNA polymerase III (ϵ)	Increased mutation rate
dnaS (*dut*)	dUTPase	Short Okazaki pieces
dnaX	DNA polymerase III (δ)	
dnaZ	DNA polymerase III (γ)	
polA	DNA polymerase I	Slow joining of Okazaki pieces uv-sensitive
polB	DNA polymerase II	Normal
lig	DNA ligase	Slow joining of Okazaki pieces
ssb	DNA-binding protein	
gyrA }	DNA gyrase	
gyrB }		
top	DNA topoisomerase I	
uvrD	Helicase II	Increased mutation rate

This system has proved very useful in the purification of the products of these *dna* genes. Fractions containing the *dnaG* gene product from wild-type cells, when added to a lysate from a mutant temperature-sensitive in *dnaG,* are able to restore DNA synthesis at the non-permissive temperature. Although the products of the *dna* genes have all been at least partially purified by this *in vitro* complementation assay, functions for all these proteins have not yet been established.

However, they have been used with some success in reconstruction experiments using fully defined components. Permealysed cells and lysates from eukaryotic cells also show a limited ability to continue DNA synthesis when provided with the four deoxyribonucleoside triphosphates, ATP and Mg^{2+}. Eukaryotic cells can be rendered permeable to deoxyribonucleotides by treatment with hypo-osmotic buffers. DNA synthesis continues in these cells for about 30 minutes before slowing down. This pattern of synthesis is probably due to completion of ongoing replicons coupled with an inability to initiate new replicons. Cell lysates show similar characteristics but isolated nuclei are much more limited in their capabilities unless supplemented with cytosol fractions.

In general, isolated nuclei appear capable of extending Okazaki pieces initiated *in vivo* but have only a limited capacity for ligating or initiating Okazaki pieces. DNA synthesis in isolated nuclei is sensitive to inhibitors of DNA polymerase a but not to inhibitors of DNA polymerase β or γ indicating a role for DNA polymerase a in this reaction. However, removal of all *soluble* DNA polymerase α activity from nuclei does not markedly reduce their DNA-synthetic capacity, possibly indicating the presence of an enzyme complex. Cell lysates do initiate the synthesis of Okazaki pieces and these have been shown to have

RNA primers 8-11 nucleotides in length. The sequence of these oligoribonucleotides is diverse. The disadvantage with lysates of eukaryotic cells is that very little is known about the nature of the DNA template or product.

As with prokaryotes, model systems have been used. Thus lysates of cells infected with SV40 or polyoma virus synthesize predominantly viral DNA *in vitro*, and soluble replication systems have also been prepared from adenovirus-infected cells and lysates made from cells infected with herpes simplex virus synthesize predominantly herpes simplex DNA in *vitro*.

Soluble Extracts

Although useful in the identification and purification of some of the factors necessary for DNA synthesis, the cellophane-disc assay is of limited value in defining the function of the various gene products and moreover it relies almost entirely on possession of temperature-sensitive mutants for every protein involved. A simplification involves the use of soluble extracts from *E. coli* free of the bacterial DNA. When cells are lysed with lysozyme the bacterial chromosome can be pelleted by centrifugation at 200 000 *g*.

The soluble preparation remaining is capable of performing specific stages of DNA synthesis when supplemented with simple DNA substrates, *e.g.* single-stranded cyclic DNA from the small phages ϕX174 or M13. Once again possession of mutants is an invaluable aid but the advantage of this system over the crude lysate is that specific steps in DNA replication may be studied and hence information on the function of particular proteins may be obtained.

Reconstruction Experiments

The ultimate aim of these experiments is to achieve the ordered replication of a complex DNA molecule such as the *E. coli* chromosome using totally defined components. Initially simpler systems have been studied. The requirements for the conversion of single-stranded phage DNA into the double-stranded replicative form (RF) and the replication of RF have been delineated and the reactions carried out using purified components. Progress was not all plain sailing for in addition to the components identified with the aid of mutants and purified by the complementation addition of soluble extract was required.

From this extract several further components have been purified for which, as yet, no mutants are known. The detailed results of these reconstruction experiments will now be considered. The simplest situation to investigate is the elongation of a primer hydrogen-bonded to a template DNA strand. When the components of such a system are known the synthesis of the primer can be investigated. Synthesis of a primer and its elongation as DNA is the reaction which occurs on the lagging strand of the replication fork, *i.e.* Okazaki piece synthesis. In order to ensure that a primer is being synthesized it is important to use a cyclic, single-stranded template molecule which has no free ends which can bend back and hybridize to the template strand. Such templates are naturally provided by the DNA of the small bacteriophage fd (M13), G4 and ϕX174. The replication of these phage in *E. coli* takes place in three stages.

(*i*) The original infecting cyclic DNA molecules (known as the viral or plus strand) is used as a template on which a complementary minus strand is synthesized (SS → RF).

(*ii*) The resulting replicative form (RF) undergoes several cycles of replication by the so-called rolling circle mechanism (RF → RF).

(*iii*) The RF molecules are used as templates from which progeny plus strands are synthesized. These are subsequently packaged into the viral coat as the mature virus leaves the cell (RF → SS).

Although of similar size the replication of the three small phage mentioned above differs in a number of significant steps which can be explained as a result of differences in DNA sequence.

MOLECULAR BIOLOGY OF THE REPLICATION FORK

Lagging-strand Synthesis

(a) Prokaryotes

In the cell single-stranded DNA is always complexed with a DNA-binding protein and such a complex is formed as soon as the DNA from the small single-stranded phages (fd, M13, G4, ϕX174) enters the host cell or is incubated with a host cell lysate. In fd there remains exposed a hairpin loop which is able to bind *E. coli* RNA polymerase. This enzyme synthesizes a short RNA which then acts as a primer for minus strand synthesis. In G4 a similar hairpin loop is exposed but this has a binding site for the *dnaG* protein— the primase. This enzyme is responsible for synthesizing the RNA primer in most prokaryotes including phage G4.

The sequence of DNA around the origin of phage G4 and the sequence of the RNA primer synthesized by the primer is shown in Fig. 6.23. Studies with phage ΦK have shown that the primase-binding site is more complex than shown in the figure and in fact two molecules of primase bind to a region of complex secondary structure extending over 100 bases having three hairpin loops. Although a primer of 28 nucleotides, in the presence of DNA polymerase as few as two ribonucleotides may be incorporated by primase before DNA polymerase takes over. In ϕX174 no binding site for either RNA polymerase or primase is exposed.

Comparison of the nucleotide sequences of the DNA of the closely related bacteriophages G4 and ΦX174 reveals a deletion in the latter covering the primase-binding site. In ϕX174 a complex is formed involving a series of proteins identified as the products of the genes *dnaB*, *dnaC* and proteins *i*, *n*, *n′* and *n″* (called X, Y and Z by Hurwitz). This complex, called the preprimosome, binds to a 55-nucleotide-long sequence which occurs once in ϕX174 DNA and contains a hairpin loop between bases 2308 and 2350. It also occurs once on each strand of the duplex Col El plasmid DNA as well as on *E. coli* DNA and this sequence when introduced into fd DNA renders this DNA capable of assembling the preprimosome. Protein *n′* is a 76 000-molecular-weight protein responsible for site selection.

A complex of proteins *n*, *n′* and *n″* interacts with DNA and this then reacts with protein *i* and a *dnaB*—*dnaC* protein complex which contains six molecules of each protein. This preprimosome complex of six different proteins interacts with primase to form the primosome and an RNA primer is synthesized. As the primase moves off in the 3′ → 5′ direction along the template strand the preprimosome translocates in the opposite direction (5′ → 3′) in a reaction activated by the DNA-dependent ATP hydrolysis catalysed by protein *n′*. At various sites the *dnaB* protein is able to affect the structure of the DNA template thereby causing another molecule of primase to interact with the preprimosome and initiate synthesis of another primer.

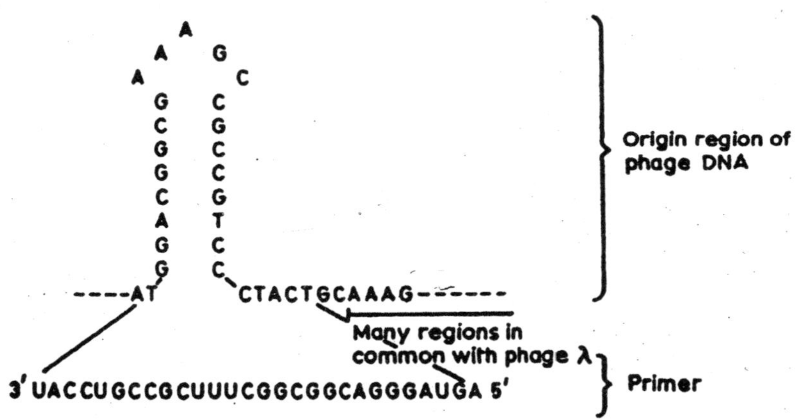

Fig. 6.23. Location of primer in phage G4 (this untranslated region is absent from ϕX174).

In all, six to eight primers are made per round of replication on ϕX174 DNA. They range in size from one to nine nucleotides long and although they initiate with pppA-Pu their internal sequence is random. Their length is in part dependent on the availability of deoxyribonucleoside triphosphates and the DNA-polymerizing enzymes. In all cases the RNA primers formed are extended by DNA polymerase III holoenzyme which catalyses the addition of deoxyribonucleotides until the growing strand reaches the next RNA primer.

Because of the limited availability of DNA polymerase III holoenzyme several primers may be present awaiting extension, and rapid transfer of polymerase occurs from a completed Okazaki piece to the adjacent primer. DNA polymerase I removes the RNA primer and completes the DNA chain in a nick-translation type of reaction. Finally the chains are joined by DNA ligase to form the cyclic duplex molecule (RF). These reactions have been recently reviewed. These reactions, as well as describing the initial (SS → RF) stages of replication of the small single-stranded phages, also represent the reaction occurring on the lagging side of the replication fork.

A similar set of reactions is catalysed by enzymes induced following phage T4 infection of *E. coli.* Seven T4-coded proteins will catalyse initiation of lagging-strand synthesis on single-stranded cyclic templates and will catalyse displacement-strand synthesis on nicked duplex DNA.

(b) Eukaryotes

In an attempt to characterize the enzymes involved in lagging-strand synthesis in eukaryotes similar closed cyclic single-stranded DNA molecules have been incubated with extracts of *Xenopus* eggs or *Drosophila* embryos. Such extracts were initially chosen as these tissues have the potential for very high rates of *in vivo* replication. These extracts, and extracts of cultured cells, catalyse the synthesis of an RNA primer which is extended to form a DNA chain of about 1000 nucleotides in length complementary to the single-stranded template. The sequence of the RNA/DNA junction is random but the primers initiate preferentially with the sequence pppA. On fractionation of the extracts it appears that single-stranded DNA-binding protein and DNA polymerase α with its associated primase activity are the only proteins required.

Leading-strand Synthesis

This reaction requires the extension of the growing daughter DNA strand and the concomitant unwinding of the duplex ahead of the point of polynucleotide synthesis. A suitable substrate to study this reaction is a nicked duplex molecule and this had been used in particular by Albert's group to investigate the proteins required for phage T4 DNA replication. On a nicked DNA duplex a complex of six enzymes is able to displace the free 5′-end at the nick and extend the 3′-OH. The proteins required include a DNA polymerase (product of T4 gene 43); the archetypal single-stranded DNA-binding protein (gene 32 protein); polymerase accessory proteins (gene 44, 62 and 45 proteins) which increase the rate and processivity of polymerase action, and a DNA-unwinding protein (gene 41) which is required to unwind the displaced strand from the template.

Another helicase is required to allow the replicating complex to pass a bound RNA polymerase molecule. In conjunction with gene-61 protein the gene-41 protein also has a primase function and the complex will synthesize and extend pentanucleotide primers on the displaced (lagging) strand. This mixture of seven enzymes is thus able to catalyse the reaction going on at the replication fork. The Okazaki pieces synthesized on the lagging side are, however, some10 000 nucleotides long, as the complex lacks the capacity to remove RNA primers and join Okazaki pieces.

RF Replication

Nicked cyclic duplexes are produced as the first step in the replication of the RFs of the small single-stranded DNA phage and it is convenient now to consider this reaction. To initiate replication of the RF requires the intervention of a specific bacteriophage-coded protein. This protein makes an endonucleolytic cut in the plus (viral) strand of the RF DNA.

The site of the cut is specific and occurs between nucleotides 4305 and 4306 of ϕX174 RF DNA, within the region coding for the endonuclease. There is evidence that the endonuclease acts preferentially on the phage DNA molecule by which it was coded. The enzyme from phage ϕX174 is called the *cisA* protein and it recognizes a site of two-fold symmetry which may be present as a hairpin structure. The unique site of action of the gene II nuclease from bacteriophage fd is between bases 5781 and 5782 which is adjacent to (24 bases 3′ on the plus strand) the unique binding site of RNA polymerase. In bacteriophage G4 the *cisA* protein acts on the opposite side of the DNA molecule (between nucleotides 506 and 507) to the primase-binding site (nucleotides 3972-3992).

The sequences recognized by the *cisA* protein of G4 and ϕX174 are identical and can be inserted into a plasmid. This makes plasmid replication dependent on the A protein and allows packaging of single-strandedplasmid DNA into preformed phage coats. Once the A protein has cut the plus strand it remains attached to a dAMP residue at the 5′-terminus by means of a phosphotyrosine bond. The 3′-end can act as a primer for plus strand synthesis by the DNA polymerase III holoenzyme which together with the primosome has remained attached to the DNA following RF synthesis.

Unwinding of the DNA is accomplished by combined action of an *E. coli*-coded ATP-dependent DNA-unwinding protein, the *cisA* protein, and the single-stranded DNA-binding protein. DNA gyrase may also help to unwind the DNA duplex ahead of the replicating fork but it is not required *in vitro*. With phage fd the gene II protein does not remain bound to the 5′-end but rather there is an additional requirement for the host *dnaA* protein. As plus strand synthesis proceeds in the 5′ → 3′ direction the 5′-end of the plus strand is peeled off the rolling minus strand and it associates with single-stranded DNA-binding protein.

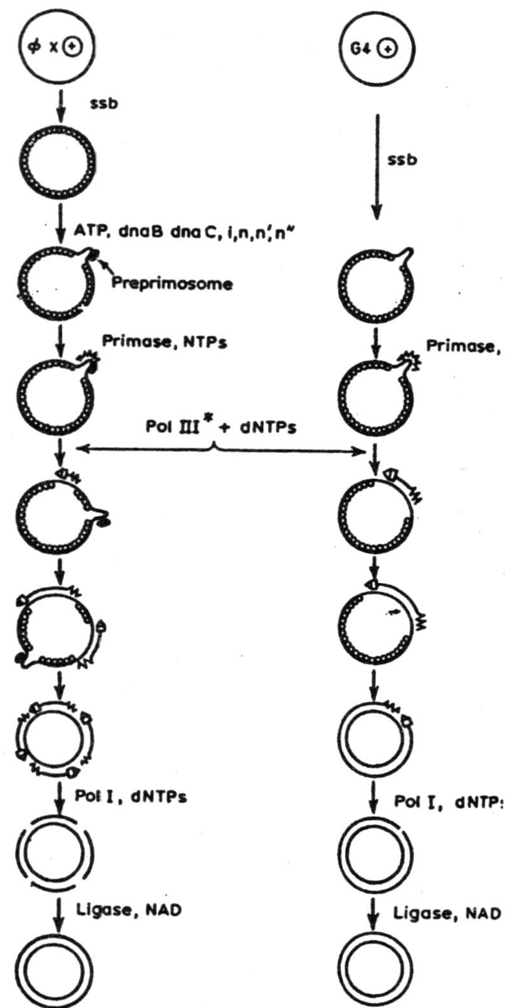

Fig. 6.24. Conversion of ϕX174 and G4 single-stranded DNA into duplex form (M13 resembles G4 except *E. coli* RNA polymerase is used in place of primase): ssb single stranded DNA binding protein; DNA holopolymerase III; primer.

The single plus strand with its associated binding protein thus resembles the initial bacteriophage DNA. In phage fd it thus remains unaltered until plus strand synthesis has gone almost a complete cycle when the RNA polymerase-binding site is exposed and minus strand synthesis is initiated. In phage G4 the primase-binding site is exposed when plus strand synthesis has gone half way round the molecule. With ϕX174 it is not certain whether or not an intact single-stranded plus strand is produced before the preprimosome reinitiates lagging-strand synthesis. However, as the preprimosome is retained on completion of SS → RF synthesis it appears likely that minus strand synthesis is initiated at several non-random sites leading to the production of Okazaki pieces.

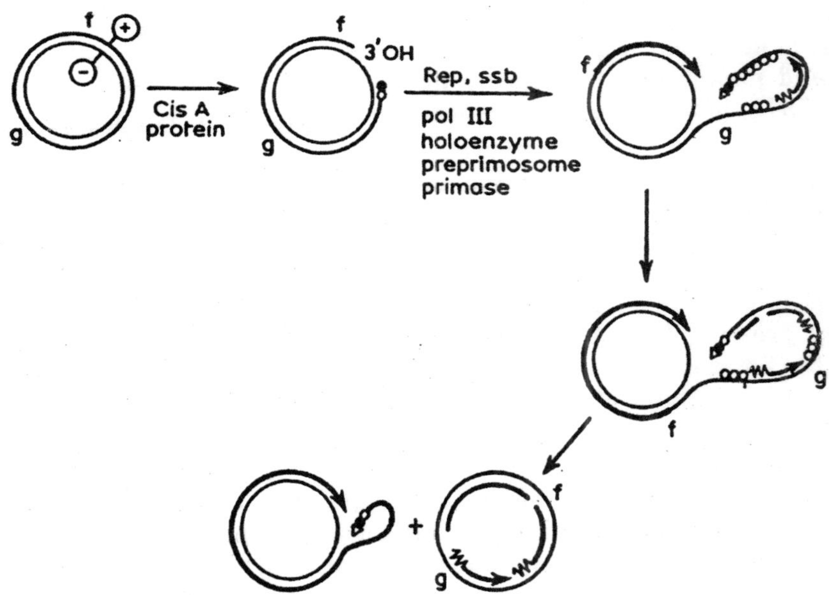

Fig. 6.25. RF replication of ϕX 174 DNA (for explanation see text). g and f correspond to the primase and RN A polymerase-binding sites on single-stranded DNA from bacteriophage G4 and fd respectively, and hence indicate the position of initiation of-strand synthesis with these bacteriophage DNAs; ssb, DNA-binding protein.

When one round of plus strand synthesis is complete the *cisA* protein at the 5′-end of the plus strand is again brought adjacent to its site of action. It now recognizes the left half of the origin sequence, nicks it, and circularizes the displaced plus strand, *i.e.* the origin sequence is both an initiation and a termination sequence. A second round of plus strand synthesis is immediately reinitiated using the same complex of *cisA* protein and rep. As the ϕX174 infectious cycle proceeds viral capsid proteins accumulate in the cell in the form of a prohead. This involves the products of viral genes F (major capsid protein), G (spike protein), H (spike-tip protein), B and D (non-structural proteins). The prohead interacts with the replicating DNA-gene A protein complex in a reaction requiring phage gene C and J proteins, and this prevents initiation of minus strand synthesis. Thus the single-stranded tail of the rolling circle is no longer available for conversion to duplex but rather is packaged into the phage prohead.

Synthesis of mature phage particles by a reconstituted mixture of purified proteins has been achieved. In the case of phage M13 the single-stranded tail becomes associated with gene V protein which is exchanged for coat proteins when the viral DNA leaves the host cell.

INITIATION OF REPLICATION - GENERAL

The results of the Meselson-Stahl experiment suggested that DNA replication would be found to be a sequential process and that successive rounds of replication would not begin at random positions on the chromosome. It is now clear that chromosomes from most viruses and prokaryotes, whether containing linear or cyclic DNA, initiate their replication at one specific site.

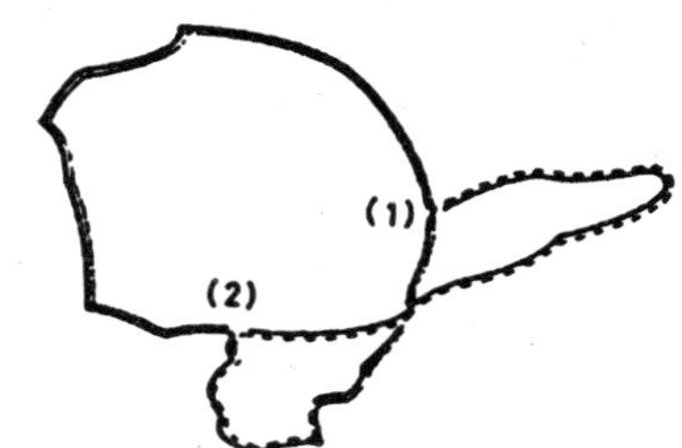

Fig. 6.26. Diagrammatic representation of an autoradiograph of the chromosome of *E. coli* labelled with tritiated thymidine for two generations [4]. The chromosome has been two-thirds duplicated and the two replication forks are indicated as (1) and (2).

For small DNA molecules, or for fragments of larger molecules, these sites can be visualized using the electron microscope when they appear as double-stranded 'bubbles'. Cairns was able to visualize the whole of the replicating *E. coli* chromosome autoradiographically by growing the bacteria for several generations in the presence of tritiated thymidine, and he showed that the chromosome exists as a continuous piece of double-stranded cyclic DNA.

Methods of Locating the Origin and Direction of Replication

That the initiation of the replication bubble occurs at a unique site has been shown by a number of different techniques for viral, plasmid, bacterial and mitochondrial DNA. Electron microscopy studies have shown that initiation of replication of the linear T7 phage chromosome occurs 17% from one end but cyclic chromosomes appear featureless in electron micrographs and so it is not immediately possible to define the position of replicating bubbles. However, Schnos and Inman were able to show that initiation of replication occurred at a particular site with reference to the partial denaturation map of phage λ and in the plasmid Col E1 and in the simian virus 40 chromosome replication is always initiated at a fixed distance from the site of action of the *Eco* RI restriction endonuclease. In most cases (Col E1 is an exception) the midpoint of the replication bubble stays fixed with respect to the ends of the molecule.

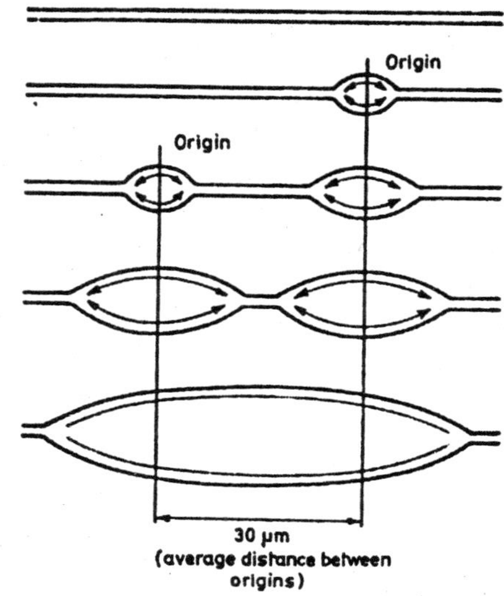

Fig. 6.27. Replication of mammalian DNA (two replicons are shown).

As the bubble gets larger it is thus clear that replication is occurring in both directions from the origin, *i.e.* replication is bidirection. As explained already that those genes located near the origin of replication of the *E. coli* or *B. subtilis* chromosome are present in more than one copy per cell in growing cells. Knowing the genetic map it is possible to measure gene copy number in growing cells relative to non-growing cells (where all genes are present in one copy per cell) and hence locate the origin of replication. Early experiments to measure gene copy number relied on measuring transforming or transducing activity and required a large panel of mutant strains.

More recent experiments measure the copy number or specific radioactivity of identifiable restriction fragments, an approach which has shown that replication of adenovirus DNA is initiated at both ends of the linear duplex molecule. Replicating DNA from eukaryotes shows a large number of replication bubbles in electron micrographs *i.e.* there is not a single origin of replication for each chromosome.

Replicons

Fibre radioautography studies suggest that DNA from their origins. Fibre radioauto-tandemly joined or 'replicons' which are separately replicated bidirectionally from their origin. Fibre radioautography involves labelling cells with tritiated thymidine and then lysing the cells in such a way that the DNA is released from the nucleus as a long fibre which attaches to a slide. On radioautography the position of the DNA is indicated by lines of grains produced by the β-particles released on tritium decay. What is seen are two parallel lines of grains whose length is dependent on the period for which the cells were exposed to the tritiated thymidine. The spacing between neighbouring lines gives the distance between replicons.

The size of replicons varies between 15 and 60 μm (40-200 kbp) in cultured mammalian cells but is considerably shorter in the very rapidly dividing cells of amphibian and insect early embryos, where S phase may last for only 20 minutes, and is much longer in slowly dividing spermatocytes. When the specific radioactivity of the tritiated thymidine is reduced after a short initial period of labelling at high specific radioactivity the density of the radioautographic grains is seen to decrease at both ends of the line of grains indicating that replication is bidirectional.

It is important in these experiments to start labelling the cells at the time of initiation of replicon synthesis otherwise gaps will be present in the middle of the lines of grains which may lead to misinterpretation of the results. With an average size of 30 μm the replicons of eukaryotes are very much smaller than the single replicon which is the *E. coli* chromosome (1300 μm). They are similar in size and may be physically identical with the loops of DNA seen in electron micrographs of dehistonized chromatin. In a mammalian cell there are typically about 60 000 replicons.

Rate of Replication

The rate of replication of DNA is dependent on the temperature and the supply of nutrients, particularly deoxyribonucleotides. The minimum time required to replicate the *E. coli* chromosome is about 40 minutes, implying a rate of synthesis of about 1700 base pairs per second. As replication is bidirectional the rate at each fork is about 850 base pairs per second or 14 μm per minute. (Compare this with the rate of transcription which proceeds at 35-40 nucleotides per second in *E. coli.*) In mammalian cells the rate of fork movement is only 0.5-1.2 μm per minute or about 60 base pairs per second. However, in eukaryotes this slower rate is compensated for by the smaller size of the replicon which is completely replicated in about 15 minutes. It is clear that only about 5% of replicons in a eukaryote cell are simultaneously active and some control must be exercised over the order in which replicons initiate DNA synthesis.

Origin Strategies

Implicit in the fact that initiation of DNA synthesis occurs at a unique site on the chromosome is the suggestion that this site has a distinctive structure which allows the binding of the replicative machinery and the unwinding of the double helix. It is considered probable that DNA synthesis is only initiated on a region of the DNA which is rendered single stranded perhaps by reason of its being involved in transcription, and specific RNA transcripts known as origin RNA (oriRNA) have been implicated. Such RNA molecules may also serve as primers for subsequent DNA synthesis. Many origin regions have been sequenced and whether present in regions coding for protein or not they contain unusual sequences, *e.g.* direct repeats, palindromes and true palindromes which may enable these regions (or the RNA derived therefrom) to assume a hairpin configuration.

The DNA may take up a hairpin or cruciform conformation in response to initiation proteins which thereby allow an RNA polymerase to synthesize a primer on which DNA synthesis may be initiated. Alternatively, as with the single-stranded bacteriophages a hairpin may be the only region of the DNA not covered by binding proteins and as such may be the only region accessible to a primase. Normally an RNA molecule, once synthesized, is released from the DNA template which returns to its double-stranded form. If the non-transcribed strand of the DNA were able to stabilize itself by forming a hairpin this would have the effect of retaining the RNA strand at the replication origin to serve as a primer for DNA synthesis.

It should also be borne in mind that rather than assuming a cruciform structure, four-stranded regions of DNA may be formed which could act as structures recognized by specific initiation proteins. Illustrates the ways some of these strategies are used but it may be helpful at this stage to point out the ways in which eplication occurs in different replicons.

(*a*) Duplex circles may replicate via a theta (θ) or Cairns structure in which the parental strands remain intact. Initiation involves production of a primer and usually relies on some mechanism to open up the duplex. The primer is extended as DNA to produce the leading strand. Lagging-strand synthesis may start almost immediately or may be delayed for a short time to produce a D-loop - or until the replication fork has traversed the whole replicon in which case synthesis leads to displacement of a single strand (displacement synthesis).

If lagging-strand synthesis proceeds through the origin region, or if separate initiation mechanisms exist for clockwise and anticlockwise replication, then replication will be bidirectional.

(*b*) Linear duplex molecules may initiate internally by similar mechanisms.

(*c*) Duplex circles often replicate by the rolling circle mechanism when initiation requires the action of a sequence-specific nuclease.

(*d*) Some linear duplex molecules initiate replication at the ends leading to displacement synthesis. Initiation may occur at both ends or a linear single-stranded molecule may be produced.

(*e*) Linear single-stranded DNA molecules can initiate by formation of a hairpin loop at the end which supplies a 3′-OH primer.

(*f*) Single-stranded cyclic DNA is converted to duplex by producing an RNA primer. This requires specific sequences on the DNA which can be recognized by the primase or primosome complex.

Positive or Negative Control of Initiation

DNA replication and, in many cases, cell division is initiated by a single event which occurs at the origin of replication of each replicon. This led Jacob, Brenner and Cuzin to propose a theory of positive control of replication similar to the induction/repression model for the control of protein synthesis in prokaryotes. This model proposes that a product of a replicon is a (protein) molecule which interacts with the origin to induce replication. The inducer molecule is *trans* acting and its concentration dictates whether or not replication will be initiated. An alternative model of Pritchard, Barth and Collins proposes negative control over initiation. They suggest that there is an inhibitor of replication present in cells.

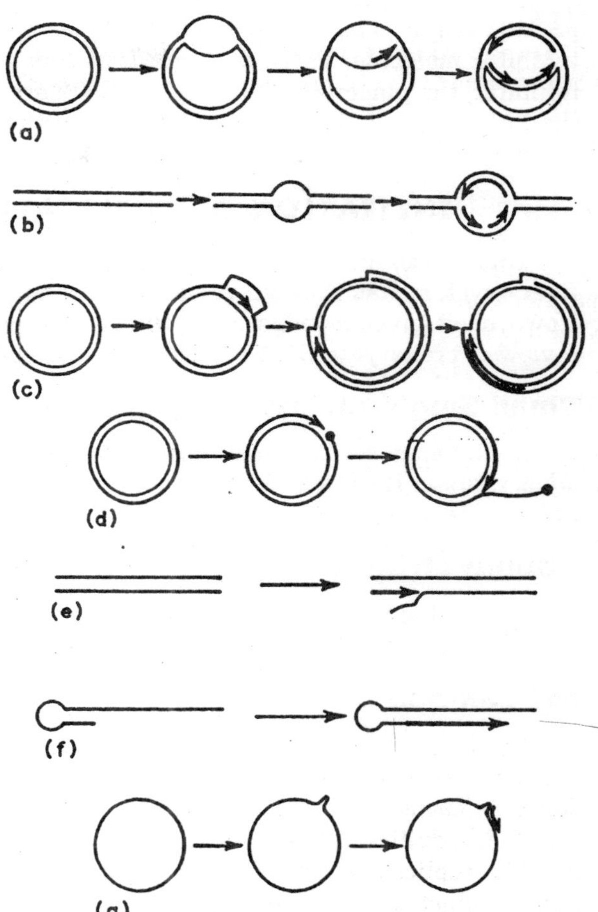

Fig. 6.28. Different ways in which initiation of DNA synthesis can occur, (*a*) and (*b*) represent Cairns (θ) structures formed by cyclic and linear duplex molecules respectively; (*c*) is a D-loop and (*d*) a rolling circle; (*e*) represents an alternative method of initiation of replication in linear duplex DNA molecules and (*f*) illustrates the mechanism of initiation in linear single-stranded DNA molecules. The initiation of replication of single-stranded cyclic molecules is shown in (*g*).

The inhibitor is synthesized by the replicon, and only when its concentration is diluted (by growth of the cells) will initiation occur. Experiments of Rao and Johnson where animal cells in the G1 phase of the cell cycle were fused with S-phase cells strongly supported the idea of a trans acting factor exerting control over initiation of DNA synthesis but did not distinguish between positive and negative control.

However, experiments involving fused plasmids strongly supported a negative control mechanism. Cabello *et al.* fused Col E1 with the related plasmid pSC101 to form plasmid pSC134 which used only the Col E1 origin and maintained the copy number of Col E1. However, in *polA* mutants Col E1 cannot grow and under these conditions pSC134 used the pSC101 origin and assumed the copy number of pSC101. These results are what would be expected of incompatible plasmids responding to identical repressers where the pSC101 origin is more sensitive to inhibition and hence is only induced in large cells in which the repressor has been diluted. Thus normally the Col E1 origin is activated first and maintains the level of repressor.

Only when the Col E1 origin is inactive does the level of repressor fall sufficiently for the pSC101 origin to be activated. Related plasmids are incompatible if they respond to the same inhibitor molecule. Thus, as a result of changes in base sequence around the target for the inhibitor, the plasmids pMB9 and pBR322 are less sensitive than Col E1 and rapidly exclude the latter. As we shall see in the next section there may be elements of both positive and negative control over the initiation of replication.

INITIATION OF REPLICATION – SPECIFIC EXAMPLES

This section considers several examples of what we know about initiation of replication. Every system has its own peculiarities and this section can in no way be comprehensive. However, it gives an idea from which some generalities can be obtained. A comprehensive review of prokaryote DNA replication has been published recently.

Small Single-stranded Phage

As it was described how replication of the duplex RF forms of the small single-stranded DNA virus is initiated. A viral-coded protein nicks one strand of the duplex molecule at a unique site and initiates replication by the so-called rolling circle mechanism.

Double-stranded Phage

Three phage (λ, T7 and T4) have been extensively studied. No common rule can be applied so each will be briefly considered.

(a) Lambda

The λ DNA molecule is linear but has sticky ends and cyclizes immediately on entry into the cell. Replication then is initiated at a unique origin and proceeds bidirectionally to give theta (θ) forms (cf. *E. coli* replication). However, at later stages of infection rolling circles are found. In addition to host functions the products of λ genes O and P are required for initiation of DNA replication. The P gene product appears to take over the function of *E. coli dnaC* gene product. It interacts with *E. coli dnaB* protein and the complex interacts with lambda gene O product and the host cell *dnaJ* and *dnaK* proteins to form a preprimosome at a specific site (ori) rendered single-stranded by transcription by host RNA polymerase. This is called transcriptional activation but a similar preprimosome can form on any single-stranded DNA in a sequence-independent manner.

Eight molecules of the O protein interact with four 19 base pair repeats within a 65-bp origin region which itself is within the gene coding for the O protein and adjacent to that coding for the P protein. This region which is necessary for phage lambda replication has been excised and propagated as a plasmid called λ*dv* which is transcribed from the P_R promoter.

This whole region has been sequenced. It has been proposed that *dnaG* primase synthesizes a 78-base RNA from a region to the left of ori. This *oop* RNA may prime DNA polymerase III-catalysed DNA synthesis. Alternatively *oop* RNA may be extended by RNA polymerase leading to synthesis of λ repressor and the state of lysogeny results where the genome becomes integrated into the host genome and the two replicate under host cell controls.

In vitro systems have been developed which will replicate λ*dv* plasmid using purified proteins. They indicate that the transfer from RNA to DNA synthesis takes place on both sides of the *ori* sequence in such a way that DNA synthesis from the two sides converges.

(b) Phage T7

Initiation of T7 DNA synthesis occurs 17% from the so-called left end of the molecule. A replication bubble forms which expands bidirectionally. When the short side is completely synthesized a Y-shaped replicating chromosome is formed.

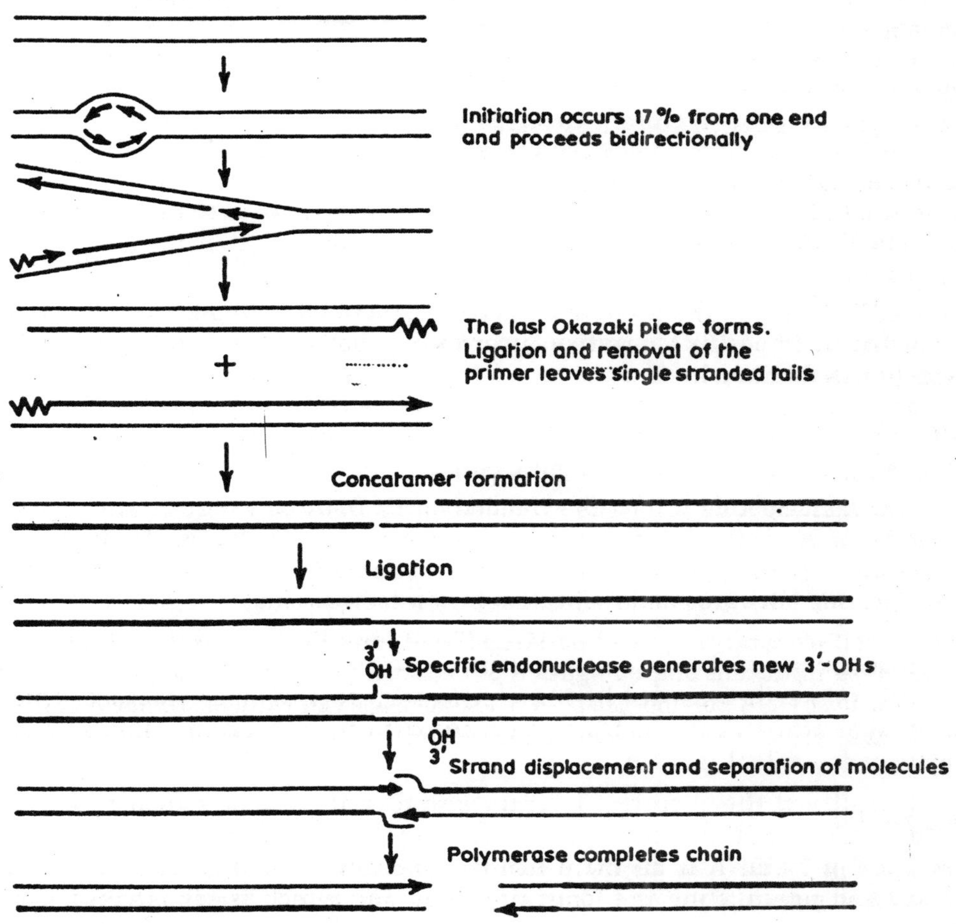

Fig. 6.29. Replication of T7 DNA.

Concatamers, *i.e.,* double length molecules, are found later in infection.

Replication requires phage genes 1, 2, 3, 4, 5 and 6. Gene 1 product is an RNA polymerase essential for the transcription of genes 2-6 and gene 2 protein inactivates the host RNA polymerase. Gene 3 protein is an endonuclease and gene 6 protein an exonuclease involved among other things in the breakdown of host DNA. There are two strong promoters for the T7 RNA polymerase in the origin region, closely followed by an AT-rich region.

Transcription causes melting of the origin region which can be detected by treatment with a single-stranded endonuclease. Within the AT-rich region there is a change-over from RNA synthesis to DNA synthesis which proceeds unidirectionally towards the right. In the presence of T7 gene 5 protein and the host protein thioredoxin which together form a DNA polymerase the T7 gene 4 protein catalyes an ATP-dependent unwinding of duplex DNA thereby promoting the DNA polymerase action. This complex is thus able to catalyse leading-strand synthesis and DNA-binding protein and gyrase further promote this action.

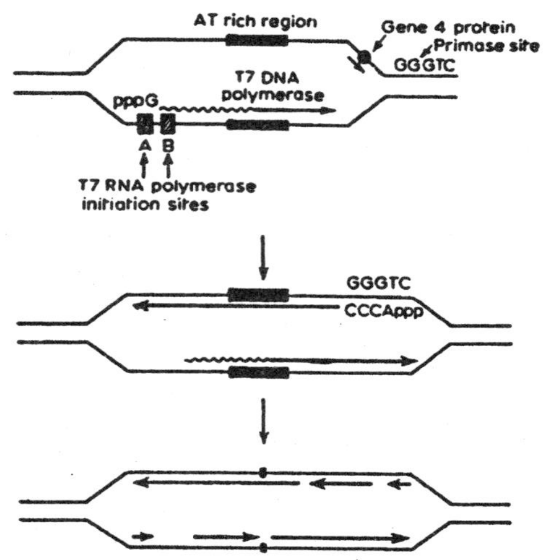

Fig. 6.30. Initiation of T7 DNA replication.

Gyrase either relieves the positive supertwist generated by replication or it may actively help unwinding by imposing a negative supercoiling ahead of the fork. Gene 4 protein has a second function in addition to its action as a DNA helicase. It also acts as a primase for lagging strand synthesis which is initiated by the synthesis of tetraribonucleotides of sequence pppACCC and pppACCA. Thus when a primase site is exposed by the rightward movement of the leading strand, synthesis of the lagging strand is initiated using the T7 gene 4 protein.

The tetraribonucleotide primer is extended by T7 DNA polymerase. DNA synthesis now proceeds left-wards on the lagging strand through the origin region. With the help of the helicase action of gene 4 protein and single-stranded DNA-binding protein a fork is now established moving leftwards and replication is now bidirectional.

Thus as well as catalysing leading-strand synthesis the gene 4-gene 5 protein complex car also catalyse initiation and elongation ol Okazaki pieces on the lagging strand. Gene 6 exonuclease or host DNA polymerase $5' \rightarrow 3'$ exonuclease can catalyse removal of RNA primers and the Okazaki pieces can be poined using either the host ligase or a ligase oded for by T7 gene 1.3.

(c) Phage T4

Replication of T4 DNA is far more complex and initiation of DNA synthesis appears to occur at several points along the long, circularly permuted, linear molecule. It has been proposed that, as with T7, primary initiation occurs following synthesis of an RNA primer but in this case it is supposed that only the most RNA polymerase can act in this manner. Mutations in T4 genes 39, 52 and 60 lead to a relay in the onset of T4 replication and the DNA synthesis that does occur is dependent on the most cell gyrose. These three genes code for in ATP-dependent to poisoners which may help to open up the duplex molecule, however, shortly after infection the host RNA polymerase is modified by interaction with the products of viral gene 33 and gene 55 and this affects promoter recognition rendering it unable to initiate DNA replication.

Initiation now becomes dependent on genes 46 and 47 which code for enzymes involved in recombination. The model which has been proposed for these secondary initiations involves recombination between partially replicated molecules which contain single-stranded regions and the circularly permuted duplex molecules. This model explains the finding of complex branched molecules later in the T4 infectious cycle. When cells of *E. coli* are infected with the T-even bacteriophages, the economy of the cells is completely altered so as to lead to the production of new phage DNA which differs from the host DNA in containing hydroxymethylcytosine in place of cytosine. These changes result in the production in the infecfed cell of a series of new and interesting enzymes.

(*i*) Within a few minutes of infection a hydroxymethylase (the product of phage T4 gene 42) appears which brings about the conversion of dCMP into hydroxymethyldeoxy-cytidine monophosphate (dHMCMP).

(*ii*) At about the same time a ***kinase*** is produced which phosphorylates dHMCMP to the corresponding triphosphate dHMCTP.

Neither of these new enzymes is found in cells infected with phage T5 which does not contain HMC.

The kinases for dTMP and dGMP are also greatly increased but not that for dAMP. This increase is due to the production of new enzymes which can be distinguished from the kinases present in the host cell prior to infection.

(*iii*) The formation of host DNA is prevented by the appearance of a pyrophosphatase (the product of phage T4 gene 56) which converts dCTP into pyrophosphate and dCMP which then acts as a substrate for the hydroxymethylase.

(*iv*) Five distinct glucosyltransferases are known to be induced after infection with T-even phages for the purpose of transferring glucose residues from uridine diphosphate glucose to the HMC of phage DNA in the proportions. The glucosylase found in T2 phage-infected cells transfers a glucose residue to HMC in the β-configuration. Two glucosylating enzymes are produced after T4 infection; one adds a glycosyl group in α-linkage to HMC while the second also adds a glucose group but in β-configuration. After T6 infection two glucosyltransferases are also produced. One adds a monoglucosyl residue to HMC in α-linkage while the other reacts with the monoglucosylated groups on HMC to add a second glucose residue, the linkage between the residues being of the β-configuration.

(*v*) A nuclease is produced which degrades DNA containing cytosine. T4 DNA which contains hydroxymethylcytosine is resistant.

In vivo the enzymes synthesizing the precursors for DNA synthesis are closely associated with the DNA polymerase and in *vitro* the complex of enzymes preferentially uses deoxynucleoside monophosphates to make T4 DNA.

Plasmids

Col El

This small double-stranded cyclic DNA molecule replicates from a unique origin but in one direction only. Theta (θ) structures are formed and initiation does not involve permanent nicking of either strand. Up to three proteins bind near the origin and bring about supercoiling of the cyclic molecule thereby exposing single-stranded regions to which *E. coli* RNA polymerase binds. An RNA transcript (RNA II) is initiated 555 base pairs upstream from the origin (–555). This transcript is not released from the origin region but a D-loop is formed with a DNA/RN A hybrid and single-stranded DNA. This hybrid is sensitive to RNase H which fragments the RNA producing several primers which can be extended by *E. coli* DNA polymerase I. This enzyme synthesizes a 6S piece of DNA (400 nucleotides) joined to a 20-long ribonucleotide primer, the whole remaining associated in the D-loop configuration.

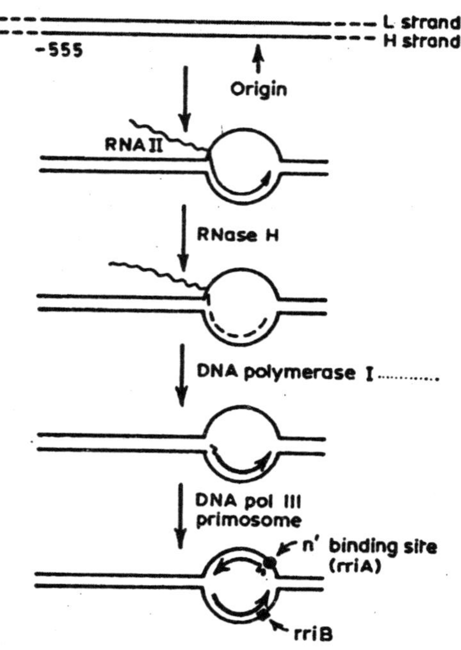

Fig. 6.31. Initiation of Col E1 replication.

The DNA polymerase III replicating complex now takes over to extend the fork. On the lagging side of the fork (on the L strand) there is an *n'* binding site (*rriA*) at which lagging-strand synthesis can be initiated. However, lagging-strand synthesis cannot pass the origin region and Col E1 replication remains unidirectional. There are two complications to this story. As well as RNA II a second RNA (RNA I) is transcribed from the region –455 to –555 from the opposite strand of the DNA. This RNA, which can assume a complex secondary structure, interacts with RNA II and either prevents its forming a stable hybrid at the origin or interferes with RNase H action in some other way. RNA I thus acts in trans to limit initiation and hence copy number of Col E1 and other compatible plasmids. A second trans acting inhibitory element exists downstream from the origin (184 to 806 bp).

The product of this region is the *rap or mm* protein (63 amino acids long) which enhances the binding of RNA I to RNA II. This leads either to premature termination of transcription of RNA II or to inhibition of processing. This protein appers to play no role in incompatibility but may regulate the balance between replication of duplex plasmid molecules (as described here) and the replication involved in the single-strand transfer reaction of conjugation. In conjugal transfer a nick is introduced into one strand (the H-strand) and the endonuclease remains bound to the 5'-side of the nick. Rolling circle replication now leads to the synthesis of a new duplex cyclic molecule and the parental H-strand can be passed to another bacterium through a conjugation tube in the form of a single-strand. Cyclization follows transfer and conversion to a duplex molecule is initiated by the binding of *n'* protein to a preprimosome recognition site (*rriB*). Thus in conjugal transfer, replication of Col E1 closely resembles replication of ϕX174' except that the order of reactions is

$$RF \rightarrow SS \rightarrow RF$$

(b) Other Plasmids

The drug-resistance plasmids (*e.g.* R1, R6 and R100) differ from Col E1 in that RNA II may act as a primer for initiation of replication but, in addition, it can be extended to form a messenger RNA from which a protein (π) is translated. π protein acts as a cis-acting initiation protein.

The origin region of pSC101 shows a similarity to that of *E. coli* and replication of this plasmid is dependent on the *dnaA* protein. In addition a plasmid-coded protein which binds to the origin region is also required. These plasmid-coded replication initiator proteins also regulate their own transcription by acting as autorepressors.

Bacteria

E. coli temperature-sensitive mutants *dnaA* and *dnaC* fail to initiate new rounds of DNA replication when placed at restrictive temperature but complete those rounds already in progress. On cooling such mutants to the permissive temperature in the presence of tritiated thymidine the first region to become radioactive is a 1.3 kilobase fragment formed by treatment of the DNA with the restriction endonuclease *Hind*III. The amount of incorporation is less in the case of *dnaA* mutants than *dnaC* mutants indicating that the *dnaA* gene product is involved several minutes before the *dnaC* gene product in the initiation process. Initiation of replication in *E. coli* is sensitive to rifampicin and chloramphenicol and a protein located in the membrane which is normally made about 15 min before initiation is not made in *dnaA* mutants.

On RNA of *E. coli* has been isolated linked to high-molecular-weight DNA. It is not made in *dnaA* mutants and is not linked to DNA in *dnaC* mutants suggesting a role for these gene products in initiating on RNA synthesis *(dnaA)* and in the change-over to DNA synthesis (*dnaC*). The origin of replication of the *E. coli* chromosome (*oriC*) has been located on a 245 bp sequence which when cloned into small Col E1-type plasmids allows them to replicate in the absence of DNA polymerase I. Such replication is bidirectional and dependent on *dnaA* protein. The *oriC* region binds with high affinity to membrane proteins. The *oriC* sequence is conserved in a number of Gram-negative species though it has been difficult closely to define the essential features. In this region some base substitutions are deleterious; others not; but deletions are harmful suggesting some of the sequences may serve as spacers.

The origin region probably provides multiple sites of interaction with replication initiation factors and must be present in a precise structure for activity. It appears, however, that initiation does not take place within the 245 bp region but just outside. In addition to the proteins required for leading- and lagging-strand synthesis, initiation shows a requirement for the products of the *dnaA* , *dnaI* and *dnaP* genes (and possibly also the *dnaJ, dnaK* and *dnaL* genes), RNA polymerase and a histone-like DNA-binding protein called HU. The *dnaA* requirement is highly characteristic and multiple copies of this protein bind to four 9-bp regions in *oriC* of sequence TTATCCACA. Initiation is severalfold more efficient in dam^+ cells indicating that DNA methylation may play a role here. A system has been reconstructed from purified proteins which has enabled initiation to be separated into two stages.

In the absence of DNA polymerase III holoenzyme an RNA primer is made but not extended. If an inhibitor of *E. coli* RNA polymerase is now added along with DNA polymerase III holoenzyme, no further primers are made but those already present can now be extended. Topoisomerase I, RNase H and HU protein are required to prevent initiation at sites other than *oriC*. In addition to RNA polymerase the production of a specific primer requires DNA gyrase and the products of the *dnaA, dnaB* and *dnaC* genes.

There are two RNA polymerase promoter sequences in *oriC* which can lead to initiation of transcription in both directions. However, there is evidence that initiation occurs first on one strand to produce a fork which moves in a counter-clockwise direction. Primase (the *dnaG* gene product) is not essential but when present it will interact with the preprimosome to synthesize a primer, *i.e.* the RNA polymerase functions to catalyse transcriptional activation. The second stage requires gyrase, single-stranded-DNA-binding protein, and the products of the *dnaB* and *dnaC* genes in addition to the polymerase III holoenzyme.

Mitochondria

In animal cells mitochondria contain a closed circular 16-kbp duplex DNA molecule which replicates unidirectionally from a unique origin. Prior to initiation the duplex undergoes partial unwinding by removal of about 48 Watson-Crick turns in the presence of a nicking closing enzyme. Initiation of DNA synthesis then occurs with synthesis of a 450-base length of DNA (the leading or heavy strand) to produce a D-looped structure. This small piece of newly synthesized 7S DNA is unstable and may turn over several times before it is extended. Extension occurs asymmetrically and unidirectionally until 60% of the heavy strand (99% in *Drosophila* mitochondria) has been synthesized, when light (lagging)-strand synthesis is initiated.

Thus although θ forms are seen in the electron microscope one loop is a single-stranded parental H-strand displaced by synthesis of daughter H-strand. The short primer RNA found at the 5′-end of human mitochondrial 7S D-loop DNA does not have a unique 5′-end probably as a result of rapid processing. This RNA, which may be synthesized by a unique RNA polymerase, is initiated in the region of a palindrome which could form a stable hairpin structure when the DNA helix is unwound. The 3′-end of 7S DNA is just downstream from a 13-bp sequence which occurs four times in mouse mitochondrial DNA but only once in human mitochondrial DNA. D-loop DNA is extended and when H-strand synthesis has proceeded two-thirds of the way round the genome a sequence is exposed which can form a stable hairpin structure.

An RNA primer is synthesized here which leads to initiation of L-strand synthesis. Both strands are now extended and the two daughter molecules separate with one being still almost half single stranded. The available evidence suggests that DNA synthesis in mitochondria is catalysed by DNA polymerase γ. In sea-urchin oocytes mitochondrial DNA synthesis is similar to that found in mouse mitochondria but duplex synthesis occurs early with multiple initiations. Yeast mitochondrial DNA contains several origins of replication. In petite strains in which a short random stretch of DNA is excised and amplified, surrogate origins are used. In one petite strain the mitochoritirial DNA contains only adenine and thymine nucleotides. Mitochondrial DNA in *Tetrahymena* and *Paramedum* is not cyclic but is a linear duplex.

In the former, replication is initiated near the centre of the molecule and 'eye'-shaped replication intermediates are seen in the electron microscope. Replication is bidirectional. In *Paramedum* replication proceeds unidirectionally from a crosslinked terminus.

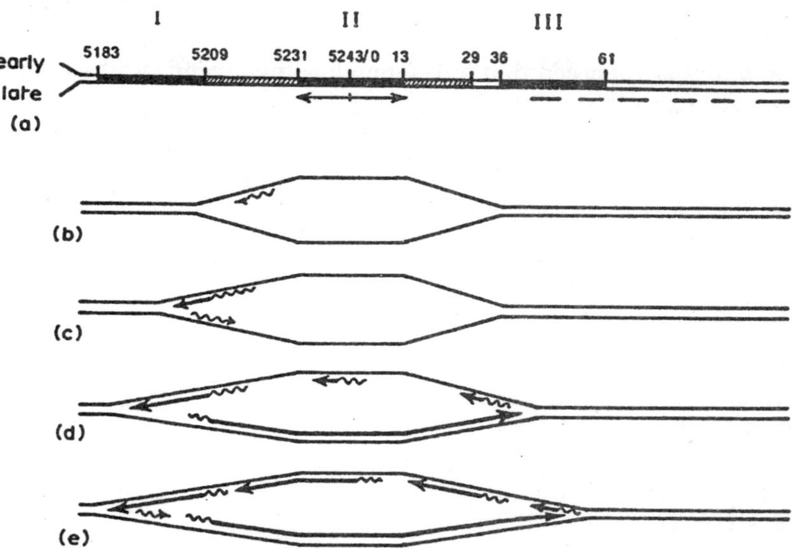

Fig. 6.32. Initiation of SV40 DNA replication: (*a*) The origin region showing the three T antigen binding sites (filled boxes), the minimal origin region (hatched box), the six 8 base pair repeats (lines under the diagram) and the palindromic region (arrows); (*b*) T antigen binding denatures the origin region and allows primase to synthesize a short primer (N^+) at one of several sites; (*c*) The primer is extended as DNA synthesis exposes primase binding sites for retrograde synthesis; (*d*) The retrograde strand is extended through the origin region to become the leading strand for late strand synthesis; (*e*) Other primers are formed on the early strand to initiate lagging-strand synthesis. Binding of T antigen to site III may also allow late strand synthesis to be initiated at the 8 bp repeats.

Double-stranded Cyclic DNA Viruses (SV40 and Polyoma)

These similar, small, animal tumour viruses contain a cyclic DNA molecule which replicates as a θ form, the parental strands remaining intact. They have been studied, not only for their own sake but as models for animal cell DNA replication. This is only poorly justified for, although the DNA is present in the nucleus complexed with histones (*i.e.* as a minichromosome) replication is not restricted to one round per cell division. Initiation occurs at a unique site and proceeds bidirectionally. The origin region is also the site of initiation of transcription which also occurs in both directions.

One of the early SV40 gene products (T antigen) is essential for initiation of replication. There are three sites in and around the origin region to which T antigen can bind. However, only one of the T antigen-binding sites (site II) falls within the 65-bp essential origin region, and the central 26-bp region of site II is palindromic. Just outside the 65-bp origin regions are six repeats of the sequence GGGCGGRR. Isolation of nascent DNA from the origin region showed initiation to occur at many sites, each time with a short (7-8 base) ribonucleotide primer. Some 70% of primers start with adenosine, the remainder with guanosine.

On the early strand (*i.e.* the template for early mRNA synthesis) DNA primase will initiate at four major sites around nucleotide 5215 between T antigen-binding sites I and II, and transfer to continuous (leading strand) DNA synthesis occurs at about nucleotide 5210 (*i.e.* at the beginning of T antigen-binding site I). There are no initiations for late-strand synthesis in the origin region and it is supposed that only when leading-strand synthesis exposes primase-binding sites does initiation of late-strand synthesis occur.

In vitro results suggest that late-strand synthesis may also be initiated in any of the six repeated sequences, removal of which reduces the efficiency of DNA replication *in vivo*. *In vitro* systems have shown that DNA polymerase a and its associated primase are involved in both initiation and continued SV40 DNA replication. The origin region of polyoma virus DNA is similar to SV40 but for initiation, in addition to large T antigen, there is a requirement for a *cis*-acting enhancer element.

Adenoviruses

These are animal viruses containing a linear duplex DNA molecule of about 36 000 bp. This replicates by a strand-displacement mechanism, initiation occurring at either end of the molecule. The adenovirus DNA contains an inverted terminal repetition extending to the very ends of the genome. In different adenoviruses the length of the terminal repeat varies from 103 to 162. Two regions in this terminal repeat are required for initiation. An AT-rich stretch from positions 9-18 is essential for binding of a 'terminal protein' (see below) and the region up to position 48 is required for binding of a host cell nuclear protein.

Covalently bound to both 5′-termini is a 55 000-molecular-weight protein. Binding is by a phosphodiester link to the β-OH of a serine residue. Initiation requires host cell DNA topoisomerase I and at least three virus-coded proteins: a single-stranded DNA-binding protein whicl may also bind to the ends of duplex molecule a DNA polymerase and an 80 000-molecular-weight 'termina protein'. The terminal protein forms a complex with the DNA polymerase and in the presence of adenovirus DNA it interacts with dCTP so that a dCMP residu becomes covalently linked to the terminal protein. This dCMP then serves as a primer to inititate displacement synthesis.

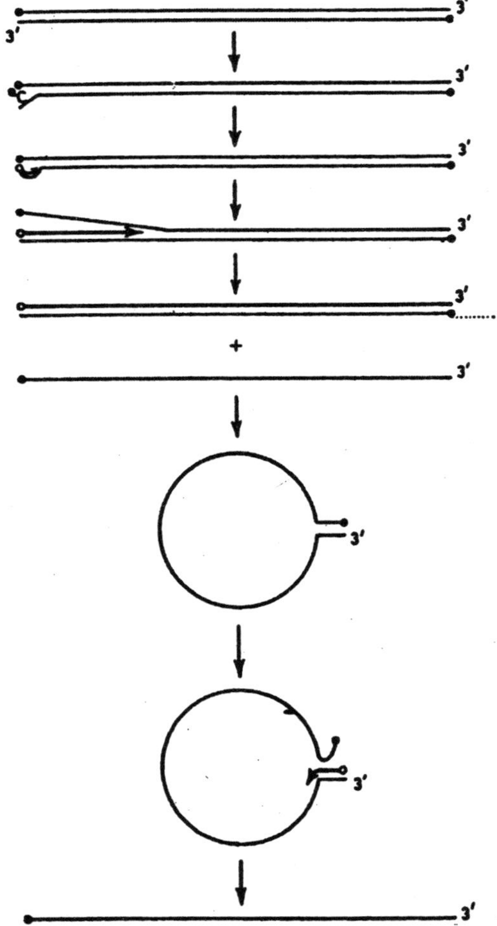

Fig. 6.33. Replication of adenovirus DNA. The terminal protein is represented by a circle (•, 55 K; O, 80 K).

Electron microscopy has been used to show that initiation usually occurs at only one end of the molecule so that one round of replication results in the formation of a duplex and a linear single-stranded molecule. However, by pulse-labelling with [^{3}H]thymidine and analysis of completed molecules it has been shown that termination occurs at both ends of adenovirus DNA showing that initiation can occur at either end and some molecules are seen in the electron microscope which are replicating from both ends. However, if a single strand of DNA is produced, the terminal repeat will allow it to form a panhandle structure to initiate conversion to a duplex DNA molecule.

At some time during infection the 80 000-molecular weight terminal protein is reduced in size to 55 000 molecular weighl but this processing does not appear to be essential. Adenovirus is not unique in adopting this strategy for initiation of replication. Thus phage ϕ29 initiates using a dAMP covalently linked to a 30 000-molecular-weight terminal protein and Pearson *et al.* have pointed out the similarity between this strategy and that adopted by ϕX174 which initiates replication of duplex circles by nicking one strand and covalently attaching a protein to the exposed end.

Yeast

Yeast cells contain a small 2μ plasmid which replicates autonomously. For this to occur the plasmid requires two proteins for which it codes and a short origin-containing region at which bidirectional replication is initiated. Other plasmids have been constructed which can also replicate in yeast provided they contain a particular region from the yeast chromosomal DNA. Such regions are known as autonomous replicating sequences (ARS) as they confer this property on the piasmid. This system has allowed the isolation of several ARS elements which all contain a very AT-rich core region which has the consensus sequence. A/TTTTATA/GTTTT/A. Similar sequences obtained from non-yeast cells will also confer autonomous replication upon yeast plasmids but we have no evidence yet that they act as origins of replication in their normal environment.

Higher Eukaryotes

As we have said earlier there are multiple origins of replication in eukaryotic cells each of which initiates once per cell cycle; yet not all replicons are simultaneously active. In the situation where genes occur in long tandem repeats it has been possible to identify an origin of replication in each repeat. For example, electron microscopy has been used to localize the centre of replication bubbles of sea-urchin ribosomal DNA to the non-transcribed spacer, and at the beginning of S phase one particular restriction fragment of the amplified dihydrofolate reductase gene in hamster cells is labelled first. In contrast, although initially subject to some controversy, it now appears clear that any DNA sequence when injected into *Xenopus* eggs will replicate under the strict cell cycle control pertaining.

Although even bacterial DNA replicates in *Xenopus* eggs, there is some evidence that sequences containing copies of the short interspersed repeated sequences (*e.g. Alu* sequences) may be replicated more efficiently. Nevertheless it is difficult to reconcile observations of specific origin sequences with the changes in size of replicons found in rapidly dividing embryos and very slowly growing cells, observations which imply that a particular origin may be used in some situations, but not in others. Relevant to this finding is the observation that origins of replication and/or replicating forks are found associated with the nuclear matrix, that residual nucleoprotein complex which remains following treatment of nuclei with high concentrations of NaCl followed by deoxyribonuclease digestion. Thus initiation may require the binding of short interspersed repeated sequences to the nuclear matrix. Certain sequences may be more susceptible to binding than others and hence may initiate earlier in S phase.

Other sequences may fail to initiate at all in slowly growing cells leading to an increased replicon size. This situation is not radically different from that found in *E. coli* where the *oriC* region is required to bind to the membrane-associated *dnaA* protein for initiation to occur.

Retroviruses

These pose a unique problem, for, they replicate by first converting their RNA genome into DNA using RNA-directed DNA polymerase (reverse transcriptase). Models of reverse transcriptase action have been proposed to explain certain unexpected observations, *i.e.*, the duplex DNA produced is longer at both ends than the viral RNA and during the reaction short molecules of both plus and minus strand DNA ('strong stop DNAs') are formed as well as two genetically identical copies of the viral RNA the virion contain; several molecules of tRNA reverse transcriptase also exhibits ribonuclease H activity (which will hydrolyse the RNA strand o an RNA:DNA hybrid) and endonucleas activity.

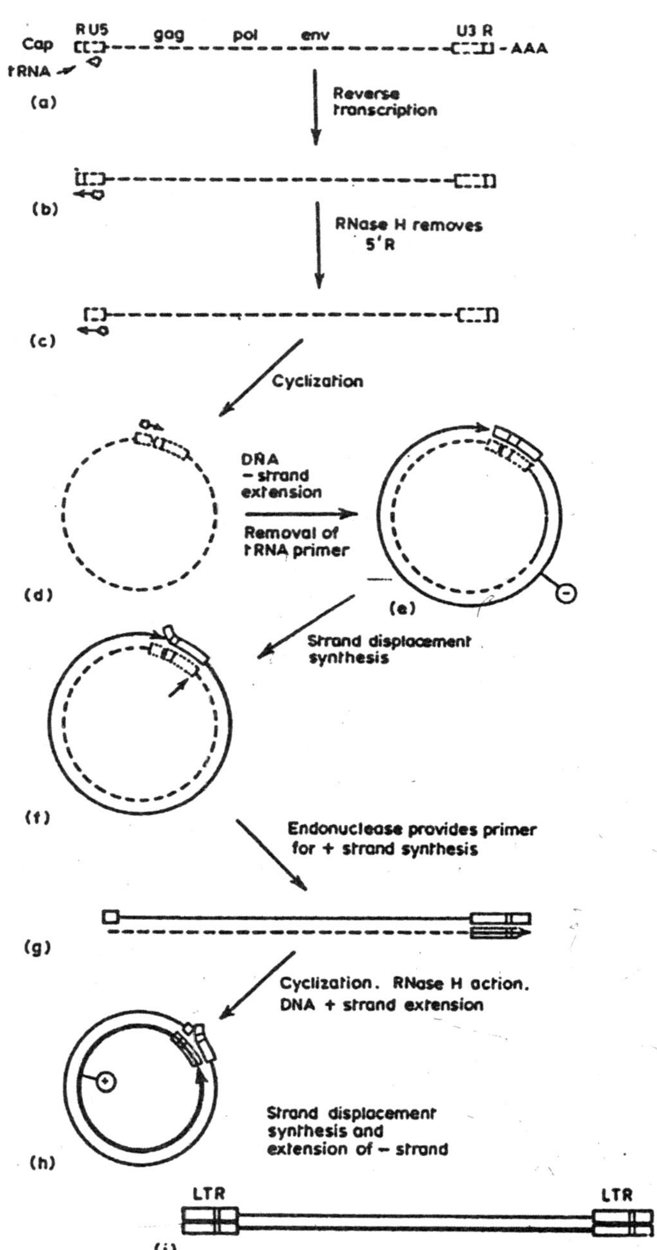

Fig. 6.34. Postulated mechanism of conversion of retroviral UNA into duplex DNA.

Avian reverse transcriptases are initiated wit $tRNA^{Trp}$ but the reverse transcription from Moloney murine leukaemia virus starts by using a $tRNA^{pro}$ which attaches to tli viral RNA at about 150 nucleotides from the 5'-end. This $tRNA^{Pro}$

serves as a primer for synthesis of minus strand DNA. Synthesis stops when the DNA strand reaches the 5′-end of the template, *i.e.,* after 100-150 nucleotides have been added. This is minus strand strong stop DNA. The RNA genome is terminally redundant and digestion of the 5′-end by ribonuclease H allows the molecule to cyclize and this provides a template for the further extension the DNA to form a copy of the whole of the viral strand.

Some strand-displacement synthesis causes the minus strand to duplicate part of the long terminal repeat sequences. It is speculated that specific endonuclease action on the viral RNA by the reverse transcriptase could now provide a primer for plus strand DNA synthesis. The plus strand strong stop DNA is extended to produce the complete duplex DNA which is usually integrated into the host cell chromosome before transcription leads to production of new viral RNA. The terminal nucleotides of the linear DNA duplex are required for integration into the host chromosome and are lost in the process. Similar conversion of RNA into duplex DNA followed by integration is believed to occur in the production of pseudogenes and in the movement of certain transposable genetic elements. The replication of the plant DNA virus cauliflower mosaic virus is also believed to involve reverse transcription from a 35S cyclic RNA transcript, and a similar enzyme has been implicated in hepatitis B virus replication.

TERMINATION OF REPLICATION

Cyclic Chromosomes

A complication arises when the two replication forks moving in opposite directions around a cyclic chromosome approach one another. Unwinding takes place ahead of the fork for a distance of about 100 bp and when the unwound regions meet the result is two gapped duplex molecules interlinked (*i.e.*, catenated) about 20 times. Replication of one of the daughter molecules appears to go to completion before the other, and a DNA topoisomerase II is required to decatenate the dimer.

Small Linear Chromosomes

These pose a different problem. A typical situation arises with T7 DNA replication. The last Okazaki piece to be made on the lagging strand might start at the extreme 3′-end of the template, but when the primer is removed this will leave a short single-stranded tail at each end of the linear molecule, a tail which cannot be replicated by any mechanism considered so far. James Watson postulated a mechanism whereby the problem might be solved. T7 DNA has an identical sequence of about 260 nucleotides at each end of the molecule, *i.e.,* the molecule is terminally redundant. Thus the unreplicated single-stranded tail of one molecule could H-bond to the similar region of another molecule giving rise to a concatamer of almost two unit lengths.

The two pieces can be joined together by ligase action and two new nicks introduced several nucleotides 5′ to the original nick. The gene 3 endonuclease is a candidate for the enzyme capable of making these specific cuts. Nick translation from the newly introduced 3′-OHs will allow the concatamer to separate into two halves and the replication to be completed. Other small linear genomes cope differently with the problem of end replication. We have seen that adenovirus DNA uses an unusual mechanism for initiation at the very ends of the DNA duplex which removes the extra stages required for termination by phage T7 DNA. Phage lambda DNA has sticky ends and forms a cyclic molecule on entering its host cell.

Telomeres

Eukaryote chromosomes are faced with the same problem as is T7 but the ends of these chromosomes are not like the ends of the small linear chromosomes. Recombination does not readily take place between the normal ends oi chromosomes showing them to be different from new ends produced by chromosome breakage (which very rapidly recombine with another piece of duplex DNA). The telomere is the terminus of a linear cellular DNA molecule with newly synthesized DNA was shown by the cleavage of nascent chromatin by micrococcal nuclease into approximately 200-bp fragments.

However, the rate and extent of cleavage of nascent chromatin is greater than that of mature chromatin indicating a somewhat different structure which requires 15 minutes to mature, *i.e.,* a time sufficient to synthesize 22000-54000 bp of DNA or approximately one replicon. Histones, as well as DNA, are made predominantly in S phase and inhibition of DNA synthesis inhibits synthesis of histones and causes pre-existing histone mRNA to break down with a much reduced half-life. In yeast, co-ordinate control of histone gene transcription and replication appears to involve the location of the promoter for the H2A and H2B genes adjacent to an ARS element.

Some histone synthesis is not dependent on DNA synthesis, however, and for instance histone H3.3 is synthesized in quiescent cells and during Gl and G2 phase together with four H2A variants and HI A and H1°. How histones associate with DNA is not clear but DNA injected into *Xenopus* oocytes rapidly assumes a nucleosomal conformation, and *in vitro* studies have implicated a heat-stable protein, nucleoplasmin, in the process. As the replicating fork progresses along a region of nucleosome-associated DNA a number of different situations might result, some of which are illustrated in Fig. 6.37.

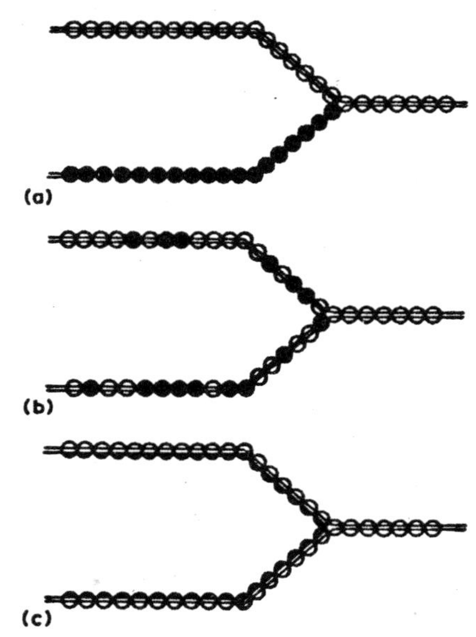

Fig. 6.37. Three possible mechanisms are shown for the distribution of nucleosomes at a replication fork. O, Parental nucleosome; •, newly synthesized nucleosome. (*a*) All the old nucleosomes remain with one of the two parental strands when they separate at the replication fork-co-operative alignment; (*b*) The old nucleosomes are dispersed at random between the two strands of DNA; (*c*) The old nucleosomes dissociate into half nucleosomes, one half going to each side of the fork.

Approaches used to distinguish between the possibilities include: (*a*) electron microscopic studies; (*b*) isolation and analysis of replicating chromatin; (*c*) crosslinking studies usually associated with attempts to separate the crosslinked species (*d*) nuclease digestion experiments and (*e*) the above approaches carried out with cells exposed to inhibitors of protein synthesis in an attempt to visualize the histone-free nascent DNA. The conclusion from these experiments is that new histones are associated with newly synthesized DNA and that the nucleosomes form by initial addition of histones H3 and H4 followed by histones H2A and H2B. Histone HI is added last and turns over quite readily. Newly synthesized nucleosomes consist predominantly of eight newly synthesized histones rather than a mixture of new and old histones.

The major controversy concerns the question whether newly synthesized nucleosomes are found adjacent to one another (as required for conservative segregation) or whether they are dispersed amongst old nucleosomes. Although some conservative segregation definitely appears to occur, the evidence is not yet strong enough to conclude that it represents the only mechanism. If segregation of nucleosomes is conservative the evidence favours the continued association of the old nucleosomes with the leading strand at the replication fork, with the newly synthesized nucleosomes being added to the lagging side. Thus DNA fork movement may be considered to occur as a series of steps caused by relocation of nucleosomes on the leading side of the fork which is always made up of duplex DNA.

At each relocation a region of about 200 bases of DNA on the lagging side of the fork is rendered single stranded providing a target for random initiation of RNA primers. For this reason the Okazaki pieces resulting have a mean size somewhat shorter than the 200 bases of nucleosomal DNA. Replicons have a length approximately equal to that of chromatin loops or the equivalent of 400 nucleosomes or about 70 turns of a solenoid. As adjacent replicons tend to replicate at the same time in the cell cycle, it is possible that a region of a chromosome becomes partially expanded and single-stranded DNA is exposed at the ends of the loops. Bidirectional replication would be initiated at such regions and hence a replicon and a loop would be the same structure.

Alternatively initiation of replicon synthesis may occur on the matrix to which the loops are attached in which case each loop would be part of two adjacent replicons. As replicon size varies dramatically in developing amphibia and insects from 5 μm long in early cleavage to up to 350 μn long during spermatogenesis chromatin structure would be expected to show parallel changes and this is indeed the case. The altered structure associated with transcribing chromatin requires between 1 and 3 minutes to reform following replication. During this time, from 18 to 54 nucleosomes will have been added to the newly synthesized DNA.

CHAPTER 7

Reverse Transcriptase Inhibitors

Long-term virological suppression of HIV replication to restore immune function is the cornerstone of HIV management. Antiretroviral regimens are becoming simpler, more potent, and better tolerated, but the development of drug resistance remains a significant problem. Despite very low rates of virological failure with the current regimens in some of the more recent clinical trials. virological failure rates are often higher in clinical practice especially in patients who are treatment experienced. Virological failure, defined as detectable plasma HIV RNA levels after initial suppression or the inability to suppress HIV RNA to undetectable levels with treatment, is influenced by many factors, including regimen potency, patient adherence, drug pharmacokinetics, and preexisting drug resistance. However, the development of drug resistance in response to treatment is a major cause and consequence of virological failure.

RESISTANCE EVOLUTION

An estimated 10^9 to 10^{10} HIV virions are produced per day in the average infected person. Coupled with the high error rate of the HIV reverse transcriptase (RT), this creates a large population of genetically distinct HIV quasispecies within infected individuals. This genetic diversity provides a large pool of HIV variants from which subpopulations can emerge in response to selective pressures, such as antiretroviral therapy. If viral replication is not suppressed to minimal levels with combinations of antiretroviral drugs, drug resistant strains will rapidly emerge.

The rate at which drug resistance develops depends on a number of factors, including the size and diversity of the viral population in a given person, the magnitude of viral replication that persists on treatment, the complexity of the genetic barrier (number of mutations) that must be overcome to allow viral replication in the presence of a given antiretroviral regimen, and the replicative fitness of the evolving resistant strains.

When treatment is stopped, resistant strains no longer have a selective advantage and wild-type strains generally become dominant. Although the resistant strains may become undetectable by conventional assays, they remain at low levels and rapidly resurface if the drugs that favor their growth are reinstituted.

EPIDEMIOLOGY

In places where antiviral therapy is readily available, antiretroviral resistance is a significant problem. The prevalence of phenotypic resistance in the United States was evaluated in late 1998 in a study designed to represent all 208,900 HIV-infected adults in the United States who received HIV treatment in urban centers. Of all patients whose HIV phenotypic resistance could be measured, 76% of patients had virus with some degree of drug resistance. Resistance to nucleoside and nucleotide analog RT inhibitors (NRTIs) was the most prevalent, found in 71% of the patients. Higher rates of drug resistance were associated with better access to treatment. Patterns of resistance generally reflect current prescribing patterns.

Resistance to the non-nucleoside RT inhibitors (NNRTIs) has steadily increased since their popularity increased in 1999. Changes in NRTI prescribing trends have altered the rates of certain NRTI resistance patterns. For example, the increased use of tenofovir and abacavir has increased the prevalence of the K65R mutation selected by both of these drugs, whereas mutations associated with zidovudine and stavudine have decreased, paralleling their decreased use or the declining proportion of patients in the population who have received single and dual nucleoside therapies containing thymidine analogs. The transmission of drug-resistant HIV is also a significant problem.

Reported rates of primary HIV resistance (resistance in patients who have never received antiretroviraltherapy) vary depending on how resistance is defined, the geographical area, and the population surveyed. The transmission rates in North America peaked at an overall estimate of 12 to 13% of patients with strains resistant to any class of drugs in 1999-2000, and are currently thought have decreased to approx 8 to 10%. Certain geographical areas, such as San Francisco, CA, have reported rates of primary resistance as high as 25%. Rates of resistance from a large European cohort of treatment-naive patients (excluding the United Kingdom) were stable at an average of 11% during 1996-2002. Rates of primary resistance reported in the United Kingdom are much higher, and climbed to as high as 26% in 2002.

MEASURING RESISTANCE

Resistance to nucleoside analogs results from changes in the nucleotide sequence of the RT gene that cause changes in the amino acid sequence of the translated enzyme. These amino acid changes alter the structure of the enzyme in regions that affect drug susceptibility, and allow the RT to function in the presence of the particular drug(s). Phenotypic drug resistance is defined as an increased concentration of drug required to inhibit in vitro growth of a virus or construct derived from a virus compared with a known wild-type (nonmutated) control.

Table 7.1. Amino Acid Abbreviations[a]

Abbreviation	*Amino acid*	*Abbreviation*	*Amino acid*
A	alanine	M	Methionine
C	cysteine	N	Asparagine
D	aspartate	P	Proline
E	glutamine	Q	Glutamate
F	phenylalanine	R	Arginine
G	glycine	S	Serine
H	histidine	T	Threonine
I	isoleucine	V	Valine
K	lysine	W	Tryptophan
L	lecucine	Y	Tyrosine

[a]The amino acid preceding the position number is the consensus wild-type amino acid and the one immediately after the position number is the substituted amino acid, *i.e.*, M184V represents a valine substitution for methionine at amino acid 184.

Phenotypic resistance is often reported as a fold change in the amount of drug required to inhibit viral replication by 50% compared with the wild-type control. The in vitro fold change in susceptibility correlates with reduced in vivo antiviral activity, but varies widely among the drugs. An increasing number of drugs have fold change cut-off data that are based on clinical trial outcomes, but these cut-off data are still needed for many of the available drug. Genotypic resistance is defined by the presence of specific mutations in the HIV RT or protease genes known to be associated with phenotypic and/or in vivo resistance to a given drug or drugs. It is determined by analysis of the nucleotide sequence of the RT and protease genes for the presence of specific genetic mutations.

The genotype is typically reported as a list of amino acid mutations of the patient virus compared with the consensus B wild-type HIV sequence. Mutations are described starting with the consensus B wild-type amino acid abbreviation (Table 1), followed by the amino acid residue number, and followed by the abbreviation of the mutant amino acid in the given sequence. For example, a valine substitution for the naturally occurring methionine at amino acid 184 is written as M184V. A predicted susceptibility profile is also reported, which is generated from algorithms that are based on the accumulated knowledge of specific amino acid changes on in vitro susceptibility and/or virological response.

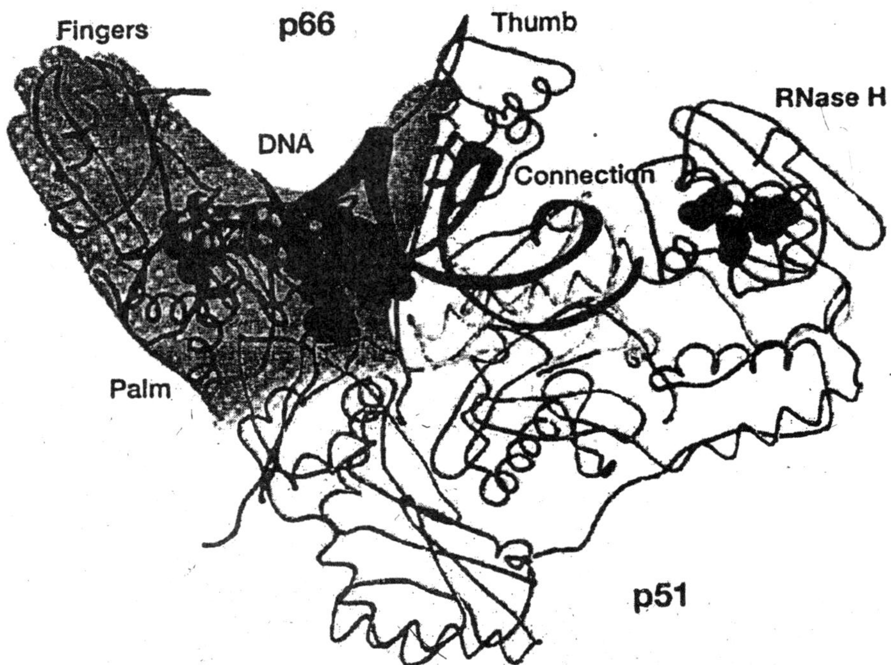

Fig. 7.1. A ribbon structure of HIV RT with the three-dimensional conformation of p66 resembling a right hand (superimposed in the background). It has a palm where the polymerase active site is contained, fingers, a thumb, a connector region that interacts with the p51 subunit to stabilize the dimer complex, and an RNase H domain. The light spheres represent dNTP bound to the active site. The adjacent darker spheres represent nevirapine bound to the NNRTI-binding pocket.

Clinical or in vivo resistance can be defined as reduced antiviral activity despite therapeutic drug levels in the blood and tissues. Phenotypic and genotypic resistance testing are both predictive of in vivo resistance, and choosing regimens based on resistance-testing results improves virological response rates in patients who have failed at least one drug regimen.

LOCATION OF NUCLEOSIDE RESISTANCE MUTATIONS

RT is a heterodimeric protein consisting of 66- and 51-kDa subunits (p66 and p51, respectively). The p66 subunit is composed of all 560 amino acids of the *pol* gene. The p51 subunit is composed of the first 440 amino acids of the *pol* gene. Although p51 contains a polymerase active site, p51 is catalytically inactive and predominantly serves as a structural base for the active p66. The three-dimensional conformation of p66 resembles a right hand. It has a "palm" containing the active site, "fingers," a "thumb," a connector region that interacts with the p51 subunit to stabilize the dimer complex, and an ribonuclease H (RNase H) domain. When incoming nucleotides bind to the enzyme, the tips of the fingers "close down" on the palm to trap the template and the developing primer strand in the active site. The major resistance mutations that are selected for by NRTIs in vivo and which have been shown by site-directed mutagenesis studies to confer nucleoside resistance, are distributed in the fingertips and the palm.

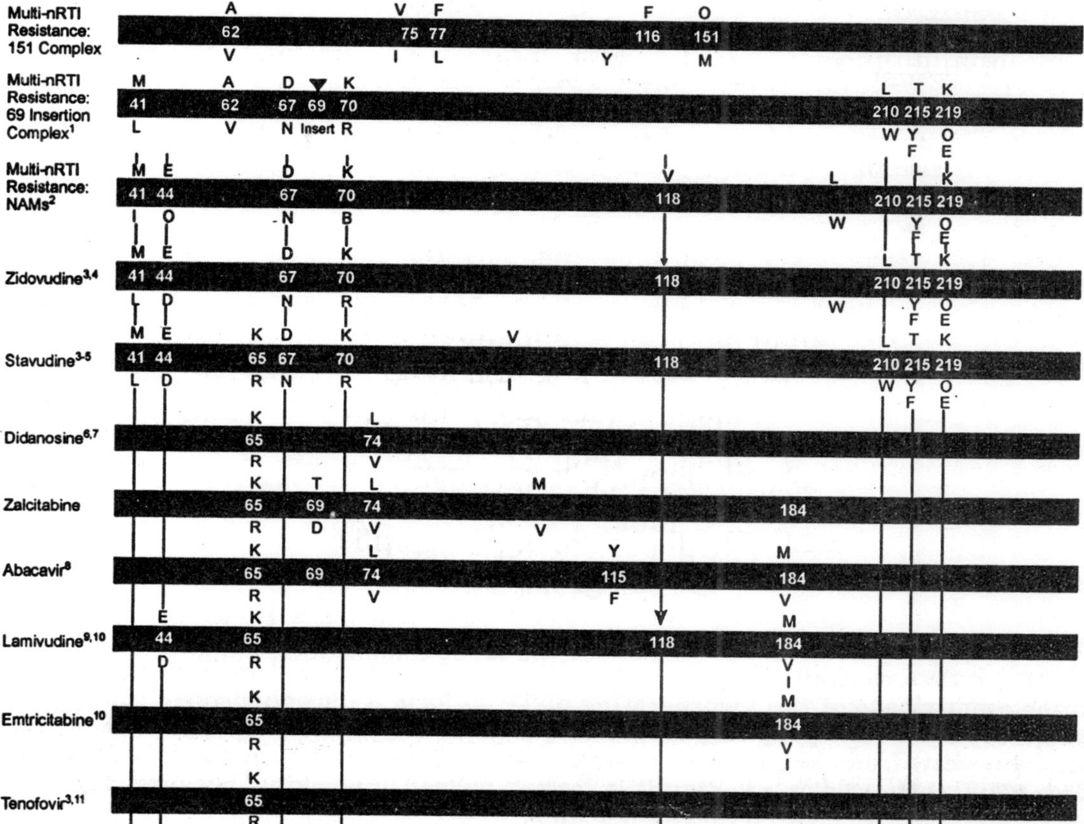

Fig. 7.2. Chart of mutations associated with nucleoside and nucleotide resistance.

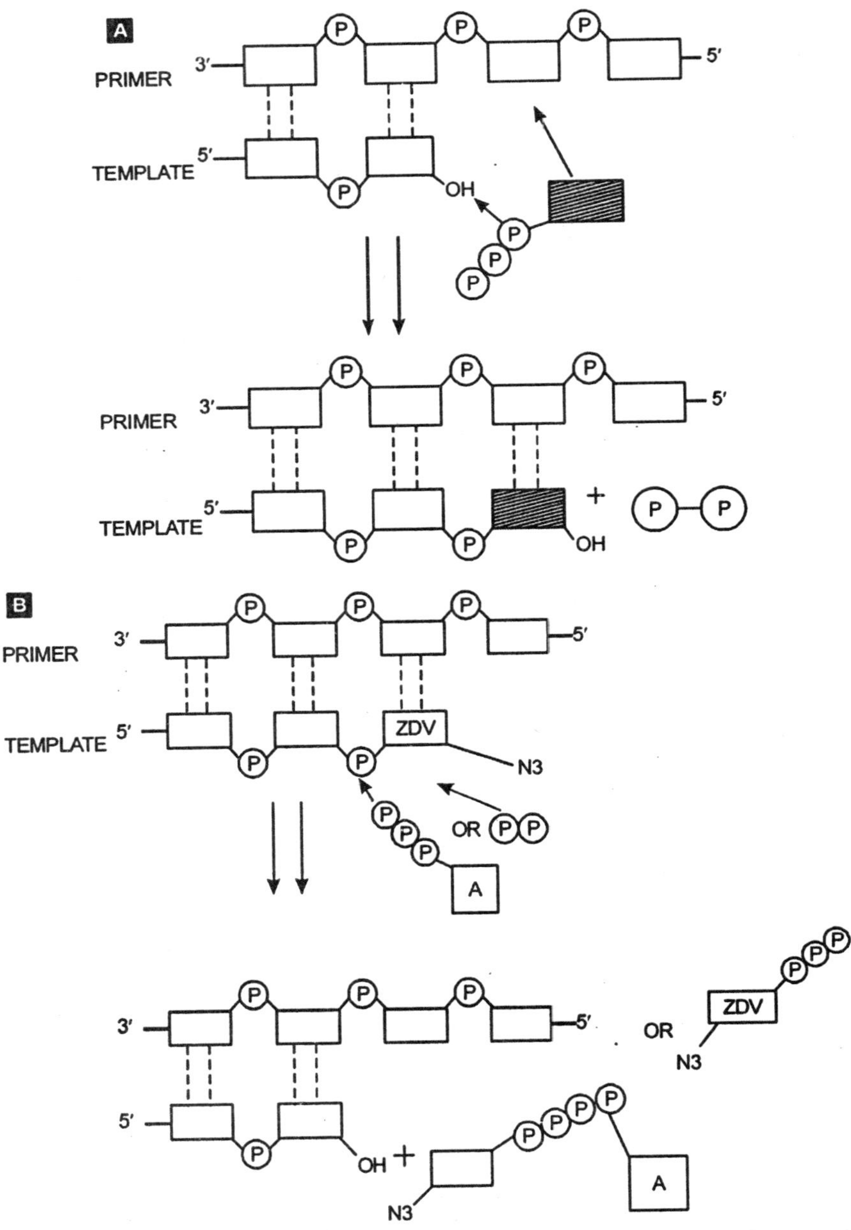

Fig. 7.3. **(*a*) DNA elongation is characterized by attack of the 3′-hydroxyl group of the primer on the α-phosphate of the incorporating dNTP to form a phosphodiester bond, releasing pyrophosphate. (*b*) Pyrophosphorolysis involves excision of the terminal nucleoside monophosphate from the primer by coupling with ATP or pyrophosphate. If the terminal base is an NRTI, this excision reaction it is termed primer unblocking, because it rescues the terminated chain so it can continue elongation.**

GENERAL MECHANISMS

HIV RT converts genomic single-stranded RNA into double-stranded DNA to be integrated in the host DNA. HIV RT has RNA- and DNA-primed DNA polymerase activity as well as RNase H activity that degrade the DNA-RNA intermediates. DNA elongation exists as an equilibrium between a dominant forward reaction and a minor reverse reaction. The forward reaction is characterized by attack of the 3′-hydroxyl group of the primer on the α-phosphate of the incorporating nucleoside triphosphate (dNTP) to form a phosphodiester bond, releasing pyrophosphate. This reaction extends the length of the primer chain by one base. This process continues to extend the primer until natural completion or until an incorporated base is a nucleoside analog, which lacks the 3′-hydroxyl group that forms a bond with the next dNTP, and therefore prematurely terminates the chain.

The reverse reaction of the RT enzyme involves excision of the terminal nucleoside monophosphate from the primer by coupling with adenosine triphosphate (ATP) or pyrophosphate. If the terminal base is a NRTI, its removal from the terminated primer will unblock the chain so it can continue extension. This is referred to as primer unblocking. In general NRTI resistance mutations cause resistance either by decreasing the incorporation of the NRTIs compared with the natural dNTPs (the forward reaction) or by increasing the rate of primer unblocking (the reverse reaction). Allosteric interference with the incorporation of the nucleoside analog is the major mechanism by which mutations M184V, Y115F, Q151M, L74V, V75T, V1181, and K65R cause resistance. The thymidine analog mutations (TAMS) at positions M41L, D67N, K70R, L210W, T215Y/F, and K219Q/E/N do not affect incorporation to any significant degree instead, they enhance the reverse reaction of nucleotide excision by enhancing the binding of ATP to the terminated primer complexed with RT.

An increase in the number of these mutations proportionally increases the degree of primer unblocking. The E44D mutation also increases primer unblocking in combination with the TAMS. Some mutations can reduce the rate of primer unblocking, thereby antagonizing this mechanism of resistance, including the NRTI-associated mutations Ml 84V, K65R, and L74V, as well as the NNRTI mutation Y181C. Although TAM-mediated excision occurs with all NRTI-terminated primers, the efficiency differs among the individual NRTIs. The efficiency is greatest with the thymidine analogs zidovudine and stavudine, and is much lower for other NRTIs, such as abacavir, didanosine, tenofovir, lamivudine, and emtricitabine.

The concentration of the next complimentary dNTP in the given environment influences the rate of excision, particularly for dideoxy NRTIs, such as stavudine and didanosine. When RT is bound to the template and the primer, the 3′ end of the primer can be in one of two positions, termed the priming (P) site or the nucleotide-binding (N) site. The nucleotide binds at the N site and, after chemical bond formation, it can translocate to the P site. If an NRTI-terminated primer translocates to the P site, the next complimentary dNTP can then occupy the N site. This forms a dead-end complex of variable stability, depending on a several factors, including the specific NRTI and next dNTP.

Excision can only occur at the N site, so this dead-end complex prevents excision. Zidovudine has a bulky azido side chain that does not favor the binding of the next complimentary nucleotide. This allows a zidovudine-terminated primer to move back and forth from N to P and from P to N, allowing excision to occur independent of the dNTP concentrations. In contrast, stavudine and other dideoxynucleosides favor the dead-end complex with the next complimentary dNTP, which keeps the 3′ end of the terminated primer in the P site and makes excision less efficient at higher dNTP concentrations are high, the TAMS confer the highest degrees of resistance to zidovudine and very small degrees of resistance to stavudine.

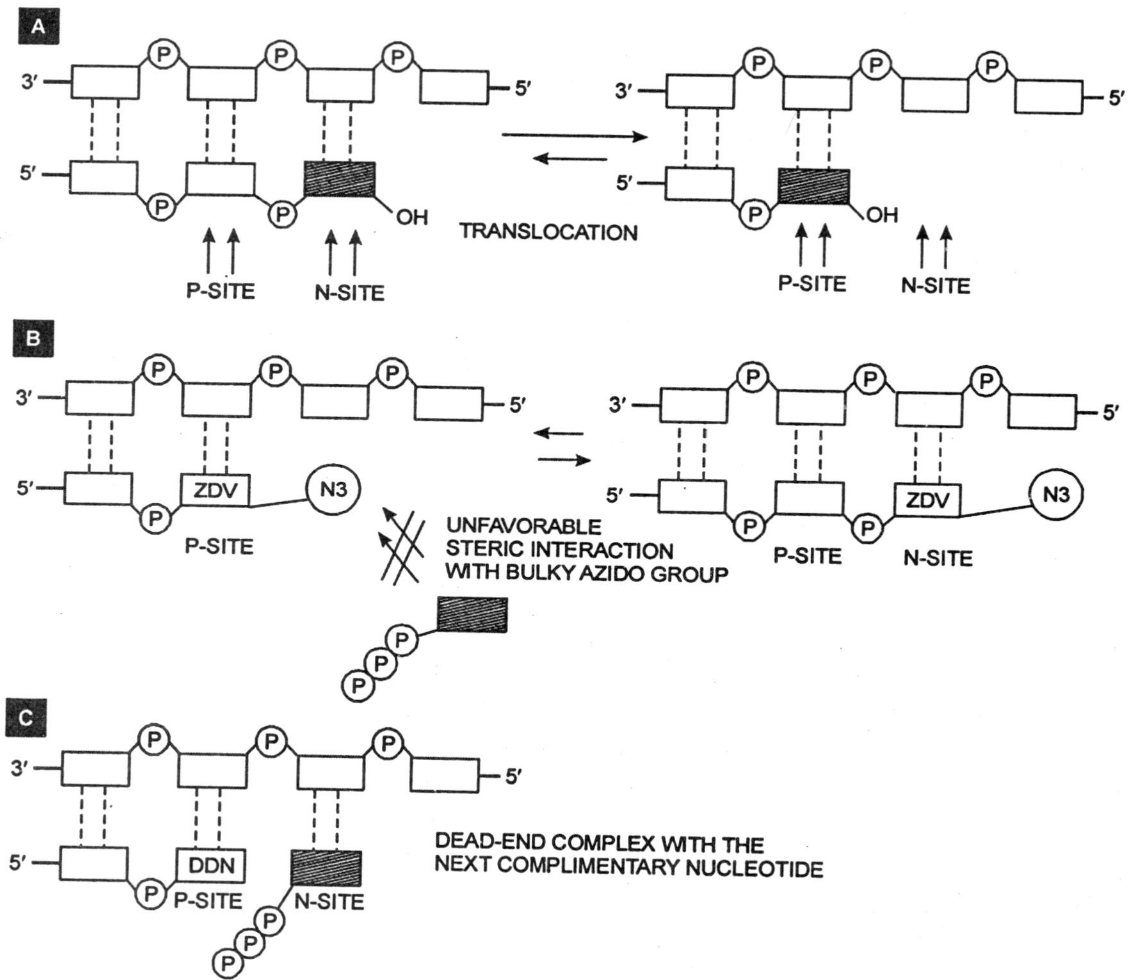

Fig. 7.4(a). (A) When RT is bound to the template and the primer, the 3′ end of the primer can be in the P site or the N site. The incoming nucleotide binds at the N site and, after chemical bond formation, it translocates to the P site. (B) Zidovudine (ZDV) has a bulky azido side chain that does not favor the binding of the next complimentary nucleotide. This allows a zidovudine-terminated primer to move back and forth from N to P and from P to N. (C) Stavudine and other dideoxy nucleotide (DDN)-terminated primers favor a stable dead-end complex with the next complimentary nucleotide, which keeps the 3′ end in the P site.

In vivo, dNTP concentrations vary, thus, TAMS cause significant clinical resistance to stavudine. Insertions in the fingers domain between amino acids 69 and 70 have been shown to destabilize the dead-end complex formed by the next complimentary dNTP, enhancing resistance mediated by TAMS and increasing the resistance to all of the NRTIs, including stavudine.

Table 7.2. Nucleoside Mutations Reported in Antiretroviral-Naïve Patients Receiving Specific Single, Dual, and Triple NRTIs or Without NNRTIs or Protease Inhibitors

Monotherapy	Mutations that arise after therapy in naive patients
Zidovudine	TAMS
Didanosine	L74V, M184V, T69D/N, K65R, rarely TAMS
Zalcitabine	T69D
Stavudine	TAMS
Lamivudine	M184V,
Abacavir	M184V, L74V, K65R, Y115F
Tenofovir	No monotherapy data
Dual-nucleoside therapy	
Zidovudine+didanosine	TAMS, Q15M complex
Zidovudine+zalcitabine	TAMS
Stavudine+didanosine	TAMS, L74V, Q15M complex
Zidovudine+lamivudine	M184V, TAMS, recently K65R
Abacavir+lamivudine	M184V, L74V, K65R, Y115F
Abacavir+zidovudine	TAMS, M184V
Didanosine+lamivudine	M184V, L74V, rarely TAMS
Tenofovir+lamivudine	M184V, K65R
Tenofovir+didanosine	K65R, L74V
Triple-nucleoside therapy	
Abacavir+lamivudine+zidovudine	M184V, TAMS, rarely L74V and Y115F
Abacavir+didanosine+stavudine	K65R, S68G
Tenofovir+lamivudine+abacavir	M184V, K65R
Tenofovir+lamivudine+didanosine	M184V, K65R

PARTICULAR MUTATIONAL PATTERNS

Table 7.2. describes the specific patterns seen with particular single NRTIs or with combinations with or without NNRTIs or protease inhibitors. The individual mutation(s) are summarized next.

Thymidine Analog Mutations

Shortly after zidovudine was first used to treat HIV, isolates from patients on prolonged zidovudine monotherapy therapy were found to accumulate characteristic RT mutations at RT amino acid positions 41, 67, 70, 210, 215, and 219. These mutations were associated with increasing phenotypic resistance to zidovudine. Similar mutational patterns were seen from in vitro passage experiments. Initial observations on relatively few patient isolates described a particular ordering of mutations that progressively increased the level of resistance. The initial mutation most often was K70R, which confers a fourfold to eightfold increase in resistance to zidovudine. This was typically followed by T215Y, the disappearance of K70R, and the subsequent evolution of M41L. D67N, K219Q/E/N, and L210W were seen, but less frequently.

More-recent observations using much larger data sets have found two different pathways of the evolution among TAMS. The first and most common pathway clusters M41L, D67N, L210W, and T215Y. The second cluster includes D67N, K70R, T215F, and K219Q/E/N. The second pathway generally displays less phenotypic resistance to all of the nucleosides than the first, perhaps because the major amino acid 215 mutation occurs later in the order of mutations. There is little data on what drives the selection of these different TAM pathways. The TAMS are selected primarily by the thymidine analogs, stavudine and zidovudine, either alone or in combination with other drugs. Stavudine and zidovudine select for TAMS at similar rates.

Although TAMS are primarily selected by the thymidine analogs, the presence of increasing numbers of TAMS confers increasing phenotypic cross-resistance to all of the nucleoside and nucleotide analogs available. TAMS also decrease in vivo activity of the nucleoside and nucleotide analogs. For this reason, some experts refer to them as nucleoside analog mutations. A 215Y/F present at baseline was predictive of increased progression to AIDS or death in patients treated with zidovudine or didanosine monotherapy. TAMS substantially reduced antiviral activity to stavudine monotherapy, didanosine monotherapy, stavudine plus didanosine therapy, and stavudine plus lamivudine therapy. The presence of TAMS, particularly T215Y/F, was associated with diminished viral load reduction in patients who had tenofovir, abacavir, lamivudine, or didanosine added to existing therapy. Incremental reductions in response were seen with increasing numbers of TAMS. Other mutation at position 215, including T215S, C, and D do not cause measurable reduced drug susceptibility in vitro, but are transitional mutations on the way from wild-type T to the two-position mutations Y or P. They are commonly called T215 revertants, because they are often seen in patients who once had T215Y/F virus, but discontinued therapy, or are more commonly found in strains of treatmentnaive patients who were infected with drug-resistant virus.

They are more fit in the absence of drug pressure than the T215Y/F mutants. Presence of these mutations in antiretroviral-naive patients was associated with an increased risk of virological failure and selection of the T215Y in one recently described cohort.

M184V

Codon 184 is a component of the catalytic active site of the RT that directly interacts with the incoming nucleotide. An M184I/V mutation causes steric hindrance that selectively reduces the incorporation of nucleosides with β- or L-ring configurations, namely lamivudine, didanosine, zalcitabine, abacavir, and emtricitabine. Lamivudine in vitro and in vivo rapidly selects for the M1841 mutation, which is replaced by the M1 84V mutation because of a fitness advantage. The M184V mutation is seen as early as 1 wk after initiation of lamivudine treatment and is present in nearly all patients after 3 mo of lamivudine monotherapy.

M184V is also selected by abacavir monotherapy, occasionally by didanosine monotherapy, by abacavir plus zidovudine combination therapy, and by all combination therapies that include either lamivudine or emtricitabine. M184V causes high-level in vitro resistance to lamivudine and emtricitabine and although M184V confers some degree of resistance to abacavir and didanosine, this mutation alone is not sufficient to significantly decrease the response to abacavir or didanosine. A recent small study also revealed that despite high levels of in vitro resistance, lamivudine may retain some in vivo activity against viruses with the M184V mutation. As mentioned in the "General Mechanisms of How the Mutations Confer Drug Resistance" section, M184V reverses TAM-associated resistance to zidovudine, stavudine, and tenofovir by decreasing the rate of primer unblocking.

The M184V mutation has been the subject of intense basic and clinical investigation. Initial studies of lamivudine monotherapy found that HIV RNA levels remain approx 0.5 $\log_{10}$ copies/mL below the pretreatment levels for 6 to 12 mo, despite the presence of high-level lamivudine resistance caused by the M184V mutatio. Laboratory studies have found that M184V reduces the viral replication capacity compared with wild-type virus in the absence of drug by decreasing the RT processivity (the relative efficiency of completing successful transcription).

In addition to decreased processivity, RTs with M184V have up to a 50-fold increase in the RT fidelity (the ability to select the correct nucleotide) compared with wild-type RT, theoretically reducing mutation rates. Clinical studies have also documented a diminished rate of acquiring TAMS when lamivudine is administered with thymidine analogs, perhaps because of the presence of the M184V mutation.

K65R

Codon 65 is in the fingers region of the RT and interacts with the γ-phosphate of the incoming dNTP. Mutations at this site confer resistance by decreasing nucleoside analog incorporation rates compared with the natural dNTPs. As mentioned in the "Epidemiology" section, the K65R mutation is increasing in prevalence, correlating with the increased use of tenofovir and abacavir. K65R was initially found to be selected by zalcitabine in vitro. It occasionally occurred in patients treated receiving didanosine or abacavir monotherapy. K65R is the first mutation selected by tenofovir in vitro, and is selected in antiretroviral-naiye patients who receive tenofovir in any dual or triple combination with lamivudine, didanosine, abacavir, or emtricitabine. K65R rarely occurs when teno-fovir is added to failing regimens in patients with preexisting TAMS.

Combination therapy with abacavir plus lamivudine can select for a K65R, but usually not when zidovudine is included as a third agent. In contrast to zidovudine, regimens containing stavudine have been reported to rarely select for K65R. In a recent study of stavudine plus lamivudine plus efavirenz in treatment-naive patients, 2 of 25 patients with virological failure developed a K65R mutation. Another study that included a combination of abacavir plus didanosine plus stavudine found that five of eight failures contained the K65R mutation, along with the unusual mutation, S68G. K65R confers partial resistance to zalcitabine, didanosine, abacavir, Tenofovir, lamivudine, emtricitabine, and possibly stavudine. K65R mutation at baseline was associated with a poor response in patients who had tenofovir added to an existing regimen, although the few patients who developed a K65R mutation after therapy in the same cohort did not have a corresponding increase in viral load toward baseline. This may result from multiple factors. Similar to M184V, K65R reduces TAM-mediated primer unblocking, resulting in a net hypersusceptibility to zidovudine. In addition, K65R reduces RT processivity and viral replicative capacity, comparable to the Ml 84V mutation.

L74V

Codon 74 interacts directly with the template and the incorporating dNTP. The mutation L74V confers resistance by selectively reducing incorporation rates of specific nucleoside analogs. L74V is the most common mutation that arises in patients receiving didanosine monotherapy and is often seen in patients receiving abacavir monotherapy. Combinations of didanosine plus lamivudine, didanosine plus emtricitabine, and abacavir plus lamivudine can select for L74V. Coadministration of zidovudine with abacavir or didanosine significantly decreases the selection of L74V, probably because L74V causes hypersusceptibility to zidovudine through diminished primer unblocking L74V has occasionally been seen in patients receiving stavudine plus didanosine. L74V confers resistance or partial resistance to didanosine, zalcitabine, lamivudine, emtricitabine, and abacavir. L74V at baseline prevented a virological response to didanosine intensification. L74V viruses are hypersusceptible to zidovudine and to tenofovir but not to stavudine. Similar to M184V, enzymes possessing the L74V mutation have less processivity than wild-type virus, conferring a replication defect in the absence of drug.

Q151M Complex

Codon 151 is in a conserved region that interacts with the incoming dNTP. Q151M interferes with NRTI binding and incorporation. Q151M develops in 5 to 10% of patients treated with thymidine analog plus didanosine combinations, and causes moderate resistance to all of the nucleoside analogs. The Q151M mutation is generally associated with mutations at positions A62V, V75I/F, F77L, and Fl16Y. Each of these mutations substantially increases the degree of resistance and/or the fitness in the presence of drug of isolates containing the Q151M mutation. Tenofovir retains some activity against HIV with this mutational complex and lamivudine may also have activity.

Codon 69 Mutants and Insertions

Codon 69 is in the fingertip region that makes contacts with the incoming dNTP. Initially, T69D mutation was associated with zalcitabine therapy, but a variety of changes at this site develop with most of the available nucleosides. T69D/N/S/A have all been shown to cause some degree of resistance to zidovudine, didanosine, zalcitabine, and stavudine in site-directed mutagenesis studies. T69D/N developed with didanosine monotherapy and was associated with clinical progression and CD4 decline in a group of 30 HIV-infected children, whereas other mutations, such as L74V and Ml 84V, were not associated with CD4 decline. Insertions at the fingers region between codons 67 and 70 (most commonly two amino acids at position 69) occur in 1 to 2% of patients extensively treated with nucleoside analogs.

These inserts are typically found in viruses that also contain TAMS. As described in more detail in the "General Mechanisms of How the Mutations Confer Drug Resistance" section, these inserts enhance primer unblocking of TAMS by destabilizing the dead-end complex formation with the next complimentary nucleotide. Insertions by themselves confer minor degrees of resistance to the nucleosides, but in the presence of TAMS, resistance is markedly increased. Isolates with insertions at codon 69 and TAMS are highly cross-resistant to all of the currently available NRTs.

E44D and V1181

Mutations E44D/A and V1181 both occur in untreated patients, but their prevalence increases in subjects who have received extensive NRTI treatment. They have generally been observed in patients receiving multiple nucleoside regimens who also have TAMS. V1181 causes resistance by decreasing the NRTI incorporation, whereas E44D enhances ATP-mediated NRTI excision. These mutations were initially presented as alternative pathway lamivudine-resistance mutations because they confer approx 10- to 15-fold resistance to lamivudine, as well as lower-level resistance to other antiretroviral agents. However, these mutations are not selected by lamivudine, and did not impair the antiviral activity of lamivudine-con-taining combination regimens in one study.

Y115F

Codon 115 directly interacts with the incorporating nucleotide. A mutation Y115F was selected in vitro and in vivo by abacavir monotherapy, but not by other NRTIs. Few data are available on the impact of Y115F on resistance to other NRTIs.

Other Mutations

An exploratory analysis from a large database of sequences from both anti-retroviral-naive and antiretroviral-experienced patients found that nine other mutations were definitely associated with NRTI therapy. These were mutations at codons 20, 39, 43, 203, 208, 218, 221, 221, 223, and 228. Nine additional mutations were associated with NRTI therapy but did not achieve statistical power after correcting for multiple comparisons. These were mutations at codons 60, 64, 104, 122, 135, 196, 200, 207, and 211.

SUMMARY

NRTIs were the first class of drugs used to treat HIV and are still included in most antiretroviral regimens. Despite the increasing number of different NRTIs available, substantial cross-class resistance exists. Further investigations are needed to determine the optimal sequencing strategies for nucleosides to maximize their use before extensive cross-resistance develops. Advancing knowl-edge of the biochemical and structural basis of resistance will aid in the design of newer compounds that are active against HIV resistant to the currently available drugs, ultimately prolonging virological suppression and life in the millions of people who are infected with HIV.

Pharmacokinetics of Reverse Transcriptase Inhibitors

The nucleoside reverse transcriptase inhibitors (NRTIs) were the first class of compounds discovered to be potent inhibitors of HIV replication and, to date, these drugs remain the backbone of antiretroviral therapy. NRTIs are essentially prodrugs, inactive in their parent form and requiring activation to exert their antiviral effects.

This chapter describes the activation of NRTIs, focusing on their intracellular kinetics, and explains how antiviral activity of NRTIs may potentially be enhanced through manipulation of activation pathways.

Reverse Transcription

The nucleic acid of HIV is RNA. To replicate, the virus must reverse transcribe its RNA into DNA. HIV reverse transcriptase (RT) catalyses the production of a DNA copy of the viral RNA using the host's endogenous deoxynucleoside triphosphates (dNTPs) as substrates. Host dNTPs within the cell are formed through either the salvage or the *de novo* pathways. The proviral DNA then enters the nucleus, where it is randomly incorporated into host chromosomal DNA by viral integrase.

Reverse Transcriptase Inhibitors

To block the reverse transcription process, a number of drugs that inhibit the enzyme RT have been developed. The NRTIs are structurally similar to the endogenous deoxynucleosides, with structural substitutions of the OH group on the deoxyribose sugar.

After entry into host target cells, NRTIs are phosphorylated by intracellular kinase enzymes to their active triphosphate derivatives (ddNTPs). NRTIs use the host's intracellular kinase enzymes, which are normally responsible for the activation of endogenous deoxynucleosides to their dNTPs as part of the replicative process of the cell's DNA.

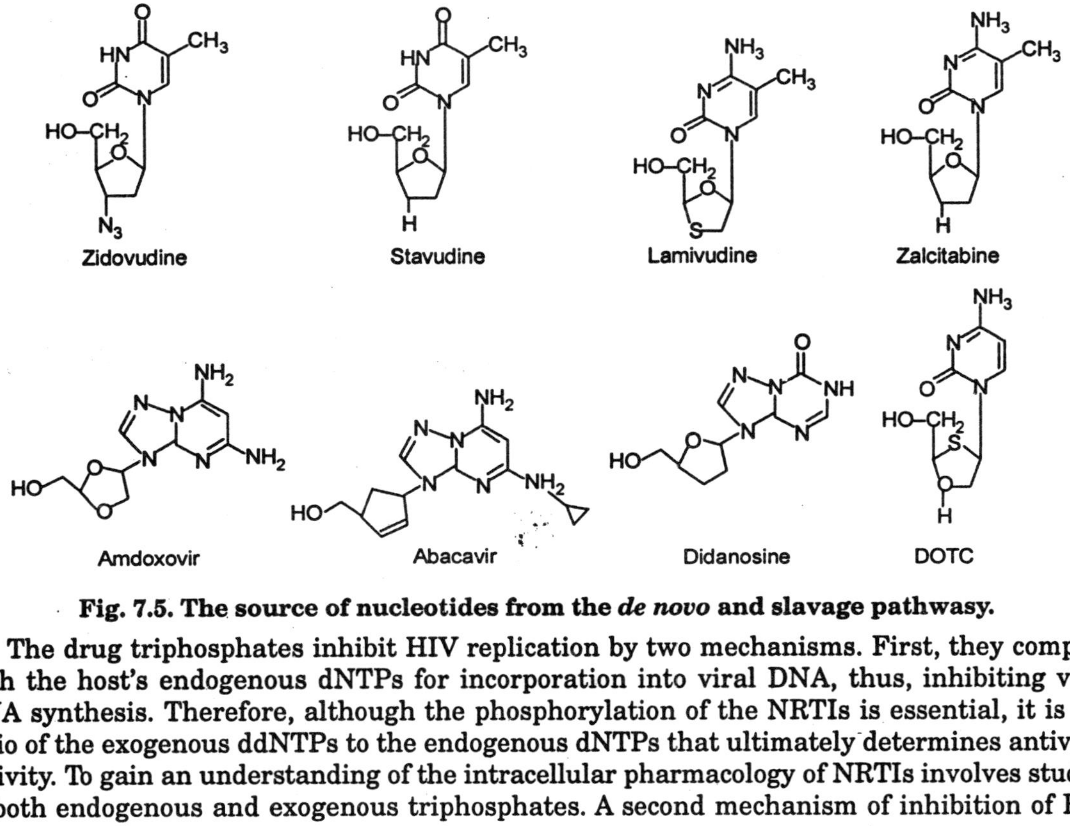

Zidovudine Stavudine Lamivudine Zalcitabine

Amdoxovir Abacavir Didanosine DOTC

Fig. 7.5. The source of nucleotides from the *de novo* and slavage pathwasy.

The drug triphosphates inhibit HIV replication by two mechanisms. First, they compete with the host's endogenous dNTPs for incorporation into viral DNA, thus, inhibiting viral DNA synthesis. Therefore, although the phosphorylation of the NRTIs is essential, it is the ratio of the exogenous ddNTPs to the endogenous dNTPs that ultimately determines antiviral activity. To gain an understanding of the intracellular pharmacology of NRTIs involves studies of both endogenous and exogenous triphosphates. A second mechanism of inhibition of HIV replication is chain termination.

The lack of a 3′-hydroxy group required to form 3′, 5′-phosphodiester linkages means that chain termination results after incorporation of NRTIs into the viral DNA. Binding of ddNTPs as chain terminators to the RT is reversible. That is, it is thought that, through mechanisms such as pyrophosphorolysis, the bound NRTIs can be removed from the proviral DNA chain. This will limit the activity and/or potency of these drugs.

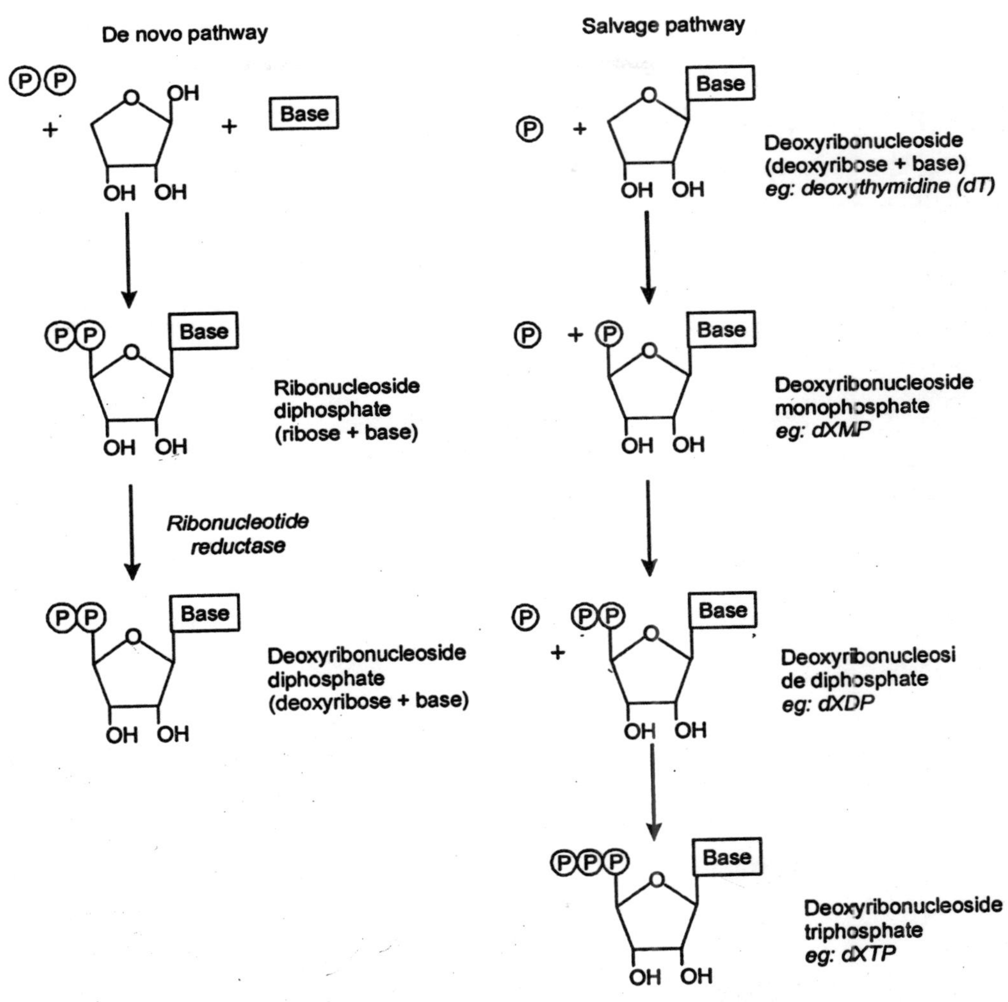

Fig. 7.6. Structure of the NRTIs showing the structural substitutions of the 3′ hydroxyl group.

INTRACELLULAR PHARMACOKINETICS OF THE NRTIs

Determining NRTI plasma concentrations is of limited benefit when attempting to relate drug concentrations to antiviral effect. Of greater significance are the studies that have measured active triphosphate in peripheral blood mononuclear cells (PBMCs) from HIV-infected patients or from cells in culture. Gives a compilation of data for all NRTIs and indicates that the triphosphate anabolites have a much longer half-life than the parent drug in plasma. Considerable research effort has been put into methodologies for determining intracellular triphosphate metabolites to optimize dosing regimens. The techniques involved are time consuming and difficult but essential to investigate pharmacokinetics at the active site. Several different approaches to triphos-phate determination have been published and all have highlighted the large variability in triphosphate levels between patients.

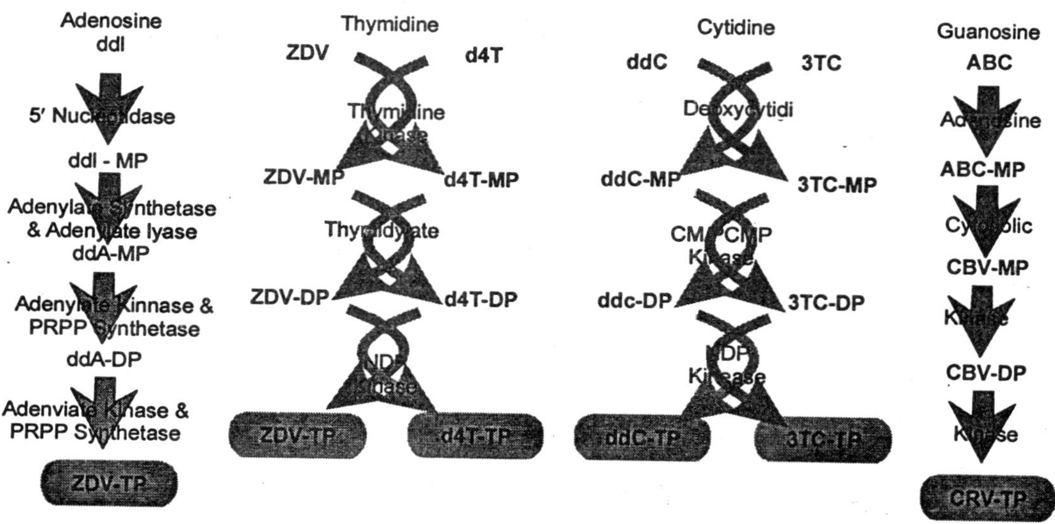

Fig. 7.7. Activation of the nucleoside analogs. NDP, nucleoside diphosphate; MP, monophosphate; and TP, triphosphate.

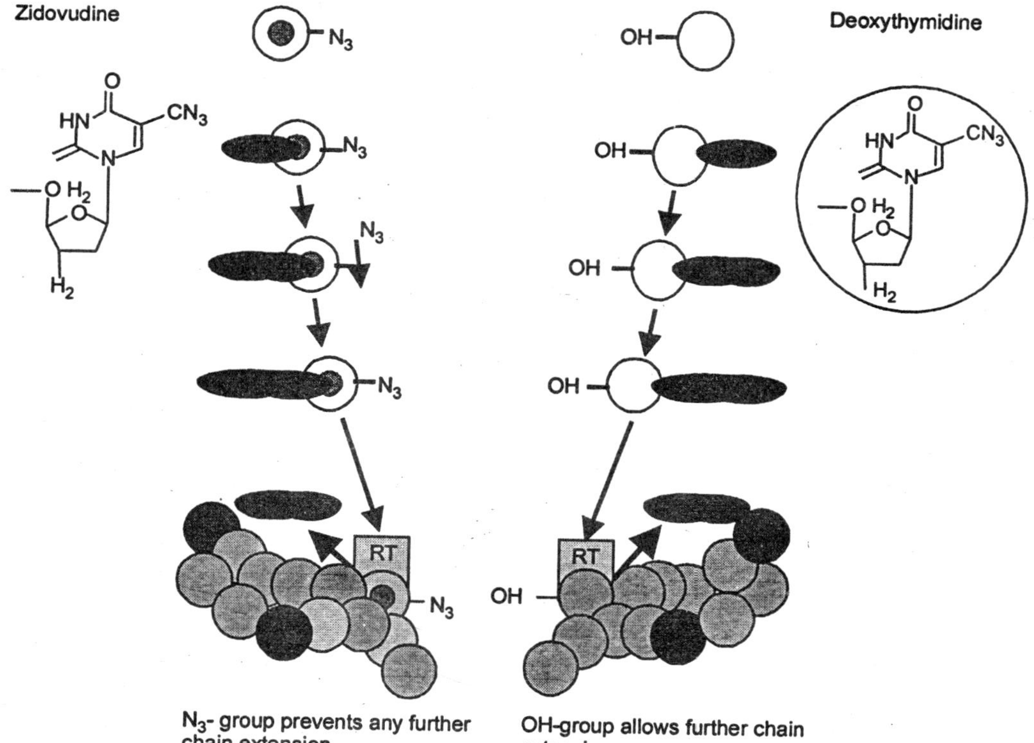

Fig. 7.8. The activation of ZDV showing the competition for endogenous kinases with endogenous deoxythymidine.

In Vitro Assays

Until recently, most data regarding the intracellular activation of the NRTIs have been obtained in vitro. The majority of in vitro studies have been performed using radiolabeled NRTIs in either immortal cell lines or isolated PBMCs. Because HIV infects both resting and activated cells, antiviral activity may not correspond to the in vitro data. Estimations of in vitro anti-HIV activity should only be used as a guide for potential clinical efficacy. Therefore, it is imperative that intracellular measurements of the NRTIs should be performed in vivo.

In Vivo Studies

One approach to measuring intracellular levels of NRTIs involves separation of intracellular phosphates by high-performance liquid chromatography (HPLC) and enzyme digestion of these metabolites before analysis of the parent drug by radioimmunoassay. Another approach, allowing quantification of zidovudine (ZDV) triphosphate (ZDVTP) and lamivudine (3TC) triphosphate (3TCTP) in HIV-infected patients, is solid phase extraction combined with mass spectrometry.

A promising direct method involving immunoaffinity and liquid chromatography coupled to tandem mass spectrometry has been recently developed for measurement of ZDVTP, allowing measurement of femtomole concentrations in PBMCs from 7 mL of blood. In addition, the latest technology, matrix-assisted laser desorption/ionization time-of-flight mass spectrometry, has been described by van Kampen et al. for analysis of ZDVTP. Methods have also been developed for determination of stavudine (d4T; 2′, 3′-didehydro-3′-deoxythymidine) and didanosine (ddI; 2′, 3′-dideoxyinosine) phosphorylated metabolites, which can be used to routinely determine both drug and endogenous triphosphate levels. Again, the method involves liquid chromatography coupled to tandem mass spectrometry (electrospray ionization).

Table 7.2. Half-Lives of the NRTIs (in Plasma) and Their Triphosphates (in Cells)[a]

Drug	*Parent drug plasma half-life (h)*	*Drug triphosphate intracellular half-life (h)*
ZDV	1.5	3.5
d4T	1	7.0
3TC	5	12
ddC	1.5	2.6
ddI	1	36
ABC	1.5	12
DAPD	1	27
DOTC	1	2.5
TDF	16 h	>60 h
FTC	10 h	39 h

[a]Data collaged from a number of studies

Template-Primer Extension Assays

Because inhibition by the NRTIs is a competition between drug and endogenous triphosphates, it is clearly preferable to also determine the endogenous dNTP levels. Both drug and endogenous triphosphates have recently been determined by template-primer extension assays. The basis of the method is similar to the mechanism of action of the NRTIs, *i.e.,* there is competition for RT between drug triphosphate and endogenous triphosphate. Using isolated HIV RT, a synthetic DNA primer, and radiolabeled dNTPs, triphosphates of 3TC, ZDV, abacavir (ABC), and (–)-β-D-2, 6-dioxalane guanosine (DXG) have been determined. Because the endogenous dNTPs are also measured, the drug-to-endogenous nucleoside triphosphates ratio can be calculated.

HPLC Methods

Early studies separated ZDV phosphates by HPLC, after enzymic removal of the phosphates. ZDV was subsequently measured by radioimmunoassay. Quantification of all three phosphates has more recently been performed with a solid-phase extraction method. After resolution of 3TC nucleotides on an anion exchange cartridge, the phosphates are removed by enzyme digestion, and the 3TC product quantified by HPLC with ultraviolet detection.

NRTIS IN CLINICAL USE

Zidovudine

ZDV is a thymidine analog that enters the cell by passive diffusion. After entry into the cell, ZDV is converted to monophosphate. The diphosphorylation step (catalyzed by thymidylate kinase) is rate limiting and results in the intracellular accunrulation of ZDV monophosphate. In vitro studies have shown that ZDVTP formation is saturated at ZDV concentrations higher than, 2 μM, but ZDVMP formation is not saturated at concentrations up to 10 μM.

Clinical studies have shown that the triphosphate metabolite of ZDV, in common with the other NRTIs, has a much longer half-life within PBMCs than the plasma half-life of the parent drug. In HIV-infected patients administered ZDV, the main metabolite in PBMCs was monophosphate, illustrating that conversion to diphosphate is rate limiting both in vivo and in vitro. Studies have shown antiviral effect to be correlated with the intracellular level of ZDVTP and the ZDVTP-to-deoxythymidine ratio vitro.

The intracellular concentration of ZDVTP was positively correlated to both the increase in CD4 cells and the decrease in plasma viral load during therapy. On the other hand, there is evidence that cytotoxicity has a better relationship with ZDV monophosphate. Interestingly, ZDV monophosphate accumulation is greater in H1V patients than in seronegative volunteers.

Stavudine

After entering the cell by passive diffusion, d4T initial phosphorylation is catalyzed by thymidine kinase. In vitro studies have shown that the majority of intracellular d4T is unphosphorylated parent drug, thereby indicating that this first enzyme conversion is rate limiting for d4T activation to its 5′-triphosphate metabolite. The intracellular concentration of d4T triphosphate provides a better measure of in vivo antiviral activity than plasma d4T levels. d4T has increased antiviral activity in replicating cells compared with quiescent cells.

After mitogen stimulation there is a 17-fold increase in the activity of thymidine kinase. This shows that thymidine analogs are preferentially phosphorylated in activated cells and have greater activity inactivated rather than quiescent cells. Recently, the first in vivo data on d4T triphosphate were obtained in a clinical trail involving 28 antiretroviral-naive HIV-infected patients who received d4T (40 or 30 mg twice daily) in combination with ddI and efavirenz. Median d4T-TP concentrations were 31 fmol/10^6 cells (range 0-99 fmol/10^6 cells).

Lamivudine

3TC is a cytidine analog that enters the cell either by passive diffusion or by the facilitated diffusion nucleoside transport process. The initial activation is by deoxycytidine kinase, followed by further phosphorylation to 3TCTP. In vitro studies have shown the predominant anabolite to be 3TC diphosphate, indicating that the rate-limiting step in 3TC activation is conversion to the active triphosphate.

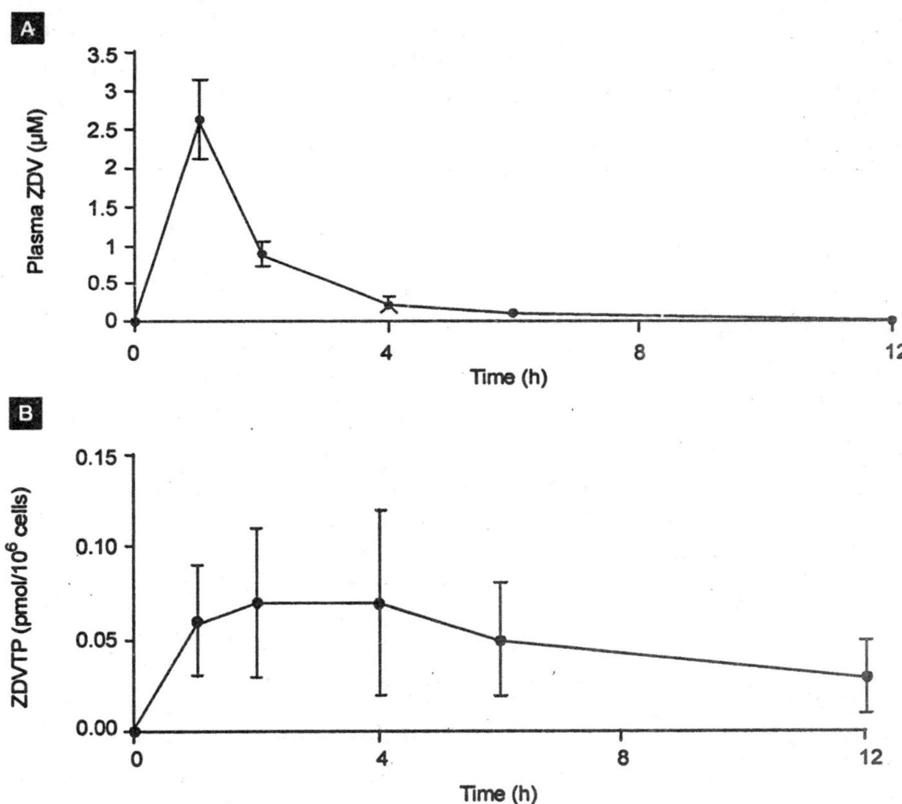

Fig. 7.9. The pharmacokinetic profile of ZDV in (A) plasma and (B) PBMCs from HIV-infected patients after oral administration of 300 mg ZDV. Data are expressed as mean ± SEM; n = 10.

The phosphorylation of this drug has shown no saturation in vitro with proportional increases in intracellular 3TC activation up to concentrations of 10 μM. A pharmacokinetic study determining the intracellular phosphorylation of 3TC in asymptomatic HIV-infected patients administered doses of 150 or 300 mg 3TC twice daily showed a less than twofold difference in the intracellular concentrations of the phosphate anabolite, despite a doubling of plasma 3TC concentrations. Recently, Yuen et. al., compared the steady-state pharmacokinetics of 3TC in plasma and 3TCTP in PBMCs in healthy subjects after 7-d courses of treatment with 300 mg 3TC once daily and 150 mg 3TC twice daily. Steady-state intracellular 3TC-TP levels were bioequivalent with respect to the area under the curve after once- and twice-daily regimens.

Zalcitabine

Zalcitabine (ddC; 2′, 3′-dideoxycytidine) is also a cytidine analog and enters the cell by either nucleoside carrier-mediated or noncarrier-mediated transporters. Activation of ddC uses the sanie intracellular enzymes responsible for 3TC phosphorylation.

No individual metabolite predominates, indicating that the initial conversion of ddC to its monophosphate is the rate-limiting step in its activation. On mitogen stimulation, a relatively small (twofold) increase in deoxycyti-dine kinase is seen, indicating that cytidine analogs are preferentially phosphorylated, and that they exert anti-HIV activity that is more potent in resting cells than in activated cells. That is, although more active ddC triphosphate is formed in the stimulated cells, the actual ratio of ddC triphosphate to endogenous deoxycytidine triphosphate is decreased. The intracellular concentrations of ddC in in vitro experiments are considerably smaller than those of 3TC when studied at the same concentrations. In addition, in contrast to 3TC, the activation of this drug is not saturated. There are no data available on ddC triphosphate levels in vivo.

Abacavir

After entry by nonfacilitated diffusion, ABC undergoes conversion by adenosine phosphotransferase to its monophosphate before catalysis to carbovir monophosphate, a guanosine analog. Further activation yields carbovir triphosphate (CBVTP), which competes with endogenous deoxy guanosine triphosphate (dGTP). Although carbovir shows potent antiviral activity, it is not directly administered because of toxicity and low bioavailability. In vitro studies have shown a linear relationship between the concentration of ABC and its active intracellular metabolite. Preliminary pharmacokinetic studies with 600 mg ABC suggest a relatively long half-life of CBVTP and support once-daily dosing.

Didanosine

ddI is an inosine analog. The mechanism of cell entry has to be fully elucidated. It is known that active transport is not involved, but the role of passive carriers is unclear. It has an unusual activation pathway, in that the enzyme primarily responsible for the first phosphorylation step is not a kinase but a phosphotransferase, with the major phosphate donor being inosine monophosphate (IMP).

After the initial phosphorylation, ddI is converted to a dideoxyadenosine analog before undergoing further activation to dideoxyadenosine triphosphate (ddATP), which competes with endogenous deoxyadenosine triphosphate (dATP) for incorporation into the proviral DNA. ddI is also metabolized by purine nucleoside phosphorylase, resulting in catabolism to hypoxanthine and a low conversion to the active ddATP. The active triphosphate anabolite has recently been determined in vivo, with median ddATP levels of 8 fmol/10^6 cells (range 0-23 fmol/10^6 cells) recorded.

Amdoxovir

Amdoxovir (DAPD; [–]-β-D-2, 6-diaminopurine dioxalane) is a guanosine analog prodrug of DXG ([2R-*cis*]-2-amino-1, 9-dihydro-9-[2-hydroxymethyl]-1, 3-dioxolan-4-yl-6H-purin-6-one).

Both DAPD and DXG are novel nucleosides displaying activity against HIV. DAPD is deaminated in vivo by adenosine deaminase to DXG. DXG is then sequentially phosphorylated to its active metabolite, DXG triphosphate, which competes with its endogenous substrate, dGTP for binding to HIV RT. Lesser amounts of DAPD triphosphate are also formed, but in vitro studies have shown that DXG triphos-phate is much more potent than DAPD triphosphate against HIV-1 RT.

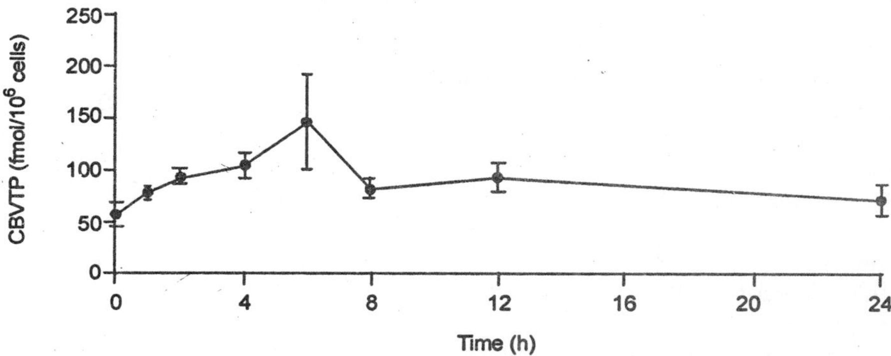

Fig. 7.10. The intracellular CBVTP pharmacokinetic profiles in HIV-infected Patients during 24 H after a dose of 600 MG ABC.

OTC (SPD754)

The novel nucleoside analog 2′deoxy-3′-oxa-4′-thiocytidine (dOTC) is a racemic mixture of two enantiomers. Both enantiomers show some activity but the (–) isomer (SPD754) shows the greater activity against HIV-1. dOTC is structurally related to 3TC. Intracellular metabolism studies showed that dOTC is phosphorylated within cells by the deoxycytidine kinase pathway and that approx 2 to 5% of dOTC is converted to triphosphate derivatives, which had measurable half-lives (2 – 3 h) within cells. Adams et al. have described the pharmacokinetics of plasma SPD754 and intracellular SPD754 triphosphate. The intracellular half-life of the triphosphate was approx 6 to 7 h.

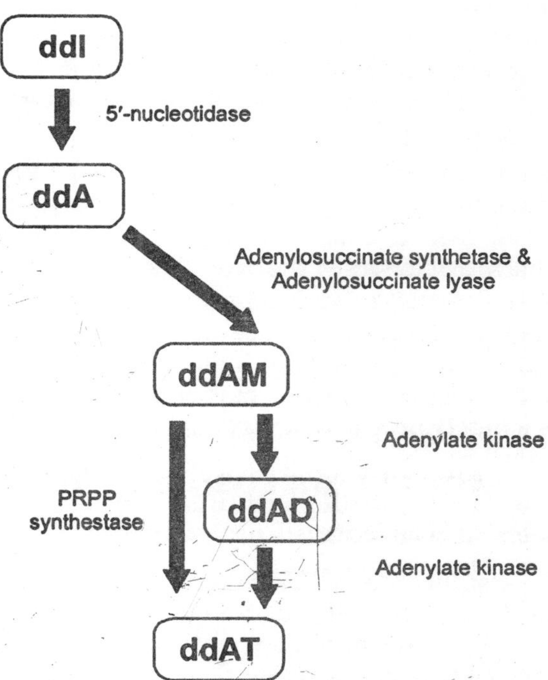

Fig. 7.11. The intracellular metabolism of didanosine. PRPP, 5'-phosphoribosyl-1-pyrophosphate.

Emtricitabine

Emtricitabine (FTC) is an analog of deoxycytidine that undergoes activation through the deoxycytidine kinase pathway. (β-L-2′, 3′-dideoxy-5-fluoro-3′-thiacytidine ([–

]FTC) is in phase III trials for HIV treatment; it is being considered for use in multidrug combination therapy of HIV-1. The half-life of FTC in plasma and cells is reported to be 7.4 h and 39 h respectively.

Nucleotide Analogs — Tenofovir

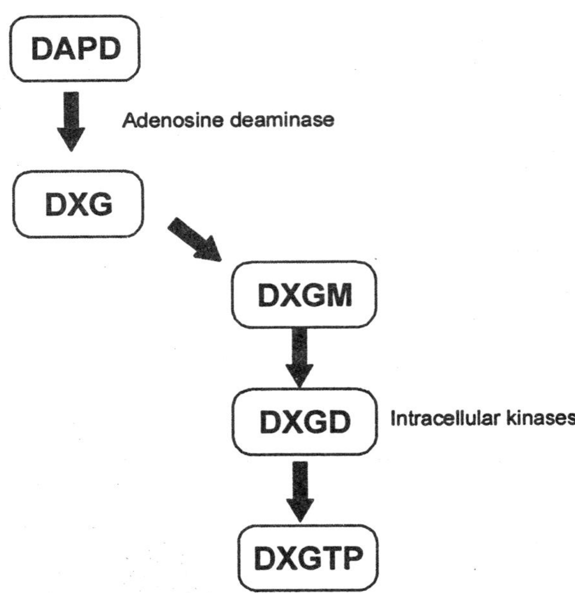

Fig. 7.12. The activation of DAPD. DXGMP, DXG monophosphate; DXGDP, DXG diphosphate; DXGTP, DXG triphosphate.

Nucleotide analogs are custom-designed drugs with a phosphate group present so that only two activation steps are necessary. Nucleoside monophosphates are not stable in vivo and, consequently, more-stable nucleotide analogs with phosphonate bonds have been designed. These phosphonate nucleotides, such as tenofovir, have poor bioavailability because of the high polarity of the phosphonate group. The attachment of labile lipophilic groups to these analogs is able to mask the polarity and allow cellular entry. Cleavage of the masking groups regenerates the monophosphate group, which is then metabolized to the active triphosphate metabolite. Tenofovir is an acyclic adenine derivative that is administered as the prodrug, tenofovir disoproxil fumarate.

After cell entry has occurred, unmasking occurs, followed by conversion to the active tenofovir diphosphate. The incorporated tenofovir anabolite acts as a chain terminator because of the lack of a 3-hydroxy moiety in the molecule. Because tenofovir diphosphate is the active form, measurement of the intracellular anabolite is necessary to fully elucidate the pharmacokinetics of the drug. The half-life of the intracellular diphosphate in vivo has recently been determined to be approx 60 h. it has also been suggested that nucleotide analogs may play an important role against HIV replication in macrophages and other nondividing cells, a unique and potentially important property that is not found with many members of the nucleoside analog class, particularly with thymidine analogs, such as ZDV and d4T. Tenofovir is a highly potent antiretroviral drug and an important addition to the armamentarium.

FACTORS INFLUENCING INTRACELLULAR PHOSPHORYLATION

Cell Activation

Because the NRTIs require intracellular activation for antiretroviral activity, influences on the enzymes responsible for activation can modulate drug activity. In vitro, cellular factors have been shown to alter antiretroviral activity. The activation state of the cell may change the relationship between drug triphosphate and endogenous triphosphate. For instance, in activated cells, the ratios of ZDVTP to deoxythymidine triphosphate (dTTP) and of d4T triphosphate to dTTP are greater than in resting cells. This is thought to be a result of changes in the cell cycle, in which increased expression of thymidine kinase is seen during S-phase.

Conversely, ddC and 3TC have been shown to have a more favorable drug-to-endogenous triphosphate ratio in resting cells. Resting cells, such as monocytes/macrophages have low endogenous dNTP levels, which results in low catalytic activity of HIV RT. Consequently, because the ratio of competition between drug and endogenous triphosphates determine activity, less drug phosphorylation may be required for the same antiviral effect.

HIV Infection

Cell culture studies have shown the intracellular phosphorylation of the NRTIs to be similar in control and HIV-infected cells. However, in vivo studies showed total intracellular phosphorylation of ZDV in PBMCs isolated from healthy volunteers to be less than that in PBMCs isolated from HIV-infected patients. Furthermore, in patients with decreased CD4 cell counts, ZDV phosphorylation was greater in resting PBMCs, suggesting that HIV infection results in an increase in activation state and in increased drug phosphorylation.

Ex vivo studies in PBMCs show that ZDV phosphorylation is greater in cells from HIV-infected patients than healthy volunteers, but in cells from HIV-infected patients there is a reduced capacity for activation by in vitro stimuli.

Long-Term Changes in Intracellular Activation—Cellular Resistance

Drug regimens fail for many reasons, including virological resistance, poor tolerability, lack of adherence, and pharmacological resistance. Possible mechanisms of pharmacological resistance to antiretroviral agents include decreased influx of drug into cells or increased efflux of drug from cells. However, a decrease in the activity of kinase enzymes responsible for the formation of NRTI triphosphates may result in subtherapeutic levels of the active metabolite and, thus, decreased efficacy of antiretroviral therapy.

Determination of drug levels in patients receiving combination therapy is required to evaluate the extent (if any) of intracellular kinase downregulation and, thus, the ability to assess whether altered phosphorylation occurs in these patients. Early studies indicated that ZDV may downregulate its own metabolism during long-term therapy. A modest inverse association between length of time on therapy and the formation of total ZDV phosphates was seen. Subsequently, the ALTIS trial suggested a difference in virological response between ZDV-naive and ZDV-experienced patients after antiretroviral therapy with d4T plus 3TC. In patients who started therapy on d4T plus 3TC, a greater decrease in plasma HIV RNA level (1.66 $\log_{10}$ copies/mL) was seen in patients who had not previously received ZDV alone or as part of a combination (0.66 $\log_{10}$ copies/mL) than in those who were ZDV experienced.

The ALTIPHAR study showed reduced d4T triphosphate and 3TCTP levels in a small number of ZDV-experienced patients who had not responded to the dual combination therapy, compared with ZDV-naive patients. This suggested that previous therapy with ZDV reduced the subsequent phosphorylation of d4T and 3TC. The decreased activation to d4T triphosphate was thought to be a result of decreased thymidine kinase activity in vivo. However, other studies failed to confirm the findings of the ALTIPHAR study.

A cross-sectional study investigating ZDV phosphorylation showed no difference between patients receiving short- and long-term therapy. A 12-mo longitudinal study investigating ZDV activation in a heterogeneous population of HIV patients observed no difference in total intracellular ZDV phosphates over time. Second, there was no decrease between activation in ZDV-naive and experienced patients. Studies also investigated whether long-term administration of ZDV altered the activation of d4T. Using ex vivo measurements, no significant differences in the activation of d4T was seen in PBMCs isolated from HIV patients with or without previous exposure to d4T. These results suggest that if a reduction in the expression of thymidine kinase occurs in vivo during long-term therapy, d4T phosphorylation is unaltered. The CHARM trial phosphorylation substudy measured triphosphates of ZDV, 3TC, and ABC to 48 wk.

The results showed no changes in either ZDV or ABC phosphorylation over time. However, 3TC phosphates at 48 wk were decreased to less than 50% of baseline (week 0). Further studies are required to investigate the consequences of this reduction. However, because drug efficacy is determined by the competition between drug triphosphate and endogenous triphosphate, it should be remembered that the efficacy of the NRTIs will be dependent on the triphosphate ratio. Determination of ratios in cohorts of patients receiving combination therapy is required to evaluate the extent of intracellular kinase downregulation and, thus, assess whether altered phosphorylation occurs in these patients.

Modulation of Intracellular Phosphorylation

As previously mentioned, ddNTPs compete with their corresponding dNTP for HIV RT and subsequent incorporation into the proviral DNA chain. Therefore, the activity of existing NRTI-containing combination therapy regimens may be improved by the inclusion of compounds that can modulate phosphorylation.

Hydroxyurea

Hydroxyurea (HU) inhibits the enzyme ribonucleotide reductase, which is responsible for the reduction of ribonucleotides to deoxyribonucleotides. Therefore, inhibition of this enzyme potentially results in decreased levels of dNTPs formed using the *de novo* pathways. However, some dNTPs are also formed using salvage pathways, which, in turn, can be affected (through feedback mechanisms) by any alteration in the *denovo* synthesis. In vitro studies have investigated the effects of HU in combination with several NRTIs.

The mechanisms behind the potential synergism of HU with these agents have been postulated. First, dATP is preferentially depleted by HU This depletion favors the incorporation of ddATP, the active anabolite of ddl, into the growing viral DNA strand. An enhancement of the anti-HI V activity of ddl, when combined with HU, has been noted in vitro, presumably because of the increased ddATP-to-dATP ratio.

Similarly, the activity of other purine analogs, such as ABC and DAPD, may be enhanced when combined with HU. However, HU does not enhance the anti-HIV activity of low-dose tenofovir disoproxil fumarate. In vitro studies of HU with pyrimidine analogs (*e.g.,* ZDV and 3TC) have demonstrated less favorable enhancements of anti-HIV activities of these agents. Increases in pyrimidine ddNTP levels have been observed, but the effects on the corresponding levels of dNTP are varied.

The differences in the effects of HU on purine and pyrimidine NRTIs may involve the reliance of purine dNTP production on *de novo* synthesis, whereas pyrimidine dNTPs are also produced using salvage pathways. To date, the majority of clinical studies investigating the effects of HU in combination with NRTIs have centered on the interaction with ddI.

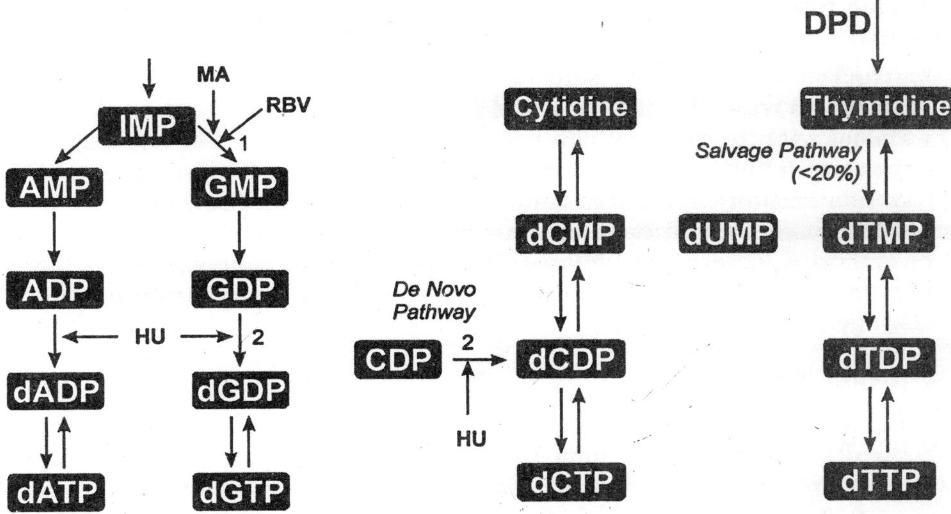

Fig. 7.137. The metabolic pathways of cellular nucleotides, showing the sites of action of mycophenolate (MA) and RBV at (1) IMP dehydrogenase; HU at (2) ribonucleotide reductase; and dipyridamole (DPD) at (3) deoxythymidine uptake.

Some studies show a clinical benefit by adding HU to ddI-containing regimens. However, other studies have been associated with a less favorable outcome. For example, a randomized trial comparing ddl plus d4T plus efavirenz with and without HU was discontinued because of excess drug toxicity in the HU-containing arm.

Results from the pharmacology substudy of the CHARM trial, a study in which patients received ZDV plus ABC plus 3TC with and without HU, demonstrated that intracellular dNTP levels were unaltered by HU. Similarly, Hamzeh et. al., observed that dATP levels were unaffected in patients receiving HU monotherapy. The discordance in data from in vitro and in vivo studies stresses the importance of determining ddNTP and dNTP levels in clinical trials.

Ribavirin

Ribavirin (RBV) is a broad-spectrum antiviral agent that inhibits IMP dehydrogenase (IMPD). This enzyme is responsible for the conversion of IMP to xanthosine monophosphate, thus, leading to a reduction in the levels of dGTP, a purine dNTP. Because dGTP is an inhibitory regulator of the reduction of pyrimidine nucleotides, the reduced levels of dGTP result in an increased pyrimidine dNTP production. In vitro studies have demonstrated that RBV reduces dGTP levels, whereas it increases the activation of purine NRTIs (*e.g.*, ddI). Furthermore, the antiviral activityof purine NRTIs is also enhanced by RBV.

Therefore, theoretically the combination of RBV plus ABC or DAPD could be advantageous. Conversely, RBV has been shown to increase the levels of pyrimidine dNTPs (dTTP and dCTP), whereas, in vitro, it inhibits the activation of pyrimidine NRTIs, such as ZDV. The effect of RBV on these pyrimidine metabolites is supported by RBV being shown to antagonize the anti-HIV activities of pyrimidine NRTIs. Clinical use of the combination of RBV with interferon or pegylated interferon is the standard of care to treat hepatitis C virus coinfection in HIV-infected patients.

However, the latest clinical data indicate that RBV does not adversely affect HIV disease control in patients receiving ZDV- and d4T-containing regimens (despite the in vitro interaction data), and does not alter intracellular concentrations of ZDVTP and d4T triphosphate, at least in the short term. There is some evidence of an effect of high-dose RBV on ZDV phosphorylation in the short term. In addition, because RBV is an IMPD inhibitor, and because it augments the inhibitory effects of the purine analog, ddl, there have been a number of reports citing clinically important ddl-related toxicities, including pancreatitis and lactic acidosis in patients coinfected with HIV and hepatitis C virus, and the recommendation is not to combine these agents.

Mycophenolate Mofetil

Mycophenolate mofetil is hydrolyzed to its active metabolite, mycophenolic acid (MPA) in vivo. Similar to RBV, MPA also inhibits IMPD, thus having similar effects on purine and pyrimidine metabolites in vitro. In vitro studies have also demonstrated that MPA can augment the anti-HIV activity of purine-based NRTIs and display activity against drug-resistant strains of HIV. For example, ABC and MPA have been observed to have a profound and synergistic anti-HIV activity in combination. However, the same study also reported no synergy or even antagonism when MPA was combined with pyrimidine NRTIs against HIV. A number of clinical studies haveTbeen undertaken, or are ongoing, to investigate the effects of MPA when used in combination with NRTIs, specifically ABG.

Margolis et al. concluded that in vivo modulation of the CBVTP-to-dGTP ratio may be achievable with doses of my cophenolate mofetil 50% lower than those used in organ transplantation. Additional studies are planned to examine the effects of MPA when administered in combination with other NRTIs (*e.g.,* DAPD or tenofovir disoproxil fumarate), on anti-HIV activity in vivo.

Other Modulators of Phosphorylation

The following compounds may also be viable candidates for use in modulating the phosphorylation of both dNTPs and ddNTPs.

Azodicarbonamide

Azodicarbonamide has been previously tested as an anti-HIV agent through its ability to expulse zinc from viral-cysteine factors and to interfere with calcium mobilization machinery. However, it shares structural analogy with HU and has been observed to inhibit ribonucleotide reductase. Unlike HU, azodicarbonamide has also been observed to inhibit thymidine phosphorylation.

Methotrexate

Methotrexate is thought to inhibit cell growth through an inactivation of dihydrofolate reductase, resulting in a disturbance of the *de novo* thymidylate and/or purine biosynthesis. Methotrexate has previously been shown to reduce intracellular dNTP levels, whereas enhancement of the phosphorylation of both d4T and 3TC has also been observed in vitro.

Fludarabine

Fludarabine (FLU) has been shown to increase the phosphorylation of 3TC as well as to reduce dCTP levels, at high concentrations, resulting in an increased 3TCTP-to-dCTP ratio. Similarly, studies by Rahn et. al., demonstrated modest increases in 3TC activation, using lower concentration of FLU. The activity of FLU has been attributed to direct and indirect effects of the active moiety of FLU, FaraATP, on deoxycytidine kinase.

Modulation of Nucleoside Transport

The sites of cellular entry of nucleosides may also act as targets for the modulation of NRTI activity. The cellular uptake of endogenous nucleosides and NRTIs is mediated by a variety of nucleoside-specific membrane transporters. These include facilitated diffusion and sodium-dependent processes. Knowledge of the abilities of the different nucleoside transporters to transport both endogenous nucleosides and NRTTs across the cell membrane may have a bearing on modulating NRTI activity For example, passive diffusion is sufficient for ZDV to reach therapeuticalw relevant intracellular levels without the need for membrane transporters.

Therefore, by inhibiting transporters used in the uptake of endogenous nucleosides, levels of dTTP can theoretically be reduced, whereas the levels of ZDVTP will be unchanged. Dipyridamole has the ability to inhibit the transport of endogenous nucleosides into the cell. This may be linked to the ability of dipyridamole to potentiate the activity of ddC in cells infected with HIV-1. However, the subject of nucleoside transporter modulation is complicated, because the expression of these transporters may be altered by theinfluence of exogenous compounds. The complex interactions between modulators and endogenous nucleoside and NRTI phosphorylation again highlight the need to determine both dNTP and ddNTP levels.

By combining a modulator with an NRTI, it may be possible to reduce the dose, or even the frequency of dosing, of the NRTI and retain efficacy. This may cause a reduction in the associated toxicity of the NRTI in question, and also have economic advantages. However, in practice, the toxicity associated with the administration of the modulator can be a major issue.

THE ROLE OF NRTI ACTIVATION

Fletcher et. al., demonstrated a positive correlation between the intracellular concentration of ZDVTP and 3TCTP, which was related to the baseline CD4 count (*i.e.,* lower CD4 count with higher triphosphates). These data, which show a correlation in the ability of HIV patients to phosphorylate NRTIs with different activation pathways, suggest that if individuals have reduced phosphorylation of one NRTI it may impact on the phosphorylation of a subsequent NRTI. Although it may be argued that this may solely reflect the activation state of the cells, other studies have shown the activity of deoxycytidine kinase to be less affected than thymidine kinase during cell activation.

Because endogenous triphosphates are formed to replicate DNA, synchronization may result in similar regulation with all the nucleoside analogs. Further studies are required in this area. In a small group of patients receiving combinations including 3TC and ABC, a weak relationship was seen between the ratios of 3TCTP to dCTP and of CBVTP to dGTP. These findings may partially explain how some patients who change from one NRTI to another have a reduced response on a subsequent NRTI.

Modulation Through Drug-Drug Interactions

Phosphorylation and clinical data clearly indicate that it is essential to co-administer two nucleoside analogs with different activation pathways, such as 3TC and d4T, rather than those with shared enzymes, *e.g.,* d4T and ZDV, to ensure maximal anti-HIV activity. ZDV and d4T are two thymidine analogs competing directly for the same cellular kinases.

Incubation of these two drugs together in vitro at equal centrations results in a decrease in activation by d4T triphosphate but activation by ZDVTP remains unaltered. This is because the first enzyme in the cascade (thymidine kinase) has a 600-fold greater affinity for ZDV than d4T. This interaction has been shown to decrease antiviral activity both in vitro and in vivo. Patients receiving ZDV plus d4T had a significantly worse response to this combination compared with patients receiving ZDV plus ddl orZDVplus ddC. Similary, studies investigating ddC phosphorylation found that 3TC inhibited ddC phosphorylation in a concentration-dependent manner at identical concentration ratios (10 and 100). 3TC also inhibits the phosphorylation of SPD754.

Interconversion of Nucleoside Triphosphates

A recent study has shown that ZDV administration in vivo results in the formation of some d4T triphosphate; intracellular concentrations of d4T triphosphate represented approx 5% of the amount of ZDVTP. The interaction between ZDV and d4T at the thymidylate kinase step suggests that the d4T metabolites are formed after this activation step, but more studies are required to investigate whether this anabolite contributes to toxicity and/or efficacy. However, in a further study, no evidence has been found for this interconversion.

Failure of Triple-Nucleoside Regimens

Recently, several triple-nucleoside regimens (ABC plus 3TC plus TDF; or ddI plus 3TC plus TDF) have been well-documented to have surprisingly high rates of virological failure. The reason for the failure of these regimens is currently unknown, but one appealing hypothesis is that there is an, as yet, unidentified intracellular interaction between two of the drugs in the regimens. However, the present data does not support this. Hawkins et. al., have presented data from a small (therefore, should be interpreted with caution) but well-designed clinical study in which intracellular CBVTP and teno-fovir diphosphate levels were determined in 15 patients.

Intracellular CBVTP and TFVDP were not altered during a period of 28 d after discontinuation of either ABC or TDF. This lack of change in intracellular phosphate anabolite was also shown in vitro when incubating cells with the drugs either alone or in combination. Considering also the data from the TONUS study it is likely that a low genetic barrier to resistance is the major cause of failure of the triple NRTIs.

CHAPTER

8 Analyzing Nucleic Acids

It was not possible in the main text of this book to describe the experimental basis of all the conclusions presented. The objective of this appendix is to try partially to remedy this deficiency by describing some of the most important general methods used in the study of nucleic acids. For some of the methods it was necessary to provide a full description earlier, and in these cases only a brief reference is made here for completeness. It should be stressed that the emphasis in this section is on the principles underlying the methods, and no attempt is made to provide laboratory protocols, which are now available in abundance, and to which reference will be made. It is inevitable that some of the methods dealt with here (especially those involving cloning), although pertinent to the work already described in this text, will in many cases be superseded in the lifetime of this edition. The reader wishing information about current methodologies is therefore advised to consult the most recent texts. However, at the time of writing, the cited references are recommended for more detailed information on principles and practice.

OCCURRENCE AND CHEMICAL ANALYSIS

Chemical Determination of Nucleic Acids in Tissues

The usual approach to the determination of nucleic acids in tissues involves fractionation to remove lipids and proteins etc., followed by assay on the basis of either phosphorus, ribose or deoxyribose, or purine and pyrimidine. The tissue is treated with cold dilute trichloroacetic or perchloric acid to precipitate nucleic acid and protein, and lipid removed by extraction with organic solvent. In the procedure of Schneider total nucleic acid (DNA + RNA) can be separated from protein by hydrolysis in hot acid, the liberated products being soluble. These latter are then assayed for ribose and deoxyribose. In the procedure of Schmidt and Thannhauser the DNA and RNA are separated by alkaline hydrolysis of the latter followed by acidification. Analysis of nucleic acid phosphorus in such fractions usually employs a colour reaction involving the formation of a phosphomolybdate complex. Following depurination deoxyribose can be determined by a colour reaction with diphenylamine. This reaction is specific for DNA. In contrast, the orcinol colour reaction employed for depurinated RNA also occurs to a lesser extent with DNA.

The spectral properties of the purine and pyrunidine bases of such hydrolysed nucleic acids may also be used as a basis of nucleic acid assay. These bases have absorption maxima in the region of 260 nm. The spectral properties of double-stranded and single-stranded DNA have already been discussed, as has the effect of pH thereon. Protein also absorbs at 260 nm, although to a much lesser extent, dependent on the content of the aromatic amino acids phenylalanine, tryptophan and tyrosine. Because the latter two of these have an absorption maximum at 280 nm it is possible to obtain an approximate estimation of total nucleic acid and protein in a mixture by employing the following formulae based on.

Nucleic acid (mg/ml) = $0.064\ A_{260} - 0.031\ A_{280}$

Protein (mg/ml) = $1.45\ A_{280} - 0.74\ A_{260}$

Typical values for the nucleic acid and protein contents of some rat tissues. These may be expressed per gram tissue wet weight, but it is often useful to express them in terms of tissue protein, which may be easily estimated.

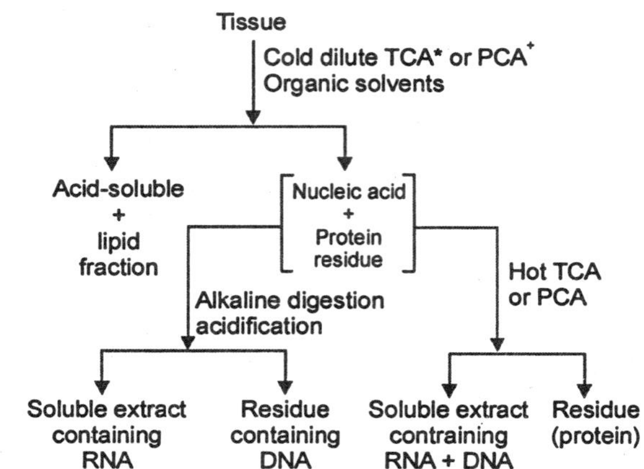

Fig. 8.1. Extraction and fractionation of nucleic acids from tissues. *TCA – trichloroacetic acid, +PCA – perchloric acid.

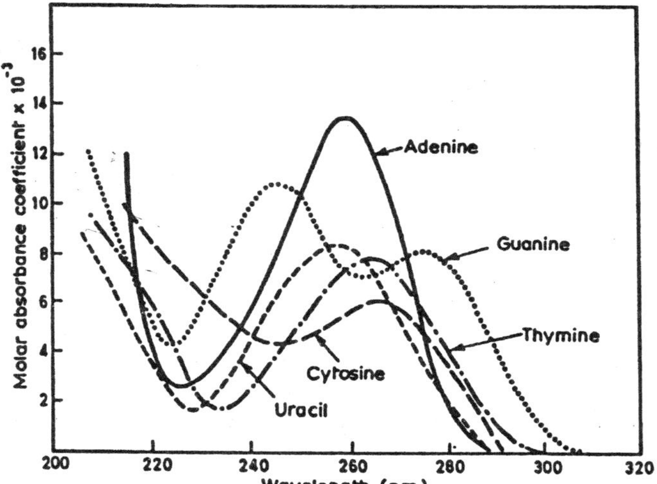

Fig. 8.2. Ultraviolet absorption spectra for purine and pyrimidine bases at pH7.

Analaysis of base Composition and Nearest-neighbour Frequency

The molar proportions of bases in nucleic acids are determined by hydrolysis, separation and spectral quantitation. Typical values for DNA are given in Table 8.3 and the variation in percentage GC content in different genomes can be found Table 8.6. The principles of determining the base composition of RNA molecules are similar to those for DNA, and values for some types of RNA (actually based on sequence analysis). Transfer RNA is not included because of the large proportion of modified bases which, in the case of rRNA have been ignored. It may be remarked that rRNA is GC-rich, this need only be true in relation to the overall base composition of the genome, rather than in absolute terms.

Table 8.1. Typical Values for the Concentrations of Nucleic Acids in Different Rat Tissues.

Tissue	*RNA (mg/g protein)*	*DNA (mg/g protein)*	*RNA/DNA*
Brain	40	15	2.7
Kidney	43	18	2.4
Liver	54	12	4.5
Skeletal muscle	62	13	4.8
Spleen	58	58	1.0
Thymus	30	93	0.3

Table 8.2. Molar Percentages of Bases in RNAs from Various Sources.

Type	*Species*	*Adenine*	*Guanine*	*Cytosine*	*Uracil*
Ribosomal**					
E.coli	16S Rrna	25.2	31.6	22.9	20.3
	23S rRNA	26.2	31.4	22.0	20.4
	5S rRNA	19.2	34.2	30.0	16.7
Rat	18S rRNA	22.5	29.2	26.5	21.8
	28S rRNA	16.6	35.5	31.6	16.2
	5S rRNA	18.3	32.5	27.5	21.7
Messenger*					
E.coli	Outer-membrane lipoprotein	28.3	24.2	22.7	24.8
Rabbit	β-globin	23.6	27.2	23.0	26.2
Rat	α-actin	21.7	26.4	30.6	21.3
Viral*					
Bacteriophage MS2		23.4	26.0	26.1	24.5
Poliomyelitis virus		29.6	23.0	23.3	24.1
Rous sarcoma virus		23.8	28.8	25.3	22.1
Tobacco mosaic virus		27.1	24.2	19.1	27.6

**Ignoring base modifications.

*Ignoring 3′ poly(A) tails.

The examples of mRNAs are rather arbitrary, and dramatic variations in the base compositions of different mRNAs are seen in the same organism. The determination of nearest-neighbour frequencies (*i.e.* the relative frequency of occurrence of the sixteen different dinucleotides, NpM) in DNA, although to a considerable extent superseded by nucleotide sequence determination still has applications.

Estimation of the Molecular Weight of DNA

Methods of estimating the size of small DNA molecules (up to 10 kbp) using *electrophoresis* on acrylamide and agarose gels have been described. Such methods can be accurately calibrated using DNA molecules of known sequence. Although measurement of the *sedimentation rate* can be used to estimate the molecular weight of an RNA molecule this method is not readily applicable to DNA molecules because of their extremely large size and asymmetric shaped.

The theory behind the calculations depends on a knowledge of diffusion coefficients but these are equally difficult to obtain either by sedimentation analysis or optical mixing spectroscopy. However, a series of empirical formulae relating sedimentation coefficient ($S°_{20,w}$) and molecular weight (M) have been derived by comparing the sedimentation rates of molecules of known size. For example, the equation for linear, double-stranded DNA is

$$S°_{20,w} = 2.8 + 0.00834\ M^{0.479}$$

On sedimentation of NDA to *equilibrium* in density gradients (usually of CsCl) the width of the band of DNA is inversely proportional to the square root of the molecular weight. As with light scattering the result is dependent on the concentration of the DNA and a series of values must be obtained and extrapolated to zero concentration. This is the most widely used, fundamental method of estimating the size of DNA molecules and is applicable to molecules of up to 10^8 daltons. *Light scattering* was widely used in the 1950s to measure molecular weights of DNA up to 5000 kbp, but only since the introduction of the laser and the production of machines to enable study of scattering at low angles has the method been extended to molecules of larger size.

However, at low angles the scattering by dust particles becomes a major problem and with large molecules different regions of the same molecule serve as scattering centres. The above methods all suffer from the disadvantage that they are badly affected by any heterogeneity in the size of the DNA molecules. It is extremely difficult to prepare high molecular-weight DNA intact from chromosomes and these methods tend to reflect the size of the smaller molecules. The problem is even greater when attempts are made to relate molecular weight to *viscosity* as this involves subjecting the DNA solution to shear forces which tend to break long DNA molecules.

However, Zimm developed a new, low-shear, viscometer consisting of a slowly rotating tube (the Cartesian diver) immersed in a DNA solution under pressure, and a relationship has been established between intrinsic viscosity (η) and molecular weight (M) of linear duplex DNA similar to that between sedimentation coefficient and molecular weight:

$$0.665 \log M = 2.863 + \log (\eta + 5)$$

A more important observation is that after the Zimm viscometer has been used to measure the size of very large DNA molecules the Cartesian diver rotates in the reverse direction for a while after the power has been switched off. This *viscoelastic* effect is caused by the stretched DNA molecules relaxing to the unstressed configuration.

The theory derived by Zimm to explain this effect has been used to measure the size of the DNA in *Drosophila* chromosomes showing the method to be applicable to DNA molecules of molecular weight up to 79×10^9 (over 10^5 kbp). The method has the major advantage that it measures the size of the largest DNA molecules in the solution. Two 'visual' methods of measuring the size of DNA molecules involve electron microscopy and autoradiography. With *electron microscopy* the Kleinschmidt technique involves spreading the DNA on grids coated with polylysine or polystyrene-4-vinylpyridine. A drop of DNA solution in 1M ammonium acetate containing 0.1 mg/ml denatured cytochrome *c* is applied to the surface of a solution of 0.2M ammonium acetate.

A film of cytochrome *c* spreads across the surface, binding to the DNA which is disentangled and spread out. A grid is touched to the surface and the sample is dehydrated in ethanol. It may be stained with uranyl acetate or by evaporation of platinum on to the surface. In the latter process the grid is rotated so as to cause the metal to pile up against all surfaces of the DNA, which stands out from the grid. Double-stranded DNA is readily visualized by this technique but single-stranded DNA (or RNA) molecules require the presence of form-amide to prevent their collapse. Even so they appear as rather wispy threads and may be stretched out to different extents.

The incorporation of proteins which bind strongly to single-stranded DNA (*e.g.* phage T4 gene 32 protein causes these regions to assume a profile thicker even than the native DNA regions and they now have a regular mass per unit length. The presence, on the same grid, of marker DNA molecules of known length (*e.g.* SV40 DNA in the relaxed circular form) allows the size of an unknown DNA molecule to be found by measuring the relative contour lengths on electron micrographs. This method am readily be used with DNA molecules of up to about 10^3 kbp, but larger molecules cannot be accommodated in a single frame. In denaturation mapping double-stranded DNA is partially denatured (usually by alkali treatment) and the unpaired bases reacted with formaldehyde to prevent their reannealing.

The duplex DNA molecule now exhibits single-stranded bubbles in the (A + T) rich regions to give a characteristic map. When DNA is labelled *in vivo* with (^{3}H)thymidine and the cells lysed on a slide to release the intact DNA molecules their position can be visualized by *autoradiography.* The radioactive DNA on the slide is covered with a layer of photographic film which is activated by the β-particles produced by the decay of the tritium to produce black grains. As with electron micrographs the size of the DNA can be obtained by measuring the contour length of the autoradiographic image. This method uses light microscopy and so the field size is much larger than with electron microscopy. Cairns applied it to the study of the *E. coli* chromosome, which is 3000 kbp long.

ISOLATION AND SEPARATION OF NUCLEIC ACIDS

Isolation and Separation of DNA

The main problems to be overcome in isolating high molecular weight chromosomal DNA from bacterial or eukaryotic cells are the removal of protein (especially deoxyribo-nucleases) and RNA and the avoidance of mechanical shearing. For eukaryotic cells a combination of pronase of protease K, a detergent and phenol extraction can be used directly to disrupt the cells and remove protein. For the cells of Gram-negative bacteria such as *E. coli* gentle lysis can be effected using lysozyme. Phenol extraction is used widely in the isolation of both DNA and RNA, which are retained in the aqueous phase while denatured protein collects at the interface between this and the phenol phase, which contains the lipids. RNA may be removed from the DNA by using pure pancreatic ribonuclease, or by isopycnic ultracentrifugation in a gradient of CsCl, which is, in any case, a useful step for further purification. In *isopycnic ultracentrifugation* a density gradient is established which encompasses the buoyant densities of the molecules to be separated.

Molecules will assume a position on the gradient corresponding to their own buoyant density, after which no further separation can occur. In this method an equilibrium is established, and the method is sometimes referred to as 'equilibrium ultracentrifugation' to distinguish it from the rate-zonal method. As already mentioned, the buoyant density of double-stranded DNA varies with the mole fraction of (G + C). In general, however, RNA has a much higher buoyant density (about 1.9 g/ml) than double-stranded DNA (about 1.7 g/ml), which, in turn, is higher than that of protein (about 1.3 g/ml).

Single-stranded DNA is of slightly higher buoyant density than double-stranded DNA (*e.g.* for a double-stranded DNA of buoyant density 1.703 g/ml, the value for the single-stranded form is 1.717 g/ml). Isopycnic ultracentrifugation is also used to separate small bacterial plasmids from chromosomal DNA. The basis of separation in this case is not differing GC contents (these are, in general, similar) but the fact that the chromosomal DNA is linear (normal methods do not extract the large circular bacterial chromosome intact) whereas the plasmids are circular. This difference is exploited by the addition of saturating concentrations of the intercalating fluorescent dye, ethidium bromide. The intercalation requires that the DNA strands be forced apart, with concomitant decrease in buoyant density and partial unwinding of the double helix. This latter process is hindered in the closed-circular plasmid molecules with the result that they bind less ethidium bromide and have a higher buoyant density than the linear chromosomal (and open-circular and linearized plasmid) molecules.

Before attempting to separate plasmid and chromosomal DNA it is advisable to enrich the preparation in plasmid DNA. The most frequently employed method for achieving this in preparative work is the use of an alkaline pH during lysis and extraction. This denatures the DNA, causing strand-separation in the case of linear molecules.

On neutralization the latter will not renature, whereas the closed-circular plasmid DNA will and can be separated from the insoluble denatured DNA. For small-scale isolation of plasmid DNA, a boiling step is usually used in place of the treatment with alkali. Ethidium/ CsCl gradients are also used in the purification of small animal viruses (*e.g.* SV40) which contain cyclic genomes. In this case the initial purification involves precipitation of most of the high molecular weight cellular DNA adsorbed onto a precipitate of sodium dodecyl sulphate to leave behind the so-called Hirt extract. Although most preparations of chromosomal DNA consist of sheared fragments, these do not in general separate on isopycnic ultracentrifugation-because the fragmentation is random and the base compositions of individual fragments are generally similar.

Exceptions do occur in the case of regions of repetitive DNA in eukaryotes which may be generated as quite homogeneous species in relatively large amounts if the fragments are of the appropriate size. Should the base composition of such a repeated DNA be markedly different from the average base composition of the chromosomal DNA, resolution from the bulk of the latter will occur. Such DNA was originally given the descriptive name 'satellite DNA' a designation which is sometimes applied to all repetitive DNA, regardless of base-composition or function. Such loose terminology is to be discouraged. It is also possible to isolate the tandemly repeated rDNA and histone genes of certain organisms by this method. These exceptions aside, separation of DNA by isopycnic ultracentrifugation is largely confined to the different chromosomal species one might find in a given cell.

Precise fragments of such DNAs can be generated by restriction endonucleases and the separation of such fragments is a common requirement in recombinant DNA manipulations. Most frequently this is achieved by horizontal *agarose gel electrophoresis*. This separates different linear DNA molecules according to size (the mobility is inversely proportional to the

$\log_{10}$ of the molecular weight), as the charge per unit length of different molecules is identical and the larger molecules will be retarded by the gel matrix.

The percentage of agarose can be increased to separate smaller-sized DNA fragments, but for very small species *polyacrylamide gel electrophoresis* (which is often used preparatively for fragments up to 1 kb) must be used. This method is also employed in rapid-nucleic acid sequencing methods in which case 8M urea is included to ensure denaturation. As agarose gel electrophoresis is frequently used to analyse preparations of plasmid DNA which may be slightly nicked, it is worth mentioning that the closed-circular supercoiled form of the plasmid migrates more rapidly than the less compact relaxed-circular form produced by single-stranded nicks.

Any linearized plasmid molecules generally migrate at an intermediate position between these two, a result that is perhaps a little unexpected. Nucleic acids may be precipitated with ethanol, 2.5 volumes in the presence of 0.3 M sodium acetate, being frequently employed.

Isolation and separation of RNA

In principle the methods used to extract RNA from cells are generally similar to those for DNA. Degradation of RNA by ribonuclease is a greater threat than that of deoxyribonuclease degradation of DNA, especially in the case of mRNA, but there are a variety of methods to counter this. The major separation problem is posed by the large number of different RNA species in eukaryotes, and to study a particular species it is necessary to separate it from the othersr. One preliminary that may be useful in particular cases is *subcellular fractionation*

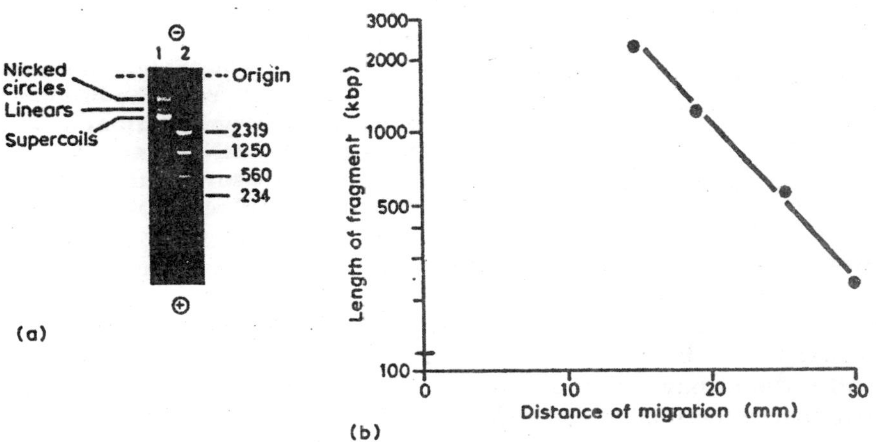

Fig. 8.3. Agarose gel electrophoresis of DNA. (*a*) Separation of: 1, different forms of DNA of plasmid pBR322; 2, fragments of DNA (lengths indicated in kbp) derived from plasmid pBR322 by double-digestion with restriction endonuclease *Bam* HI and *Bgl* I; (*b*) Plot of length of DNA fragment (log scale) against distance of migration (linear scale) of data from (*a*) 2, illustrating linear relationship.

Thus the nuclei may be removed by low-speed centrifugation (700 g for 5 min) after gentle disruption of the cells in the presence of sucrose (to preserve the osmolarity), the mitochondria may be removed by further centrifugation (10000 g for 10 min), and the ribosomes separated from soluble RNA species such as tRNA by ultracentrifugation (*e.g.* 100 000 g for 90 min).

It may of course be necessary further to purify the subcellular fraction in which one is interested, and methods for nuclei, nucleoli mitochondria and ribosome are available. It may be remarked that the most appropriate method for isolating an organelle, such as the nucleus, for subsequent RNA extraction (where purity is of greatest importance) may be quite different from the method which will yield the most biologically active material. One of the most widely used methods of separting RNA molecules is *rate-zonal ultracentrifugation employing sucrose density gradients*. In this method the separation of RNA is based primarily on size (in principle, molecular shape can also have an influence, but with RNAs differences in shape make no significant contribution to the separation).

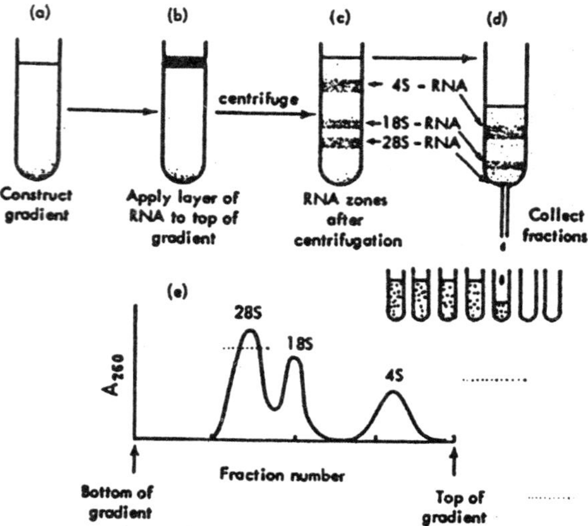

Fig. 8.4. Rate zonal centrifugation of RNA through a sucrose density gradient. A sucrose density gradient is constructed in a centrifuge tube (*a*) and the RNA solution applied as a layer on top (*b*). During ultracentrifugation the main components of the RNA separate into zones, primarily on the basis of molecular weight (*c*). These zones may be recovered by puncturing the bottom of the tube and collecting different fractions in separate tubes (*d*). The separated RN As may be visualized and quantitated by measurement of the absorbance at 260 nm (*e*). Steps (d) and (*e*) may be conveniently combined by pumping the gradient through the flow-cell of a recording spectrophotometer.

It should be emphasized that the density range of sucrose solutions does not extend to that of RNA species and, in contrast to isopycnic ultracentrifugation involving CsCI gradients, no equilibrium is reached. The main purpose of the sucrose gradient is to prevent diffusion of the zones of individual species that are separated from the narrow zone applied initially to the top of the gradient. Isopycnic ultracentrifugation in gradients of dense sucrose (often employing a discontinuous or 'step gradient') may be used to separate 'free' ribosomes from those bound to the membrane of the endoplasmic reticulum. Rate-zonal ultracentrifugation in sucrose density gradients is also used to separate different size classes of polyribosomes and to separate the large and small subunits of ribosomes after their dissociation. Rate-zonal ultracentrifugation is clearly in-appropriate for the separation of different species of *tRNA,* which are broadly similar in size.

Methods involving separation on the basis of charge and hydrophobicity differences (*e.g.,* chromatography on BD-cellulose and RPC5) may be employed for this purpose. The major problem to be overcome in the isolation of *mRNA* is purification from other species of RNA, which are approximately 20 times more abundant. Affinity chromatography using oligo(dT)-cellulose or poly(U)-Sepharose-materials that bind the poly(A) tails of mRNAs—is the basis of the separation of mRNA from other species of RNA. The minimum size of poly(A) tail required for binding to poly(U)-Sepharose is smaller than that for oligo(dT)-cellulose (about 6-10 residues compared with 15 residues), and although the latter material is in more widespread use, the former has applications to mRNAs having very short poly(A) tails.

A mixture of different mRNAs can be separated to a certain extent on the basis of size, either by sucrose density gradient centrifugation or, on a smaller scale, by agarose gel electrophoresis-basically as for DNA but generally under denaturing conditions. Both agarose gel electrophoresis and polyacrylamide gel electrophoresis have been applied to other types of RNA. The isolation of individual mRNA species is usually achieved by recombinant DNA techniques.

HYBRIDIZATION OF NUCLEIC ACIDS

The denaturation and renaturation of homoduplex DNA molecules has been dealt already where the effects of base composition, temperature and ionic strength were mentioned. The application of DNA renaturation to determining the copy number of different portions of eukaryotic genomes (*C_0t value analysis*) has already been described. An important technique in molecular biology is the formation of heteroduplexes between different DNA molecules or between DNA and RNA. The application of electron microscopic visualization of such heteroduplexes formed in solution to the determination of the positions of introns (*R-loop mapping*) has already been described and the electron microscopic analysis of heteroduplexes can be most useful in a variety of other situations.

However, for most routine work heteroduplexes are detected by the radioactivity of one of their components (methods for the radioactive labelling of nucleic acids – 'probes'- are described. It is operationally convenient to have one component of the hybridizing pair immobilized on a solid 'paper' support in order that the hybrids may be easily visualized by autoradiography. Before the introduction of heteroduplex formation involving immobilized nucleic acids analysis involved the time-consuming inconvenience of multiple solution-hybridizations, when the experimental protocol involved identifying which component of DNA fractionated by agarose gel electrophoresis hybridized to a particular radioactively-labelled probe.

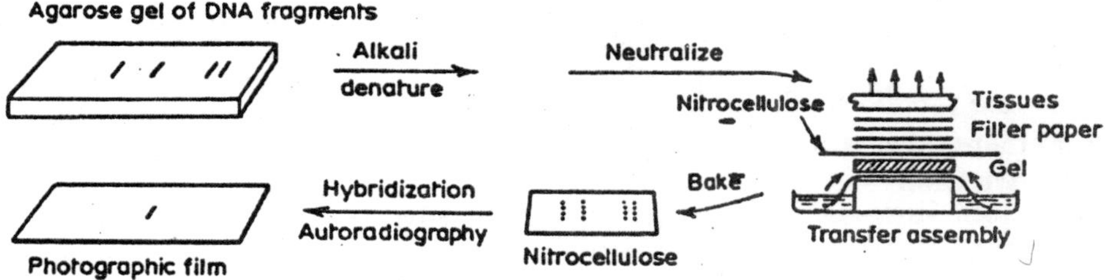

Fig. 8.5. Southern blotting. Fragments of DNA are fractionated by electrophoresis in agarose. the DNA denatured in alkali and, after neutralization, transferred (blotted) from the gel to nitrocellulose paper by capillary action at relatively high ionic strength in the transfer assembly shown. The transferred DNA is baked on to the nitrocellulose, after which it can be hybridized in solution with a radioactive probe and the hybridizing bands visualized by autoradiography. In the hypothetical example illustrated only one of the original four bands has hybridized to the probe.

To overcome this problem Southern devised a method of transferring the DNA from the fragile gel to nitrocellulose paper by means of capillary action. The DNA must be denatured with alkali before transfer, and afterwards it is baked at 80°C (in a vacuum oven to prevent it igniting) in order to fix the DNA permanently to the nitrocellulose. This method of transfer is commonly called *Southern blotting*, after its originator. The nitrocellulose paper is usually hybridized with the radioactive DNA probe in a minimum volume of solution in a sealed polythene bag at 68°C and conditions of relatively high ionic strength that promote heteroduplex formation. (If 50% formamide is included the hybridization is performed at 42°C.) The membrane is washed under suitable conditions, dried, and subjected to autoradiography. This method is most easily applied to detecting which restriction fragment of a larger piece of cloned DNA contains sequences homologous to a particular probe.

Fig. 8.6. Examples of results of Southern blotting experiment, (*a*) Cloned DNA from a mouse/bacteriophage lambda recombinant or (*b*) chromosomal DNA from mouse liver, were digested with (different) restriction endonudeases, subjected to electrophoresis on agarose gels and Southern blotting performed. Hybridization was to a (^{32}P)-labelled mouse actin cDN A clone. S: ethidium bromide stained gels photographed under illumination with ultraviolet light. A: autoradiographs of the nitrocellulose.

However, the sensitivity of the method is such that it can detect individual fragments in the continuum produced by restriction enzyme digestion of total genomic DNA and in fact the method was originally applied to uncloned genomic and sub-genomic DNAs. Such genomic Southern blotting has become a routine tool in screening for human genetic disorders by the detection of restriction fragment length polymorphisms for full details of which the reader is directed elsewhere. The conditions of washing nitrocellulose membranes (nylon membranes may also be used) depend on the degree of homology between the two members of the heteroduplex. (The homology need not be 100% as hybridization may be between a variety of related but non-identical sequences.)

When the two members of the heteroduplex are identical the washing is usually performed 12°C below the melting temperature (T_m), which in a solution 0.2 M in Na^+ is related to the mole percentage of (G + C) by the equation:

$$T_m = 69.3 + 0.41\,(G + C)$$

The effect of ionic strength (μ) on the melting temperature is given by the equation.

$$T_{m_2} - T_{m_1} = 18.5 \log_{10} \frac{\mu_2}{\mu_1}$$

Thus, to allow for the fact that the T_m of duplex DNA decreases by 1°C per 1 – 1.5% mismatch the 'stringency' of the hybridization can be decreased by lowering the temperature of the wash and/or by increasing the ionic strength of the washing buffer. Southern's method for the transfer of DNA from gels to paper has been extended to RNA, where it has acquired the somewhat ridiculous name of *Northern blotting*. (There is a third point on the compass, 'Western blotting', the name sometimes given to the electrophoretic transfer of proteins from polyacrylamide gels to nitro-cellulose.)

The methodology of Northern blotting is different from that of Southern blotting because RNA does not bind to nitrocellulose paper under conditions in which DNA does, and is hydrolysed by alkali. To overcome this problem, transfer can be made to diazobenzyloxymethyl (DBM) paper, which binds RNA (and DNA). RNA can, in fact, be transferred to nitrocellulose paper from agarose gels provided that it is denatured (*e.g.* by formaldehyde). Hybridization of immobilized nucleic acids on paper supports is not restricted to material transferred from agarose gels. It can be used for detecting recombinant DNA in bacterial colonies or bacteriophage plaques after transfer from petri dishes or for multiple samples of total cellular RNA applied directly to the nitrocellulose (the so-called *Dot-blot* technique).

METHODS OF LABELLING NUCLEIC ACIDS

There is a variety of circumstances in which the detection of DNA is only possible if the DNA is labelled in some way. Furthermore, certain techniques require DNA labelled specifically at one end. By far the most common method of achieving this is with radioactivity (usually ^{32}P; less commonly ^{35}S or ^{3}H), and the discussion below will be exclusively in terms of this. However, non-radioactive labelling methods are available. Currently the most prominent of these involves the use of biotin-labelled derivatives of dNTPs which are detected by using an antibiotin antibody. A variety of systems has been devised to visualize this, for example with a second antibody coupled to an enzyme such as horseradish peroxidase that catalyses a colour reaction.

General Labelling Methods

When the objective of labelling a nucleic acid is merely that of allowing its detection, methods that cause the incorporation of radioactivity throughout the molecule are suitable, as well as the specific end-labelling methods have been described. Historically, RNA and DNA were labelled from (^{3}H), (^{14}C) or (^{32}P)-labelled precursors *in vivo,* but methods involving the incorporation of radioactive label *in vitro* are more convenient for recombinant DNA work and produce material of higher specific activity. The most widespread method of labelling double-stranded DNA is by the method of *'nick translation'* using a(^{33}P)dNTPs [46]. In this method nicks are introduced into the DNA with DNase and the 5'-phosphate ends produced can serve as substrates for the 5′ to 3′ exonucleolytic activity of *E. coli* DNA polymerase I, which at the same time repairs the gaps by addition of the a(^{33}P)dNTPs to the 3'-OH end of the 'nick'.

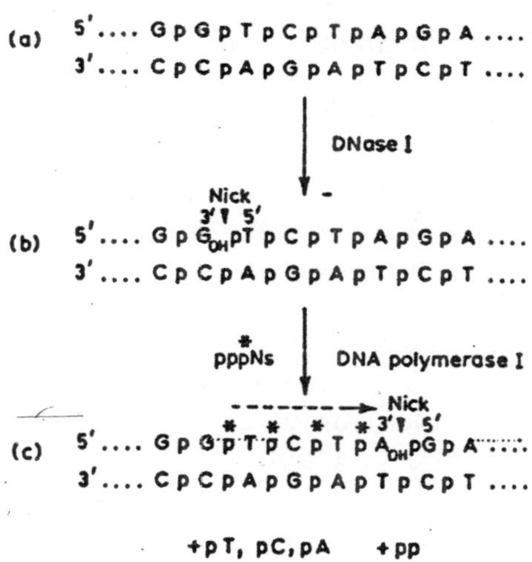

Fig. 8.7. Nick translation. Double-stranded DNA(a) is treated with a low concentration of pancreatic DNase producing occasional nicks with 3'-OH groups (*b*) on which DNA polymerase can commence polymerization. The use of a (^{32}P)dNTPs produces a radioactive phosphodiester bond p and the 5′ → 3′ exonuclease action of the enzyme successively cleaves further non-radioactive phosphodiester bonds which are replaced by radioactive ones (*c*). In the course of this replacement the position of the nick undergoes vectorial translation, as indicated.

Another replacement synthesis method involving T4 DNA polymerase has certain advantages over nick-translation, but is less widely used. Single-stranded labelled DNA probes may be prepared by copying the DNA after cloning into the single-stranded phage vector M13. More recently a method of producing highly radioactively-labelled RNA copies of a piece of DNA has been introduced, which involves transcription with bacteriophage SP6 RNA polymerase from the DNA cloned into a special vector that puts it under the control of a bacteriophage SP6 promoter. Another method of producing very high specific activity DNA probes also involves copying but does not require preliminary subcloning.

The Klenow fragment of DNA polymerase is employed and random oligonucleotides are used as primers with the denatured DNA as template. It is possible to label mRNA directly *in vitro* by certain end-labelling methods but a copying method is generally preferred because it produces a more stable product of higher specific activity. This involves the use of α(^{32}P)dNTPs and reverse-transcriptase to produce a (^{32}P)-labelled single-stranded cDNA copy. The labelling of tRNA *in vitro* is important in sequence determination in cases where insufficient material is available for spectral analysis and it is not possible to label with ^{32}P *in vivo*. As well as enzymic end-labelling methods (sge-betew) chemical labelling using (^{3}H)borohydride has been employed.

End-labelling Methods

These are particularly important in preparing DNA for sequence analysis by the method of Maxam and Gilbert and for certain other analytical purposes. For these purposes the label must be covalently incorporated and the products must be homogeneous. The DNA to be labelled is generated by cleavage with a restriction endonuclease and the method of labelling depends on whether flush ends, or 'sticky' ends (in which there is a 5′-P or 3′-OH overhang) are produced. Ends with a 5′-P overhang are easily labelled by either of two methods, in one of which the 5′-P is removed by alkaline phosphatase and replaced by the γ-phosphate of γ(^{32}P)ATP in a reaction catalysed by bacteriophage-T4 polynucleotide kinase.

In the other the Klenow fragment of *E. coli* DNA polymerase I is usually used to fill the gap in the strand with the recessed 3′-OH end, using an appropriate α(^{32}P)dNTP, although bacteriophage T4 DNA polymerase may also be used. (The 5′ to 3′ exonuclease of the complete *E. coli* enzyme would result in degradation of the template for this reaction.) Thus these two methods complement each other in labelling different strands. The polynucleotide kinase method may be adapted for labelling the 5′-P of flush ends by effecting local denaturation of the latter, but cannot be applied efficiently where there is a 3′-OH overhang.

The 3′ to 5′ exonuclease and 5′ to 3′ polymerase activites of the Klenow fragment of DNA polymerase I can be used to lable flush ends or 3′-OH protruding ends by a replacement synthesis method, although bacteriophage T4 DNA polymerase (which has a more powerful 3' to 5′ exonuclease activity and also no 5′ to 3′ exonuclease) is more effective. Currently, the most effective method of end-labelling 3′-OH overhangs for sequencing involves the use of terminal transferase, but with α(^{33}P)2′,3′dideoxyATP to ensure addition of only a single nucleotide. A number of methods for end-labelling are available when the nucleic acid is only to be used as a probe.

With DNA, extension of 3′-overhanging ends using terminal transferase and α(^{32}P)dNTP is an alternative to the use of DNA polymerase. For mRNA, extension of the 3′-end with *E. coli* poly (A) polymerase can be used for labelling with α(^{32}P)ATP, or α(^{32}P) cordecypin (3′-deoxy ATP) if the addition of only a single nucleotide is required. An alternative to covalent incorporation as a method of end-labelling, useful in certain restricted circumstances, is hybridization of a labelled fragment to suitable single-stranded regions. In the case of eukaryotic mRNAs, (^{3}H)poly(U) can be hybridized to the poly(A) tails.

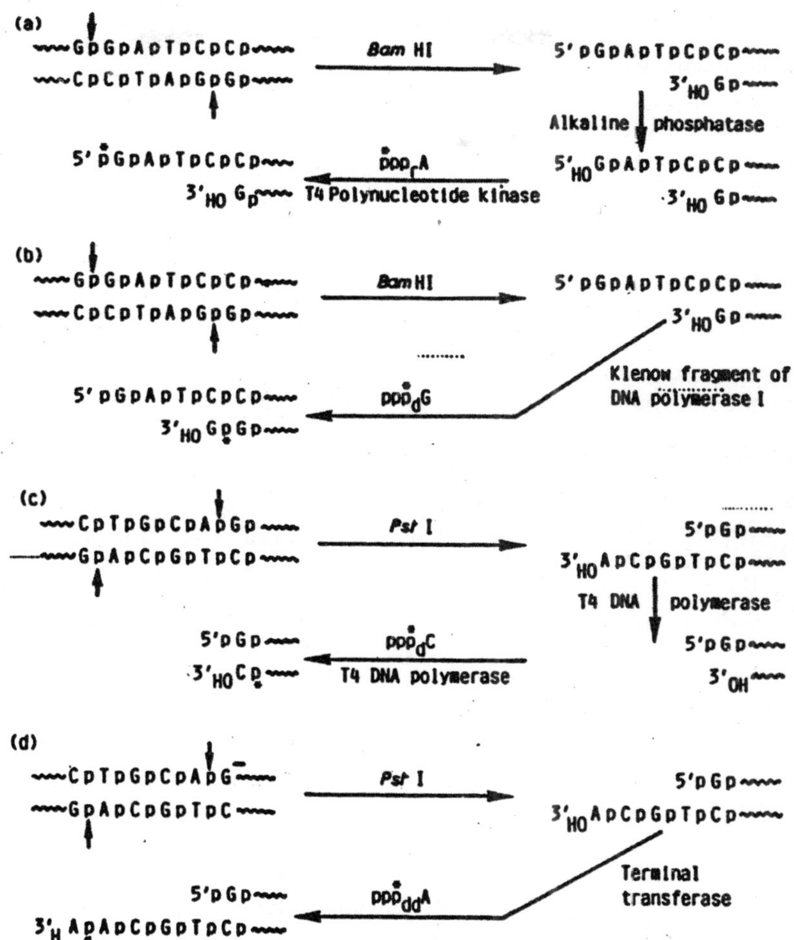

Fig. 8.8. Methods of end-labelling DNA fragments, (a) Use of polynucleotide kinase and γ(^{32}P) ATP for 5′-labelling of cohesive end with 5′overhang produced by restriction endonuclease such as *Bam* HI; (b) use of Klenow fragment of DNA polymerase I and α(^{32}P)dNTPs for 3′-labelling of cohesive end with 5′-overhang produced by restriction endonuclease such as *Bam* HI. In this case any of the α(^{32}P)dNTPs could have been used for labelling if the other non-radioactive dNTPs had been included to allow complete fill-in; (c) use of T4 DNA polymerase and α(^{32}P)dNTPs for 3′-labelling of cohesive ends with 3′-overhang produced by restriction endonuclease such as *Pst*/I. An excess of the other three non- radioactive dNTPs prevents the exonuclease activity proceeding further; (d) use of terminal transferase and α(^{32}P)ddATP for 3′-labelling of cohesive ends with 3′-overhang produced by restriction endonuclease such as *Pal.* Methods (a), (c) and (d) may also be applied to blunt ends produced by restriction endonucleases such as *Sma* I, and method (d) can be applied to ends with a 5′-overhang. N.B. The other end of the fragment (not illustrated) will also be labelled if it is similar, and a second restriction endonuclease digestion or strand-separation will be required to obtain a fragment of DNA labelled at only one end.

DETERMINATION OF NUCLEIC ACID SEQUENCES

Determination of DNA Sequences

There are two main methods currently in use for sequencing-large fragments of DNA. These methods (that of Maxam and Gilbert and that of Sanger) both involve the generation of a ladder of fragments of different sizes but with one common end. The basis of these methods is to generate specific sets of radioactively labelled fragments, each set terminating at a particular base (in the ideal case), and the use of high-resolution polyacrylamide gels allows fragments differing by only a single nucleotide to be resolved and the sequence to be deduced. In both cases cloned DNA is normally used, although the direct application of the Maxam and Gilbert method to genomic DNA has been described.

A third method of sequencing (the 'Wandering Spot' or 'Mobility Shift' method), although no longer used for large DNA molecules, still finds application to small oligonucleotides (especially those produced synthetically, which are difficult to sequence by the other methods. An outline of earlier methods, now superseded, can be found in.

(a) The Method of Maxam and Gilbert

The essence of this method is the chemical fragmentation of either single- or double-stranded DNA by base-specific reactions. The reactions most commonly used are those absolutely specific for guanine residues and for cytosine residues respectively, and those that are specific only for purines or pyrimidines, respectively. Cleavage at guanine residues is effected by methylation with dimethyl sulphate at the N7 position, leading to instability of the glycosidic linkage which is then hydrolysed by piperidine, followed by β-elimination of both phosphates from the sugar. Purine nucleotide linkages are hydrolysed with acid, again followed by piperidine treatment.

Pyrimidine residues are hydrolysed by hydrazine a reaction which, in the case of thymine, can be inhibited by 2 M NaCl, thus allowing specific cleavage at cytosine residues. (Under the conditions normally used, cleavage does not occur at 5-methylcytosine residues, causing a gap in the ladder.) Conditions of chemical cleavage are generally adjusted to try to obtain a single scission per DNA molecule. Even so, each scission would produce two fragments. In order to visualize on polyacrylamide gels only fragments of increasing length emanating from one end of the DNA, a single end of the DNA must be radioactively labelled using one of the methods described. Such methods, in fact, generally label both ends of a piece of duplex DNA, and fragments with a single-labelled end are generated by restriction endonuclease digestion of this labelled DNA followed by separation on polyacrylamide gels.

Less commonly, single end-labelled molecules are obtained by strand separation of the DNA. As sequence data accumulate, the use of computer programs to handle them and identify restriction endonuclease sites greatly facilitates sequencing by the Maxam-Gilbert method.

(b) The Sanger Dideoxy Method

The essence of this method is the primed synthesis of partial copies of the DNA to be sequenced, with random base-specific premature termination of the copying producing ladders of different length fragments terminating in each of the four different bases. The copying is catalysed by the Klenow fragment of *E. coli* DNA polymerase I, which requires a primer and that the DNA to be copied be in the single-stranded form. (The Klenow fragment, lacking the 5′ to 3′ exonuclease, is used to prevent attack on the 5′-end of the primer.

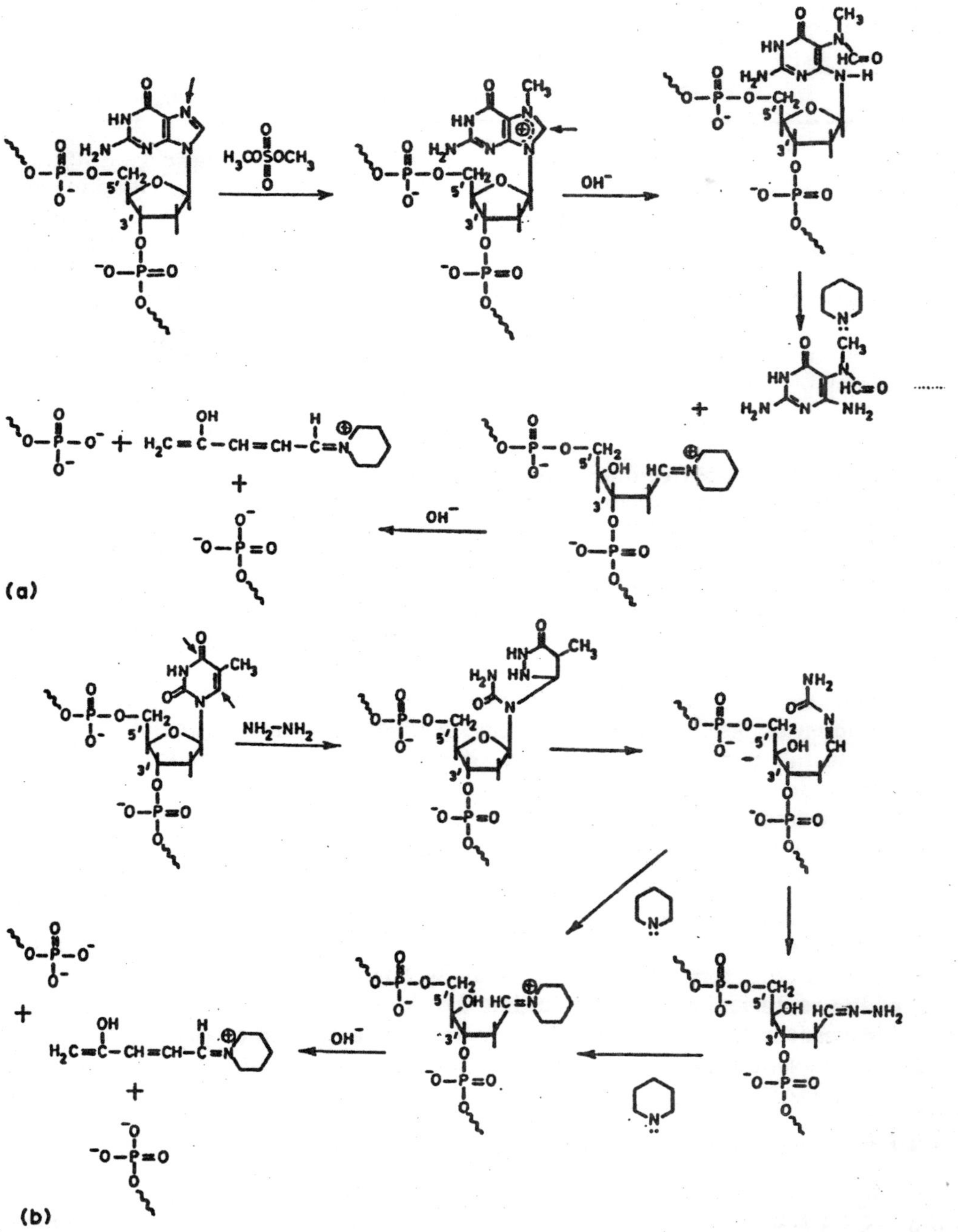

(a)

(b)

Fig. 8.9. Base-specific chemical cleavage reactions used in the sequencing method of Maxam and Gilbert. (*a*) Dimethyl sulphate reaction for guanine residues; (*b*) Hydrazine reaction for pyrimidine residues (thymine illustrated).

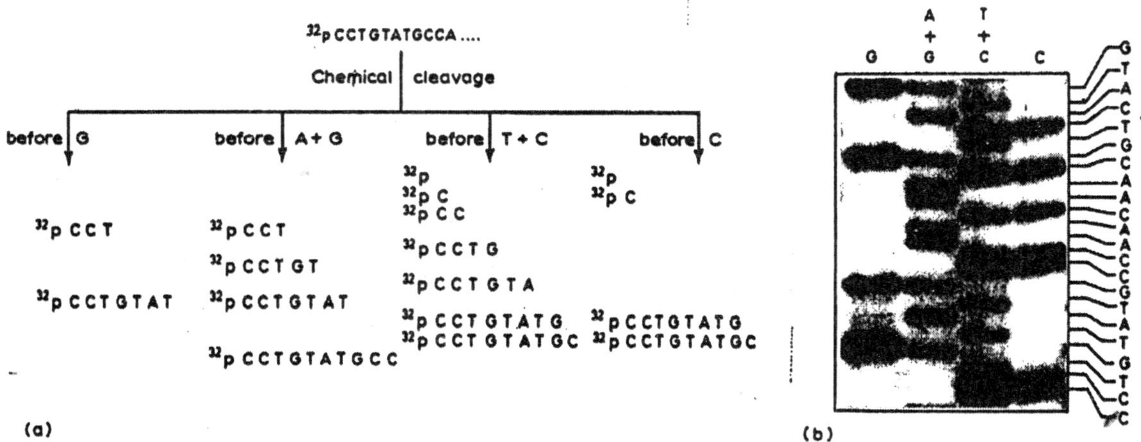

Fig. 8.10. Example of DNA sequence determined by the method of Maxam and Gilbert. (*a*) The ^{32}P-end-labelled fragments derived from chemical cleavage of 5′-end-labelled DNA. These are set out in order of increasing size to allow comparison with (*b*) the autoradiograph of the part of the gel on which these (and longer fragments) were separated on the basis of size. It can be seen that bands appearing in both G and G + A tracks are assigned to G whereas those appearing only in the G + A track are assigned to A. C and T are similarly differentiated. (This is, in fact, a simplified example, and the extreme 5′-fragments are normally quite difficult ot read.)

The copies are made radioactive by inclusion of an α(^{32}P)dNTP or a α(^{35}S)dNTP in the reactions, and base-specific chain termination is effected by the addition of the appropriate. 2′ 3′dideoxynucleotide triphosphate (ddNTP), which, lacking a 3′-OH, cannot be extended further. Initially, the widespread application of this method was prevented by the requirement for a single-stranded template and the need for a separate oligonucleotide primer for each piece of DNA to be sequenced. However, these limitations were overcome when the single-stranded bacteriophage, M13, was adapted for use as a sequencing vector.

The DNA to be sequenced may now be subcloned into one of a variety of adjacent sites in the double-stranded replicative form of the bacteriophage vector, and can be sequenced in the easily prepared single-stranded DNA using oligonucleotide primers complementary to regions of the bacteriophage DNA flanking the cloning sites. This development has led to a dramatic increase in the use of the Sanger method. The random subcloning of fragments from the DNA to be sequenced has resulted in the need for computer programs to order the sequences obtained.

(c) The Wandering-spot Method

The Maxam-Gilbert and Sanger methods described above, although extremely powerful for large restriction fragments of DNA, are not well suited to smaller oligodeoxynucleotides. The sequence of these can be determined following end-labelling by the chromatographic separation of a non-base-specific partial enzymic digest of the oligodeoxynucleotide in which all the partial digestion products having the same labelled end are represented. Digestion is by venom or spleen phosphodiesterase and two-dimensional thin layer homochromaography on DEAE-cellulose is employed. The shortening of the oligodeoxynucleotide by a single nucleotide causes a change in mobility that is characteristic of the nucleotide removed and which allows the sequence to be read.

Determination of RNA Sequences

Before the advent of rapid DNA sequencing methods considerable effort was expended on developing techniques for sequencing RNA. The availability of partially or totally base-specific endonucleases allowed employment of approaches formally rather similar to those used for sequencing proteins, but requiring different methods for separating and analysing the oligo-ribonucleotides. Although the sequencing of a DNA copy must now be the method of choice for sequencing any RNA, this approach may not be applicable to small RNAs or oligonucleotides, and in any case does not identify modified nucleotides. In these cases (especially for tRNAs) direct methods (including the 'wandering-spot' method, mentioned above) are still employed. Rapid sequencing methods involving basespecific cleavage and separation of fragments on polyacrylamide gels have been introduced for 5′-end-labelled RNA.

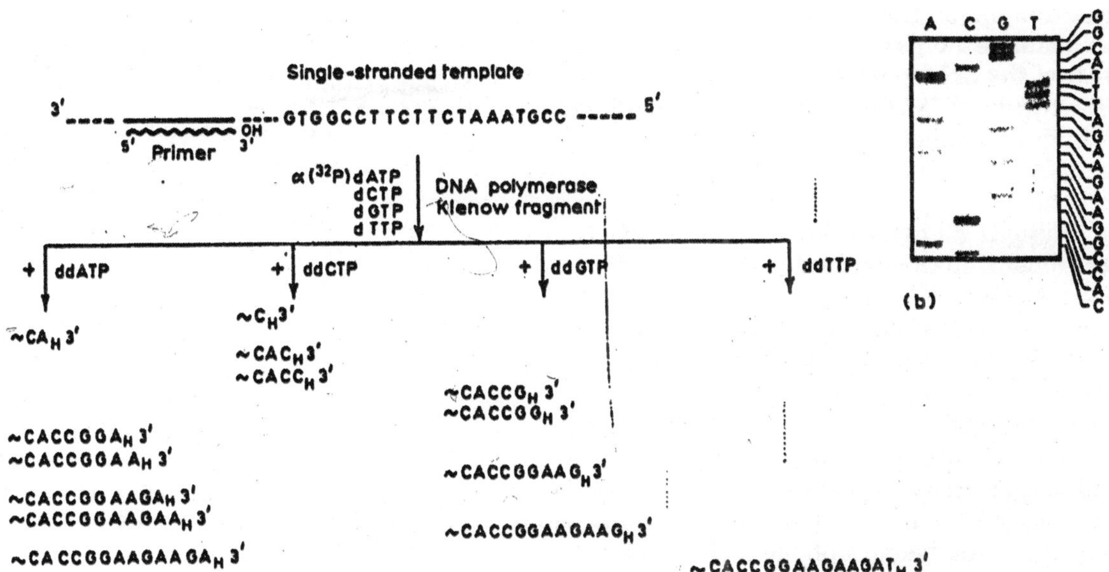

Fig. 8.11. Example of DNA sequence determined by the dideoxy method of Sanger. (*a*) The various products of extension of the primer (~) by DNA polymerase are set out in order of increasing size. The terminating ddNTP is represented as N_H. Chains are labeled internally by the incorporation of the α(^{32}P)-or α(^{35}S)dATP; (*b*) autoradiograph of the part of the gel on which these (and longer) species were separated on the basis of size. Note that the sequence obtained is the complement of the original priming strand.

The cleavage is effected enzymically with ribonucleases the following reactions being currently in use:

Ribonuclease T_1 : cleaves 3′ of G

Ribonuclease U_2 : cleaves 3′ of A
Ribonuclease Phy M : cleaves 3′ of A and U
B. *cereus* ribonuclease : cleaves 3′ of U and C

In other respects the method and the interpretation of the results are formally similar to that of Maxam and Gilbert.

RESTRICTION MAPPING OF DNA

When studying specific regions of genomic or cloned DNA it is necessary to have some sort of 'map' to differentiate one area from another. There are two different types of map. A *genetic map* can be constructed (in suitable organisms) in which the positions of functional genes contributing to particular phenotypes can be interrelated by studying the frequency of genetic recombination between normal and mutated alleles. A *physical map*, on the other hand; can be constructed in the absence of genetic data or knowledge of any function the DNA might have, and uses purely physical techniques. These include the formation of heteroduplexes between standard DNA molecules and the DNA under investigation, and restriction mapping, the subject of this section. A restriction map indicates the positions along a piece of DNA at which there are recognition sites for particular type II restriction endonucleases.

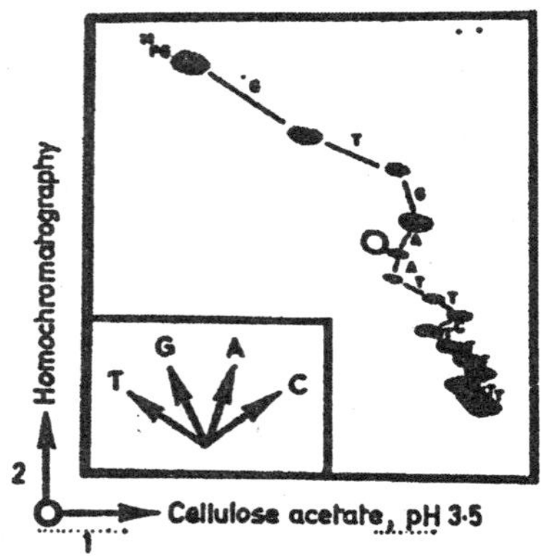

Fig. 8.12. Example of oligodeoxynucleotide sequence determined by the 'wandering spot' method. A partial pancreatic DNase digest of GGTGAATTCTTTCTT labelled with ^{32}P at its 5′-end was subjected to two-dimensional separation. The inset shows the effect of removal of the four different 3′-end deoxynucleotides on the mobility of the resulting fragment. The identity of the 5′-nucleotide is determined by comparison with standards. The circle adjacent to the second A indicates the position of a marker dye. Adapted from, with permission.

The method in its simplest form involves digesting the DNA separately and in combination with different restriction endonucleases, separating the resulting fragments by agarose (or sometimes polyacrylamide) gel electrophoresis, visualizing these under ultraviolet lignt after staining with ethidium bromide, and estimating their size in relation to standards. If there are not too many recognition sites for the enzymes it is possible to deduce the position of these. It can be appreciated that for enzymes with many recognition sites it is both difficult and tedious to construct a restriction map by this procedure.

In the case of large pieces of DNA (*e.g.*, those cloned in bacteriophage lambda or cosmid vectors most restriction endonucleases will have many recognition sites and it is difficult to circumvent this problem by choice of enzyme. In any case it is often useful to have a restriction map for enzymes that cleave a piece of DNA at many .positions. An alternative approach was devised involving end-labelling of the DNA and *partial* digestion with *single* restriction endonucleases. In essence the strategy is similar to that for Maxam and Gilbert DNA sequencing using a single base-specific reaction, except that analysis of the larger fragments usually employs the standard agarose gels. Although the pattern of stained fragments is complex, the autoradiograph of the dried gel reveals a ladder of fragments, the increasing sizes of which represent increasing distances from the labeled end.

A modification of this strategy, for application to DNA cloned into bacteriophage or cosmid vectors, replaces end-labelling by hybridization of a specific (^{32}P)-labelled oligonucleotide to either the right or left cohesive end of the molecule. This avoids the need for secondary cleavage and gel fractionation to produce a fragment of DN A labelled at one end.

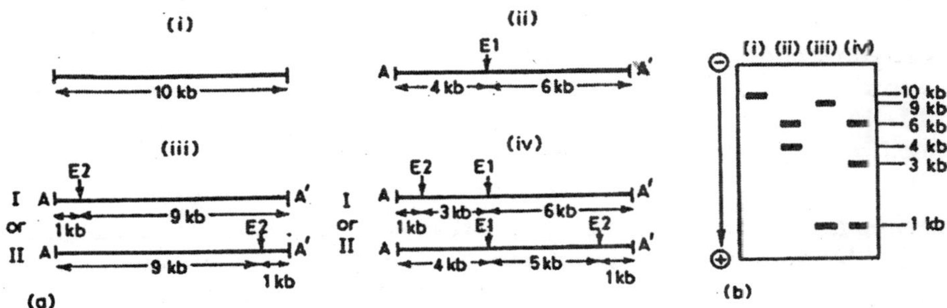

Fig. 8.13. Example of restriction mapping. It is desired to determine the relative positions of the recognition sites for two restriction endonucleases, El and E2, along a hypothetical 10 kb fragment of DNA. (*a*) If the 10 kb fragment (*i*) is cleaved by El to give 4 kb and 6 kb fragments thė two ends of the DNA can be distinguished as A and A' (*ii*). This then defines two possible positions of cleavage (I and II) of a second enzyme E2, which generates 1 kb and 9 kb fragments (*iii*). The two alternatives I and II may be distinguished by the different sized fragments they predict in the case of double-digestion with E1 and E2 (*iv*); (*b*) diagram of the results of agarose gel electrophoresis of (*i*)-(*iv*). The result in lane (*iv*) is only consistent with alternative I, and hence (*iv*)/I represents an (albeit limited) restriction map of the 10 kb fragment.

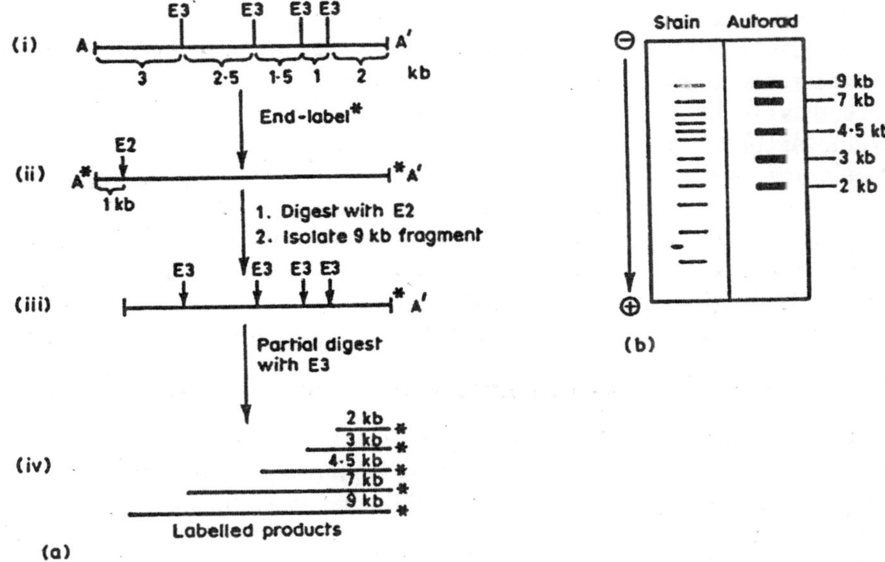

Fig. 8.14. Example of partial digestion restriction mapping. (*a*)(*i*) The objective is to determine the relative positions of the four recognition sites for the restriction endonuclease E3 on the hypothetical DNA fragment. This is end-labelled (*ii*), digested with enzyme E2, which is known to cleave close to one end, and the Targer (9kb) fragment isolated (*iii*). This is subjected to partial digestion with E3 giving a mixture of end-labelled fragments together with non-labelled fragments not illustrated (*iv*); (*b*) diagram of the results of electrophoresis of the partial digestion products. The distances of the E3 recognition sites from end A' can be read off the autoradiography ladder using molecular size markers. This allows the restriction map of (*a*)(*i*) to be deduced.

CLONING DNA

The Principles

The study of individual mRNAs and of particular regions of genomic DNA were hampered historically by two problems. The first was the difficulty of devising chemical methods to isolate one particular nucleic acid species from, perhaps, tens of thousands of chemically similar distinct species. The second was that, even if a chemical separation method had been found, the amount of nucleic acid isolated would in most cases have been too small to allow study. The cloning of DNA employs a biological, rather than a chemical, strategy to overcome both these problems. If one wishes to isolate a single bacterium from a million others one spreads them out on an agar plate so thinly that each is separated from its neighbour.

Each bacterium gives rise to a visible colony which is, in fact, composed of identical bacteria derived from the division of the original one. Such an assembly of genetically identical individuals is called a *clone*. If the bacterium that gives rise to that clone harbours a particular mutation that distinguishes it from the wild type one could say that one had cloned the mutant DNA. Even though this mutant DNA originally existed as a single chromosome in minute amount, one could obtain large quantities of the DNA by using the colony to seed a bulk liquid culture. The basis of the purification and isolation of an mRNA or a chromosomal DNA segment using a biological strategy is to introduce them into individual bacteria so that they may be cloned in an analogous way to the hypothetical mutant chromosomal PNA described above. There are several problems to be overcome before this strategy can be realized in practice.

The foreign DNA must be presented as part of a molecule that can replicate along with the bacterial (or other host) chromosome. Such vehicles for introducing the DNA are called *vectors* and the chimeric DNA produced by the insertion of the DNA to be cloned is a *DNA recombinant*. The DNA must be inserted into the vector, or the mRNA be converted into a form in which it can be inserted, and the vector must be introduced into the bacterium, the process of *transformation*. Methods must be available to distinguish the transformed bacteria from the untransformed ones, and bacteria containing recombinant DNAs from those containing non-recombinant DNAs from those containing non-recombinant vector (*selection*). Finally it is necessary to be able to identify which one (if any) of thousands of clones contains the particular DNA of interest *(screening)*. The ways in which these problems may be overcome are illustrated below, almost exclusively in relation to *E. coli* as a biological host. For details of cloning in other bacteria, or in eukaryotic animal or plant cells, the reader is directed elsewhere.

The Construction of Recombinants and their Introduction into Bacteria

As already stated, the cloning of a piece of foreign DNA in *E. coli* requires its introduction into a suitable vector. There are three main types of vector in use for primary cloning-plasmids, bacteriophage lambda and cosmids-and these will be discussed in turn, together with the way that DNA for cloning into them is prepared. It is worth mentioning at the outset that in all these vectors foreign DNA is inserted at specific sites by the use of the highly-specific restriction endonucleasesr Although it is possible in theory to clone foreign DNA into vectors using cleavage by non-specific endonucleases such as DNase I, in actual practice it was the discovery of restriction endonucleases that made it practicable both to clone and subsequently analyse DNA. It was not only the specificity of restriction endonucleases that was important, but the fact that many of them give rise to DNA fragments with cohesive ends that can be far more easily joined with DNA ligase than can fragments with blunt ends.

Cloning in Plasmid Vectors and Bacteriophage M13

Plasmids are self-replicating double-stranded circular DNA molecules found in certain bacteria. Especially useful are those that occur in multiple copies per cell. Natural plasmids have been extensively modified to serve as cloning vehicles, the most well known of which is pBR322. The essential features are an origin of replication, unique restriction endonuclease sites into which foreign DNA may be inserted without disabling the plasmid, and selectable markers for the presence of the plasmid in a bacterial cell. These latter are genes endowing resistance to the antibiotics tetra-cycline and ampicillin (a derivative of penicillin). For clarity they are abbreviated as Tc^R and Ap^R although the Ap^R gene is referred to by its correct genetic designation *(bla)*. In the simplest case a fragment of foreign DNA generated from a larger molecule by digestion with a restriction endonuclease (*e.g. Bam* HI) is ligated with DNA ligase to pBR322 DNA that has been digested with the same enzyme, generating the chimeric molecule. To enable the plasmid DNA to enter the *E. coli* cells (*transformation*) these are pretreated with $CaCl_2$ which, together with a short heat shock, appears to act by altering the structure of the cell wall. Transformed cells are selected by plating the bacteria on agar containing the appropriate antibiotic (ampicillin in this case) which will prevent the growth of cells that do not contain any plasmid.

Although re-ligation of the linearized plasmid to itself (but not to foreign DNA) can be prevented by pretreating it with alkaline phosphatase, in many cases some of the transformed bacteria will contain plasmid with no insert. Colonies of such bacteria may be distinguished by the fact that they will also grow on tetracycline, unlike those containing foreign DNA inserted in the *Sam* HI site. (A 'replica' plate of the bacterial colonies is prepared for testing on tetracycline; the colonies of interest corresponding to those that will not grow on tetracycline being maintained on the ampicillin 'master' plate.

If, however, it had been necessary to clone into a restriction site such as *Eco* RI, which is not located in an antibiotic-resistance gene, the presence of inserts can only be confirmed by physical analysis of the plasmid DNA. The problem of identifying transformed bacteria containing plasmids with inserted DNA is overcome in the more recently developed pUC vectors, which also have the advantage of containing a large number of useful cloning sites. These vectors contain the ampicillin-resistance gene of pBR322 to allow selection of transformants, in addition to the *lac* operator and promoter regulatory regions and the N-terminal portion (α-fragment) of the β-galactosidase gene.

If the plasmid *lac* gene is induced with IPTG (isopropylthio-β-galactoside) in an *E. coli* strain with a β-galactosidase gene lacking the N-terminal portion (actually carried on the F'episome), complementation of the two peptides will occur giving enzymically-active β-galactosidase. This can be detected by the blue colour produced when 'X-gal' (5-bromo-4-chloro-3-indolyl-β-D-galactoside) is present as a substrate. In the pUC vectors the extreme N-terminal region of the α-fragment is interrupted by a region of multiple cloning sites. This does not alter the reading frame or result in inactivation of the α-fragment of β-galactosidase. However, in most cases the insertion of foreign DNA into these sites will cause inactivation (so-called 'insertional inactivation').

Thus if the transformed bacteria are plated on agar containing IPTG and X-gal (as well as ampicillin) the white colonies will generally contain plasmid with inserts, whereas the blue colonies will generally not. In Fig. 8.17 illustrated the cloning vector M13mp18, a member of the family of M13mp vectors, for which the *lac* fragment of the pUC vectors was initially engineered.

Although this is technically a single-stranded phage vector, the replicative form illustrated is formally analogous to a plasmid (the ability to form 'plaques' replaces the ampicillin resistance). It can be used in a similar way, although, as already mentioned, the main purpose of cloning DNA fragments into M13 vectors is to facilitate the determination of their sequences by the Sanger dideoxy method. This is possible because the single-stranded form of the recombinant can easily be generated and, as cloning with different enzymes is into the same region, the same primers (corresponding to flanking regions of the *lac* Z′ gene) can be used with different recombinants. One of the major uses of plasmid vectors has been for *cloning cDNA* - DNA copies of mRNA.

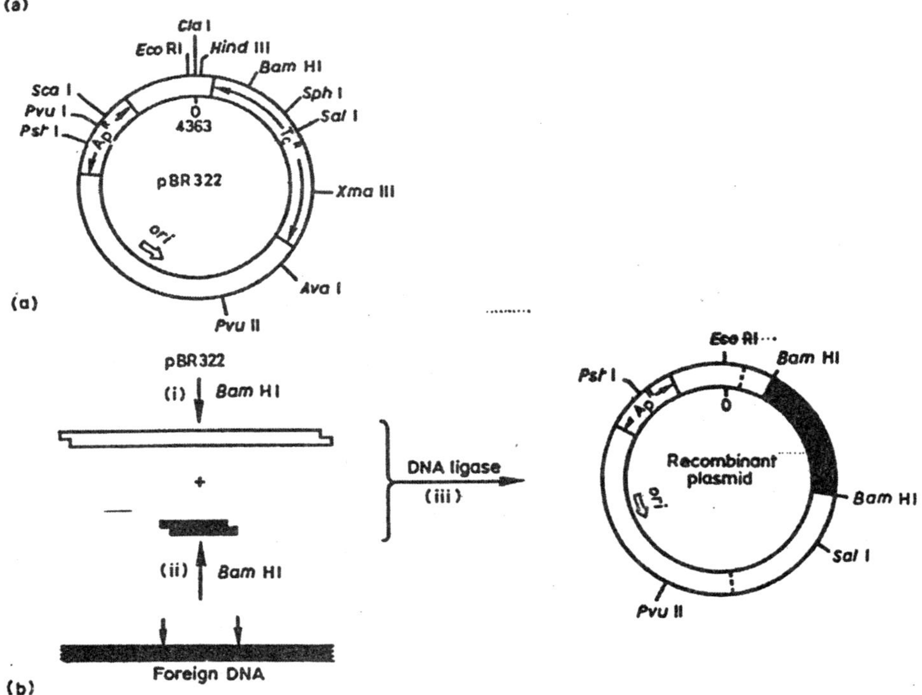

Fig. 8.15. Cloning cDN A. Poly(A)$^+$ mRN A is primed with oligo(dT) (*i*) and a cDNA copy synthesized (*ii*). After and tetracycline resistance (Tc^R), the origin of replication (*ori*), and the positions of some of the restriction endonucleases that have a single recognition site in this DNA: (*b*) to clone into the *Bam* HI site of pBR322 the DNA is digested with this enzyme (*i*), as is the foreign DNA (*ii*), and these are ligated together (*iii*), among the products of the ligation being the desired recombinant plasmid shown. Note that although this latter still retains the gene for ampicillin resistance, the gene for tetracycline resistance has been inactivated by the insertion of the foreign DNA.

We shall therefore now briefly describe this procedure as it applies to polyA-containing (poly(A)$^+$) mRNA. The poly(A)$^+$ mRNA is usually purified on an affinity column of oligo(dT)-cellulose and reverse-transcribed using retroviral reverse transcriptase with oligo(dT) hybridized to the poly(A) tail of the mRNA as a primer. This results in a single-stranded cDNA, complementary to the mRNA. For a number of years the usual procedure for synthesizing the second strand relied on the self-priming that occurs when (after removing the mRNA by alkali or heat denaturation) the single-stranded cDNA folds back on itself. *E. coli* DNA polymerase I was generally used to generate a double-stranded cDNA by copying the first strand, and S1 nuclease had to be used to destroy the loop and digest any protruding ends on the double-stranded cDN. The ends of such double-stranded DNA must be modified so that they can anneal to the termini of a plasmid linearized by restriction endonuclease digestion.

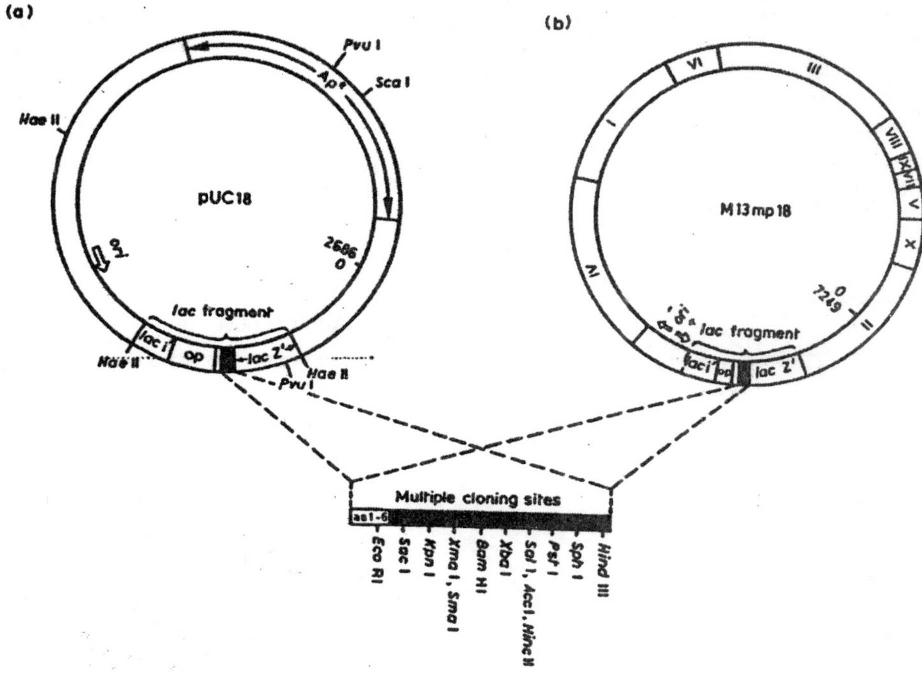

Fig. 8.16. pUCIS and M13mpl8. (*a*) The plasmid vector pUC18. This is derived from a fragment of pBR322 into which a fragment of the *lac* gene has been inserted at *Hae* II sites. The multiple cloning sequence interrupts the *lac Z'* (truncated β-galactosidase) gene. The *Pvu* I and *Sea* I sites of the original Ap^R gene areshown, but note that the Pit I site has been removed so that there is only a single *Pst* I site, as for the other sites in the multiple cloning region; (*b*) the double-stranded replicative form of single-stranded phage vector M13mpl8. The *lac* fragment has been inserted into the phage M13, the genes of which are indicated I-X.

One way of achieving this is to 'blunt-end' ligate 'linkers' to them using DNA ligase. Such linkers are double-stranded oligodeoxynucleotides containing a restriction endonuclease site which is cleaved after the ligation. This generates 'cohesive ends' that may be specifically ligated to the corresponding cohesive ends in the digested plasmid. (This is more efficient than blunt-end ligation directly into a blunt-ended site. Blunt-end ligation is much less efficient *per se* than ligation of fragments with cohesive ends. However, the high molar concentration of the small linkers can drive this reaction in a manner impossible with the larger molecules.) An alternative strategy-perhaps more widely used has been to add homopolymeric tails of complementary oligonucleotides to the 3′OHs in both cDNA and inearized plasmid using the enzyme terminal transferase.

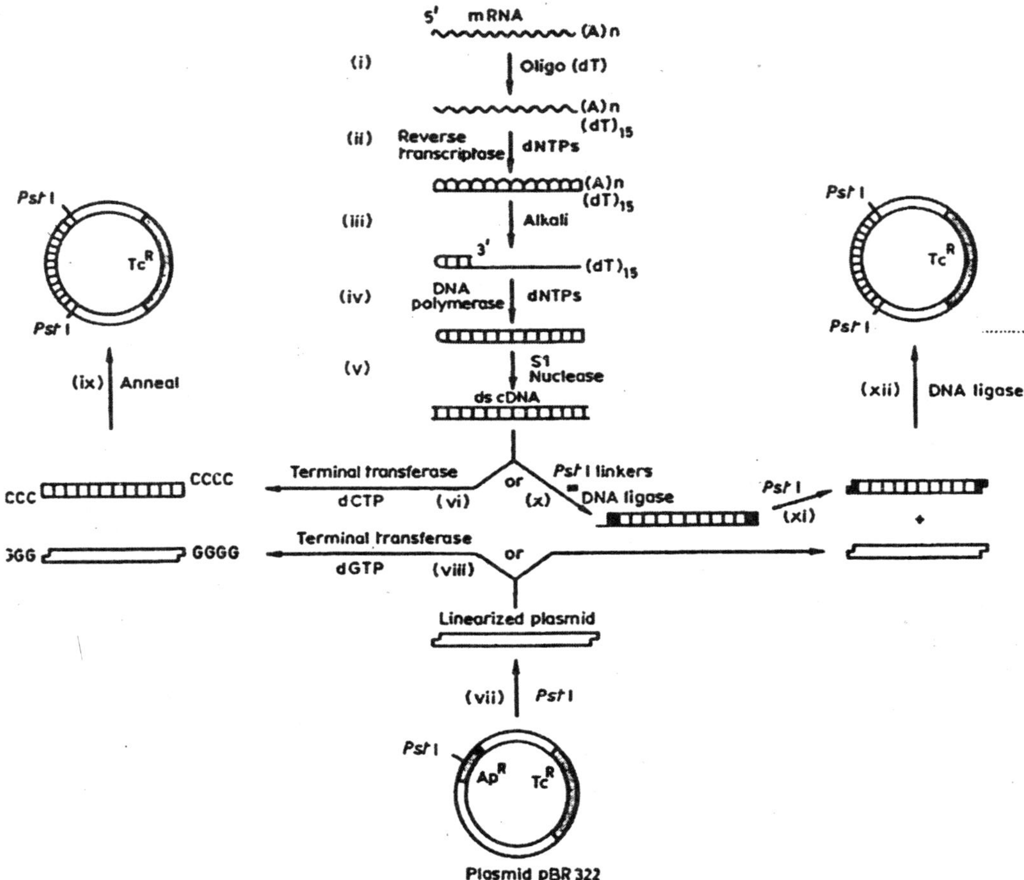

Fig. 8.17. Cloning cDNA. Poly(A)⁺ mRNA is primted with oligo(dT) *(i)* and a cDNA copy synthesized *(ii)*. After removing the mRNA *(iii)* this self-primes second-strand synthesis *(iv)* and the loop and any ragged ends are then removed with nuclease S1*(v)*. One strategy is now to tail the double-stranded cDNA with dCTP using terminal transferase *(vi)*, add this to linearized plasmid *(vii)* that has been tailed with dG *(viii)* and allow these to anneal *(ix)* to form the recombinant. Alternatively, linkers can be ligated onto the cDNA *(x)*, these cleaved with the appropriate restriction endonuclease *(xi)* and then ligated *(xii)* to the plasmid linearized with the same enzyme. Other linkers may be used for step *(x)* but for tailing *(vi)* and *(viii)* a site such as *Pst* I, that on cleavage gives the necessary 3′-OH overhang for terminal transferase is required.

The cDNA and vector are 'tailed' with complementary oligonucleotides (generally C and G), and, after allowing these to anneal to one another, the hybrid can be used to transform *E. coli* without the need for prior ligation (which is subsequently effected by bacterial DNA repair mechanisms). *PstI* is usually used as the cloning site in this strategy: not only does it generate the necessary 3′OH overhang for terminal transferase, but its recognition site and cleavage position (CTGCA′G) are such that 'tailing' with G will regenerate it. The use of S1 nuclease in cDNA cloning has the disadvantage that it destroys the extreme 5′ end of the cDNA.

For this reason it is increasingly being superseded by methods that allow production of full-length cDNA copies of mRNAs. The problem is clearly how to prime the second strand. One solution has been to 'tail' the single-stranded cDNA with oligo(dC), allowing the second strand to be primed with oltgo(dG). However, a more sophisticated solution, with an additional advantage of producing full-length cDNA with high efficiency, has been devised by Okayama and Berg. The basic strategy is to use a plasmid vector tailed with oligo(dT) at one end as primer for the first cDNA strand, and, after tailing the free 5′ end of the cDNA with oligo(dC), replacing the other end of the vector (which was also tailed with C) with an adaptor DNA previously tailed with oligo(dG) to allow completion of the second strand.

The interested reader should consult the original reference for the full details of how this is achieved in practice. However, one aspect of this rather sophisticated method has been adapted to allow almost complete second-strand copying of a simple complex of mRNA and first strand cDNA. This is the use of *E. coli* RNase H to nick the mRNA in the hybrid, and DNA polymerase (the Klenow fragment or the enzyme from bacteriophage T4) to perform replacement synthesis of DNA from the 3′OHs generated. Finally, it should be mentioned that cDNA copies of poly(A)$^-$ mRNAs can be obtained by polyadenylating with *E. coli* poly(A) polymerase, or, more usually, by priming with a random mixture of oligonucleotides (cf. the use of oligonucleotides to prime labelling although in this case DNA corresponding to the 3′ end of the mRNA may be lacking.

Furthermore, in the method of *primer-extension* a single-stranded oligonucleotide fragment derived from the 5′ end of a cDNA clone that is less than full length may be used as a primer in an attempt to obtain a cDNA clone-corresponding to the area that is lacking (cf. the use of this method in mapping transcripts.

Cloning in Bacteriaphase Lambda and Cosmid Vectors

The plasmid and bacteriophage M13 vectors described above are most suitable for the cloning of relatively small pieces of DNA (usually up to about 3 kb), either as cDNA copies of mRNAs or as portions of larger pieces of cloned DNA (subcloning). However, other vectors are required when the objective is to prepare a eukaryotic genomic library—a collection of clones in which the whole DNA of the genome of a particular eukaryotic organism or cell is represented. The main reason for this is the large size of eukaryotic genomes. For example, to be 99% certain that a library of 15 kb inserts is representative of the 3×10^6 kbp mouse genome, 10^6 clones are required.

The reason for the unsuitability of plasmid vectors for such a library is the limited efficiency of the transformation of calcium-treated cells by DNA. This is true even with inserts of 1 or 2 kb, but the efficiency falls dramatically as the size of insert increases. The use of derivatives of bacteriophage lambda as vectors for constructing genomic libraries rests on the fact that a large portion (approximately 16 kb) of the central region of the 49 kbp bacteriophage genome codes for lysogenic and other functions that are not required for the lytic infective cycle of the bacteriophage. Thus if this region is replaced by foreign DNA the bacteriophage will still be viable. Furthermore, after constructing DNA recombinants of bacteriophage and foreign DNA it is possible to 'package' them into phage particles with high efficiency *in vitro,* using an extract from infected cells containing the necessary enzymes.

Cells may then be infected with the packaged recombinant, resulting in plaques. The size of DNA that can be cloned depends on the construction of the vector (see below), but is governed by the fact that there is a size restriction of 38.5-52 kb for the DNA to be packaged into phage heads. In general, bacteriophage lambda vectors can be classed as 'insertion' or 'replacement' vectors. In *insertion vectors* there is a single unique restriction site in thecentral non-essential region into which the foreign DNA is ligated. The non-essential region is engineered so that the size of the vector DNA is towards the lower limit of packageability, and this type of vector can therefore accept foreign DNA up to a size that will produce a recombinant near the upper limit of packaging.

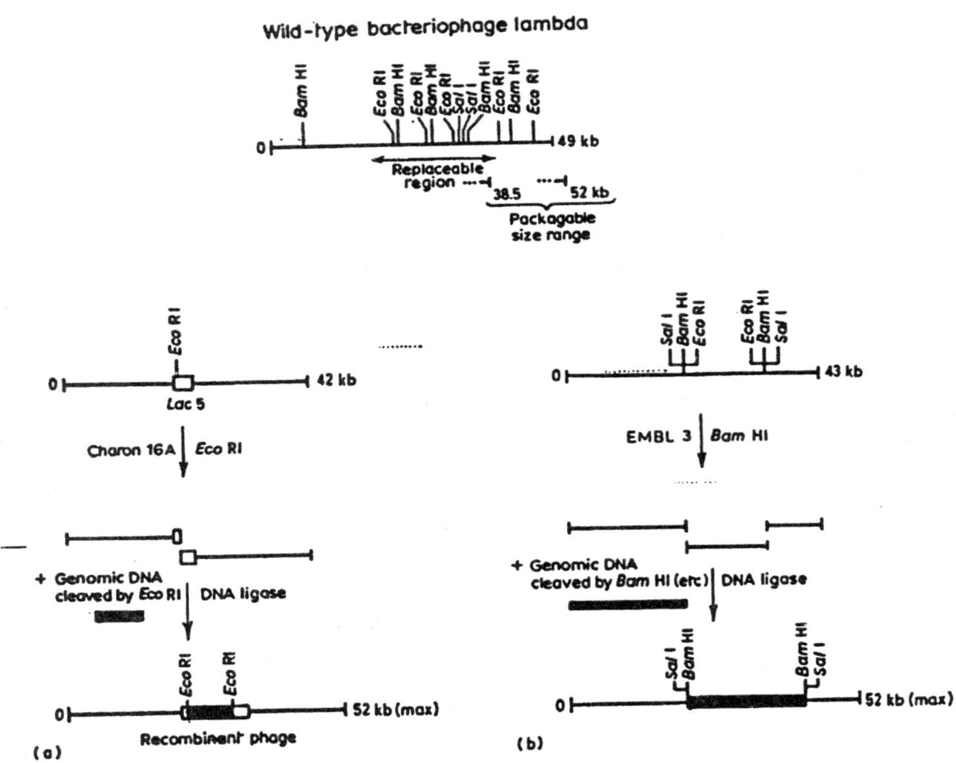

Fig. 8.18. Cloning vectors derived from bacteriophage lambda, (*a*) Charon 16A, an example of an insertion vector. *Lac* 5 is a fragment of the *E. coli lac* region containing all of the β-galactosidase *(lac* Z) gene and associated control elements. Only the restriction site for *Eco* RI is shown; (*b*) EMBL 3, an example of an insertion vector. The central fragment may be digested with *Eco* Rl to prevent it being re-ligated to the '*Bam* arms'. If genomic DNA digested with *Sou* 3A rather than *Bam* HI is used (see text) the recombinant will generally lack the *Bam* HI sites indicated. However, the insert can be recovered using *Sal* I. Only the desired products of ligation are illustrated.

In Charon 16A this is approximately 10 kb. Different vectors employ different genetic devices for identifying recombinants but it will suffice to mention that Charon 16A employs as a cloning site a portion of the *E. coli lac* region containing the complete β-galactosidase gene *(lac* Z) and upstream operator and promoter regulatory elements, allowing colour selection of recombinant phage plaques using X-gal when plated on *E. coli* with a deletion in the *lac Z* gene. Lambda *replacement vectors* are generally employed when it is necessary to clone larger pieces of foreign DNA. An advanced vector of this type, EMBL, is illustrated in Fig. 8.19. This has been engineered so that restriction endonuclease digestion with *Bam* HI, *Eco* Rl or *Sal* I cuts out a non-essential piece of DNA of 14 kb, which can be replaced by foreign DNA (up to about 23 kb) with compatible cohesive ends.

To prevent religation of lambda 'centre' when using replacement vectors, the lambda 'arms' can be purified first. However, vectors such as EMBL 3 allow simpler methods to be employed. It may be mentioned that the compatible cohesive ends generated in the fragments of genomic DNA need not be (and generally are not) produced using the same restriction endonuclease as used in cleaving the vector. Thus, cleavage of DNA with the enzyme *Sou* 3AI (which has the 4-base recognition sequence 'GATC) produces identical cohesive ends to cleavage with *Bam* HI (which has the 6-base recognition sequence G'GATCC).

If high molecular weight chromosomal DNA is subjected to partial digestion with *Sou* 3AI one is more likely to obtain clonable fragments of any particular area of the genome than by complete (or partial) digestion with *Bam* HI, for which there are fewer recognition sites. Because of the large size of the introns in many eukaryotic genes, a particular lambda genomic clone may not contain the whole of a given gene of interest. It is then necessary to perform the operation known as *chromosome walking,* in which a fragment of unique sequence DNA near the extremity of the clone is used as a probe to re-screen the genomic library for an overlapping clone containing the adjoining area of the genome. An alternative to chromosome walking (or a way of speeding it up) is to construct the genomic library in a *cosmid vector* (*e.g.* pHC79) which can accommodate larger pieces of DNA than lambda replacement vectors.

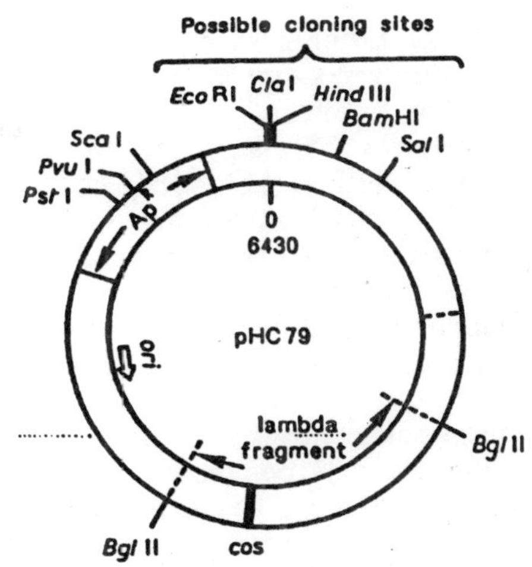

Fig. 8.19. Cosmid vector pHC79. This simple cosmid vector contains the ampicillin resistance gene, origin of replication and useful cloning and manipulation sites of pBR322, together with a *Bgl* II fragment of circularized lambda DNA containing the *cos* site recognized by the lambda packaging system, and another fragment of DNA. The cleaved vector is ligated to large restriction fragments of genomic DNA under conditions that allow the formation of linear assemblies containing insert flanked by two molecules of linearized cosmid to provide the two *cos* sites needed for packaging *in vitro*.

Cosmids are based on bacteriophage lambda but retain only the cohesive ends (*cos*) of the bacteriophage, two of which are sufficient to allow packaging *in vitro*. This may allow insertion of up to 40 kb of foreign DNA, but the packaged recombinant DNA is no longer infective after it penetrates the host cell. For this reason the cosmid vector has an origin of replication and antibiotic resistance gene that allows it to behave like a plasmid, and selection of transformed colonies can be made on antibiotic plates. Further discussion of cloning in cosmids is beyond the scope of this book.

The Identification of Recombinant DNA Clones

The objective of cloning DNA is, as already stated, to isolate a particular piece of BNA or mRNA copy in a pure state. However, the methods described above will, in general, produce a large number of clones which together represent many different mRNA copies or genomic segments. The solution to the problem of how to identify clones of the DNA of interest depends on its nature. We shall consider the most common case, that of DNA which encodes a protein, or part thereof. One might be able to identify this protein immunologically, by the biological properties it possesses, by its amino acid sequence or merely by its electrophoretic behaviour in polyacrylamide gels. In order to use these properties to identify the clone of interest it is usually necessary to express the genetic information in some way (random sequencing to search for cDNA clones corresponding to abundant proteins of known sequence an exception here). In general, direct expression may not occur and indirect expression is necessary. By indirect expression is meant the expression of the mRNA after selection by DNA-RNA hybridization.

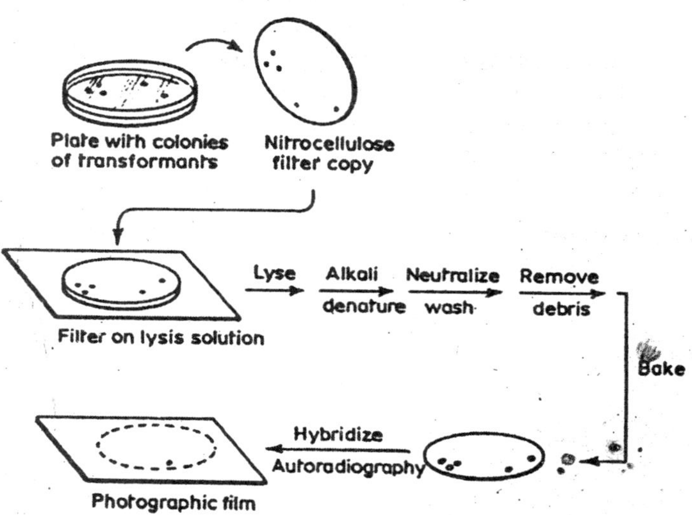

Fig. 8.20. Colony hybridization. Plasmid DNA in transformed bacteria is immobilized on nitrocellulose and hybridization of this to a suitable ^{32}P-labelled probe (see text) allows identification of colonies carrying plasmids with inserts of interest. In the diagram the number of colonies has been greatly decreased for clarity.

The cloned DNA is immobilized on a nitrocellulose filter, and an mRNA preparation (consisting of a mixture of species) from an appropriate tissue or cell is passed through. Only the mRNA complementary to the particular immobilized DNA clone will form a hybrid and be retained on the filter. It can then be released (washed off at low ionic strength) and translated in a cell-free system. This is known as *hybrid-selection* or *hybrid-release translation*. Alternatively, the denatured cloned DNA can be mixed with the mRNA preparation, the mixture added to the cell-free system and the loss of a particular product detected (*hybrid-arrest translation*). In both cases the product may be identified immunologically or electrophoretically as appropriate or, in particular cases, by its biological activity.

Screening a large number of clones by these methods is a formidable task, and it is usual to try to reduce this number by some means. For example, one can fractionate the mRNAs initially on sucrose density gradients and clone only those mRNAs in the relevant size range. If an antibody is available enrichment of the mRNA by immunological isolation of polysomes containing nascent peptides of the protein can be attempted. Preliminary screening is often employed by comparing the hybridization of (^{32}P)-labelled single-stranded cDNA probes derived from mRNA populations enriched with or depleted of the species of interest. (This is possible if the mRNA is differentially expressed in certain tissues or circumstances.) This *'colony hybridization'* is due to Grunstein and Hogness who devised a method of transferring duplicate copies of colonies from a 'master' agar plate to nitrocellulose filters (replica plating) on which the plasmid DNA could be immobilized after lysis of the cells.

If the sequence of the protein, or part thereof, is known it is possible to screen the colonies with a (^{32}P)-labelled *oligonucleotide probe* corresponding to the nucleotide sequence predicted by a part of the amino acid sequence. (Indeed, such an oligonucleotide may be used to prime cDNA synthesis as an initial cloning strategy. Because of the degeneracy of the genetic code such nucleotide sequences cannot be predicted with certainty. It is necessary to select sequences containing amino acids of minimal codon degeneracy, to consider what bias in codon usage might exist in the organism from which the mRNA is derived and to employ sufficient different oligonucleotides of sufficient length to allow for mistakes.

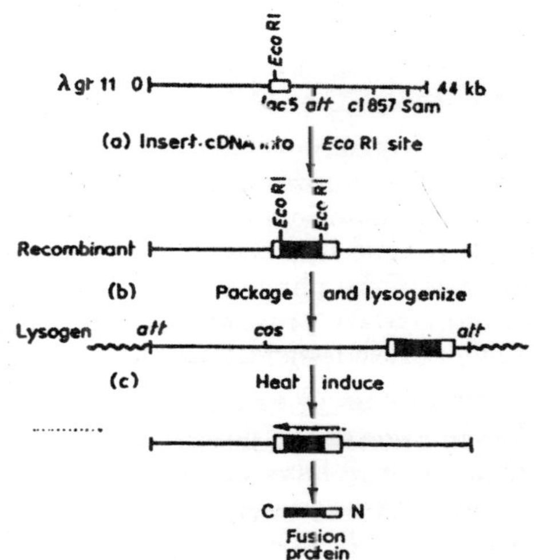

Fig. 8.21. Expression vector λgt11. Relevant differences from the vector Charon 16A are indicated: the *att* site and other functions required for lysogenic integration into *E. coli* (absent in Charon 16A are present, cI857 indicates a is mutant in the cl gene responsible for maintaining lysogeny, Sam indicates an amber mutation in the *S* gene coding for one of the cell lysis functions, (*a*) The vector is cut with *Eco* RI and ligated to a library of cDN A to which *Eco* RI linkers have been attached; (*b*) the recombinants are packaged *in vitro* and transfected into an *E. coli* strain carrying a mutation *(hflA)* that ensures lysogeny rather than lysis; (*c*) subsequent heat induction causes release of the lambda recombinants, but these do not lyse the bacteria so that, in one case out of six, colonies contain large amounts of fusion proteins comprising the N terminal portion of the β-galactosidase gene joined to part of the protein encoded by the cDNA.

Nevertheless, this method is extremely powerful and has been increasingly employed as protein microsequencing technology has improved. Identification of cloned DNA by direct expression of the protein product has the major advantage that thousands of clones can be screened directly. This is particularly useful for low-abundance mRNAs for the protein product of which no sequence data are available but against which antibodies have been raised. The problem with direct expression in *E. coli* is that a bacterial promoter and Shine and Dalgamo sequence are required. In the case of bacterial genes this is no problem, and genetic selection can often be effected by using suitable auxotrophs, but eukaryotic DNAs lack such sequences.

Although analogous genetic selection in eukaryotic cells can be effected in favourable instances where mutants are available, the more general solution to this problem has been to put the eukaryotic DNA under the influence of bacterial regulatory signals. In fact the reader may have realized that if an incomplete eukaryotic cDNA clone (*i.e.,* lacking sequences corresponding to the 5′ untranslated portion of the mRNA) is cloned into the pUC or M13 vectors there is a one-in-six chance of this being inserted in the right orientation and with maintenance of the reading frame, and of this leading to the synthesis of a fusion protein comprising the N-terminal amino acids of β-galactosidase and a part of the protein corresponding to the cDNA clone. Such an expressed fusion protein would most likely have antigenic determinants that would be recognized by a polyclonal antibody to the eukaryotic protein. If the objective is to screen a cDNA library by expression of a fusion protein in *E. coli* it is better to construct the library in a purpose-built expression vector rather than in the pUC or M13 vectors.

Such expression vectors have features that maximize expression of the fusion protein to increase the chances of immunochemical detection. One such vector that is currently in wide use is the bacteriophage lambda insertion vector, λ*gt*11. This has an *Eco* RI cloning site in the *lac Z* gene like the vector Charon 16A but has a mutation *(cI* 857) that allows thermal induction of the phage recombinants after lysogeny into a suitable *E. coli* host strain. Another mutation (S100) prevents lysis of the cells, resulting in the accumulation of large quantities of fusion protein inside the bacterial colonies.

The protein is eventually made accessible to the antibody by treatment of the bacteria with chloroform. In describing methods of screening recombinant DNA clones eukaryotie genomic clones have so far been ignored. This is because it is completely impracticable to apply indirect expression methods (mRNA selection) to massive genomic libraries, and direct expression in *E. coli* is generally impossible. In fact genomic clones for a particular protein are normally identified by plaque hybridization methods, analogous to colony hybridization but using the (^{32}P)-labelled cloned cDNA insert as a probe. An alternative approach using genetic selection by recombination between the plasmid with the cDNA insert and the corresponding bacteriophage lambda or costnid clone has been devised for more rapid screening of genomic libraries.

ANALYSIS AND MANIPULATION OF CLONED DNA

Once a nucleic acid sequence has been cloned an investigator may wish to subject it to a wide variety of manipulations. In this section we restrict ourselves to describing the principles underlying the most common of these.

Mapping of RNA Transcripts

After cloning a piece of genomic DNA thought to encode a protein it is generally necessary to determine *in vivo* precisely which parts of the DNA are represented in the final processed RNA transcript. The positions of intron splice sites and the points at which transcription of mRNA starts and finishes are usually determined by the technique known as *nuclease S1 mapping*. The principle of this method is to form a DNA/RNA hybrid between the mRNA encoded by the cloned DNA and the complementary single strand of the latter. This is possible by judicious choice of hybridization temperature, because of the greater stability of the DNA/ RNA hybrids compared with the original DNA/DNA hybrid. The endonuclease S1 is used to digest the non-hybridized, single-stranded regions of the DNA, leaving the regions of DNA complementary to the mRNA intact.

The original method of characterizing the undigested DNA was by size after electrophoresis on alkaline agarose gels (to hydrolyse the RNA) and identification by hybridizing to an appropriate (^{32}P)-labelled probe. A more precise determination of the transcription boundaries can be obtained if an appropriate (^{32}P-endlabelled restriction fragment of the cloned DNA is used in the hybridization. This can then be subjected to electrophoresis on a polyacrylamide sequencing gel alongside suitable standards, often produced by Maxam/Gilbert chemical cleavage of the restriction fragment in question. In determinmg the position of introns it can be helpful to employ exonuclease VII addition to the nuclease S1.

Exonuclease VII will digest single-stranded ends protruding from a mRNA/DNA hybrid but, unlike the endonuclease S1, cannot digest a single-stranded intron looped out from such a hybrid. While discussing the topic of mapping RNA transcripts it is worth mentioning a situation that may arise with cDNA clones. As such clones may not reflect the full extent of the mRNA to its 5′-end one may wish to determine how far it lies from the position corresponding to the 5′-end of the cDNA. The technique of *primer extension* utilizes a small restriction fragment near the 5′ end of the cDNA clone which is denatured and then used to prime further copying of the mRNA with reverse transcriptase.

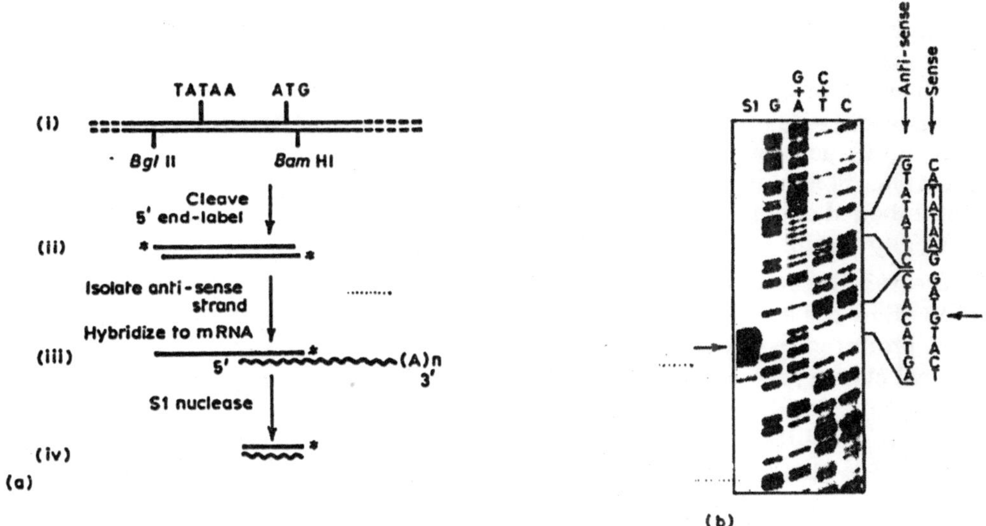

Fig. 8.22. Example of nuclease Si mapping to determine transcriptional start point, (*a*) A cloned fragment of genomic DNA, hybridizing to a particular mRNA has been sequenced and the likely initiation codon and TATAA box identified (*i*). A restriction fragment that is likely to contain the transcriptional start point is isolated from this region and 5′-end-labelled (*ii*), the anti-sense strand isolated and hybridized to mRNA (*iii*) and the single stranded regions of the DNA (and RNA) not involved in the hybrid digested with SI nuclease leaving the hybrid (*iv*); (*b*) the size of the protected DNA in (*iv*) is determined on an acrylamide sequencing gel. Maxam-Gilbert sequencing reactions are performed on the undigested anti-sense strand of (*iii*), and electrophoresed in parallel, allowing exact definition of the transcriptional start point in the sequence (the G indicated by the arrow). The position of the TATAA box is also indicated on the sequencing gel.

If the primer is (^{32}P)-labelled at its 5′ end and non-radioactive dNTPs are used in the extension reaction, the size of the product can be determined by polyacrylamide gel electrophoresis, as for nuclease SI mapping. Clearly, this technique is limited by the ability of the reverse transcriptase to overcome any secondary structure barriers that may exist at the 5′ end of the mRNA.

Identification of Regions of DNA that Interact with Proteins

The regulation of gene expression is achieved through the interaction of proteins with DNA, hence when a particular piece of genomic DNA has been cloned it is often of interest to identify the regions of DNA involved in such interactions. One general approach is to form a complex between the DNA and protein and then subject this to digestion with bovine DNase. The region of DNA in contact with the protein is protected by the latter from digestion. Originally it was necessary to isolate this protected fragment and determine its structure; quite a difficult undertaking. This was much simplified by the adaptation of the Maxam-Gilbert sequencing strategy to the problem in the technique known as *DNA footprinting*. In this approach a fragment of double-stranded DNA is generated, radioactively end-labelled at one end only, as for Maxam-Gilbert sequencing.

The protein is bound to the fragment and the complex subjected to digestion with DNase I under conditions that on average result in a single endonucleolytic cut per molecule (cf . the use of similar conditions for chemical cleavage). The DNA is then isolated and subjected to electrophoresis in the type of denaturing polyacrylamide gel used for sequencing. If no protein is present during the DNase digestion a ladder of end-labelled oligonucleotides encompassing all possible sizes is produced on autoradiography. However, when the DNA-protein complex is subjected to DNase treatment no scission occurs in the protected region and no end-labelled oligonucleotides terminating in this region will be generated. This appears as a hole or gap on the sequencing gel.

Unprotected DNA subjected to chemical cleavage is subjected to electrophoresis on the same gel to allow the easy identification of the region of the DNA protected. When comparing the ladders produced by chemical cleavage and cleavage with DNase I it must be remembered that chemical cleavage produces oligonucleotides with a 3′-phosphate group whereas DNase I produces oligonucleotides with a 3′-OH group. This causes a difference in charge, and hence electrophoretic mobility, which must be taken into account. Reagents other than DNase I have been used to cleave DNA, or to change its sensitivity to subsequent cleavage.

Neocarzinostatin (a double-strand specific endodeoxyribonuclease) has been used in place of DNase I because it generates products with 3′-phosphate groups that allow direct comparison with the chemical cleavage ladder. Chemical methylation and 5-bromodeoxyuridine substitution with subsequent cleavage by ultraviolet light have been used to address the role of specific purines and thymine residues, respectively. More recently the development of sequencing in vivo has allowed the footprinting technique to be extended to the study of interactions between uncloned DNA and protein *in vivo*. One approaches to irradiate nuclei with ultraviolet light, producing pyrimidine dimers. The resulting saturation of the 5, 6-double bond allows ring opening through reduction with $NaBH_4$.

Alternatively, it is possible to apply the use of dimethyl sulphate for the methylation of purines to intact systems such as *E. coli* cells and mammalian nuclei. This section should not be concluded without mentioning that the most precise identification of the features of DNA-protein interactions has been achieved by X-ray crystallography of complexes formed between proteins and chemically synthesized oligonucleotides corresponding to their binding sites.

Allocation of Cloned Genes to Specific Chromosomes

Three methods in which cloned genes can be assigned to specific chromosomes have been described in this book. These are the formation of somatic cell hybrids, hybridization to fixed chromosomes *in situ,* and hybridization to DNA from individual chromosomes separated by a fluorescence-activated cell sorter. A fourth method, recently developed for the mouse, is useful for determining whether two genes are close together or distant, and will be described here as it also illustrates the use of restriction fragment length polymorphisms (RFLPs). This approach combines recombinant DNA with classical mouse genetics. The classical genetic approach to study the chromosomal 'linkage' of two gene loci is to determine whether markers co-segregate or segregate independently during genetic crosses between two different inbred strains that show polymorphism for the marker.

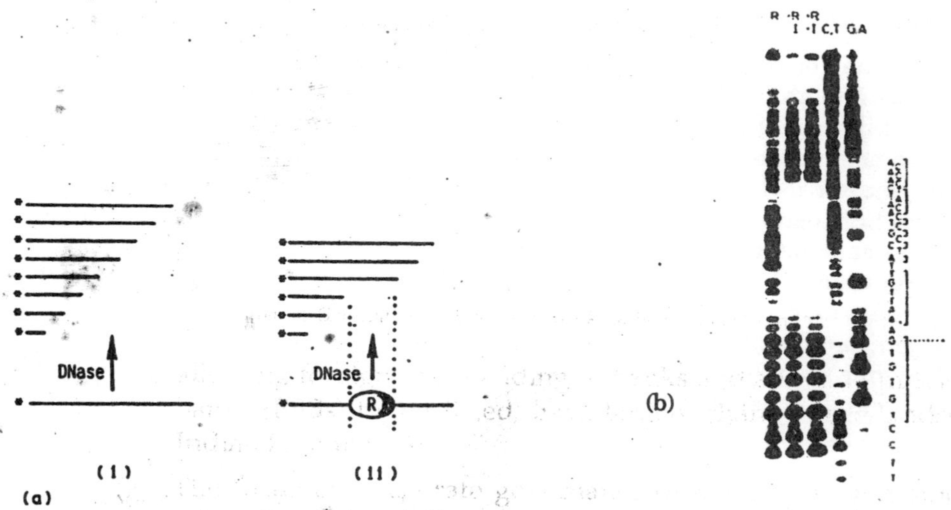

Fig. 8.23. Example of DNase I footprinting. (*a*) Partial digestion of a piece of end-labelled DNA to which a protein (R) is bound (*ii*) results in the absence of the end-labelled fragments cleaved in this region but found in the digest (*i*) of the unprotected DNA; (*b*) the results of separation of the products of such an experiment on a polyacrylamide sequencing gel. The example is with *lac* operator DNA and the *lac* represser (R). I is IPTG (0.3 m), which does not prevent the binding of a mutant represser used in this study. C,T and G,A represent the results of Maxam-Gilbert C + T and A -H G reactions on the undigested end-labelled fragment.

In higher organisms externally observable phenotypes associated with polymorphisms are relatively infrequent; however, genetic analysis has been aided by the more numerous cases in which a polymorphism results in differences in electrophoretic mobility of an easily identifiable enzyme. (The assignment of certain markers to particular chromosomes is possible when they are the site of a visible chromosome translocation. Other markers can be assigned by linkage to these.) To assign cloned genes to particular chromosomes the above principle is taken one step further by directly observing the genotype rather than the phenotype. Advantage is taken of polymorphisms in the nucleotide sequence of the gene that affects sites for restriction endonucleases. Such polymorphisms will produce differentsized restriction fragments in the two cases and these can be visualized by hybridization of a radioactive probe to a genomic Southern blot of the DNA.

The best chance of finding suitable RFLPs in mouse is by comparing an inbred strain of the common laboratory mouse, *Mus musculus domesticus,* with an inbred strain of the distantly related *Mus spretus.* Crosses between these strains have been used to study the linkage of mouse myosin genes. The use of RFLPs in the prenatal diagnosis of human hereditary diseases and the detection of carriers falls outside the terms of reference of this Appendix. Nevertheless, the utility of RFLPs generated by the mutation responsible for a particular genetic defect should be obvious. In addition, however, RFLPs outside the disease locus (which may be unknown) can be diagnostically useful if they show strong genetic linkage to this locus. The genetic analysis needed to use such RFLPs, although understandably more complex, is in principle similar to that just described.

Expression in Eukaryotic Cells

Although bacterial systems are extremely well suited to cloning eukaryotic genes they cannot be used to study their transcription. In this section we shall discuss briefly some of the systems that can be used to detect expression of cloned eukaryotic genes and study factors important in their regulation. One can classify systems for expression in eukaryotes into three categories: those in which the exogenous DNA is neither replicated nor integrated into the host genome but is subject to *transient expression,* those in which the DNA is *stably integrated* into the host genome, and those in which the DNA is expressed on a *eukaryotic vector* based on an animal virus. This last category, although of considerable interest, is too wide to consider here, especially as expression in eukaryotic vectors is not yet a technique generally applied to cloned genes. (Yeast plasmid vectors are unfortunately unsuitable for expression of the genes of higher eukaryotes. This is because of fundamental differences in the signals for transcription and processing. One way of obtaining transient transcription of a cloned gene is *microinjection* into the nucleus of frog oocytes.

The large size of the nucleus in the oocyte facilitates this operation, but it is still a specialized technique requiring special equipment and considerable practice. More generally accessible is the technique of *calcium phosphate co-precipitation* of DNA on to tissue culture cells which take up the precipitate, apparently by a process of phagocytosis. The DNA remains in the cells for several days and expression is usually studied about 48 h after transfection. In transient expression experiments the problem is how to detect the relatively small amounts of gene products of the transcripts and distinguish them from the gene products of the host cell.

In particular instances specific antisera may be available, but a more general approach is possible when, as is often the case, it is the regulatory rather than the coding regions of the gene that are being studied. In such cases constructs are made in which the 5′-non-coding and untranslated regions of the gene are linked to the coding region for the bacterial enzyme chloramphenicol acetyltransferase (CAT). *CAT assays,* as they are called, are extremely sensitive and are performed using [^{14}C]-chloramphenicol as a substrate, separating the acetylated chloramphenicol products from unreacted substrate by thin-layer chromatography. The calcium phosphate co-precipitation-technique also results in the stable integration of DNA into the chromosomal DNA of a small proportion of the cells.

If such 'transformed' cells can be selected from the vast majority of untransformed cells, cell lines can be obtained in which to study the expression of such stably-integrated exogenous genes. This is achieved by co-precipitation with a piece of DNA containing a selectable marker, as it transpires that the form of integration is as long tandem concatamers and inclusion of a second DNA into the co-precipitate results in its inclusion in the integrated unit. Many experiments have employed the herpes simplex viral thymidine kinase (tk) gene as a selectable marker. Cellular thymidine kinase is not required for the normal synthesis of dTTP from

CDP, but allows salvage of thymidine produced by breakdown of DNA by converting it to dTMP. It is possible to select tk^- mutants by growing cells on bromo-deoxyuridine (BrUdr), which can only be incorporated into DNA in tk^+ cells in which thymidine kinase converts BrUdr to BrdUMP.

Ultraviolet light will fragment the DNA of those cells in which thymine has been replaced by bromouracil, killing them, and hence allowing selection of spontaneous tk^- mutants. Tk^- cells that have been transformed to tk^+ by the integration of exogenous DNA can be selected by growth on 'HAT medium containing hypoxanthine, aminopterin and thymidine. Aminopterin inhibits the activity of dihydrofolate reductase, a key enzyme in both purine and pyrimidine biosynthesis. Hypoxanthine can act as a preformed precursor of purines, but thymidine can only act as a precursor of pyrimidines if thymidine kinase is present. Other selectable markers have been developed that do not require mutant cell lines such as tk^-.

Such dominant selectable markers include: dihydrofolate reductase (resistance to methotrexate), the multifunctional CAD enzyme of uridine biosynthesis (resistance to *N*-phosphonacetyl-*L*-aspartate) *E. coli* xanthine-guanine phosphoribosyltransferase (resistance to mycophenolic); and bacterial neomytin phosphotransferase (resistance to neomycin). A limitation of systems in which cloned genes are stably integrated into tissue culture cells arises when the major interest is in the tissue-specificity of expression of a particular gene.

A technique in which this problem is overcome involves the microinjection of cloned DNA into fertilized mouse eggs to produce so-called *transgenic mice*. The foreign DNA is integrated at such an early stage that it is generally present in all daughter cells, and certainly in the progeny of those mice that pass it on through the germ line to the next

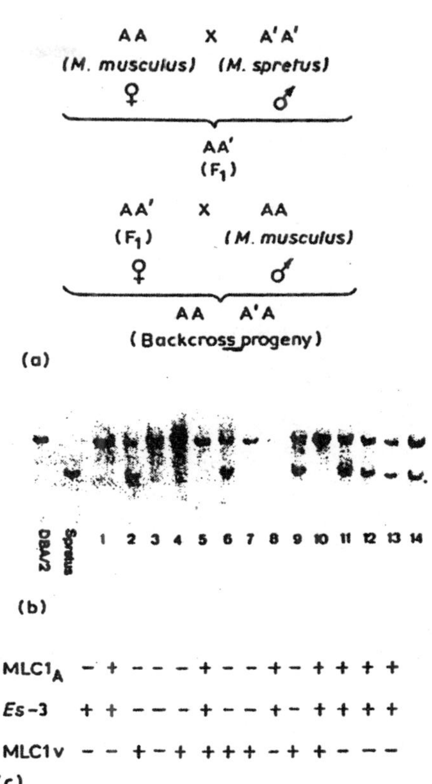

Fig. 8.24. Allocation of genes to mouse chromosomes and study of linkage using restriction fragment length polymorphisms, (*a*) The mating strategy is illustrated. Because the F_1 male progeny are infertile the females must be back-crossed to one of the parental strains to allow the segregation of the two alleles A and A' to be studied; (*b*) an example of the segregation of RFLPs for the MLCI A (myosin alkali light chain, cardiac atrial form) gene. The results of *Bam* HI digestion and electrophoresis of the DNA of the backcross offspring are shown. The different sized fragments in *Spretus* and DBA/2 (a *Musculus* strain) are formally equivalent to alleles A and A'; (*c*) comparison of the segregation of genes for MLCU. MLClv (myosin alkali light chain, cardiac ventricle form) and *Es-3* (an esterase enzyme exhibiting electrophoretic polymorphism) the gene for which is known to be located on chromosome 11. Homozygotes have been marked-, and heterozygotes +. These results illustrate the lack of linkage between the genes for $MLC1_A$ and MLClv, and that MLCU lies on chromosome 11. (More offspring than illustrated were, in fact, analysed.

generation. It is thus possible, using suitable constructs, to study whether a particular tissue-specific gene is expressed in a tissue-specific manner in its new location, and to investigate the regions of DNA which confer tissue-specificity of expression.

Mutagenesis in Vitro

In studies of cloned genes of the type conclusions regarding the importance of regulatory regions of DNA can be tested and extended if it is possible to produce mutations in the regions of interest. Such mutations are difficult to produce in higher eukaryotes *in vivo,* but powerful techniques are available to produce them in cloned DNA *in vitro.* The technique of mutagenesis *in vitro,* especially when directed to specific nucleotides (see below) can also be applied to the coding regions of genes the products of which it is possible to express, purify and subject to structural or functional studies. A variety of strategies has been used to produce mutations in cloned DNA.

Discussion here will be restricted to two of the currently most widely used and powerful techniques, and the reader interested in a thorough treatment of this topic is directed elsewhere. *Deletion mutagenesis* is frequently performed using the enzyme *Bal* 31 nuclease. This enzyme has exonuclease activity against double-stranded DNA. Hence if a plasmid containing the cloned gene is linearized by restriction endonuclease digestion at a point in the region of interest, *Bal* 31 nuclease will digest it in a way that produces a deletion emanating from this point in both directions.

The size of the deletion is determined by the time allowed for digestion, after which the blunt ends (treatment with the Klenow fragment of *E. coli* DNA polymerase and dNTPs is employed to fill in any overhangs) are ligated to linkers cleaved, and ligated together with DNA ligase. The most powerful technique at present available to produce mutations at specific sites is the so-called *site-directed mutagenesis using synthetic oligonucleotides.*

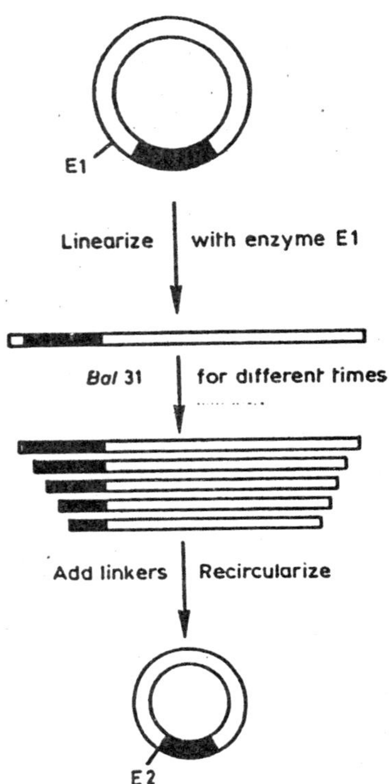

Fig. 8.25. Use of nuclease *Bal* 31 to generate deletions. A hypothetical plasmid (*a*) is illustrated in a region of which (solid shading) it is desired to obtain deletions. A neighbouring restriction endonuclease site is used to linearize the plasmid (*b*) and this is digested for different times generating a spectrum of deleted molecules (*c*). These are recircularized using linkers for a restriction endonuclease, E2, giving plasmids such as (*d*).

For this the DNA must be subcloned into a bacteriophage M13 or related vector that allows the generation of a single stranded circular recombinant DNA. An oligonucleotide (*e.g.,* 10-12 mer) is synthesized, corresponding to the target of mutation and surrounding nucleotides, except that it contains the desired base-change. This is annealed to the single-stranded circular DNA, which is then converted to the replicative double-stranded form *in vitro* using the Klenow fragment of DNA polymerase I and DNA ligase. When *E. coli* are transfected with this DNA, phage containing both wild-type and mutant single-strands will result and give rise to 'plaques'.

These can be distinguished by blotting on to nitrocellulose and hybridization to the (^{32}P)-labelled oligonucleotide originally used, under conditions that enable a perfect match (the desired mutant) to be distinguished from a single mismatch (the wild type). A modification of the procedure just described (and in fact preceding it historically may be used to generate short, specific deletions of a type hardly possible with *Bal* 31 nuclease. In this case the oligonucleotide synthesized is a hybrid of the sequences directly flanking the area to be deleted. This can hybridize to the single-stranded circular DNA if the area to be deleted loops out, and the desired mutant second strand can then be synthesized *in vitro*.

CHEMICAL SYNTHESIS OF OLIGONUCLEOTIDES

The importance of chemically synthesized oligonucleotides in cloning cDNAs in labelling DNA as primers for sequencing in M13 for producing mutagenesis *in vitro* and in studying specific interactions between nucleic acids and proteins has already been mentioned. Custom synthesis of oligonucleotides is nor available commercially, as are machines for 'in house' synthesis. The chemistry involved derives from that developed by Khorana, the application of which culminated in the synthesis of a tRNA gene. The methods Khorana used have now been superseded, and therefore the principles of the two methods at present in greatest use will be outlined here. Further details can be found elsewhere. In both methods successive diester bonds are 11 formed between the 5'-OH group of one nucleotide derivative and 3'-OP of a second nucleotide derivative.

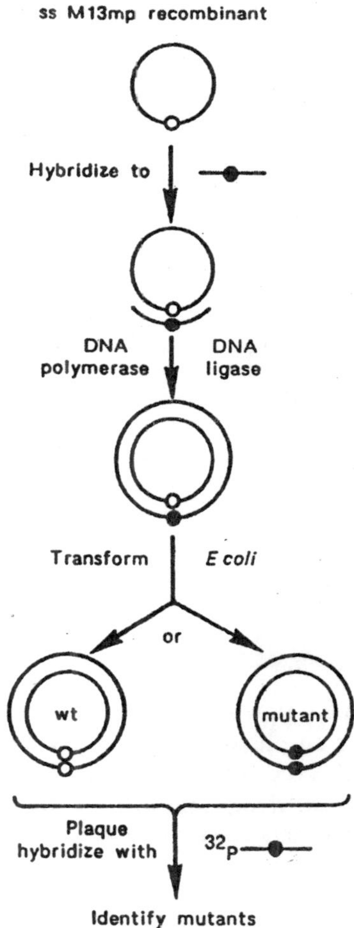

Fig. 8.26. Site-directed mutagenesis. The base to be altered is shown as an open circle and the mismatch (mutation) in the chemically synthesized oligonucleotide as a solid circle.

The chemical environment of the P determines the chemistry of the condensation and is different in the two cases. Both methods share the common feature that other potentially reactive groups elsewhere in the molecules are 'protected' by reversible chemical modification, and such protected nucleotide derivatives are used as the starting building blocks. The groups protecting the bases are not removed until the end of the overall synthesis, and the same is true of the protecting group for the P-OH. However, as the oligonucleotide chain is chemically synthesized in a 3' → 5' direction, the 5'-position (protected in both methods by a dimethoxytrityl group) must be deprotected after the addition of each monomer to allow further reaction. In order to simplify the synthetic procedure the first nucleotide is directly linked to a solid support (*e.g.* silica gel) packed in a column. This allows the excess reagents to be washed off the column after each step. It is usually necessary to block or 'cap' any unreacted 5′-OH groups after each step, using acetic anhydride.

At the end of the synthesis the oligonucleotide must be chemically released from the column and deprotected. In the the *phosphotriester* method the nucleotide with the reactive 3′-moiety is a derivative of the nucleoside 3′-monophosphate in which one P-OH is blocked by a 2-chlorophenyl group, and hence is in fact a phosphodiester. It is generated by a suitable base from a precursor in which the remaining hydroxyl on the P is blocked by a β-cyanoelhyl group, and reaction in the presence of a coupling reagent and catalyst results in the formation of a phosphotriester. The 5′-O dimethoxytrityl protecting group is removed by suitable acid treatment and another activated monomer added to allow the next round of condensation.

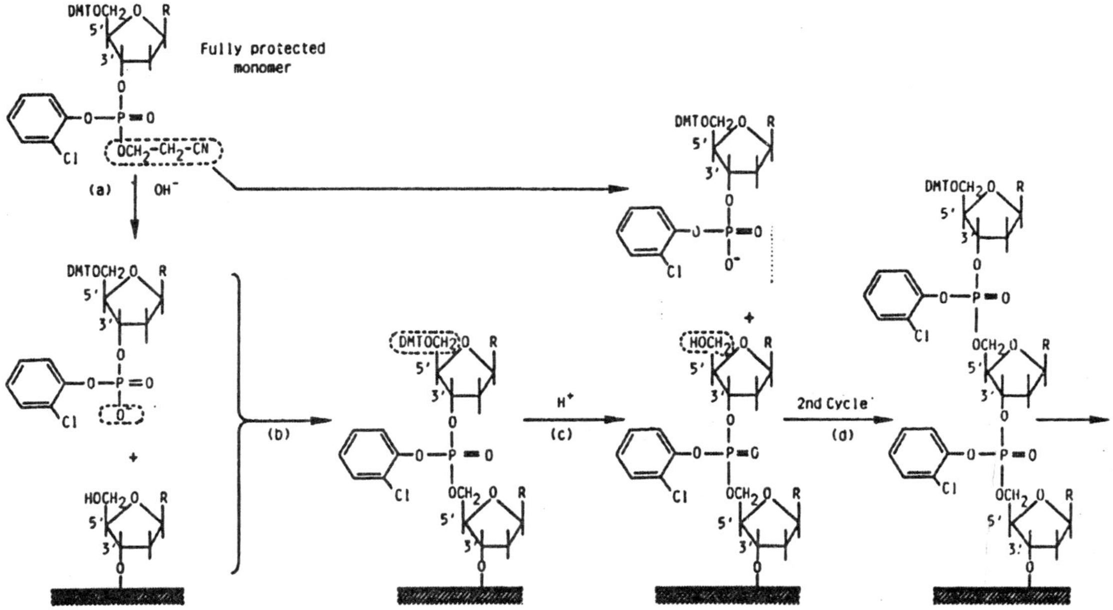

Fig. 8.27. The phosphotriester method of deoxyoligonucleotide synthesis. (*a*) The fully protected monomer is unblocked at the 3′-phosphate position by treatment with a base such as triethylamine allowing (*b*) condensation with a second molecule of monomer attached *via* its 3′-O to a solid support and having its 5′-OH unprotected. The dinucletoide produced must be (*c*) treated with acid to remove the dimethoxytrityl group (DMT) protecting the 5′-OH before (*d*) condensation with another molecule of 3′-activated monomer can occur. *R* = benzoyladenine, benzoyl cytosine, isobutyryl guanine, or thymine.

At the end of the whole reaction sequence the 2-chlorophenyl groups are removed by tetramethylguanidine-aromatic aldoxime in aqueous dioxane, the acyl blocking groups of the base by concentrated ammonium hydroxide, and the dimethoxytrityl group by 80% acetic acid. Removal of the oligonucleotide chain from the solid support depends, of course, on the nature of the linkage. Ester linkage via succinyl groups is quite common, in which case cleavage occurs at the same time as deprotection of the basest. The phosphotriester method is useful for synthesizing oligonucleotides containing up to about 50 bases, but is limited by the fact that each condensation takes about 1 h to go to completion, even in the presence of catalysts.

The *phosphite triester* method employs a more reactive form of the phosphorus which allows the condensation to be completed in about 2 min, and has been used to synthesize oligonucleotides containing more than 150 bases. The 5′-OH of the monomer is protected with the same blocking group as in the phosphotriester method but the reactive 3′-component is a dialkyl phosphoamidite (one of a variety of the possible dialkyl substitutions of the N is illustrated) and the product is a phosphite triester, which must be oxidized (*e.g.*, with aqueous iodine) to a stable phosphotriester before the next synthetic step. A methyl group is usually employed to block the hydoxyl on the P. This so-called phosphoamidite method is now widely regarded as the method of choice for oligonucleotide synthesis, and most automatic synthesizers ('gene machines') use this chemistry.

CELL-FREE SYSTEMS FOR TRANSCRIPTION AND TRANSLATION

A number of systems for studying the *transcription* of cloned genes in eukaryotic cells are mentioned. All of these are quite sophisticated and involve the use of intact cells. For many purposes the availability of cell-free systems which would allow transcription of exogenous DNA would offer considerable advantages; however, progress in this area has been slow. The importance of systems in which to assay the *translation* of eukaryotic mRNAs has also been indicated in relation to cloning and the study of jirotein synthesis in general.

Although the intact cell system of microinjection into frog oocytes (or eggs) has the advantage that translation may proceed for several days cell-free systems are of greater utility to the non-specialized laboratory. Such systems are not as simple to establish in eukaryotes as in *E. coli* (see below) and activity is easily lost during cell fractionation. Necessary components of the system may be lost with nuclei and mitochondria or adhere to intercellular structures. Furthermore, the disruption of the cytoskeleton may be a factor contributing to loss of activity. It is therefore perhaps not coincidental that the two most widely used systems for cell-free translation are derived from rather specialized cells. By far the most popular system is that derived from rabbit reticulocytes using hypotonic lysis followed by low-speed centrifugation to remove the cell membranes. This *rabbit reticulocyte lysate* has been further modified to allow more efficient utilization of exogenous mRNA.

The endogenous globin mRNA is degraded by the Ca^{2+}-dependent micrococcal nuclease that can subsequently be inactivated by chelation with EGTA. Such nuclease-treated lysates are available from several commercial sources, obviating the need for the user to maintain and bleed his own anaemic rabbits. The other cell-free system still in general use is that derived from the post-microsomal supernatant of wheatgerm, which has an intrinsically low conten of endogenous mRNA7. One limitation of the reticulocyte lysate and wheatgerm systems (but not of the frog oocyte) is the lack of endoplasmic reticulum for correct processing of translation products that contain signal peptides. This may be overcome by the addition of a microsomal membrane fraction, *e.g.*, from dog pancreas. Prokaryotic cell-free systems for translation were developed very early on; the ribosomes, tRNAs and soluble factors for translation of exogenous mRNA being present in the supernatant after centrifugation of disrupted cells at 30 000 g for *c* 30 min. This system was originally developed by Nirenberg and Matthaei in their work on the genetic code and was subsequently developed for use in studying the translation of the stable mRNAs of the RNA coliphages.

The low endogenous mRNA activity of such systems is a reflection of the very short half-lives of bacterial mRNAs, which make it very difficult to isolate bacterial mRNAs and study their translation directly. To overcome this problem a cell-free system for translation was devised in which the mRNA could be continually generated by simultaneous transcription of the gene. This coupled transcription-translation system was originally applied to genes cloned by more traditional genetic techniques using specialized transducing phages.

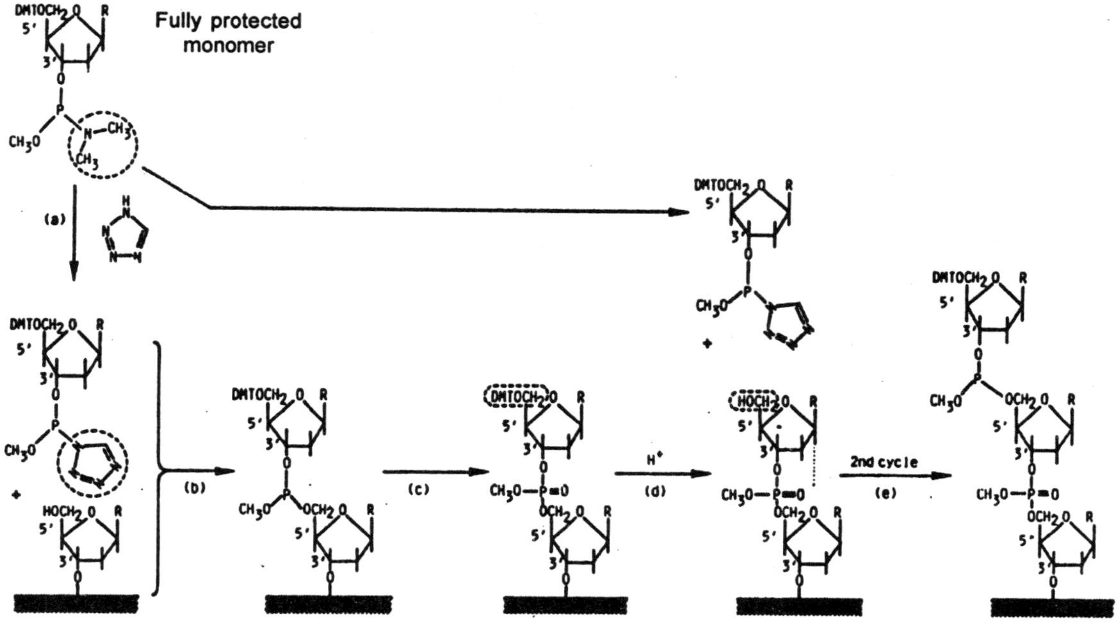

Fully protected monomer

Fig. 8.28. The phosphate trimester method of deoxyoligonucleotide synthesis. (*a*) The protected phosphoramidite monomer is activated in the 3′-P position by a suitable weak acid (*e.g.* tetrazole) allowing condensation (*b*) with 5′-activated monomer attached to a solid support. The phosphate is then oxidized to a phosphate (*c*). Removal of the 5′-protecting DMT group of the dinucleotide with, for example, dichloroacetic acid (*d*) prepares this for the next cycle of condensation (*e*) R is as in Fig. 8.27.

Table 8.3. Some Inhibitors of DNA Synthesis

Category	*Examples*
1. Reagents interacting with DNA*	
(a) Alkylating agents	Nitrogen and sulphur mustards
	Dimethyl sulphate
	NNMG and NMS
	Mitomycin C
(b) Intercalating agents	Ethidium bromide
	Propidium diiodide
	Acridine dyes
	Actinomycins

Category	Examples
	Anthracenes
	Benzpyrene
	Adriamycin
(c) Intertwining agents	Netropsin
	Distamycin
2. Analogues of bases, nucleosides etc.*	5-fluorouracil
	6-azauracil
	8-azaguanine
	6-mercaptopurine
	diazaoxynorleucine
	azaserine
	5-bromodeoxycytidine
	5-fluorodeoxycytidine
	2-aminopurine
	Cytosine β-1-D-arabinoside
	Adenine β-1-D-arabinoside
	Dideoxynucleosides
	Aphidicolin
	Aminopterin
	Amethopterin
	5-fluorodeoxyuridine
	Hydroxyurea
	Acyclovir
	2′ azido-2′-deoxynucleosides
3. Inhibitors of topoisomerases	Coumermycin
	Novobiocin
	Nalidixic acid
	Oxolinic acid
4. Inhibitors of cell division	Colcemid
	Colchicine
	Vinblastine
	Vincristine

* These may be mutagens and/or agents which interfere with replication or transcription.

** These may interfere with nucleotide biosynthesis or, following incorporation into DNA, block further chain elongation or transcription.

THE USE OF INHIBITORS IN THE STUDY OF GENE EXPRESSION

In the course of this book reference has been made to the use of antibiotics and other inhibitors of macromolecular synthesis. For the convenience of the reader the most common of these are summarized in Table 8.3 to 8.5. Many inhibitors of transcription and translation are specific for eukaryotic or prokaryotic systems, and this specificity. Most inhibitors of DNA synthesis do not discriminate between eukaryotes and prokaryotes; however, the dideoxynucleosides, on conversion to the triphosphates, do not inhibit eukaryotic DNA polymerase *a*. In contrast, a fairly specific inhibition of eukaryotic DNA polymerase a is obtained with aphidicolin. Further details may be found elsewhere.

Table 8.4. Some Inhibitors of RNA Synthesis.

Inhibitor	*Remarks*
Rifampicin and Streptovaricin	Bind to β subunit of *E. coli* RNA polymerase, inhibiting initiation
Streptolydigin	Binds to β subunit of *E. coli* RNA polymerase, inhibiting elongation
Actinomycin D	Inhibits both prokaryotic and eukaryotic transcription by complexing with deoxyguanosine residues. Hence transcription of rRNA is more sensitive than that of mRNA
α-amanitin and other fungal amatoxins	Inhibits eukaryotic RNA polymerase II specifically (also RNA polymerase III at very high concentration)
Cordycepin (3′-deoxyadenosine)	Inhibits poly(A) polymerase responsible for 3′-polyadenylation of eukaryotic mRNAs
Dichlororibofuranosyl benzimidazole	Inhibits appearance of eukaryotic mRNA in the cytoplasm, although only partially inhibiting synthesis of hnRNA

Table 8.5. Some Inhibitors of Protein Synthesis.

Inhibitor	*Prokaryotic/eukaryotic*	*Remarks*
Streptomycin	P	Binds to 30S subunit to cause misreading and (at higher concentrations) inhibition of initiation
Paromomycin	P	Inhibits initiation. Resistant mitochondria have altered small rRNA
Kasugamycin	P	Inhibits initiation. $m_2^6Am_2^6$ A sequence of 16S rRNA involved
Pactamycin	P, E	Inhibitis initiation, was used to determine gene order of proteins derived from picornaviral polyproteins
Neomycin	P, E	Causes miscoding and inhibits initation and elongation. Used as a selectable marker in transfection of eukaryotic cells
Edeine A	P, E	Inhibits initiation at either mRNA-or initiator tRNA-binding step
Aurintricarboxylic acid	P, E	Inhibitis initiation by preventing binding of mRNA to ribosome
Showdomycin	P, E	Inhibits initiation at the stage of ternary complex formation
Fluoride ion	(P?) E	Inhibits initiation in intact cells
Puromycin	P, E	Inhibits elongation by binding to A-site and reacting with peptidel-tRNA
Tetracycline	P	Inhibits elongation by blocking binding of aminoacyl-tRNA to the A-site on the 30S subunit
Chloramphenicol	P	Inhibits elongation at peptidel transferase. Resistant mitochondria have altered large rRNA

Inhibitor	*Prokaryotic/eukaryotic*	*Remarks*
Erythromycin	P	Inhibits elongation at transpeptidation step
Sparsomycin	P, E	Inhibits elongation at peptidyl transferase step
Spectinomycin	P	Inhibits elongation at transpeptidation. Resistant ribosomes have altered protein S5
Thiostrepton	P	Inhibits elongation, preventing binding of EF-G-GTP complex to ribosome
Fusidic acid	P, E	Inhibits elongation, preventing release of EF-G-GDP complex from ribosome. Resistant bacteria have altered EF-G
Kirromycin	P	Inhibits elongation, preventing release of EF-Tu GDP complex from ribosome
Kanamycin	P, E	Inhibits elongation and causes misreading
Cycloheximide	E	Inhibits elongation and initiation, 'freezing' ribosomes on polysomes. Resistant yeast have altered 60S subunits
Emetine	E	Inhibits elongation at translocation step. Resistant hamster cells have altered protein S14
Diphtheria toxin	E	Inhibits elongation by inactivating EF-2 by ADP ribosylation
Ricin and abrin	E	Inhibits elongation, affecting 60S rRNA
Colicin E3	P	Specific nuclease for site on 16S rRNA
α-sarcin	(P), E	Specific nuclease for site on 23S and 28S rRNA in ribosomes, although inactive against intact *E. coli.* Cleaves many sites in naked rRNAs
Guanylyl methylene diphosphonate and Guanylylimidodiphosphate	P, E	Non-hydrolysable analogues of GTP
Trimethoprim	P	Prevents formation of fMet-tRNA by inhibition of synthesis of N^{10}-formyl-H_4-folate
5-fluorotryptophan	P, E	Prevents activation of $tRNA^{Trp}$ by competitive inhibition of synthetase
Norvaline	P, E	Prevents activation of $tRNA^{Val}$
Ethionine	P, E	Causes synthesis of abonormal proteins with ethionine instead of methionine
O-methyl threonine	P, E	Causes synthesis of abnormal proteins with O-methyl threonine replacing threonine

E. COLI GENES RELEVANT TO NUCLEIC ACID METABOLISM

Much of the study of nucleic acid metabolism has been conducted in *E. coli,* and many of the components involved were identified by mutations in their genes. Because of this, and in view of the widespread use of specialized *E. coli* strains among non-geneticists involved in recombinant DNA work, we felt it might be useful to provide a summary of the names given to some selected genetic loci of *E. coli.* A few words of caution are necessary regarding Table 8.6. Because genetic loci were often defined before the corresponding gene product was known, there exist alternatives to some of the names listed. In some cases names have merely been rationalized to conform to a pattern such as *dna, rpo, rps;* and older nomenclature based on, *e.g.*, antibiotic resistance may still be found especially in description of genotypes.

Thus bacteria with a totally inactive gene for ribosomal protein S12 may be designated *rps*L$^-$, whereas bacteria with the mutation in the *rps*L gene that conveys dependence on the antibiotic streptomycin are generally designated Sm^D, the gene being originally called strA. Also worth mentioning are the *sup* genotypes: These define mutations in tRNA genes (*e.g.* *sup*F is in *tyr*T) which cause a particular termination codon to be read as an amino acid codon (suppression). For fuller details of these and other gene symbols the reader is advised to consult the *E. coli* linkage map published periodically in *Microbiological Reviews.*

Table 8.6. Some *E. coli* Genes Relevant to Nucleic Acid Metabolism.

Gene symbol	*Explanation*
ada	Repair enzyme for removal of methyl groups
*ala*S	Alanyl-tRNA synthetase (similarly: *glnS*, *pheS*, etc.)
*ala*T (U, V etc.)	Alanyl-tRNAs (similarly: *glnT*, U, V etc.)
*att*λ	Integration site for bacteriophage lambda (similarly *att* ϕ80, etc.)
cca	tRNA nucleotidyl transferase
dam	DNA adenine methylase
dcm	DNA cytosine methylase
*dna*A (B, C etc.)	Enzymes of DNA biosynthesis
dut	(*dna*S) dUTPase
*end*A	DNA endonuclease I
*fus*A	Protein synthesis factor, EF-G
gpp	Guanosine pentaphosphatase
*gyr*A, B	(*cou*, *nal*A) DNA gyrase (two subunits)
hfl	High frequency of lysogenization by bacteriophage lambda
hsd M, R, S	Host restriction/modification system (methylase, endonuclease and specificity components of *EcoB* and *Eco*K in B and K strains, respectively).
*inf*A, B, C	Protein synthesis factors, IF-1, –2 and –3
*lac*A, I, Y, Z	Components of lac operon
lep (B)	Singal peptidase I
*lex*A	Repressor protein involved in response to DNA damage
lig	(*dna*L) DNA ligase
*lsp*A	Signal peptidase II (specific for lipoprotein)
*mut*D	(*dna*Q) DNA polymerase III (ϵ)
*mut*H, L, R, S	Uncharacterized proteins involved in DNA repair
*nrd*A, B	(*dna*F) ribonucleotide reductase (two subunits)
*nus*A, B	Proteins involved in termination of transcription
*nus*C, D, E	= *rpo*B, *rho* and *rps*J, respectively
*ori*C	Origin of NDA replication

Gene symbol	*Explanation*
phr	photolyase
pnp	polynucleotide phopshorylase
*pol*A, B, C	DNA polymerases I, II and III (*dna*E)
*prl*A-D	Uncharacterized proteins affecting protein export
*prm*A, B	Ribosomal protein methylases
*pur*A-M	Enzymes of purine biosynthesis
*pyr*A-I	Enzymes of pyrimidine biosynthesis
*rec*A	Protein involved in recombination and repair
*rec*B, C	DNA exonuclease V (two subnits)
*rec*F	DNA exonuclease VIII
*rel*A	Stringent factor
rep	DNA helicase
rho	Transcription termination factor rho
*rim*B-L	Ribosomal protein modification enzymes (methylases, acetylases etc.)
rna-rnh	Ribonucleases I-III, D, E and H
*rnp*A, B	Ribonuclease P (protein and nucleic acid components)
*rpl*A-Z, rpmA-H	50S subunit ribosomal proteins L1-L34
*rpo*A-D	RNA polymerase subunits α, β, β' and σ, respectively
*rps*A-U	30S subunit ribosomal proteins S1-S21
*rrf*A-E, G, H	5S rRNA genes of operons *rrn*A-H
*rrl*A-E, G, H	23S rRNA genes of operons *rrn*A-H
*rrn*A-E, G, H	rRNA operons
*rrs*A-E, G, H	16S rRNA genes of operons *rrn*A-H
*sbc*B	DNA exonuclease I
*sec*A, B, Y	Uncharacterized proteins affecting protein export (*sec*Y = *prl*A)
*spo*T	Guanosine tetraphosphatase
ssb	DNA-binding protein
*thy*A	Thymidylate synthase
tdk	Thymidine kinase
*top*A	topoisomerase I
*tra*I	Helicase I (plasmid encoded)
*trm*A-D	tRNA methylases
tsf	Protein synthesis factor EF-Ts
*tuf*A, B	Protein synthesis factor EF-Tu (two genes)
ung	Uracil-DNA-glycosylase
*uvr*A-C	AP endonuclease (three components)
*uvr*D	Helicase II
*uvr*E	Protein involved in mismatch repair
*xth*A	DNA exonuclease III
*xse*A	DNA exonuclease VII

CHAPTER 9

Resistance to Inhibitors

The prevalence of non-nucleoside reverse transcriptase (RT) inhibitor (NNRTI) resistance is increasing, especially as the number of patients who have been treated with this class of agents has grown. The three Food and Drug Administration-approved NNRTIs, efavirenz, nevirapine, and delaviridine, have been shown to be potent additions to the antiretroviral (ARV) armamentarium. In many studies, these compounds have been shown to be effective components of highly active ARV therapeutic (HAART) regimens; results have been similar to or better than the comparator protease inhibitors (PIs). Resistance to the NNRTIs occurs more frequently than resistance to other classes of ARVs and is a serious limitation to their use.

The use of these agents in monotherapy and in partially suppressive regimens rapidly selects for resistance, and this resistance precludes the use of the other agents in this class. When considering these agents in newly infected or ARV-naive patients, the rate of transmitted resistance, especially in areas in which NNRTIs have been used extensively, may ultimately limit the usefulness of this class of therapeutics. As the incidence of transmission of NNRTI resistance increases, resistance testing for recently infected patients should be strongly considered. The potency of these agents must be balanced against the rapid development of resistance.

RESISTANCE ASSAYS

Defining resistance to NNRTIs is straightforward with both genotype and phenotype assays. A number of genotypic changes have been associated with resistance to NNRTIs. Primary resistance mutations often confer high-level (>30-fold) phenotypic resistance, so that genotype and phenotype assays yield concordant resistance interpretations. Occasional viral isolates are encountered in NNRTI-naive patients that have low-level phenotypic resistance (<10-fold change [FC] in 50% inhibitory concentration [IC_{50}]). These isolates do not have the primary resistance mutations often found with higher levels of resistance (*i.e.,* K103N). These low level phenotypic shifts are caused by polymorphisms in the genotype and not by primary resistance mutations. Clinical response to treatment with NNRTI-based regimens in patients with low-level phenotypic resistance has not been reduced. Occasional mutations associated with NNRTI resistance do not confer phenotypic resistance suggesting that some issues in resistance testing and interpretation remain unresolved.

RESISTANCE PROFILES OF NNRTIS

NNRTIs bind noncompetitively to the p66 subunit of RT in a hydrophobic pocket near the active site. Nevirapine has been studied with RT in crystal form. It lies on top of a hairpin motif that contains aspartic acid at positions 185 and 186. Tyrosine at positions 181 and 188 are in contact with nevirapine. This partially explains why the Y181C mutation confers resistance to nevirapine. Nevirapine binding may indirectly affect the conformation of aspartic acid residues at the active site, or binding may prohibit movement of different domains of the protein in relation to each other, thus inhibiting the function of the RT enzyme. There are several single point mutations that are associated with significant decreases in susceptibility to NNRTIs, and combinations of mutations may further reduce susceptibility.

The most widely recognized NNRTI-resistance mutation is K103N. This mutation confers resistance to all three NNRTIs. NNRTIs bind the RT enzyme at a site distinct from the active site, generally between codons 100 to 110 and 180 to 190. Nucleoside RT inhibitors (NRTIs) bind at the active site, and because of the different binding sites of the two drug classes, there is little overlap in the mutations that confer NRTI and NNRTI resistance. One exception may be the Q145M mutation. This mutation may be responsible for drug recognition and processing, and has been shown to confer resistance to both NNRTIs and NRTIs.

Although one study suggested that zidovudine augmented the development of K103N, other data found no association between the use of AZT or the presence of AZT mutations on the emergence of K103N. In fact, NRTI-resistance mutations have been associated with NNRTI hypersusceptibility. Mutations at other sites may confer resistance to NNRTIs. An example is the Y318F mutation. When site-directed mutants containing only the Y318F mutation were constructed, significant resistance to delavirdine (41-fold decrease in susceptibility) was seen, whereas a less than threefold decrease in susceptibility to nevirapine or efavirenz was seen.

However, in clinical isolates, the presence of Y318F when Y181C was also present conferred an additional 3.3-fold increase in resistance to efavirenz; when Y318F was present with K103N, there was a significant increase in resistance (from 15-fold for K103N alone to 43-fold when both mutations were present). Detecting which mutations confer resistance to specific drugs is made more difficult because mutations selected by serial passage of virus with sub-inhibitory drug concentrations in vitro may be different from those seen in vivo. There are several hypotheses for this; among them is viral fitness.

The mutations selected for in vitro may confer significant fitness disadvantages, limiting their emergence in vivo. Certain mechanisms of resistance may provide a marked fitness disadvantage in vivo that is not seen in vitro because of the absence of host interactions in the in vitro model.

Efavirenz

In vitro studies with efavirenz have demonstrated that resistance occurs rapidly. Passage of virus in MT-2 cells resulted in mutations V179D, L1991, and Y181C. After 24 passages, there was greater than 1000-fold reduced susceptibility to the drug. In another study, highly resistant virus containing the double mutations L1001 and K103N was selected for in several passages. The rapid development of resistance in vitro correlates with the rapid development of resistance seen clinically, especially when efavirenz is administered alone or in suboptimal regimens. When efavirenz was administered as monotherapy, an initial decline in HIV RNA was followed by virological rebound to baseline within 6 wk, and rebound was accompanied by resistance.

Of the overall study of 135 patients, a subgroup of 19 patients were randomly selected to evaluate drug resistance. All isolates were found to be sensitive to efavirenz before the start of therapy. All five patients receiving the high dose (1000 mg/d) had resistance (defined as greater than eight-fold change in susceptibility in this study) to efavirenz; four of five patients receiving 300 mg/d had resistance; and two of six patients receiving 50 mg/d had resistance. This suggests that exposure to therapeutic levels of efavirenz as monotherapy induces resistance, whereas ineffective doses do not. In a study of patients failing an efavirenz-containing regimen, either efavirenz plus indinavir or efavirenz plus zidovudine plus lamivudine, K103N (which causes resistance to all three available NNRTIs) was the most common NNRTI-resistance mutation, occurring in 90% of patients. VI081 and P225H were also frequently observed, primarily in viruses that also contained other mutations that conferred resistance to NNRTIs.

Mutations that often emerge during treatment with nevirapine or delavirdine, such as V106A, Y181C, and Y188C, were rare in this study. Phenotypic susceptibility to efavirenz may seem to be retained with certain primary NNRTI-resistance mutations (*i.e.,* Y181C). When mutations K103N and Y188L are present in a particular viral isolate, high-level reduced susceptibility to efavirenz is likely. In one study that demonstrated this, viral isolates from patients undergoing dual combination therapy with efavirenz plus indinavir or from patients failing a zidovudine plus lamivudine-containing regimen, to which efavirenz was added, were analyzed.

Mutations K103N, Y188L, and G190S/E were associated with reduced susceptibility and failure to respond to efavirenz. Secondary mutations V1061, V1081, Y181C, Y188H, P225H, and F227L caused little resistance to efavirenz alone, although they significantly enhanced in vitro resistance when combined with other mutations. Some isolates from patients who had failed nevirapine or delavirdine therapy, with Y181C, Y188C, or V106A mutations that lacked K103N or Y188L mutations retained phenotypic susceptibility to efavirenz in vitro. The clinical relevance of this apparently retained susceptibility is unclear.

Continued therapy with an NNRTI could rapidly lead to development of additional mutations, such as K103N and Y188L. Because the clinical response to efavirenz with the Y181C mutation alone is probably suboptimal *(see* "Cross-Resistance"), there is an apparent discordance between the phenotypic profile and genotype pattern for efavirenz. However, recognition that higher-level reduced phenotypic susceptibility (>10-fold) to any NNRTI probably precludes response to any currently available member of the class would improve the interpretation of the phenotype assay for patients with Y181C as the sole mutation.

Nevirapine

In vitro studies with nevirapine demonstrate that resistance occurs rapidly, in as little as one passage. One study found Y181C to be the most common single mutation occurring in vitro. Another study demonstrated that, by the fifth passage of nevirapine, mutations V106A, Y181C, and G190A were present and were associated with a greater than 100-fold reduction in susceptibility. An assay of in vitro fitness demonstrated that, in the presence of nevirapine, resistant virus was more fit than wild type, with Y181C being the most fit. A surprising finding of unclear significance was that in two separate assays conducted by the same group, virus containing the Y181C mutation was more fit than wild-type virus in the absence of drug.

The clinical relevance of this finding is yet to be established. However, anecdotal evidence suggests that NNRTI resistance, both genotypic and phenotypic, persists long after NNRTI therapy is removed (R. Haubrich, personal observation). This argues in favor of a relative lack-of- fitness impairment of NNRTI resistance-associated mutations. Resistance to nevirapine has recently been shown to be influenced by perturbations in its interaction (and hence binding energy) with the RT enzyme at positions Y181 and Y188. Efavirenz seems to be less influenced by these interactions. Crystal structures of efavirenz complexed with RT containing the Y181C mutation, and nevirapine complexed with RT containing the Y181C mutation and complexed separately to RT containing the Y188C mutation were recently published.

With efavirenz and the experimental agent, UC-781, only small rearrangements were seen in the inhibitor binding pocket when compared with nevirapine. This represents a structural mechanism for the observation that Y181C confers significant resistance to nevirapine, whereas it only confers a low-level reduced susceptibility to efavirenz (in the absence of other mutations. Several clinical studies of nevirapine monotherapy have shown that resistance develops quickly in this setting. In one study, 20 patients with CD4 counts of greater than 500 cells/mm^3 received oral nevirapine monotherapy at 400 mg once daily.

Some patients received a dose-escalating regimen of nevirapine at 200 mg/d for the first 4 wk, followed by 400 mg/d afterwards. Seventeen of the patients completed the study, and 16 of them had samples available for resistance assays using virus grown up from peripheral blood mononuclear cells. In 13 of 16 subjects, virus was isolated and phenotypic resistance detected as early as 4 wk in 1 subject, 6 wk in another subject, 8 wk in 3 subjects, and 12 wk in 8 subjects. In the remaining three patients, virus became detectable at 20, 24, and 60 wk. There was no correlation between baseline level of RNA and development of resistance. In another nevirapine monotherapy study, seven patients received nevirapine (one patient was also receiving zidovudine) as part of a double-blind, placebo-controlled trial.

As early as 1 wk after starting nevirapine therapy, the Y181C mutation was detected. It was detected in six of seven patients receiving nevirapine at 2 wk. The detection of Y181C was correlated with a rise in plasma viral load. The extremely rapid emergence of resistance in the ART-naive patients (except for the one patient who received zidovudine) suggests that point mutations conferring NNRTI resistance preexist in a proportion of virions within a patient, and that exposure to nonsuppressive drug pressure leads to rapid replacement of the wild-type population with a population containing the resistant mutation. In addition to mutations at codon 181, two patients had K103N mutations. Also detected were K101E, G190A, K101R, K102Q, Q182R, Y183D, T107A, K102E, and VI 081 mutations. The plasma HIV RNA levels returned to baseline values within the 28-d observation period in two of the seven subjects.

The remaining five patients had viral loads between 0.7 $\log_{10}$ copies/mL and 1.32 Iog_{10} copies/mL below baseline at 28 d. Another study was performed to determine whether there was a difference in the development of resistance in patients with lower viral loads in the presence of nevirapine monotherapy. In 16 of the 17 patients for whom drug-resistance data were available, all isolates were genotypically resistant to nevirapine at 12 wk. Drug susceptibility and RT mutations were analyzed in 167 isolates from 38 patients treated with nevirapine, either alone or with zidovudine.

Resistance to nevirapine occurred rapidly in all patients, in as little as 1 wk, regardless of whether the patient received nevirapine alone or in combination. The presence of zidovudine altered the pattern of mutations, but did not prevent resistance. Mutations at residues 103, 106, 108, 181, 188, and 190 were found. Individual clones were sequenced from selected patients. The presence of different viral subpopulations was suggested by the presence of mixtures of mutations at several positions. Y181C was the most common mutation seen in the absence of, zidovudine.

Even though Y181C was prevented with the administration of zidovudme alternative mutations (such as Y181 Y, Y181S, and Y181H), which confer nevirapine resistance, were noted in place of the Y181C mutation. In a retrospective cohort of 88 patients who failed a nevirapine plus PI-containing regimen, resistant isolates to nevirapine were found in 92% of patients; the isolates were cross-resistant to efavirenz (68%) and delavirdine (73%) in most cases. The Y181C mutation was seen in 76% of mutants. A combination of the Y181C and the K103N mutations was also common (23%). Of the eight isolates carrying only the single mutation Y181 C, 29% retained phenotypic susceptibility to efavirenz. Resistance to NNRTIs in this study was correlated with baseline resistance to PIs.

Delavirdine

In vitro, delavirdine selects for P236L, which rarely occurs in vivo. In a fitness study, the P236L mutant was shown to be less fit than the K103N mutant, which may partially explain the infrequent occurrence of this mutant in vivo. Delavirdine resistance is also associated with cross-resistance to other NNRTIs. In a study of delavirdine administered as monotherapy, phenotypic resistance was found in 28 of 30 subjects within 8 wk. K103N and Y181C were the most common mutations. P236L, which confers delavirdine resistance, but hypersusceptibility to nevirapine and efavirenz was seen in less than 10% of the 30 patients studied. In another delavirdine monotherapy study, resistance was noted to emerge rapidly and, for this reason, the study was discontinued at 8 wk. K103N and Y181C were the mutations most commonly observed.

CROSS-RESISTANCE

Cross-resistance among NNRTIs is broad and largely prevents sequential use of agents of this class. Cross-resistance is associated with the K103N mutation, which is among the most common mutations arising with drugs of this class. Some mutations, such as Y181C, confer resistance to delavirdine and nevirapine, but only low-level reduced phenotypic susceptibility to efavirenz. Whether patients with Y181 C, when it is present as the sole RT mutation, can respond to efavirenz is of considerable interest. The degree of cross-resistance among NNRTIs was further illustrated in a study of 5000 clinical samples from a large database.

Seventy-nine percent of patients who had phenotypic NNRTI resistance to at least one drug had resistance to all three approved NNRTIs. Twelve percent of patients were resistant to both delavirdine and nevirapine, but sensitive to efavirenz. Six percent of patients were susceptible only to delavirdine, and 3% of patients were resistant to nevirapine but sensitive to delavirdine and efavirenz (genotypes not specified). Although many patients with previous NNRTI exposure experience broad cross-resistance, some do not, and interest in sequencing the NNRTI-resistant has been evaluated in small studies.

One study used genotypic data to determine whether a switch from a failing nevirapine regimen to efavirenz would provide additional activity from the NNRTI class. High-level cross-resistance to efavirenz was seen in 41% of nevirapine-treated patients (38% with K103N mutations, 3% with Y188L mutations) and probable resistance in 17% of patients (a variety of mutations, including Y181C). There was a correlation between shorter treatment durations with nevirapine and response to efavirenz. In this study using genotypic resistance, 2 of 12 patients with a single nevirapine-resistance mutation had a response (defined as an undetectable HIV RNA level for 3 mo) to efavirenz, whereas 35 of 67 patients without any known resistance mutations had a response. None of the patients with the Yl 81 C mutation had a response.

This study illustrates that NNRTI cross-resistance was common, and after treatment failure of one NNRTI, the likelihood of another one retaining antiviral activity was low. To evaluate the sequential use of NNRTIs, a prospective study of 47 patients who were failing a nevirapine-containing regimen assessed the impact of changing to a regimen containing efavirenz plus abacavir plus a new PI. NNRTI-resistance mutations were seen in 79% of patients, mainly at codons 181 (49%), 103 (40%), and 106 (19%). Sixty-two percent of patients had phenotypic resistance to efavirenz (>10-fold decreases in susceptibility).

All strains with the K103N mutation demonstrated reduced phenotypic susceptibility to efavirenz, but only 20% of specimens harboring solely the Y181C mutation had phenotypic resistance to efavirenz (>10 FC). By 24 wk, 38% of patients taking the efavirenz-containing regimen had at least a 1 $\log_{10}$ copies/mL decrease in their viral load, with 19% of patients having an undetectable viral load. Most of the responses (1 $\log_{10}$ copies/mL viral load decrease) were seen in the small number of patients with no evidence of genotypic or phenotypic NNRTI resistance (7/10 patients responded). Factors associated with failure of the efavirenz-containing regimen were phenotypic resistance, presence of the K103N mutation, and a previous longer period of nevirapine therapy (mean of 288 vs 170 d).

Although this data suggests the activity of efavirenz after nevirapine failure, it is far from conclusive. Cross-resistance between the three available NNRTIs can be explained by their binding to the same site of the RT. A mutation altering the binding properties of one of these drugs also effects the other members of the class. The sequential use of these agents in salvage therapy is still under study, but does not seem promising. It is not clear what, if any, fitness cost the virus must pay to maintain the resistance to NNRTIs, or whether, in a multidrug salvage regimen, recycling these drugs has a role.

LOW-FREQUENCY NNRTI-RESISTANT MUTATIONS

The assessment of NNRTI resistance can be complicated because of failure to detect NNRTI-resistance mutations. Patients with previous NNRTI therapy who subsequently change to new regimens that do not contain an NNRTI may lose the NNRTI-resistance mutations. If an NNRTI is reintroduced in a subsequent regimen, this absence of NNRTI mutations on the genotype may not predict a good treatment response. In ACTG 398, patients with a history of NNRTI treatment were less likely to achieve an HIV RNA level reduction to fewer than 200 copies/mL than those with no history of NNRTI treatment, even if they did not have evidence of NNRTI resistance. A potential explanation for the inability of the genotype to predict response in these NNRTI-experienced salients could be the relative insensitivity of standard assays to low-frequency mutations.

To explore this hypothesis, Mellors examined virus from 11 NNRTI-experienced and 12 treatment-naive patients using single-genome sequencing and other more-sensitive techniques. Using these methods, patients with a history of NNRTI treatment were more likely to have low-frequency resistant variants detected than treatment-naive patients (6/11 vs 2/12, respectively) and had greater numbers of viral clones with resistant variants. In the NNRTI treatment-experienced patients who ultimately failed the new NNRTI-containing regimen, the viral isolates were genetically related to the baseline minority viruses found by single-genome sequencing. This data suggests that previous treatment (if virological failure occurred on the regimen) with an NNRTI is likely to preclude the salutatory use of this class of agents in a future regimen, even if the resistance assay fails to detect evidence of NNRTI resistance.

STIMULATION OF REPLICATION

A newly characterized mutation found after NNRTI therapy seems to stimulate viral replication (in vitro) in the presence of NNRTIs. The M230L mutation was observed in four patients, either alone or in combination with other known NNRTI-resistance mutations. The combination of M230L and other NNRTI-resistance mutations seemed to markedly reduce the susceptibility of all NNRTIs (>250-fold decreased susceptibility). Interestingly, the virus isolated from three of the four patients had 50 to 100% greater levels of replication in the presence of NNRTI than in the absence of drug. Thus, certain NNRTI concentrations seemed to stimulate viral growth. This was not observed in the one patient who had M230L combined with Y181C mutations. These data sug-gest that, for certain patients, continued use of NNRTIs after virological break-through could be deleterious. Further work is needed to clarify these findings.

NNTRI-RESISTANT VIRUS

Transmitted NNRTI resistance is well-documented and is becoming a major problem. High-level resistance can be defined as a 10-fold decrease in susceptibility of the virus to a particular drug. The rate of high-level phenotypic NNRTI resistance in patients recently infected with HIV-1 has been estimated at 7.1% in a retrospective cohort in 10 major cities in infections occurring between 1999 and 2000. Although this high prevalence of resistance is not found in every population studied, it is concerning, especially given the broad cross-resistance among the NNRTIs.

In an earlier primary infection study by the same team, 3 of 141 total patients had high-level phenotypic resistance (>10-fold reduced susceptibility) to one or more ARVs. Two of these patients had resistance to both PIs and NNRTIs, although the presence of known mutations conferring NNRTI resistance was not documented. Thirty-six patients had intermediate susceptibility to one or more classes of ARVs, including NNRTIs. Additionally, the rate of resistance to the different NNRTIs was further broken down: 9 of 24 patients with reduced NNRTI susceptibility had resistance to both nevirapine and delavirdine, and only 2 patients had reduced susceptibility to all 3 NNRTIs.

In other studies, the rates of transmitted NNRTI resistance were less. If the trend of increasing transmission of NNRTI resistance continues, empiric use of NNRTIs may be problematic without resistance testing. Even if resistance is transmitted, the resistance assay may fail to detect the resistance. In untreated, therapy-naive patients, in the absence of selective drug pressure, the virus may revert to wild type, and a routine resistance test may not detect transmitted resistant viral subpopulations. There are emerging data, however, that contradict this conventional wisdom. One study showed that, in four patients with recent infection, mutations conferring resistance to NNRTIs persisted for at least 13 mo.

A study in five patients with zidovudme resistance in the setting of acute infection showed that resistance persisted for up to 1 yr in three of the five patients, whereas the other two patients had reversion to wild-type virus at 1 yr. These data suggest that it may be possible to use resistance assays in treatment-naive patients, even when it is unclear when transmission of resistance occurred. More studies are needed to determine the value of resistance testing in chronically infected ARV-naive patients. There is also evidence to suggest that drug resistance is detectable m chronically infected patients. In a study of 230 samples of chronically infected, ARV-naive patients, 6% of patients had high-level NNRTI resistance.

In an observational study, 281 chronically infected ART-naive patients were enrolled. Genotypic resistance to any ARV agent, using the *Visible Genetics* algorithm, was seen in 8.9% of patients, with 5.3% of patients having NNRTI resistance. The K103N and Y18IC mutations were the most frequent NNRTI-resistance mutations. This finding also argues for considering resistance testing before the initiation of ARV therapy, even in chronically infected patients.

Low-Level NNRTI Resistance

In treatment-naive patients, low-level phenotypic NNRTI resistance (FC between 2.5 and 10) has been noted. This low-level resistance does not pre-clude the use of efavirenz. In one case-control study, baseline resistance was determined for 50 patients who had virological failure on an efavirenz-containing regimen (either with indinavir or zidovudine plus lamivudine) and compared these patients with 50 patients who had long-term suppression on the same regimens. Entry criteria precluded previous use of NNRTIs. Thirty-four percent of baseline samples showed greater than fourfold phenotypic resistance to one or more NNRTIs.

Low-level (4- to 10-fold) phenotypic resistance was found in baseline samples of 26% of treatment failures and in 34% of treatment successes. Thus, there was no difference in the rate of low-level NNRTI resistance in failing vs successful treatment, suggesting that low-level NNRTI resistance did not influence virological outcome. High-level (>10-fold) resistance was seen in five of six subjects who subsequently failed their efavirenz-containing regimens. Several of these six patients had the K103N or Y181C mutations at baseline, consistent with undisclosed previous NNRTI therapy or transmitted resistance. Another study demonstrated that low-level reduced susceptibility (4- to 10-fold) to NNRTI was not associated with a significant risk of virological failure.

In this study of NNRTI-naive patients, a significant level of reduced sensitivity to NNRTIs in the 279 patients was observed. Greater than fourfold reductions in susceptibility to delavirdine were seen in 24% of patients *(n* – 60), to nevirapine were seen in 6% *(n* = 15), and to efavirenz were seen in 6% *(n* =15). Analysis of time-to-viral suppression or time-to-viral rebound for these patients treated with a nevirapine-based regimen were not associated with low-level reduced susceptibility to the NNRTI. In the absence of previous use of NNRTIs, low-level reduced phenotypic susceptibility should not be a con-traindication to using an NNRTI.

TREATMENT OF NNRTI-RESISTANT VIRUS

Because the concentration of NNRTI in plasma is many fold greater than the IC_{50}, investigation into whether higher levels of NNRTIs may be able to overcome certain resistance mutations has been considered. One study evaluated 40 patients who had virological breakthrough (HIV RNA levels >50 copies/mL after being <50 copies/mL) while taking a nevirapine plus two-NRTI regimen. After efavirenz was substituted for nevirapine without changing the NRTI in the regimen, 43% (17/40) of the patients had an undetectable viral load or a 1 $\log_{10}$ copies/mL reduction in viral load. For the 31 of 40 patients from whom genotypes were obtainable at baseline, 25 of 31 had NNRTI-resistance mutations (codons 103, 181, and 190).

In patients without NNRTI-resistance mutations, five of six regained viral suppression. In those subjects with NNRTI mutations, viral suppression was seen only in those patients with efavirenz levels greater than 2 μg/mL (therapeutic levels are generally considered to be >1.4 μg/mL). One-third of subjects with efavirenz levels greater than 2 μg/mL had a response (undetectable viral load or >1 $\log_{10}$ copies/mL decrease in viral load) to efavirenz, but details of how many patients had higher levels, which specific mutations correlated with response, or how many had viral loads of fewer than 50 copies/mL after switch were not provided. The benefit of this strategy remains speculative at present.

NNRTI RESPONSE WITH NRTI RESISTANCE

The ability of a regimen with two NRTIs plus an NNRTI to achieve and maintain undetectable viral load levels is dependent on the potency of the NRTI component. If one or more of the nucleosides are impaired because of previous treatment and resistance, response to the NNRTI-based regimen may be sub-optimal. Switch studies, in which a patient with HIV RNA of fewer than 400 copies/mL taking a current PI-based regimen, has an NNRTI substituted for the PI, have evaluated this concept. In a prospective study involving 34 NNRTI-naive patients with 6 mo of an undetectable viral load on a PI-containing regimen, the PI was changed to nevirapine.

Before the successful PI-containing regimen, 12 patients had previous treatment with NRTIs, whereas 22 patients were treatment naive. Forty-one percent of the patients with previous NRTI exposure and none of the treatment-naive patients experienced virological failure. All patients with virological rebound had virus with evidence of NNRTI mutations, a K103N mutation in three patients, and Y181C and G190A mutations were seen in one patient each. The K101E mutation was seen in conjunction with other NNRTI-resistance mutations. NRTI- but not Pi-resistance mutations were seen in those patients whose treatment failed. The NNRTI mutations occurred in patients who also had NRTI mutations detected during rebound, suggesting that archived resistant virus (NRTI resistance, in this case) played a significant role in treatment failure.

Analysis of proviral DNA from archived specimens obtained during virological suppression taking PI therapy in four of the five patients with subsequent failure did not demonstrate NRTI-resistance mutations; however this does not preclude the existence of low levels of archived NRTI-resistance mutations, because the methods used to detect the mutations may not have been sufficiently sensitive. These data suggest that previous NRTI therapy may impair virological response to a regimen containing an NNRTI plus two NRTIs.

NEW NNRTIS AND RESISTANCE

Exposure to NNRTIs, in the presence of ongoing viral replication, leads to resistance to the NNRTI class of agents, and this resistance can occur quickly. New NNRTIs are needed that will circumvent existing patterns of NNRTI resistance and provide additional therapeutic options to patients who have resistance to NNRTIs. Novel NNRTIs are in development, some of which may be active in the presence of the most common, class-limiting NNRTI mutations, especially K103N or Y181C mutations. The rational approach to designing NNRTIs that retain activity against virus harboring common NNRTI imitations, which will not themselves select for new mutations, is a challenge.

One method would be to select a compound that induces mutations that confer a significant decline in viral fitness. W229 is such a site. It resides in the "primer grip" portion of RT. A mutation at this site, such as W229F or W229Y, has been shown to result in a virus with a significantly lowered DNA polymerase activity. Although this is a potential target of newer generation NNRTIs, no compounds in development take advantage of this strategy. There are several new NNRTIs in clinical trials: calanolide A, capravirine, and TMC-125.

Calanolide A is a natural product isolated from tropical plants from the genus *Calophyllum* which has activity against HIV strains with the Y181C mutation. It has enhanced activity against strains with mutations conferring zidovudine resistance. This compound does not have good activity against strains carrying other NNRTI-associated mutations, such as those occurring at sites L100, K103, T139, and Y188. TMC125 is another NNRTI in human clinical trials. In vitro, it demonstrates potent activity against strains with the K103N or Y181C mutations, but has poor activity against strains with both mutations. It also seems to select for resistance more slowly than the currently available, first-generation NNRTIs.

In a phase Ha study, 16 patients who had a treatment failure on an efavirenz- or nevirapine-containing regimen had TMC125 substituted for their current NNRTI for 7 d. The median log change in viral load was –0.89 $\log_{10}$ copies/mL. Patients in this study had virus with multiple NNRTI mutations, including LI001, K103N, Y181C, Y188L, and G190A/S. At least prelim-inarily, TMC125 is a promising candidate for those patients who have had treatment failures with the first-generation NNRTIs. Capravmne is another NNRTI currently in phase I to II clinical trials. In preliminary data, it seems to be a potent inhibitor of RT, and has activity against virus harboring the K103N mutation.

A phase I study in which capravirine was administered as monotherapy for 10 d demonstrated 1.2 to 1.7 $\log_{10}$ copies/mL decreases in HIV RNA levels in the monotherapy treatment arms, compared with a 1.7 $\log_{10}$ copies/mL decrease in the nelfinavir plus zidovudine plus lamivudine control arm. This study did not report data regarding resistance. In an ongoing phase II study involving PI-naive patients who had previously failed an NNRTI-containing regimen (plasma HIV RNA levels >2000 copies/mL), more than 50% of the 50 patients for whom data was available attained plasma HIV RNA levels of fewer than 400 copies/mL when administered a regimen of capravirine plus nelfinavir plus two NRTIs. Capravirine seemed to be well-tolerated and efficacious. Thus, capravirine seems to be an effective and well-tolerated NNRTI, which, similar to TMC-125, has activity against HIV strains that are resistant to currently available NNRTIs.

RESISTANCE AND VIRAL FITNESS

The effect of various NNRTI mutations on viral fitness has been evaluated in several studies. One study used site-directed mutagenesis to construct viruses with NNRTI-resistance mutations and compared replication capacities and phenotypic susceptibilities with wild-type virus. The following commonly occurring mutations had no effect on viral replication capacity: K103N, Y181C/I, Y188C/H/L, and G190A. Several mutations, V106A, G190C/S, P225H, M230L, and P236L, conferred substantial reductions in viral fitness. Double mutants had varying degrees of altered replication capacity, indicating that the replication capacity of virus with more than one NNRTI mutation cannot be predicted based on the replication capacity of single mutants.

However, the accumulation of multiple mutations has been shown to cause a significant reduction in fitness. Because virus with NNRTI-resistance mutations may persist long after the NNRTI-selective drug pressure is removed, whether these mutations cause fitness impairment remains controversial. Several common mutations (K103N, Y181C, and YI88C/H/L) do not reduce fitness, but other mutations do confer a significant fitness cost. Another study evaluated the replication capacity of the P236L mutation that occurs readily in vitro, but infrequently in vivo in the presence of delavirdine. K103N occurs much more frequently in the setting of delavirdine therapy, and it was hypothesized that virus with the K103N mutation is more fit than virus with the P236L mutation.

This was borne out in one in vitro study, which noted a twofold to threefold decreased p24-antigen production in the presence of delavirdine in P236L-mutated virus compared with the K103N mutation. The P236L mutant demonstrated decreased rates of ribonuclease H cleavage that were presumed to be the cause of the differential fitness. Differential rates of ribonuclease H cleavage were also noted in another study demonstrating that a virus harboring the V179D mutation was more fit than one with the Y181C mutation, which, in turn, was more fit than one with the V106A mutation. Whether the reductions in fitness, caused by NNRTI-resistance mutations, have clinical consequences has not been determined.

HYPERSUSCEPTIBILITY

Recent observations based on phenotypic susceptibility assays have revealed that some virus isolates have increased susceptibility to NNRTIs. For these isolates, the IC_{50} of the patient virus is lower than the IC_{50} of the reference (wild-type) virus. The FC in IC_{50} would be less than one for these viruses.

This phenomenon has been termed hypersusceptibility. The clinical implications and durability of NNRTI hypersusceptibility have been evaluated in clinical trials. In a large analysis of 17,000 sequentially received plasma samples, specimens were analyzed for phenotypic as well as genotypic resistance. Hypersusceptibility to efavirenz, delavirdine, and nevirapine were detected in 10.8%, 10.7%, and 8% of the specimens, respectively. Hypersusceptibility (defined as FC <0.4) was most common in NNRTI-naive/NRTI-experienced patients.

A subanalysis of NNRTI-naive patients was undertaken to elucidate the effect of NRTI mutations on hypersusceptibility. Among the samples, 331 patients who were NNRTI naive/NRTI naive and 447 who were NNRTI naive/NRTI experienced were identified. The prevalence of hypersusceptibility to delavirdine, efavirenz, and nevirapine was 5%, 9%, and 11%, respectively, in the NRTI-naive group; whereas it was 29%, 26%, and 21%, respectively, in the NRTI-experienced group. Mutations at several sites in RT were significantly associated with NNRTI hypersusceptibility, predominantly at codons 41 44, 67, 69, 74, 75 118, 184, 210, 215, and 219. Another study looked at the clinical relevance of NNRTI hypersusceptibility in a prospective cohort.

In this study, hypersusceptibility was defined as a FC in IC_{50} of less than 0.4. NNRTI hypersusceptibility was present in 29% of the cohort of 177 patients. A longer duration of NRTI therapy and a reduced NRTI susceptibility were correlated with efavirenz hypersusceptibility. NNRTI therapy was begun in 106 patients. The mean change in HIV RNA level after 6 mo was 1.2 $\log_{10}$ copies/mL for patients with hypersusceptibility, as compared with –0.8 $\log_{10}$ copies/mL for those without hypersusceptibility ($p = 0.016$). Differences remained statistically significant at 12 mo. Multiple linear regression models confirmed that NNRTI hypersusceptibility was a significant independent predictor of the magnitude of response to NNRTIs ($p < 0.02$). Efavirenz hypersusceptibility was evaluated in a multidrug salvage protocol.

This study treated patients with single vs dual PI plus NRTI plus efavirenz. Phenotype assays were retrospectively performed on a subset of 139 randomly selected patients. Previous NNRTI treatment was positively associated with virological failure (HIV RNA level >200 copies/mL at weeks 24 and 48), whereas NNRTI hypersusceptibility reduced the likelihood of virological failure at week 48 (odds ratio, 0.16; $p < 0.01$). Hypersusceptibility to delavirdine was also shown to be a significant independent predictor of virological response.

Several methods to determine a hypersusceptibility cut-point found an FC value near 0.4 to be the best predictor of response. Thus, clinical studies have demonstrated the relevance of NNRTI hypersusceptibility for NNRTI-naive patients. Because hypersusceptibility is more likely to occur in NRTI-experienced patients with the presence of NRTI mutations, whether hypersusceptibility should influence the initial choice of regimens (PI vs NNRTI) remains controversial.

IMPACT OF RESISTANCE

NNRTI resistance in the developing world could be a significant problem in the future. Because HAART is not available to the majority of infected people worldwide, the use of NNRTIs in resource-poor settings has been largely restricted to small cohorts. The ability to obtain resistance assays is also a limiting factor in the effective use of this class of agents. Additionally, especially in parts of West Africa, because of the prevalence of HIV-2, this class of agents has limited usefulness. Currently, nevirapine has been most widely used in the developing world for the prevention of mother-to-child HIV transmission.

Single doses of nevirapine, administered to the mother and newborn at the time of delivery, have been shown to significantly reduce HIV transmission. In this setting, a fully suppressive regimen (*i.e.*, two NRTIs plus one NNRTI) is not provided because of cost and other logistical considerations. Given the efficacy of single-dose nevirapine, the use is justifiable, but resistance after this single dose has been reported. The half-life of the NNRTI drugs could lead to detectable levels of the medications for days to weeks after a single dose is administered. This persistent exposure to suboptimal NNRTI concentrations has lead to the development of resistance. In the HIV Network for Prevention Trials (HIVNET) 012 study in Uganda, 111 treatmentnaive pregnant women received a single 200-mg dose of nevirapine, and neonates received a single dose within 72 h after birth. Thirty-three infants were infected despite prophylaxis.

Genotypes were performed in 32 of the transmitting mothers and in 72 non-transmitting mothers. Nevirapine-resistance mutations were detected in 7 (22%) of the transmitting women, and 11 (16%) of the nontransmitting women. K103N was the dominant mutation in 90% of samples with a mutation. Y181C was seen in six samples. The women in HIVNET 012 had a follow-up genotype at 12 to 18 mo after therapy, which showed a return to wild-type virus in all of the patients. Seven of the infected infants also had genotypes performed, the K103N mutation was detected in one and the Y181C mutation was detected in two infants.

Interestingly, the infected infants had different mutations than their mothers; this difference suggests that the nevirapine-resistance mutations were caused by the selective pressure of the nevirapine treatment of the infants, rather than transmission of virus with resistance mutations. Additional larger studies have confirmed high rates of development of NNRTI resistance after a single dose of nevirapine administered to prevent HIV transmission from mother to baby.

The rate of nevirapine resistance was 39% in 456 women who received nevirapine to prevent transmission in a study in South Africa. Nevirapine resistance was more likely in women with a lower baseline CD4 cell count, a higher baseline HIV RNA level, and in women who took more than one dose of nevirapine. In some studies, single-dose nevirapine was administered in addition to background zidovudine; the addition of zidovudine to the prophylactic regimen did not prevent the emergence of nevirapine resistance. The impact of development of NNRTI resistance, after a single dose of nevirapine, on the subsequent response to a nevirapine-containing HAART regimen is a critical concern regarding the use of this strategy to prevent HIV transmission from mother to infant.

The potential benefit to the infant could be diminished if maternal NNRTI resistance emerges and reduces the response to a new regimen containing an NNRTI. Even if the NNRTI mutations in the plasma revert to wild type with time, archived virus, present in lymph nodes or proviral DNA, could lead to the rapid reemergence of resistant populations if therapy with an NNRTI is introduced. Because the World Health Organization guidelines and country programs recommend the use of an NNRTI-based regimen for initial therapy in resource-limited settings, data concerning this question are important. A recent study in Thailand suggests that previous exposure to single-dose NNRTI and development of NNRTI mutations can reduce response to an NNRTI-containing regimen.

In a study of treatment of women who had received single-dose nevirapine for transmission prophylaxis, 18% of 90 tested women had major NNRTI-associated mutations (K103N, G190A, or Y181C). All women subsequently received an NNRTI-based regimen for treatment of their HIV. After 6 mo of dosing, 85% (n = 27) of women not exposed to nevirapine had HIV RNA levels of fewer than 400 copies/mL, whereas the response rates for exposed women who did not have mutations was 80% (n = 97), and the rate for women with at least one mutation was 68% (n = 50; p = 0.003 in a test for trend). The time after nevirapine exposure was also a factor in response to treatment, those exposed to nevirapine longer than 6 mo before initiating an NNRTI-based regimen were more likely to respond whether or not they had previously had a NNRTI mutation.

Clearly, exposure and resistance from single-dose NNRTI prophylaxis is an important determinant of subsequent response to an NNRTI-containing regimen. Strategies to *g* overcome this limitation of preventing mother-to-child transmission will need to be carefully considered because the role of nevirapine prophylaxis yields such significant benefit in preventing HIV transmission to the baby. NNRTIs have little activity against the RT of HIV-2. HIV-2 is prevalent in west Africa. It is estimated that the pol genes (which contain RT) share a 60% homology between HIV-1 and HIV-2.

Despite the difference between these two strains, the RT enzymes seem similar in overall structure. One study compared in vitro susceptibilities of the three available NNRTIs to HIV-1 (III_b) and the two different strains of HIV-2 (ROD and EHO). All three NNRTIs demonstrated significantly reduced activity to HIV-2. For example, delavirdine had an IC_{50} of 0.01 μM against HIV-1, and an IC_{50} of 2.1 to 47.3 to the two strains of HIV-2. To have any significant activity in human serum, the concentrations of the three available NNRTIs will have to be upwards of 50 times greater than those achieved by current dosing regimens.

One group hypothesized that NNRTI resistance in HIV-2 could be reversed by inducing a mutation at position 188. Indeed, an L188Y mutation led to an HIV-2 that became sensitive to delavirdine and efavirenz. One study evaluated patients with HIV-2 who were treated with a nevirapine-containing regimen. Seven patients were treated with a nevirapine-containing HAART regimen, and had no significant decrease in viral load. Four patients received an indinavir-containing regimen and had undetectable viral loads at 4 mo of therapy. Several studies have looked at *de novo* NNRTI resistance of HIV-1 in differ-ent geographic regions, where different HIV clades may predominate.

No significant differences in drug susceptibility were noted in HIV-1 isolates across regions or clades. There are limited data on the pattern of resistance mutations that develop when patients with nonclade B virus are treated and fail a NNRTI-based regimen. As resistance assays become more available, studies can be performed to determine whether patients with non-B virus fail NNRTI regimens with different mutational pathways than patients with clade B virus. These data indicate that NNRTIs should not be used to treat patients with HIV-2, and that NNRTI susceptibility does not vary significantly among clades.

Monitoring HIV resistance will be important in resource-poor countries that have-limited access to ARVs, because suboptimal regimens may be the only affordable options. Additionally, patients may attempt to share their medications with infected family members, hoping that some benefit may be obtained. As ARVs are introduced in these settings, resistance may become a major problem.

CONCLUSIONS

Resistance to the NNRTIs occurs rapidly in nonsuppressive regimens and results in cross-resistance to all of the currently approved agents in the class. The increased prevalence of NNRTI resistance in newly infected patients provides rationale to obtain a resistance assay before initiating therapy in these patients. Current resistance testing guidelines recommend the use of resistance testing for ARV-naive patients with recent HIV infection (often defined as infection within the past 12 mo). The routine use of resistance tests in chronic HIV infection has recently been revised to recommend testing in patients who were infected with in the past 2 yr, but testing for those infected for an unknown duration or greater than 2 yr remains controversial and awaits further data. Nower NNRTIs are targeted to inhibit virus that exhibits current patterns of resistance, and are hoped to benefit patients with multidurg-resistant infections. NNRTIs with differing resistance profiles will be especially useful if they do not induce resistance as rapidly as the currently available agents.

CHAPTER 10 Peripheral Neuropathy

Peripheral neuropathies are the most common neurological syndromes associated with HIV infection. The spectrum of neuropathic syndromes is broad, and neuropathies may be encountered during any stage of the infection. Some disorders of the peripheral nerves in patients with HIV infection are presumed to be caused by pathological factors resulting from the virus itself, whereas others may result from responses of a competent immune system to viral antigens, opportunistic processes unleashed when immune deficiency has occurred, or as adverse effects of therapies for AIDS-related diseases.

To provide a context for understanding the toxic neuropathies, these other HIV-related neuropathies are briefly described in this first section. Inflammatory demyelinating polyneuropathies (IDP) are immune-mediated reactions directed at myelin peptides, which often present in association with a viral infection. They may occur acutely (Guillain-Barre syndrome or acute inflammatory demyelinating polyneuropathy [AIDP]) or follow a chronic fluctuating or progressive course (chronic inflammatory demyelinating polyneuropathy [CIDP].

Because they result from competent immune responses directed at antigens that are either derived from an invading pathogen or exposed by damage induced by the resulting inflammatory response, these syndromes tend to occur earlier in the course of HIV infection, even at the time of seroconversion. IDP is notable for a predominant picture of motor weakness and reflex loss, which can progress rapidly, compromising respiratory muscles and cranial nerve function as well as limb strength. Autonomic features may be present at onset, including tachycardia, blood pressure elevation, diaphoresis, and sphincter disturbances. In HIV infection, the clinical picture is indistinguishable from other settings, except for the presence of a moderate lymphocytic pleocytosis accompanying the elevation of protein levels in cerebrospinal fluid, in contrast to the usual increased protein in the absence of cells (albumino-cytological dissociation).

Electrophysiological studies are indicative of a demyelinating process, demonstrating severe slowing and blockage of nerve conduction velocities (NCV) and dispersion of compound action potentials. Variable degrees of axonal involvement may be seen and may have prognostic significance when present. Fortunately, the response to conventional therapy with plasmapheresis or pooled human immune globulin transfusions is usually successful. Mononeuntis multiplex is a multifocal process affecting multiple nerves. Inflammatory cells may be found in the epineurial and endoneurial perivascular spaces, however, vascular destruction is not seen.

Table 10.1. Neuropathies Associated With HIV Infection.

IDPs
- Acute IDP (Guillain-Barre syndrome)
- Chronic IDP

Monomeuritis multiplex
- Inflammatory—vasculitic
- Opportunistic (cytomegalovirus)

Polyradiculitis
- Cytomegalovirus, Varicella-zoster, lymphomatous meningitis, syphilis

Dermatomal zoster radiculitis—may affect multiple dermatomes

DSP

Toxic neuropathies
- Dideoxynucleosides (clinically DSP)
- Acute neuromuscular weakness syndrome associated with hyperlactatemia and d4T
- Other toxic neuropathies causes by treatment agents
  - Hydroxyurea, vincristine, INH, dapsone, thalidomide, etc

Diffuse infiltrative lymphocytosis syndrome

The distribution of clinical features reflects the specific cranial and/or peripheral nerves involved. This type of neuropathy is characteristically asymmetric; affecting both motor and sensory functions, and may occur during the period of symptomatic HIV disease. It may occur because of vasculitis as part of a multiorgan systemic syndrome. In patients with AIDS, multiple mononeuropathy has been found in association with cytomegalovirus infection. An acute polyradiculopathy syndrome may be encountered in patients with AIDS, most commonly resulting from cytomegalovirus infection, although cases caused by varicella-zoster virus, lymphoma, and syphilis have also been reported.

Typical features include an ascending motor weakness with reflex loss and sphincter disturbances, usually urinary retention. Painful paresthesias in the perineum and lower limbs are common, and sensory levels or Babinski signs may indicate involvement of the spinal cord in addition to the nerve roots. Diagnosis is usually made by cerebrospinal fluid analysis, which commonly reveals a polymorphonuclear pleocytosis and evidence of specific viral DNA when tested with polymerase chain reaction amplification techniques. Although the clinical picture superficially resembles an IDP, electrophysiological testing demonstrates features of an axonal process, including diminished action potential amplitudes and increased recruitment patterns. Treatment with specific antiviral agents may be successful when initiated early in the course of the opportunistic infection.

Additional HIV-associated neuropathies include varicella-zoster virus ganglio-neuritis, which presents with pain in a dermatomal distribution, similar to that affecting patients without HIV. In some patients, multiple segments will be affected, and occasionally the syndrome can be seen in the absence of the characteristic vesicular rash. Among the rarer neuropathic entities is the diffuse infiltrative lymphocytosis syndrome, which is a painful axonal sensory-motor neuropathy that may be either symmetric, or multifocal and asymmetric. Nerve biopsies reveal an angiocentric infiltration of CD8 lymphocytes in the epineurium. Peripheral blood CDδ lymphocytosis and infiltration of salivary glands are characteristic.

Dorsal root ganglionopathies producing sensory ataxia, acute motor-sensory axonal neuropathies, and neuropathies related to nutritional deficits are also reported in HIV-infected patients. The most common of the HIV-associated neuropathies, however, is the distal sensory polyneuropathy (DSP), which is seen most often in patients with HIV infection that is more advanced, and causes neurological symptoms in up to one-third of such individuals. Typicalclinical symptoms include numbness beginning in the soles of the feet followed by burning or shooting pains, pin and needle-like paresthesias, and variable aching and cramping pains. Symptoms occur symmetrically, gradually ascend into the legs, and may result in sufficient loss of sensation to cause impairment of balance (sensory ataxia).

Eventually, numbness and paresthesias involve the distal upper extremities as well. Ankle reflexes are typically reduced or lost early, and reduction of other reflexes may follow. Characteristically, motor weakness is not a prominent complaint, although examination may disclose modest weakness in the small muscles of the feet and hands, especially the toe extensors. Pathologically, DSP is a dying-back axonopathy, affecting both small and large nerve fibers. Because conventional NCV and EMG studies are relatively insensitive to pathology in the smallest nerve fibers, they may be unrevealing in early stages of symptomatic neuropathy, despite severe pain and evidence of distal sensory loss. Alternatively, some individuals without neurological complaints or findings may be found to have evidence of neuropathy using neurophysiological testing.

Quantitative testing of sensory thresholds (QST), a computer assisted determination of perception dependent on patient response to repeated stimulus presentation at various intensities, may be more sensitive to small fiber involvement. Such testing may reveal abnormalities in individuals without clinical findings or complaints, sometimes foreshadowing the evolution of symptomatic DSP. In a study using QST, the pain was associated with diminished thresholds to mechanical noxious stimulation and increased responses to supramaximal mechanical stimuli. In contrast, although thermal detection thresholds were also increased in subjects with neuropathy, the levels were similar in subjects with HIV-associated neuropathy, regardless of whether the neuropathy was painful or not.

This selective relationship of pain to mechanical receptors was interpreted as indicative of a peripheral pain mechanism selectively involving a class of nociceptors responsive to pressur. Assessment of intraepidermal nerve fiber density in skin biopsy specimens has recently been used in clinical trials as a means of quantifying loss of small nerve fibers in patients with polyneuropathies. The extent of epidermal small fiber loss was shown to correlate with measures of pain intensity in a recent clinical therapeutic trial of nerve growth factor for HIV associated polyneuropathy. The pathological process causing DSP is not completely understood at present.

Direct HIV infection of nerves is not seen, although occasional macrophages containing HIV genetic material are found in biopsies of affected nerves and some biopsies also show lymphocyte infiltration in the perineurium. A current theory is that the neuropathy results from local secretion of proinflammatory cytokines by activated macrophages, resulting in cytokinemediated neurotoxicity. Recent studies have suggested that the prevalence of DSP is related to both CD4 count and plasma viral load, although epidemiological studies comparing periods before and after the common use of highly active antiretroviral therapy (HAART) have shown conflicting effects on the prevalence of DSP, possibly in relation to an increasing frequency of adverse effects of components of that therapy.

TOXIC NEUROPATHIES

Early clinical studies of dideoxynucleoside reverse transcriptase inhibitors were complicated by a painful neuropathy, which proved to be a dose limiting toxicity. It has been estimated that approx 10% of patients receiving zalcitabine (ddC) or stavudine (d4T) and 1 to 2% of those receiving didanosine (ddl) will have to discontinue therapy because of a painful neuropathy. Affected individuals developed burning or shooting pains, or pins and needles paresthesias in the soles of the feet and toes associated with numbness and sometimes with aching and cramping pains in the feet and calves. The pain and paresthesias worsened over time, gradually ascending up the legs.

On detailed neurological examination, impairments of distal sensation to pain, temperature, light touch and vibration, variable impairment of balance as a result of the sensory loss, and allodynia (hypersensitivity to normally nonpainful stimuli), were accompanied by loss or reduction of ankle reflexes, and variable weakness of the small muscles in the feet. Symptoms could precede abnormal examination findings, and were remarkable for their resemblance to the HIV-associated DSP.

The initial symptoms might be aching or neuralgic pain or paresthesias, depending on the dosage of the neurotoxic agent. When the dideoxynucleoside treatment was discontinued, the majority of individuals experienced gradual improvement in the neuropathic symptoms, however, this improvement was often preceded by a period of up to 3 months following cessation of the medication during which symptoms persisted or might even worsen, a phenomenon referred to as coasting. EMG and NCV could be normal, depending on the severity of the neuropathy, but demonstrated features of an axonal neuropathy when abnormalities were present.

QST was more sensitive than nerve conduction velocities in one study evaluating both methods. Clinical trials have identified a number of factors associated with increased risk for the development of painful neuropathies in association with dideoxynucleoside therapy, in addition to the parameters of drug dose and administration. These include lower CD4 lymphocyte levels in some but not all studies, previous history of neuropathy from other cause, excessive use of alcohol, exposure to potentially neurotoxic medications such as isoniazid or vincristine, neuromuscular or neuropathic symptoms at onset of therapy, diabetes mellitus, or weight loss before entry.

Most of these factors are either causes of neuropathy in their own right, or are associated with increased risk of HIV DSP. Not all individuals who develop paresthesias or pain go on to develop clinical neuropathy. In a retrospective study of subjects at one center participating in a large clinical trial, 28% of individuals receiving either ddl or ddC experienced transient numbness, paresthesias, or pain, but did not develop signs of toxic neuropathy. Criteria have been suggested for attribution of a compatible clinical syndrome to a toxic neuropathy. These criteria emphasize the occurrence of symptoms after exposure to the agent and their improvement after its discontinuation.

A subacute progression is also considered suggestive of a toxic neuropathy, because HIV-associated DSP is typically slowly progressive. Involvement of the upper extremities is more common in DSP than in toxic neuropathy, but has been associated with nudeoside exposure as well. The descriptions of toxic DSP were derived from experience in early clinical trials of dideoxynucleoside agents, and reports of clinical anecdotes. It is immediately evident from these descriptions that HIV-associated DSP and painful toxic neuropathies commonly associated with exposure to dideoxynucleoside reverse transcriptase inhibitors are remarkably similar in their clinical symptoms and examination findings. This leads to the obvious question of whether evolution of such symptoms during treatment with one of these agents represents unmasking or aggravation of an occult DSP, or represents a primary effect of the potentially neurotoxic agent itself. In many of these early trials, subjects either did not undergo detailed screening for prevalent polyneuropathy at entry or were entered with symptoms suggestive of preexisting neuropathy. In the next section, we review this data for each of the relevant nucleoside analogs.

Table 10.2. Reported Risk Factors for Toxic Neuropathy With Dideoxynucleoside Therapy

Dose and regimen of dideoxynucleoside
Combination therapy with multiple dideoxynucleosides or with hydroxyurea
Lower CD4 levels
Hyperlactatemia
History of peripheral neuropathy before dideoxynucleoside exposure
Exposure to other potentially neurotoxic drugs
Heavy alcohol use
Diabetes mellitus
Diminished creatinine clearance
Vitamin B_{12} deficiency
Weight loss during dideoxynucleoside therapy

Table 10.3. Criteria for Diagnosis of Dideoxynucleoside-Associated Toxic Neuropathy

Symmetrical pain and/or paresthesias in distal lower extremities—may ascend over time Diminished distal sensation to vibration, pain, and/or light touch on neurological examination—may have distal contact hypersensitivity (allodynia)
Normal motor tone and normal strength, except for mild weakness in small distal muscles[a]
Absent or diminished ankle reflexes
Onset of symptoms associated with exposure to dideoxynucleoside agents and improvement associated with their discontinuation or dose reduction, allowing for a period of persistence or intensification within the first 3 mo after the change in regimen
Elevated serum lactic acid levels[b]

[a] The recently described acute neuromuscular syndrome associated with hyperlactatemia and d4T therapy is characterized by acutely or subacutely progressive motor weakness and areflexia. The clinical pattern superficially resembles Guillain Barre syndrome rather than the more commonly seen distal sensory polyneuropathy *(see* text).

[b] Elevated serum lactic acid levels have recently been found to discriminate distal sensory polyneuropathy caused by dideoxynucleosides from that caused by HIV.

DIDEOXYNUCLEOSIDE THERAPY

Zalcitabine

Early phase I trials encountered painftil neuropathy as a major dose-limiting toxicity in ddC therapy. Yarchoan and colleagues administered doses of ddC ranging from 0.03 mg/kg to 0.25 mg/kg, every 8 h, to 20 subjects with AIDS or AIDS-related complex (ARC) and CD4 lymphocyte levels fewer than 350 cells/mm^3. A reversible painful neuropathy syndrome occurred in 10 of the participants, beginning between 6 and 14 wk after onset of therapy. This syndrome was characterized by aching pain in the feet followed by burning paresthesias 2 to 3 wk later, with impairment of light touch, temperature, vibration, and position senses, and diminished ankle reflexes on neurological examination. Three subjects also developed mild weakness of the toe and foot extensor muscles.

After discontinuation of ddC, worsening of neuropathic symptoms continued for up to 5 wk before improvement. Four weeks after discontinuing continuous ddC, one patient with toxic neuropathy who was treated with an alternating regimen of 200 mg zidovudine (ZDV) every 4 h for 1 wk and 0.03 mg/kg ddC every 4 h for the next week, experienced improvement in neuropathic symptoms that continued during the subsequent 5 wk.

Electromyography (EMG) and nerve conduction velocity (NCV) studies were obtained in seven of the ten subjects who developed neuropathy at baseline, or within the first 2 wk of therapy. Five of these patients had normal test results and two patients showed a preexisting neuropathy, one axonal and slowly progressive (HIV-associated DSP?), the other a polyneuropathy attributed to diabetes mellitus.

Ten additional studies were performed after the onset of neuropathic symptoms, which began concurrently with treatment in nine subjects, and 4 wk after discontinuation of ddC in the 10th patient. NCVs revealed markedly decreased or absent sural nerve action potentials in the lower extremities, and, in addition, diminished sensory action potentials from ulnar nerves in the upper extremities, with preserved distal latencies, of seven subjects in whom these tests were performed. EMGs demonstrated abnormal spontaneous activity, increased polyphasic potentials, and abnormal interference patterns. These features are consistant with an axonal neuropathy. Merigan and colleagues conducted a dose-finding study in 61 subjects with AIDS or ARC, excluding individuals with "significant neurological disease."

All subjects administered 0.06 or 0.03 mg/kg ddC every 4 h developed painful neuropathy. The time-to-onset was dose related, with the higher-dose group developing neuropathic symptoms in a mean of 7.4 wk, and the 0.03-mg/kg group developing neuropathic symptoms in a mean of 8.2 wk. Individuals receiving 0.01 mg/kg ddC every 4 h also developed painful neuropathy, which occurred in 80% of these subjects in a mean of 11.8 wk. Symptoms escalated with onset over weeks and involved the distal upper extremities in half of the cases. QST revealed abnormalities of vibratory perception coincident with the earliest svmptoms.

Even the group receiving the lowest dose of 0.005 mg/kg ddC every 4 h had a small incidence of neuropathy, which occurred in 17% of the patients, evolving more insidiously during a mean of 17.5 wk. Improvement on discontinuation of ddC was dose related, with the two lower dose groups experiencing complete recovery within 6 mo. Improvement was slower in the higher dose arms (requiring up to 1 yr), and incomplete in several subjects receiving the highest dose who continued to experience pain and sensory loss after 1 yr. Fifty-two of these subjects receiving ddC every 4 h at 0.06 mg/kg, 0.03 mg/kg, 0.01 mg/kg, or 0.005

mg/kg, were evaluated for toxic neuropathy using serial clinical and electrophysiological monitoring combined with symptom questionnaires.

Criteria for neuropathy were specified as at least two symptoms of moderate intensity on the questionnaire, or one symptom of severe or moderate intensity plus an objective abnormality on either the neuro-logical examination or the electrophysiological testing. In the two higher-dose arms, all subjects developed neuropathy with a mean onset of 7.7 wk. QST for vibration became abnormal before the appearance of clinical symptoms, with a mean onset of 7.3 wk. In the intermediate 0.01 mg/kg-dose arm, all subjects developed similar but less intensely painful symptoms, evolving more gradually with a mean onset at 9.3 wk.

Vibratory QSTs became abnormal in 70% of this group, coincident with the clinical symptoms. NCVs were obtained in only 20% of the total cohort. They revealed abnormalities in all subjects from the higher-dose arms after the development of clinical dysfunction. NCVs were mildly abnormal in 40% of the intermediate dose cohort. In the low dose group, one subject developed mild but transient neuropathic symptoms 26 wk after onset of therapy. Another developed asymptomatic abnormalities on vibratory QST and a loss of ankle reflexes, however, no further change occurred despite continued therapy with ddC. One might reasonably question whether this subject developed HIV-associated DSP rather than a toxic neuropathy.

In contrast, however, when ddC was discontinued in symptomatic participants, improvement occurred in 83% of patients, with a mean time-to-onset of improvement of 19 wk, 15 wk, and 11 wk for the high dose, intermediate dose, and low-dose groups, respectively, consistent with a toxic etiology. In an open-label study conducted in individuals taking ZDV for at least 48 wk, ddC administered in a dosage of 0.75 mg every 8 h was associated with symptoms of painful neuropathy in 25% of 59 subjects. The pain was characterized as moderate or severe in 17% of patients, and 10% of patients withdrew from the study because of the neuropathic symptoms.

Subjects entering the trial had CD4 counts of fewer than 200 cells/mm^3 and no known peripheral neuropathy, although the screening methodology is not elaborated in the report. A regression analysis found that neuropathic symptoms occurred more frequently in subjects with CD4 levels of fewer than 50 cells/mm^3 and in subjects with diminished creatinine clearances of less than 110 mL/min. Another open-label trial cornpared courses of 200 mg ZDV plus ddC at either 0.01 mg/kg or 0.03 mg/kg every 4 h alternating either weekly or monthly, 0.03 mg/kg ddC at administered on alternate weeks, and 200 mg ZDV every 4 h administered either on alternate weeks or continuously, in a cohort with a median CD4 level of 142 cells/mm^3 (mean, 164 ± 139 cells/mm^3) and no known previous neuropathy.

Neurological screening methods included a base line evaluation for neuropathy, a neuropathic symptom questionnaire administered every 2 wk, a focused examination for neuropathy every 4 wk, and a vibratory QST every 8 wk. The frequency of painful neuropathy in subjects on ddC was related to the dose. Among individuals in the three 0.03 mg/kg-dose arms combined, neuropathy occurred in 34% of patients compared with 15% of patients in the two 0.01 mg/kg-dose arms. Of patients on ZDV alone taken either continuously or on alternate weeks, 17% of patients developed neuropathy, which was comparable to the results for those taking lower-dose ddC on a weekly (10%) or monthly (21%) alternating regimen, and those taking higher dose ddC on the monthly alternating regimen (14%).

In subjects alternating ddC and ZDV weekly, 46% developed neuropathy, compared with 14% of those alternating ddC and ZDV monthly. In those who developed neuropathy, 68%

experienced the onset of symptoms within 24 wk of starting therapy. This cohort had relatively low CD4 levels at entry and it is not noted whether the time-to-onset of neurological symptoms differed between those on ddC and those taking only ZDV, or whether any increase in CD4 level in response to therapy had an independent relationship with the occurrence of neuropathy. A large community-based, open-label clinical trial compared 500 mg/d ddl with 2.25 mg/d ddC in subjects with up to 300 CD4 cells/mm^3 who had failed therapy with ZDV.

Peripheral neuropathy was an exclusion criterion for entry, and occurred during the study at a rate of 22.1 per 100 patient years with ddl compared with 45.1 per 100 patient years for ddC, a statistically significant twofold increased risk in the ddC arm. A subsequent study evaluated the frequency of toxic neuropathy in a clinical trial using ddC dosed at 2.25 mg/d (~0.03 mg/kg) alone, or in combination with ZDV, in subjects with CD4 counts of up to 200 cells/mm^3 or up to 300 cells/mm^3 with concurrent systemic symptomas of HIV. At one site, 76 sub-jects entered into this multicenter clinical trial were evaluated with serial neuro-logical examinations, neuropathy symptom questionnaires, and vibratory QST.

Individuals who reported moderate or severe symptoms of peripheral neuropathy on the initial questionnaire, or who had moderate deficits on examination (undefined) at screening were excluded. Operational criteria for toxic neuropathy included at least one distal sensory symptom and at least one objective sign on examination, persisting for at least 72 h, and appearing while the subject was receiving ddC. During a follow-up period of up to 73 wk, the incidence of neuropathy was 34% in the two groups receiving ddC compared with 4.3% in the group receiving ZDV alone, an increased risk of 7.9 times for the ddC groups. The mean time-to-onset of neuropathy was 16 wk, but varied widely, from 1 to 51 wk.

In the entire multicenter cohort, the rate of moderate or worse neuropathy was 22% in the ddC group and 23% in the combination arm, compared with 13% for the ZDV group alone. In addition, in the multicenter cohort, for those entering with CD4 lymphocyte counts greater than 150 cells/mm^3, moderate or worse neuropathy was significantly associated with ddC therapy alone (16%) or in combination (21%), compared with ZDV alone (9%). However, there was no significant association with toxic neuropathy for ddC therapy, alone or combined, compared with ZDV alone, for patients with CD4 cell levels of fewer than 150 cells/mm^3 at entry. Although there seemed to be a trend favoring a relationship to ddC, this result could also reflect a greater risk of HIV associated DSP in patients with reduced CD4 levels.

Interestingly, the rate for neuropathies characterized as severe in the parent study was similar in all groups, at 4% for ddC alone and 6% for combination therapy or ZDV alone. In the smaller single center cohort, a case-control study comparing baseline characteristics found only diabetes mellitus to be a significant risk factor for neuropathy, whereas weight loss during the study showed a trend suggesting a relationship but did not reach statistical significance. No effect of CD4 level was demonstrated in this smaller sample in contrast to the total cohort. Development of moderate or worse neuropathic symptoms was managed by dose interruption until improvement occurred, followed by resumption of treatment at half of the original dose.

Blum and colleagues observed subiective symptomatic improvement within an average of 10 wk in 67% of their subjects after dose reduction, but no reversal of abnormal findings on neurological examination. The persistent neurological abnormalities could represent either residual damage from a toxic neuropathy or emergence of HIV-associated DSP during the study.

Didanosine

Some early dose-escalation studies with ddl were limited by a painful neuropathy similar to that seen with ddC. Lambert and colleagues encountered symptoms of neuropathy in almost one-third of a cohort of 37 subjects with CD4 levels of fewer than 400 cells/mm^3, who received doses of ddl greater than 12 mg/kg/d, but in only 1 of 15 subjects who received smaller doses. Subjects were not screened for preexisting neuropathy at entry. The investigators also noted an increased risk of neuropathy with total cumulative dose greater than 2 g/kg. The time-to-onset of symptoms was dose related and ranged from 55 to 201 d after initiation of therapy. Withdrawal of ddl resulted in resolution or improvement of pain and paresthesias during 4 to 8 wk in seven of eight affected individuals, three of whom tolerated rechallenge with ddl at 10 mg/kg/d for 4 to 12 wk without recurrence.

A subsequent follow-up, of these subjects and several additional individuals taking 250 to 750 mg/day of ddl, for a mean of 34 wk (2-72 wk) revealed neuropathy in two additional subjects, both taking ddl at 750 mg/d. One developed symptoms at 30 wk. The other, who entered with what was described as mild peripheral neuropathy symptoms (HIV-associated DSP?), developed intensified neuropathic discomfort at 18 wk. After an 8-wk suspension of ddl, however, the patient was able to continue treatment at a reduced dose of 500 mg/d. In contrast, a large multicenter trial comparing ddl with continued ZDV in individuals previously taking ZDV for at least 16 wk and having either AIDS, AIDS related complex (ARC), and fewer than 300 CD4 cells/mm^3 or fewer than 200 CD4 cells/mm^{3}without systemic symptoms, found no difference in the rates if of symptomatic neuropathy between the groups.

Neuropathy at entry was not noted to be an exclusion criterion, and it appears that only symptomatic neuropathic complaints were evaluated. Perhaps of some importance, CD4 counts rose in the two ddl-treatment arms but not in the continued-ZDV arm. In another dose-escalation study, a cohort of 58 subjects with AIDS or ARC and a median CD4 count of 47 cells/mm^3 (range, 4-267 cells/mm^3) was followed for a median of 8.9 mo (with 28 subjects followed for longer than 1 yr).

Yarchoan and coworkers encountered painful neuropathy in 15 subjects. Twelve of these individuals received doses greater than 9.6 mg/kg/d. Preexisting neuropathy was not evaluated at entry. A subsequent report from this group described a 5-yr follow-up of 72 subjects, including individuals from this study and others from phase I studies who were subsequently switched to doses of 200 or 250 mg ddI twice daily. The authors reported 11 cases of painful neuropathy during this period, affecting approx 15% of the cohort. Of these, four patients continued treatment without dose change, two were successfully rechallenged at the same dose after temporary suspension of therapy, two tolerated rechallenge at a lower dose, and three discontinued ddl and switched to ZDV.

No evaluations for persistent neurological abnormalities in those who experienced pain were described. In an open label, dose escalation study of ddl administered once daily to 34 individuals with CD4 cell levels fewer than 400 cells/mm^3 (mean, 125 ± 110 cells/mm^3 at entry), followed for an average of 15.8 wk (2-56 wk; median, 12 wk), severe painful neuropathy was reported in only one of six individuals receiving the highest dose, of 30.4 mg/kg ddI. After suspension of ddI, partial resolution was noted during the next 8 wk. An interesting aspect of this study is the observation of 11 subjects entering with grade 1 neuropathy (paresthesias not requiring treatment), 8 had resolution of their symptoms during the trial, whereas the other 3 were unchanged.

During the trial, CD4 levels rose to greater than 200 cells/mm^3 in the majority of the subjects, suggesting that neurological improvement of the baseline neuropathy (HIV-associated DSP?) might have occurred because of a therapeutic response induced by the ddl regimen. Evaluation of subjects from a single site participating in another multicenter ddl dose escalation study identified toxic neuropathy, confirmed by a neurologist, in 10 (23%) of 44 subjects who complained of pain or numbness or were noted to have abnormalities on study examinations. Neuropathic symptoms improved or resolved after suspension of ddl. Toxic neuropathy was defined as the occurrence of distal sensory symptoms in the lower extremities in with reduced or absent ankle reflexes and distal sensory impairments on examination, appearing in relation to ddl therapy and improving on its withdrawal.

One subject maintained at half of the original dose had resolution of the neuropathic symptoms during 6 mo follow up, whereas four others tolerated rechal lenge after suspension of therapy either without recurrence or with mild paresthesias. In these 10 subjects, the mean CD4 level at entry was 114 cells/mm^3 (0-212 cells/mm^3) and improved by an average of 49 cells/mm^3 (range, 0-170 cells/mm^3) in seven patients, whereas it worsened by a mean of 38 cells/mm^3 (range, 11-75 cells/mm^3) in three other patients. The Alpha trial, a large, double-blind European study, used two dose regimens of ddI, 10.5 mg/kg/d and 3.3 mg/kg/d.

Peripheral neuropathy at entry was not an exclusion criterion, except in French centers. Only neuropathy leading to study discontinuation was reported, with rates of 7.8% in the higher dose group and 5.9% in the lower-dose arm. An open label, community based study comparing ddl with ddC in 467 subjects with CD4 levels of up to 300 cells/mm^3 reported the development of neuropathy in 21.6% ofsubjects taking ddl at 500 mg/d.

Stavudine

Painful neuropathy was a dose-limiting side effect in early studies of d4T, as with the other dideoxynucleosides. In a series of 41 subjects with CD4 cell levels of fewer than 400 cells/mm^3, 55% of patients developed symptoms of painful neuropathy. The time-to-onset ranged from 5 to 40 wk. Painful neuropathy was related to both dose and dosing frequency. Each increase of 1 mg in the dose unit was associated with a twofold increased risk of neuropathy. Dosing twice daily was associated with a fourfold increased risk of neuropathy compared with administering the same total dose as a thrice daily regimen. It appears from the report that no screening for neuropathy at baseline was incorporated into the entry criteria.

Of 20 subjects developing painful neuropathies, 4 were noted to have used alcohol excessively, 1 was taking INH, and 1 was taking amphotericin. Five additional subjects reported myalgias at baseline. Some subjects tolerated rechallenge at reduced dose. The maximally tolerated dose was determined to be 2 mg/kg/d. Skowron reviewed early trials of d4T and found painful neuropathy to be the principal dose-limiting side effect.

In a combined analysis, the rate of neuropathy at doses of 0.5 and 1.0 mg/kg/d was 21 per 100 person-years compared with 66 per 100 person-years for doses of 2 mg/kg/d. In addition to the factors noted in the Browne study existing neuropathy at entry was also found to be a risk factor for subsequent neuropathic adverse effects. In a phase II study of subjects with CD4 levels of up to 250 cells/mm^3, painful neuropathy was dose- and treatment duration-dependent and occurred up to 48 wk after initiation at 2.0 mg/kg/d. The rate of painful neuropathy was 17 per 100 person years at a dose of 0.5 mg/kg/d *h* compared with 41 per 100 person-years at a dose of 2 mg/kg/d.

Resolution of pain and tolerance to rechallenge with a lower dose of d4T was noted in 55% of subjects who were followed for a median of 32 wk. A large, double blind, multicenter trial compared 40 mg d4T twice daily with 200 mg ZDV thrice daily in a cohort of 882 subjects with CD4 levels from 50 to 500 cells/mm^3, and previous exposure to ZDV for at least 6 mo. Individuals with a history of peripheral neuropathy were excluded. The rate of painful neuropathy requiring dose modification in the d4T group was 12%, compared with a rate of 4% in patients taking ZDV.

Neuropathy was more common in those with AIDS at baseline, regardless of treatment assignment, with rates of 25% in the d4T group and 12% in the ZDV group, compared with 10% and 3%, respectively, for those without AIDS at entry. Of those receiving d4T who developed neuropathy, 63% resolved symptomatically within a median of 17 d after suspension of the medication (range, 4-57 d) and tolerated resumption of therapy at reduced dose for a median of 23 wk (range, 1-120 wk). Differences in neuropathy rates for those with AIDS, in addition to toxic neuropathy associated with the antiretroviral therapy may affect these results. A different pattern of neurotoxicity has recently been described in association with d4T therapy, which superficially resembles the acute inflammatory polyneuropathy, or Guillain-Barr'e syndrome, described in the first section of this chapter. This new neuromuscular syndrome has been attributed to hyperlactatemia related to dideoxynucleoside reverse transcriptase inhibitor therapy, particularly with d4T.

Individuals with this syndrome develop profound progressive motor weakness during several weeks, which may be associated with ataxia and numbness. The cranial nerves may be involved, resulting in facial weakness, and involvement of the respiratory muscles may lead to ventilatory failure. In contrast to typical Guillain-Barr'e syndrome, in which the primary pathology is demyelinating, electrophysiological studies in this syndrome reveal features of an acute axonal neuropathy. In some instances, neurological symptoms were preceded by symptoms of lactic acidosis syndrome, including malaise, nausea, emesis, and abdominal cramping. Elevated plasma lactate levels have been documented in some of the relatively few case reports thus far reported.

Recovery in those individuals who do improve may be prolonged up to 1 yr or longer. Cases have been described in pediatric patients. Recently, a patient presenting with the Miller-Fisher variant of Guillain-Barre syndrome consisting of the clinical triad of ophthalmoplegia, ataxia, and areflexia evolving in association with hyperlactatemia and after discontinuation of anti retroviral therapy including d4T was reported. As this chapter was being finalized, a report appeared describing a retrospective collection of 69 cases, including some from previous reports, meeting criteria of new acute or subacute neuromuscular limb weakness associated with lactic acidosis.

Cases were classified for diagnostic certainty based on the presence of confirmatory diagnostic studies and the exclusion of confounding causes of neuromuscular weakness. In 88% of the cohort, the dideoxynucleoside used was d4T, with a median duration of 10.5 mo of therapy (range, 1-71 mo). No association between serum lactate level and duration of d4T use was found. Of interest is the finding that cases were almost equally likely to occur in women as in men. Approximately one-third of the cases were associated with symptoms of lactic acidosis syndrome. An important finding was that the neuromuscular weakness syndrome occurred despite previous discontinuation of the antiretroviral therapy in 36% of the cases (median, 14 d). Respiratory failure requiring ventilator support occurred in 23% of the cohort.

Mortality was notable at 16% of the patients and was marginally associated with levels of serum lactate greater than 6 mmol/L, although levels greater than 10.08 mmol/L showed a strong trend to associate with death. Most cases evaluated more extensively were sensorimotor polyneuropathies, although both axonal and demyelinating pathologies were seen in the limited nerve biopsies obtained. Four of 15 muscle biopsies revealed evidence of mitochondria! dysfunction, and 3 revealed inflammatory infiltrates. The full spectrum of this syndrome, its underlying pathophysiology, and its relationship to nucleoside therapy are only beginning to be understood as this is written. Similarly, although isolated case reports have described improvement in association with various treatments no consistently effective therapy has yet been identified.

TOXIC NEUROPATHY WITH COMBINED AGENTS

Several reports have evaluated cohorts or described neuropathy in individuals receiving combinations of dideoxynucleoside agents, or combination dideoxynucleoside agents with hydroxyurea (HU). An early case report concerned an individual participating in a clinical trial alternating ddC and ZDV monthly. Neurological findings at the onset of therapy were not described. After 16 wk of treatment, the patient developed paresthesias, followed by burning pain 4 wk later, at which time the ddC was discontinued. Two weeks later, he reported the neurological symptoms to be the same. The following week, another physician initiated ddl treatment, and 1 wk after that, the neurological symptoms worsened with ascending sensory impairments in the legs and increased pain. The ddI was stopped, and the symptoms resolved 1 mo later.

The authors suggested that ddl could potentiate ddC neurotoxicity. The combination of ddI and d4T was evaluated in a cohort of 52 treatmentnaive subjects with a mean baseline CD4 count of 330 cells/mm^3. During 24 wk, nine subjects developed neuropathic symptoms (17%). Two of these withdrew and the other seven continued therapy. In a 24 wk extension completed by 30 subjects, 1 additional subject withdrew because of painful neuropathy and 2 other subjects developed minor paresthesias. The authors noted that toxic neuropathy was the most common adverse event in the trial, but that only three subjects discontinued therapy because of it, and concluded that the combination was safe and well tolerated. In a dose finding, double blind study comparing the effect of combinations of ddI and d4T on plasma viral load, there were 13 adverse events in 86 subjects followed for a median of 349 d (1-728 d).

Patients with peripheral neuropathy and previous treatment with neurotoxic drugs were excluded, and most subjects were treatment naive, with CD4 cell counts greater than 200 cells/mm^3 (mean, 343 cells/mm^3). It seems, from the report, that determination of neurotoxicity was based solely on subject complaints and only two subjects were reported to develop neuropathyas a serious adverse event (2.3%). The addition of HU to dideoxynucleosides has been shown to increase neurotoxicity.

A cohort of 144 subjects with CD4 levels of 200 to 500 cells/mm^3 were randomized to receive ddl and d4T with or without HU. Seventy two subjects were initially randomized to the three-drug arm and 12 additional subjects switched to that arm after 12 wk. After 2 yr, only 25% of subjects remained in the trial, most withdrawing because of lack of efficacy or adverse events. Neuropathy occurred in 35% of the group taking all three drugs, compared with 15% of those taking only the ddl and d4T,a statistically significant difference. In a retrospective database review of 1116 patients treated with various regimens containing ddl, d4T, and HU in one large center, the risk of neuropathy was calculated for the agents alone and in combination.

Inclusion criteria for neuropathy required sensory symptoms in the lower extremities or all four extremities, beginning, recurring after resolution, or (if preexisting) worsening after the onset of therapy with the agents. The mean CD4 count was 270 cells/mm^3. The risk of developing neuropathy was related to lower CD4 count, history of non-drug-asso-ciated neuropathy, and, weakly, with age greater than 40 yr. Adjusting for these factors yielded a relative risk compared with ddl alone of 1.39 for d4T, 2.35 for ddl plus HU, 3.5 for ddI plus d4T, and 7.8 for ddI plus d4T plus HU. The crude incidence rate for ddl alone was 6.8 per 100 person-years of treatment. Another retrospective review found a frequency of peripheral neuropathy 27% of a small group of subjects administered ddI plus d4T in conjunction with HU, compared with 10% in a group administered ddl plus d4T.

Although the trend did not reach statistical significance, this report is consistent with the findings in the two studies described above. A single case report describes a patient with neuropathic toxicity attributed to lamivudine (3TC). This individual experienced onset of paresthesias while taking ddC plus ZDV. His symptoms improved after discontinuing the ddC, however, distal sensory impairment and mild distal weakness persisted, with electrophysiological findings consistent with a polyneuropathy.

Eighteen months later, he started taking 3TC, and, within 3 wk, he experienced recurrent painful dysesthesias. These symptoms improved within 4 wk after the 3TC was discontinued, but recurred on rechallenge, consistent with a toxic exposure aggravating the preexistent neuropathy.

PATHOPHYSIOLOGY OF TOXIC NEUROPATHY

The mechanism of nucleoside induced neurotoxicity is not completely understood at the present time. Current hypotheses emphasize toxic effects of nucleoside reverse transcriptase inhibitors on mitochondria, and tissue specific vulnerabilities based on energy requirements. Nucleoside analogs can replace nucleotide bases and can be incorporated into newly synthesized DNA. In contrast to other forms of DNA synthesized from nuclear chromosomal genetic material, mitochondrial DNA (mtDNA) is inherited from the cytoplasm of the oocyte and replicates by random segregation rather than recombination.

Replication of mtDNA requires nuclear coded DNA polymerase-γ. In mtDNA, there are few introns and limited self-repair mechanisms, thus, mutations are more likely to affect coding sequences altering protein production, and dysfunctional mutations are more likely to persist with subsequent replication. This results in mixtures of mitochondria containing wild-type and mutant DNA in the same cells in varying proportion, referred to as heteroplasmy. The impact of heteroplasmy on a particular tissue may be a threshold effect, dependent on the severity of the resulting defect in oxidative phosphorylation in the abnormal mitochondria and the relative metabolic demands of the tissue for energy production.

The hypothesis, as articulated by Lewis and Dalakas, suggests that manifestations of nucleoside toxicity reflect:

1. Subcellular availability and abundance of nucleoside analog in the target tissue.
2. The ability of the cellular thymidine kinases to use the nucleoside analog as a

competitive alternative substrate for monophosphorylation and, subsequently, triphosphorylation.

3. The ability of the triphosphate nucleoside analog to inhibit DNA polymerase-γ either by acting as an alternative substrate for incorporation into mtDNA via competitive inhibition resulting in aberrant chain extension, or by noncompetitive incorporation and premature chain termination.
4. The relative metabolic requirements of the tissue for oxidative phosphorylation.

It has been suggested that depletion of the mtDNA induced by nucleoside analogs results in the depletion of enzymes encoded by the mtDNA, leading to impaired oxidative phosphorylation. An analogy has been drawn to multiple symmetric lipomatosis, a hereditary condition in which mtDNA mutations and deletions impairing oxidative phosphorylation complex IV have been identified. This condition results in hypertriglyceridemia, insulin resistance, and fat deposition abnormalities similar to those associated with HAART, and is associated with peripheral neuropathy in almost all cases. In vitro studies have demonstrated the relative potency of nucleoside agents to inhibit mtDNA in a T-lymphoblastoid cell line, CEM, to be ddC greater than d4T greater than ZDV greater than ddI.

The concentration of nucleoside analog required to inhibit mtDNA replication was less than that required to inhibit cell growth at 4 d for ddC, ddI, and d4T, but not for ZDV. In a neuronal cell model, using PC 12 cells derived from rat pheochromocytoma and stimulated to differentiate into neurons by incubation with nerve growth factor, ddC and ddI inhibited production of mtDNA and neurite generation at pharmacological concentrations, whereas ZDV and 3TC did not. In contrast, d4T inhibited neurite generation at concentrations that did not affect mtDNA synthesis, despite the ability of d4T to impede activity of DNA polymerase-γ, and despite the presence of d4T triphosphate in the mitochondria, suggesting a different mechanism of neurotoxicity.

The authors suggested that a repair mechanism, perhaps the 3′5′ exonuclease present in DNA polymerase-γ, might offset toxicity of d4T to a degree. Absence of phosphorylated 3TC in the isolated mitochondria was consistent with the hypothesis that the inability to transport phosphorylated 3TC accounted for its lack of toxic effect on mtDNA. The lack of neurotoxicity of 3TC may also lie in its ability to act as a substrate for the integral 3′5′ exonuclease activity of DNA polymerase-γ, making incorporation less likely. A consequence of mtDNA polymerase-γ dysfunction is a reduced capacity of the mitochondria to synthesize uridine nucleotides, which are integral to the synthesis of oligosaccharides on membrane glycoproteins and glycolipids.

Cytidine nucleotides, derived from uridine nucleotides, are involved in phospholipid synthesis. In PC 12 cultures, the addition of uridine and pyruvate can prevent ddC-induced reduction of mtDNA replication. Nucleoside analog neurotoxicity has been demonstrated in animal models. Axonal degeneration and reduction in fiber size, myelin degeneration manifested by myelin splitting and edema, and abnormal mitochondria in Schwann cells were seen in rabbits treated witn ddC. It was not possible to determine from the study whether the axonal changes were a primary toxic effect or were secondary to the myelin damage.

In rats treated with ddI, sections of sciatic nerve taken at various durations after onset of exposure demonstrated myelin toxicity manifested by edema and myelin splitting at 15 wk. Concentrations of ddI used resulted in peak plasma levels five to six times those in humans administered therapeutic doses. A recent study analyzed morphological changes in sural nerve

Subjects were treated with 1500 mg acetyl-L-carnitine twice daily for a median of 14 mo (range, 5-33 mo) and followed both clinically and by serial skin biopsies taken from the distal leg. At baseline, biopsies in patients with DSP showed reduced total innervation of epidermis, dermis, and sweat glands in comparison with five healthy controls. Selective staining for small sensory (type C and Aδ) fibers revealed an 80% reduction in epidermis and a 25% reduction in the dermis. After 6 mo of therapy with acetyl-L-carnitine, the total epidermal and dermal nerve fiber innervations reached or exceeded normal levels, whereas the fibers around sweat glands increased by 75%. Selective small fiberjnnervation reached 40% of control in the epidermis and 175% in the dermis.

Further increases were seen on subsequent biopsies in those continuing therapy for longer periods. Clinical neuropathic symptoms improved in 76% of the subjects and did not change in the others. No significant changes in CD4 or CDS lymphocyte counts or HIV plasma viral load occurred during the study period and, thus, did not affect results. In a brief trial, 10 of 16 patients with painful DSP receiving 500 to 1000 mg daily of acetyl-carnitine intramuscularly or intravenously reported improvement in pain symptoms after 3 wk. Five subjects reported no change in pain and one reported worsened pain. No large controlled clinical trial of acetylcarnitine therapy in HIV-associated or nucleoside-associated neuropathies has been published, to date.

MANAGEMENT OF NEUROPATHY

Complaints of pain or paresthesias in the lower extremities of a patient taking ddI, ddC, or d4T suggests the possibility of a toxic neuropathy and warrants detailed evaluation. Compared with HIV-associated DSP, toxic neuropathies appear more abruptly and progress more quickly, thus, an individual who develops such progressive symptoms within several months after the introduction of a potentially neurotoxic antinucleoside agent is more likely to have a toxic neuropathy. The diagnosis, however, can be confirmed only if symptoms resolve on discontinuation of the agent. Because some subjects in clinical trials experience only transient paresthesias minor symptoms may not necessarily require immediate change in therapy.

Furthermore, in some instances, it may be preferable to continue the agents despite minor neuropathic symptoms because of a robust antiretroviral response. A detailed neurological examination to assess the extent of sensory and reflex impairment, and a detailed neurological history to pursue other causes of neuropathy is an important measure, which should be obtained at the onset of symptoms and is also a reasonable screening measure before institution of dideoxynucleoside therapy. Other potential treatable causes for neuropathy, such as vitamin B_{12} deficiency, diabetes mellitus, thyroid disease, syphilis, Lyme disease, or exposure to other neurotoxic substances or medications should be sought.

Electrophysiological studies may be helpful in centers with the capability to do both QST and conventional EMG-NCV studies, however, routine EMG-NCV studies may be unrevealing untilafter clinical signs appear. Serial monitoring of the neurological symptoms and examination is indicated in any individual experiencing incident neuropathic symptoms or abnormal findings on examination while taking dideoxynucleosides. In the event the symptoms progress or are severe enough to limit function, and if other antiretroviral treatment options are available, consideration should be given to suspending therapy with the nucleoside analog to allow the neuropathy to improve.

The patient should be counseled about the possibility of symptoms worsening in the first months after discontinuation of the presumed offending agent (coasting). If substantial improvement occurs, rechallenge with the agent at a lower dose is reasonable after 2 to 3 mo if no alternative therapy is available. Neuropathic pain may be intermittent and tolerable for some patients and require only nonsteroidal antiinflammatory agents for relief. For many individuals, the pain is both continuous and severe enough to limit function. The World Health Organization has published a three-step analgesic ladder to provide a systematic and rational approach to pain management.

Although developed to treat cancer-related pain, the same approach is useful for addressing neuropathic pain associated with HIV and dideoxynucleoside therapy. Initially, adjuvants such as tricyclic agents (amitriptyline and others) or anticonvulsants (gabapentin, lamotrigine, carbamazepine, and others) are used in combination with nonopioid analgesics. In the second step, a second adjuvant with a differing mechanism of action is added. Many patients with moderate neuropathic pain can be controlled at this level; however, for those who prove refractory to these measures, step three adds opioid analgesics, which should be used in long-acting formulations (fentanyl, oxycodone, and others). A recent study comparing the analgesic efficacy of lamotrigine to placebo for HIV-associated neuropathic pain found self-reported benefit in subjects taking dideoxynucleoside agents, but not in those who were not using neurotoxic treatment regimens.

A randomized trial of amitriptyline and mexiletine failed to show benefit for either drug compared with placebo in subjects with painful HI V-associated neuropathies and varying exposure to dideoxynucleosides. Individuals with sensory ataxia caused by loss of postural sensation may require support for safe ambulation. Such patients are at increased risk of falling in the dark, when changing posture or direction, or when descending stairs. In contrast to the patient complaining of distal paresthesias, urgent evaluation is required for the individual who complains of acute motor weakness. These types of neuropathies can progress rapidly to paralysis and, potentially, to respiratory failure.

Plasma lactate levels should be drawn, and dideoxynucleoside agents discontinued, particularly if the neurological symptoms are associated with gastrointestinal symptoms. EMG-NCV studies may help to differentiate IDPs from axonal or mixed neuropathies that are more characteristic of toxic syndromes or opportunistic disease. In a patient with progressive weakness, hospitalization and serial monitoring of neurological and respiratory function is indicated until stability or sufficient recovery occurs to allow independent function. Cerebrospinal fluid samples should be obtained to exclude opportunistic infections or lymphoma.

Ventilatory support should be initiated for subjects with deteriorating respiratory status when vital capacity falls to 1 L, and prophylaxis for venous thrombosis should be initiated. Physical and occupational therapy to maintain range of motion and prevent contracture formation should be started early in individuals anticipated to have prolonged weakness. No specific therapy for the acute neuromuscular syndrome associated with nucleoside analogs and lactic acidosis has been shown to be effective. In one review of reported cases of pathological syndromes attributed to nucleoside-associated lactic acidosis, some individuals who improved were treated with cofactors such as thiamine, riboflavin, L-carnitine, prostaglandin E, or coenzyme Q. When neuropathic weakness improved, it did so very slowly, taking up to 1 yr after discontinuation of the nucleoside analog agent.

IMPLICATIONS FOR FUTURE CLINICAL TRIALS

Current data on the toxicity of dideoxynucleoside analogs in humans has been drawn largely from subject complaints during clinical treatment trials and from case reports. No large carefully controlled prospective neurological trial focused specifically on nucleoside-associated neuropathy in HIV has been performed, to date. In many clinical treatment trials, previous neuropathy was not an exclusion factor at entry. In others, detailed neurological examination to determine the prevalence of asymptomatic neuropathy at entry, and to confirm the origin of symptomatic complaints during the study, were not incorporated into the protocols. As a result, errors in attribution of symptoms or in the recognition of occult neuropathy may have occurred. Simpson and colleagues reanalyzed data from a large clinical trial of ZDV, ddl, ddC, and placebo in four combinations.

Using criteria for DSP as described above, they reevaluated diagnoses made at the treatment sites. In 63% of reported cases of neuropathy, the site investigators were unable to determine the cause of the symptoms. On review, Simpson and colleagues were able to classify approximately half of these as DSP, and, therefore, potentially treatment related. Of subjects classified as having treatment related neuropathy, 27% did not have typical features of DSP as reflected in the criteria used by the reviewers, and were, therefore, considered unlikely to have medication-related toxic neuropathy. These findings suggest that data from studies not incorporating detailed baseline and serial neurological assessments may both underestimate or overestimate true treatment-related neurotoxicity.

Furthermore, in addition to modifying the frequency of DSP, successful HAART might also modify the time-to-onset and rate of progression of neuropathies induced by nucleoside analog therapy, confounding the temporal characteristics currently used to discriminate HIV-associated DSP from toxic neuropathy. The question of whether painful neuropathies appearing in association with nucleoside therapy represent exacerbation of a previously occult HIV-associated distal sensory neuropathy, or whether they arise *de nnvn* as a result of nucleoside toxicity is of significant clinical importance.

Neuropathy in advanced HIV infection seems to be common, and not infrequently asymptomatic. Marra and colleagues examined 226 subjects drawn from a large AIDS clinical trials unit and found DSP in 21%. Of the subjects with DSP, 71% had signs of neuropathy on examination but no symptoms. Individuals with DSP were more likely to have received dideoxynucleosides, they had more advanced HIV infection by Centers for Disease Control classification, and they had a trend toward lower CD4 counts. Schiffito and associates determined the frequency of DSP in a cohort selected for cognitive symptoms and CD4 counts of fewer than 300 cells/mm^3 or fewer than 200 cells/mm^3.

They found symptomatic DSP in 35% of subjects, and asymptomatic DSP in an additional 20% of subjects. In this cohort evaluations, 72% of subjects without neuropathy at entry developed incident DSP during a period of 30 mo, with a 1 yr incident of 52%. Among those with no neuropathy or asymptomatic neuropathy at entry, the 1-yr incidence of symptomatic DSP was 36%. Only 23% of the subjects were taking dideoxynucleosides and, interestingly, neither these subjects nor those with asymptomatic neuropathy were more likely to develop symptomatic DSP during the study. In a series of 251 HFV- infected subjects undergoing clinical and neurophysiological assessments, Tagliati and coworkers found evidence of DSP in 38% of the entire cohort, and in 48% of subjects with AIDS.

Of subjects with clinical features of DSP, 19% had normal electrodiagnostic studies, whereas, of those without clinical findings, 28% had abnormalities on the neurophysiological assessments suggestive of DSP. The limitations of previous data and the frequency of asymptomatic polyneuropathy demonstrated in these studies underscore the need for any future clinical trial using potentially neurotoxic therapies, including nucleoside analogs, to incorporate a sufficiently detailed neurological screening at baseline and during the observation period to accurately assess and attribute incident neuropathy, and to determine risk factors for the emergence of incident neuropathy. In some recent clinical trials incorporating dideoxynucleoside treatment regimens, fewer subjects have been noted to develop symptomatic painful neuropathy.

Additionally, some individuals seem to experience improvement of neuropathic symptoms in association with dideoxynucleoside therapy, further confounding the determination of frequency and etiology of painful neuropathies in patients with advanced HIV infection who are treated with potentially neurotoxic therapies. Factors contributing to the emergence of treatment-related neuropathic toxicity need to be elucidated. Do host factors, such as genetic susceptibilities, predispose to the development of toxic neuropathy? Are there particular interactions among the multiplicity of drugs taken by HIV-infected patients and the multiorgan pathologies, particularly those affecting the liver or kidneys, which combine to increase the toxicity of a standard therapeutic dose of a dideoxynucleoside in a given individual? Are the beneficial effects of successful therapy on DSP partially masking the neurotoxicity of dideoxynucleoside therapy? The risk of acquiring HIV-associated DSP s appears to be related to both increased plasma HIV viral load and decreased CD4 count.

Successful treatment with current HAART may have the capacity to improve peripheral nerve function, as demonstrated in a study using serial QST assessments, and as reported in an earlier study of ddl, described above. Thus, HIV-associated DSP may become less prevalent with successful retroviral suppression and control of other, as yet unidentified factors, offsetting an increased frequency of neuropathy induced by dideoxynucleoside components of HAART regimens. This offsetting effect was suggested in a retrospective study of a German cohort, in which the prevalence of DSP in a 2-yr period before the introduction of HAART was 42.5%, with a suspected rate of toxic neuropathy of 20.4%. In the following 2-yr period, during which subjects were placed on HAART in large numbers, the prevalence of DSP was reduced to 34.4%; however, the proportion of neuropathies attributed to nucleoside analog toxicity increased to 31.2%. Recently, a small prospective study assessed serum lactate levels in patients with DSP, comparing those taking d4T to those without exposure to dideoxynucleosides, as well as to HIV-infected individuals without neuropathy taking d4T and non-HIV-infected controls.

An elevated serum lactate level was found to be 90% sensitive and 90% specific in discriminating d4T-related toxic DSP from HIV-associated DSP. If results of this study are confirmed, the presence of a surrogate marker should contribute to the understanding of the pathogenesis of toxic DSP, and perhaps clarify the potential contributions of both the dideoxynucleoside- and HIV-related pathologies to the natural history of DSP. Future controlled clinicaltrials which include longitudinal assessments both of symptomatic and asymptomatic neuropathy from onset of antiretroviral therapy, and epidemiological studies of the natural history of HIV-associated neuropathies in relation to HAART therapy to identify factors associated with their occurrence, may help to clarify which subjects are at risk of severe toxic neuropathies and provide prognostic indicators to facilitate modification of therapeutic regimens.

Perhaps serum lactate levels will prove useful in this regard. It may be possible, with better understanding of the pathophysiology of these syndromes, to identify adjunctive therapeutic agents which protect against the emergence of toxic neuropathies, as has recently been suggested for acetylcarnitine. In addition to detailed serial clinical assessments, better measurement tools may be required, including electrophysiological tests that are more sensitive, such as QST assessments, and pathological measurements, such as quantification of intraepidermal and, perhaps, dermal nerve fiber density obtained by skin biopsy. Treatments relieving neuropathic pain may not result in improvement of clinical, electrophysiological, or pathological measures in neuropathy, as was seen in the recent clinical trial of nerve growth factor.

The natural history of such painful neuropathies relieved, but not resolved, after withdrawal of dideoxynucleosides, awaits elucidation in future studies. Will these prove to be aborted toxic neuropathies with incomplete resolution, or incident HIV-associated DSP exacerbated by a superimposed agent with some neurotoxicity, perhaps synergistic with primary neuropathological processes? Thus, most importantly, studies to improve our understanding of the mechanisms of toxicity in both HIV-associated DSP and nucleoside-associated toxic neuropathies are essential to provide the basis for future therapies that specifically target the pathophysiological mechanisms of these syndromes.

Index

G

H

K

M

N

O

P

biopsies taken from patients with ddC-induced toxic neuropathy, in comparison with biopsies from patients with HIV-associated distal sensory neuropathy, who either never received any antiretroviral therapy or received only ZDV, and from individuals with non-HIV-associated axonal neuropathies.

Dalakas and colleagues found abnormalities in mitochondria in axons and Schwann cells of subjects receiving ddC more frequently than in controls. No unique features were noted on light microscopy; however, using electron microscopy, abnormal mitochondria with interrupted inner morphology, enlarged size, osmophilic deposits between the inner and outer limiting membranes, and diminished cristae were noted in all three groups, with different degrees of severity. In the ddC-associated neuropathies, 55% of axonal mitochondria were abnormal, compared, with 28% in HIV-associated DSP and 9% of the other axonal neuropathies.

In Schwann cells, 47% of the mitochondria were abnormal in the ddC-associated neuropathies, compared with 5% in the HIV-associated DSP and 14% in the other neuropathic controls. Competitive polymerase chain reaction amplification was used to quantify mtDNA in the nerves, and demonstrated significant depletion of mtDNA in the ddC-treated group compared with the neuropathic controls. Results were considered to be supportive of the ability of ddC to induce a toxic mitochondrial neuropathy by depletion of neural mtDNA, consistent with the selective inhibition of DNA polymerase-γ demonstrated injieural cell lines. The potential role of acetyl-carnitine deficiency as a result of mitochondrial damage in toxic neuropathy was suggested by Famularo and colleagues, after a study in which they demonstrated diminished serum levels of acetylcarnitine but not total carmtine, in 12 subjects with symptoms of painful neuropathy while taking ddI, ddC, or d4T in comparison with 10 subjects takingjidl and 11 subjects taking ZDV, who had no symptoms of neuropathy.

Acetyl-L-carnitine is a transport molecule for longchain fatty acids into the mitochondria, where they undergo β-oxidation, generating adenosine triphosphate, and donate the acetyl group to acetyl coenzyme for subsequent participation in Krebs cycle reactions, generating additional adenosine triphosphate. The authors hypothesized that acetylation of carnitine is important to normal function of peripheral nerve, facilitating binding of nerve growth factor and enhancing neural repair, and that lower levels in serum reflected defects in carnitine metabolic pathways caused by mitochondrial injury. In contrast to Famularo's findings, in a sample of 232 subjects drawn from a clinical treatment trial of nerve growth factor for HIV-associated neuropathy, Simpson and colleagues found an increased proportion of subjects with abnormally low levels of total and free carnitine among the study subjects compared with normal HIV-negative controls (no HIV controls without neuropathy were assessed), but found no association between the levels of total or free carnitine, or acetylcarnitine levels and the severity of the peripheral neuropathy as measured by clinical symptoms, QST, or nerve fiber density in skin biopsy specimens.

No relationship of carnitine levels to current or recent use of dideoxynucleoside agents, as compared with subjects with no use or remote use (6 mo or longer before entry), was found. There was no difference in any of the neuropathy measures studied between subjects with decreased levels compared with subjects having normal levels of carnitine. A possible explanation for the discrepancies between these studies could lie in differences between tissue and plasma levels of carnitine compounds. A recent report describes results of open label acetyl-carnitine treatment in 21 subjects with nucleoside reverse transcriptase inhibitor associated toxic neuropathies.